FRANCISCAN GEOLOGY OF NORTHERN CALIFORNIA

Editor
M.C. BLAKE, JR.
U.S. Geological Survey
Menlo Park, California

November 5, 1984

Published by
The Pacific Section
Society of Economic Paleontologists and Mineralogists
Los Angeles, California
U.S.A.

Pacific Section
Society of Economic Paleontologists and Mineralogists

Pacific Section SEPM Officers - 1984

Pacific Section
Society of Economic Paleontologists and Mineralogists

Pacific Section SEPM Officers - 1984

For copies of this volume, write to:
Treasurer, Pacific Section S.E.P.M.
P.O. Box 10359
Bakersfield, CA 93389

Printed by Comet Reproduction Service
Santa Fe Springs, CA 90670

Contents

EDGAR H. BAILEY
1914-1983

INTRODUCTION AND DEDICATION TO EDGAR BAILEY

M. C. Blake, Jr.
U.S. Geological Survey, 345 Middlefield Road
Menlo Park, California 94025

This volume is dedicated to Edgar Bailey, who studied the Franciscan[1/] throughout his long and productive career and, just as importantly, inspired numerous other geologists to do the same--including many of the contributors to this volume.

Bailey's work on the Franciscan began in 1937, when he studied the rocks on Catalina Island for his doctoral dissertation at Stanford University (1941). Although this study was primarily a mineralogic investigation of the glaucophane schists and related metamorphic rocks, he mapped a thrust fault separating overlying high-grade amphibolites from underlying glaucophane-bearing Franciscan rocks and referred to the chaotically deformed Franciscan rocks as melange. This was nearly 30 years before melange became a household term among workers in deformed orogenic belts.

His wartime studies with the U.S. Geological Survey (from 1941 to 1947) were concentrated on mercury deposits and associated Franciscan rocks in California. They resulted in numerous publications and maps, including his monumental New Almaden report, U.S. Geological Survey Professional Paper 360 (Bailey and Everhart, 1964). That report contains some of the best descriptions of Franciscan lithology and structure ever published, and the three-dimensional geometry afforded by the deep mine workings allowed documentation of the intrusive origin of the Franciscan serpentinite bodies.

In 1957, Bailey began mapping the Skaggs Springs 15-minute quadrangle in Sonoma County, selected after he had spotted the Cazadero ultramafic mass from a commercial airplane. Numerous other geologists, including Jim Berkland, Clark Blake and Carl Wentworth, participated in this study and, as a result of Bailey's inspiring guidance and teaching ability, went on to graduate studies in other areas of Franciscan and related rocks.

During this same period, Bailey, Porter Irwin, and Rollin Stevens developed a technique for staining graywacke specimens to determine their detrital K-feldspar content. They found (Bailey and Irwin, 1959) that the varying K-feldspar content of the Franciscan of northern California could be used to subdivide it and that the K-feldspar content of most Franciscan rocks differed markedly from that of coeval rocks of the Great Valley sequence. This staining technique was applied in great detail during mapping of the Skaggs Springs quadrangle, and the resulting units served as a model for subsequent regional mapping of the entire northern San Francisco Bay region (Blake and others, 1971, 1974; Fox and others, 1973; Sims and others, 1973).

Bailey's mapping of the Skaggs Springs quadrangle also stimulated several important ancillary studies in this quadrangle. Most important was the work on glaucophane-bearing metamorphic rocks of the Ward Creek area by Bob Coleman, Don Lee, and others (Coleman and Lee, 1962, 1963; Lee and others, 1963a, b, 1966), during which metamorphic aragonite was discovered and the very high pressure under which Franciscan rocks had been metamorphosed was confirmed.

In 1960, Bailey, Porter Irwin, and David Jones began preparation of a report that would collect and summarize all that was then known concerning the Franciscan and related rocks in California. This comprehensive paper was published in 1964 as California Division of Mines and Geology Bulletin 183 and is still the definitive Franciscan reference. In this report, Bailey first defined the terms "blueschist" and "blueschist facies," which are now used throughout the world. Later, Bailey collaborated with Bob Garrison on a study of Franciscan limestones (Garrison and Bailey, 1967), and with Harold Wollenberg on an investigation of the radioactive-element content of Franciscan graywackes in northern California (Wollenberg and others, 1967).

In 1968, several papers were published describing what has now come to be known as plate tectonics. This theory, in general, states that the surface of the Earth is made up of various interacting plates that are created at spreading centers, such as the Gorda Ridge, slide past one another along transform faults, such as the San Andreas, and are consumed in trenches, such as the one presumed to lie north of the Mendocino Escarpment. The application of this theory to Pacific coast geology and, in particular, to the origin of the Franciscan assemblage was prompt. Already in 1964, Bailey, Irwin, and Jones, in the last paragraph of their Bulletin 183, suggested that the Franciscan rocks might have been underthrust beneath the Great Valley sequence, with attendant formation of such high-pressure minerals as jadeite, in what is now called a subduction zone. Subsequent papers (Hsu, 1968; Bailey and Blake, 1969; Page, 1969; Hamilton, 1969; Dickinson, 1970; Ernst, 1970) using plate-tectonic theory postulated that the Franciscan rocks were deposited in a trench and subsequently deformed, and then metamorphosed in a subduction zone beneath the Sierran-Klamath arc.

Bailey, together with David Jones, began in 1968 to map the Colyear Springs quadrangle, located in northern California at the junction between the Klamath Mountains, Great Valley, and Coast Ranges provinces. During their study, Bailey and Jones discovered a depositional contact between basal sedimentary rocks of the Great Valley sequence and underlying mafic volcanic rocks, gabbro, and serpentinite. The implications of this discovery were numerous and resulted in the paper "On-Land Mesozoic Oceanic Crust in California Coast Ranges" (Bailey and others, 1970). Other maps and reports dealing with the complex junction in the Colyear Springs quadrangle were published by Jones and others (1969), Bailey and Jones (1973), and Jones and Bailey (1973).

[1/]Franciscan rocks are currently most commonly referred to as the Franciscan Complex or Franciscan assemblage (see Bailey and others, 1964; Berkland and others, 1972).

in Blake, M.C. Jr., ed., 1984, **Franciscan Geology of Northern California:** Pacific Section S.E.P.M., Vol. 43, p. 1-4

From 1971 to 1973, Bailey, with the help of Deborah Harden, prepared a map and report showing the mineral resources of the San Francisco Bay region and several other papers were published by Bailey and Blake in 1973 and 1974 dealing with the geochemistry and origin of the Coast Range ophiolite, and by Jones, and others (1978) describing the distribution of Franciscan-like subduction complexes along the entire west coast of North America.

Twenty years have now passed since Edgar Bailey, Porter Irwin, and David Jones published their now-classic volume on the Franciscan rocks of the California Coast Ranges. Within this time, the plate-tectonic revolution has further stimulated study of the Franciscan such that more than 400 papers have already been published on various aspects of Franciscan geology.

The present volume draws together a number of important contributions on the Franciscan in the region of Bailey's principal interest, the northern Coast Ranges. Papers dealing with the Franciscan rocks in the San Francisco Bay region are presented first, followed by regional and topical studies elswhere in northern California. These papers demonstrate the continuing challenge of the Franciscan as a topic for study, for our ideas are still changing rapidly.

Ed is no longer with us; he died on July 28, 1983, after a long bout with cancer. The absence of his gentle manner, soft smile, and cheerful whistle has created a void that can never be filled.

REFERENCES

Bailey, E. H., 1941, Mineralogy, petrology, and geology of Santa Catalina Island, California: Stanford University, Stanford, California, Ph.D. thesis.

Bailey, E. H., and Blake, M. C., Jr., 1969, Tectonic development of western California during the Late Mesozoic (in Russian): Geotektonika, pt. 3, p. 17-30; pt. 4, p. 24-34.

-----1973, Ophiolite, an indicator of ancient plate boundaries in western California (abs.): Abstracts of Papers, International Symposium on Ophiolites, Moscow, p. 95-97.

-----1974, Major chemical characteristics of Mesozoic ophiolite in the Coast Ranges of California: U.S. Geological Survey Journal of Research, v. 2, p. 637-656.

Bailey, E. H., Blake, M. C., Jr., and Jones, D. C., 1970, On-land Mesozoic ocean crust in California Coast Ranges, in Geological Survey research, 1970: U.S. Geological Survey Professional Paper 700-C, p. C70-C81.

Bailey, E. H., and Everhart, D. L., 1964, Geology and quicksilver deposits of the New Almaden district, Santa Clara County, California: U.S. Geological Survey Propessional Paper 360, 206 p.

Bailey, E. H., and Harden, D. R., 1975, Map showing mineral resources of the San Francisco Bay region, California--present availability and planning for the future: U.S. Geological Survey Miscellaneous Investigations Map I-909, 12 p, scale 1:250,000.

Bailey, E. H., and Irwin, W. P., 1959, K-feldspar content of Jurassic and Cretaceous graywacke of the northern Coast Ranges and Sacramento Valley, California: American Association of Petroleum Geologists Bulletin, v. 43, p. 2797-2802.

Bailey, E. H., Irwin, W. P., and Jones, D. L., 1964, Franciscan and related rocks, and their significance in the geology of western California: California Division of Mines and Geology Bulletin 183, 177 p.

Bailey, E. H., and Jones, D. L., 1973, Preliminary lithologic map, Colyear Springs quadrangle, California: U.S. Geological Survey Mineral Investigations Field Studies Map MF-516, scale 1:48,000.

Berkland, J. O., Raymond, L. A., Kramer, J. C., Moores, E. M., and O'Day, Michael, 1972, What is Franciscan? American Association of Petroleum Geologists Bulletin, v. 56, p.2295-2302.

Blake, M. C., Jr., and Bailey, E. H., 1973, California Coast Range ophiolite--Ocean ridge, marginal basin, or island arc environment? (abs.): Eos (Transactions of the American Geophysical Union), v. 54, no. 11, p. 1219-1220.

Blake, M. C., Jr., Bartow, J. A., Frizzell, V. A., Jr., Schlocker, Julius, Sorg, Dennis, Wentworth, C. M., and Wright, R. H., 1974, Preliminary geologic map of Marin and San Francisco Counties and parts of Alameda, Contra Costa and Sonoma Counties, California: U.S. Geological Survey Miscellaneous Field Studies Map MF-574, scale 1:62,500.

Blake, M. C., Jr., Smith, J. T., Wentworth, C. M., and Wright, R. H., 1971, Preliminary Geologic Map of western Sonoma County and northernmost Marin County, California: U.S. Geological Survey open-file map, scale 1:62,500.

Coleman, R. G., and Lee, D. E., 1962, Metamorphic aragonite in the glaucophane schists of Cazadero, California: American Journal of Science, v. 260, p. 577-595.

-----1963, Glaucophane-bearing metamorphic rock types of the Cazadero area, California: Journal of Petrology, v. 4, pt. 2, p. 260-301.

Dickinson, W. R., 1970, Relations of andesite, granites, and derivative sandstones to arch-trench tectonics: Review of Geophysics and Space Physics, v. 8, p. 813-860.

Ernst, W. G., 1970, Tectonic contact between the Franciscan melange and the Great Valley sequence, crustal expression of a Late Mesozoic Benioff zone: Journal of Geophysical Research, v. 75, p. 886-902.

Fox, K. F., Jr., Sims, J. D., Bartow, J. A., and Helley, E. J., compilers, 1973, Preliminary geologic map of eastern Sonoma County and western Napa County, California: U.S. Geological Survey Miscellaneous Field Studies Map MF-483, scale 1:62,500.

Garrison, R. E., and Bailey, E. H., 1967, Electron microscopy of limestones in the Franciscan Formation of California, in Geological Survey research, 1967: U.S. Geological Survey Professional Paper 575-B, p. 94-B100.

Hamilton, Warren, 1969, Mesozoic California and the underflow of the Pacific mantle: Geological Society of America Bulletin, v. 80, p. 2409-2430.

Hsu, K. J., 1968, Principles of melanges and their bearing on the Franciscan-Knoxville paradox: Geological Society of America Bulletin, v. 79, p. 1063-1074.

Jones, D. L., and Bailey, E. H., 1973, Preliminary biostratigraphic map, Colyear Springs quadrangle, California: U.S. Geological Survey Minerals Investigations Field Studies Map MF-517, scale 1:48,000.

Jones, D. L., and Imlay, R. E., 1969, Structural and stratigraphic significance of the Buchia zones in the Colyear Springs-Paskenta area, California:

U.S. Geological Survey Professional Paper 647-A, 24 p.

Jones, D. L., Blake, M. C., Jr., Bailey, E. H., and McLaughlin, R. J., 1978, Distribution and character of Upper Mesozoic subduction complexes along the wsst coast of North America: Tectonophysics, v. 47, p. 207-222.

Lee, D. E., Coleman, R. G., and Erd, R. C., 1963a, Garnet types from the Cazadero area, California: Journal of Petrology, v. 4, p. 460-492.

Lee, D. E., Thomas, H. H., Marvin, R. F., and Coleman, R. G., 1963b, Isotopic ages of guaucophane schists from the area of Cazadero, California. art. 142 of Short papers in geology and hydrology: U.S. Geological Survey Professional Paper 475-D, p. D105-D107.

Lee, D. E., Coleman, R. G., Bastron, H., and Smith, V. C., 1966, A two-amphibole glaucophane schist in the Franciscan Formation, Cazadero area, Sonoma County, California, in Geological Survey research, 1966: U.S. Geological Survey Professional Paper 550-C, p. 148-157.

Page, B. M., 1969, Relation between ocean floor spreading and the structure of the Santa Lucia Range, California (abs.): Geological Society of America Abstracts with Programs, v.1, no. 3, p. 51-52.

Sims, J. D., Fox, K. F., Jr., Bartow, J. A., and Helley, E. J., compilers, 1973, Preliminary Geologic map of Solano County and parts of Napa, Contra Costa, Marin, and Yolo Counties, California: U.S. Geological Survey Miscellaneous Field Studies Map MF-484, scale 1:62,500.

Wollenberg, H. A., Smith, A. R., and Bailey, E. H., 1967, Radioactivity of Upper Mesozoic graywackes in the northern California, California: Journal of Geophysical Research, v. 72, no. 16, p. 4139-4150.

Coherent metagraywacke and metasiltstone of the Yolla Bolly terrane, Pacheco Pass, California. Photo by E. H. Bailey, 1980.

TECTONOSTRATIGRAPHIC TERRANES OF THE SAN FRANCISCO BAY REGION

By

by M. C. Blake, Jr., D. G. Howell, and A. S. Jayko
U.S. Geological Survey
345 Middlefield Road
Menlo Park, CA 94025

ABSTRACT

Seventeen tectonostratigraphic terranes have been differentiated within the Franciscan Complex, the Great Valley sequence, and the Salinian block in the San Francisco Bay region. The Franciscan terranes include subduction or collision complexes (Yolla Bolly, Pickett Peak, and Burnt Hills), fragments of ocean-floor or ocean-islands (Marin Headlands, Nicasio Reservoir, and Permanente), and graywacke-rich terranes of uncertain origin (Coastal, Alcatraz, Rio Nido, San Bruno Mountain, and Novato Quarry). All of the Franciscan terranes are tectonically enveloped in a melange known as the Central terrane.

The three terranes of the Great Valley sequence are lithologically less diverse than the Franciscan terranes, but contain notable differences from one another. The basal part of the Del Puerto terrane consists of the Coast Range ophiolite of inferred island-arc origin, overlain by arc-derived tuffs and sedimentary rocks. In the Elder Creek terrane, the Coast Range ophiolite has a more oceanic nature and is overlain by mafic breccia that was derived entirely from the ophiolite. In the Healdsburg terrane, a basal ophiolite is overlain by thick (3000 m) conglomerate consisting largely of continentally-derived rhyolitic clasts.

West of the San Andreas fault, the terranes of the Salinian block include the Salinia terrane (continental fragment) with its high-grade metamorphic-plutonic basement and the Point Arena terrane, which is probably floored by oceanic crust, but which includes Cretaceous strata of both continental and oceanic origin.

Analysis of the present distribution of the terranes, the nature of the faults that bound them, and the timing of unconformities and overlap assemblages places some constraints on the nature and timing of accretion. It appears that some of the older parts of the terranes of the Great Valley sequence were accreted to North America in the Early Cretaceous. Accretion of many of the Franciscan terranes followed in the middle Cretaceous, Late Cretaceous, and early Cenozoic, accompanied by low-angle, imbricate thrust-faulting and in some terranes, by high P/T metamorphism. Accretion of Franciscan rocks ceased with the initiation of the San Andreas transform fault system. The Salinia and Point Arena terranes were moved to their present positions along strike-slip faults concurrent with northward dispersion of Franciscan and Great Valley sequence terranes along the Hayward, Calaveras, and related fault strands.

INTRODUCTION

The San Francisco Bay region contains three different, fault-bounded complexes that overlap in age (fig. 1). The Jurassic and Cretaceous Great Valley sequence, the Jurassic to early Tertiary Franciscan Complex, and the Salinian block which includes middle Cretaceous plutonic rocks overlain by Upper Cretaceous and younger sedimentary rocks. Concepts concerning the origin and tectonic history of these rocks have evolved rapidly since about 1977. During the decade prior to that, nearly all of the region had been remapped or at least reevaluated in terms of plate tectonics (see the Introduction to this volume for a historical sketch). By 1977, however, it was clear that simple plate-tectonic models did not explain the complex pattern of fault-bounded units of roughly the same age that exhibited different stratigraphy, structural and metamorphic histories. At about this same time, the concept of tectonostratigraphic terranes was introduced to describe similarly complex relationships in British Columbia (Monger and Ross, 1971), the Klamath Mountains (Irwin, 1972), and Alaska (Jones and others, 1978). Stated simply, a tectonostratigraphic terrane is defined as a fault-bounded geological entity of regional extent, characterized by a geologic history that is different from its neighbors. Using this concept as a working hypothesis, we identified and mapped a number of distinctive terranes in California (Blake and others, 1982), many of them in the San Francisco Bay region. The present paper presents further stratigraphic, structural, sedimentologic and metamorphic data for each of the Bay area terranes. The amount of detailed information available varies greatly from terrane to terrane, as shown in the following section.

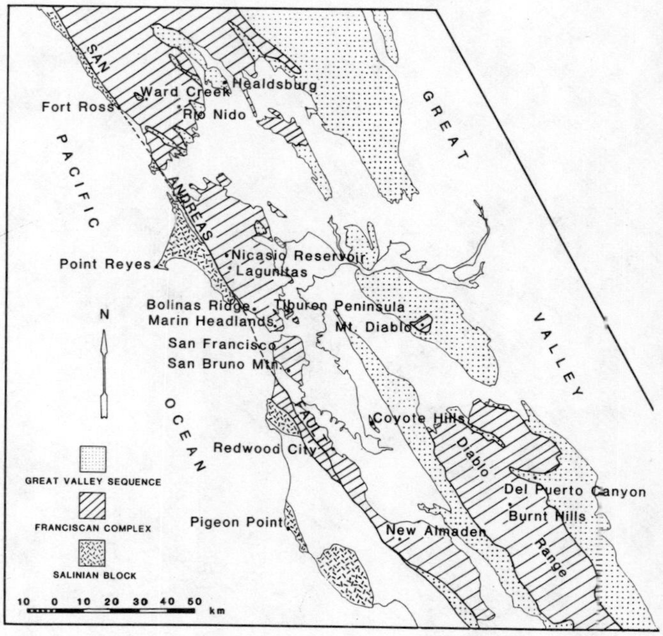

Figure 1. Index map showing the three major geologic units of the San Francisco Bay region and various localities referred to in text.

in Blake, M.C. Jr., ed., 1984, **Franciscan Geology of Northern California:** Pacific Section S.E.P.M., Vol. 43, p. 5-22

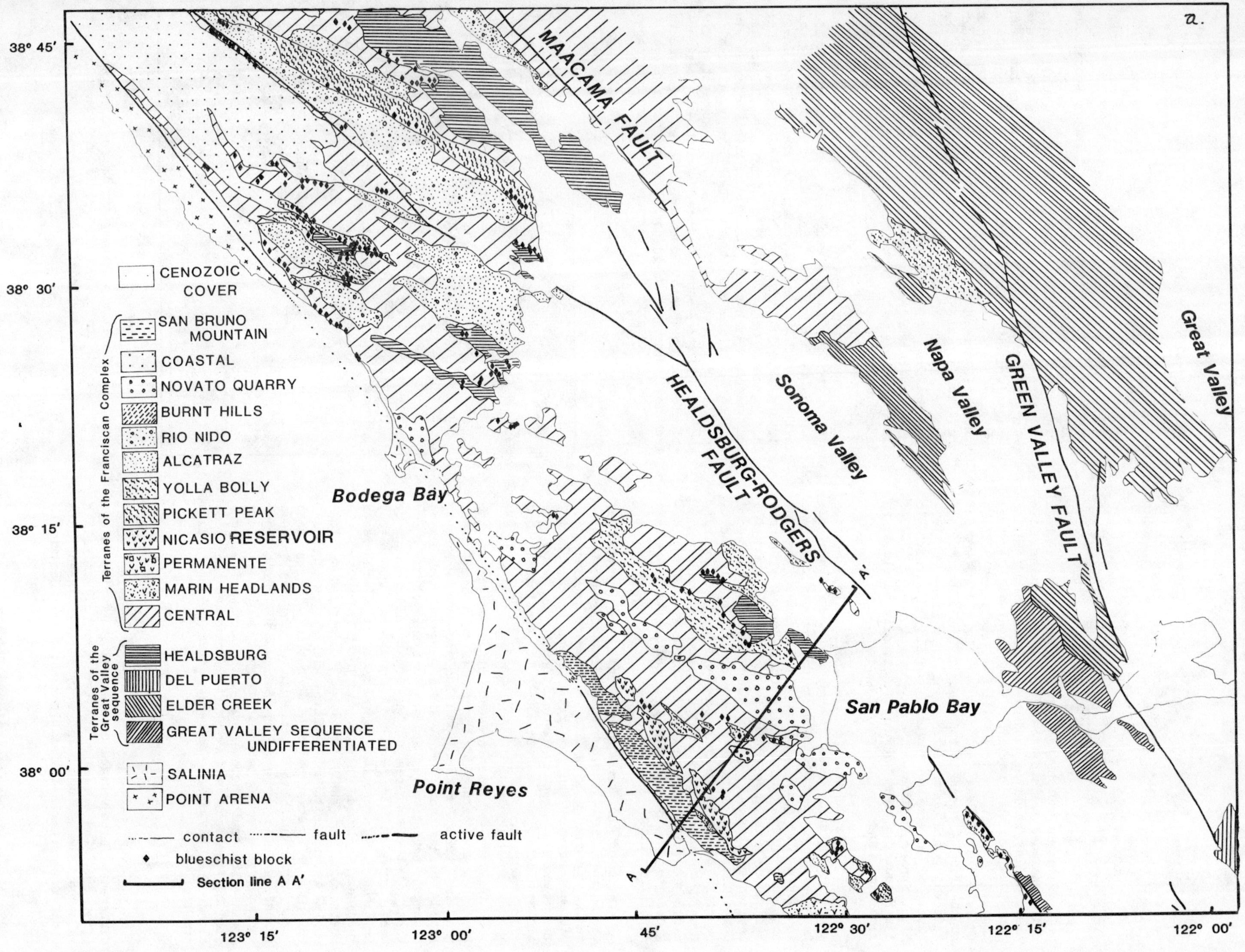
a.
MAACAMA FAULT
HEALDSBURG-RODGERS FAULT
GREEN VALLEY FAULT
Great Valley
Napa Valley
Sonoma Valley
Bodega Bay
Point Reyes
San Pablo Bay
A
A'
CENOZOIC COVER
Terranes of the Franciscan Complex
SAN BRUNO MOUNTAIN
COASTAL
NOVATO QUARRY
BURNT HILLS
RIO NIDO
ALCATRAZ
YOLLA BOLLY
PICKETT PEAK
NICASIO RESERVOIR
PERMANENTE
MARIN HEADLANDS
CENTRAL
Terranes of the Great Valley sequence
HEALDSBURG
DEL PUERTO
ELDER CREEK
GREAT VALLEY SEQUENCE UNDIFFERENTIATED
SALINIA
POINT ARENA
contact
fault
active fault
blueschist block
Section line A A'
38° 45'
38° 30'
38° 15'
38° 00'
123° 15'
123° 00'
45'
122° 30'
122° 15'
122° 00'

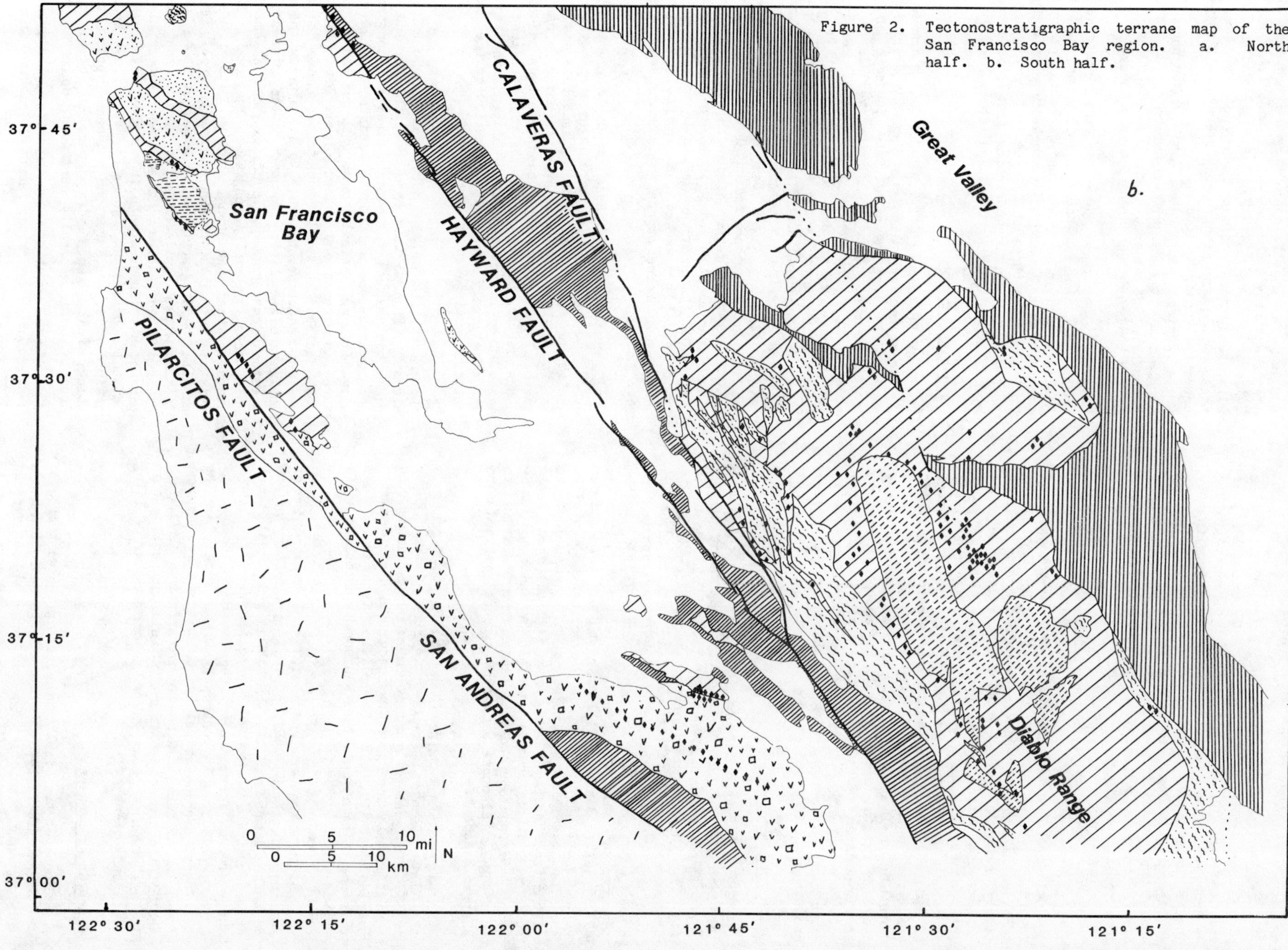

Figure 2. Tectonostratigraphic terrane map of the San Francisco Bay region. a. North half. b. South half.

DESCRIPTION OF TERRANES

Figure 2 shows the tectonostratigraphic terranes of the San Francisco Bay region. We emphasize that each terrane: 1) is characterized by a stratigraphic record that differs from the records of nearby, partly coeval terranes; 2) is fault-bounded; and is abruptly juxtaposed with contrasting terranes (no transitional zones are presently recognized); and 3) records a deformation and metamorphic history different from those of its neighbors.

Great Valley Sequence Terranes

Along the eastern margin of the Coast Ranges and in fault slices within them, several distinctive terranes have been identified within the unit collectively referred to as the Great Valley sequence (Bailey and others, 1964; Bailey and others, 1970). This sequence consists of the Coast Range ophiolite, representing oceanic crust and/or island arc basement of Middle Jurassic age (Hopson and others, 1981), and overlying well-bedded sedimentary rocks of Late Jurassic and Cretaceous age. In general the rocks lack the stratal disruption and metamorphism seen in the Franciscan terranes, contain more abundant fossils, and nearly everywhere structurally overlie Franciscan rocks along the Coast Range thrust (Bailey and others, 1970).

Del Puerto Terrane

The Del Puerto terrane (fig. 2) crops out along the east side of the Diablo Range from Mount Diablo south to a termination west of Bakersfield (south of fig. 2). The basal part of this terrane consists of a dismembered ophiolite that includes a thick accumulation of siliceous volcanic rocks (keratophyre and quartz keratophyre) in the upper part. This ophiolite sequence probably represents an island arc (Evarts, 1977, 1978; Blake and Jones, 1981). The Upper Jurassic Lotta Creek Formation overlies the ophiolite and is overlain by mudstone of Late Jurassic (Tithonian) and Early Cretaceous (Albian) age. These rocks are disconformably overlain by sandstone, mudstone, and conglomerate that ranges from Cenomanian to as young as Paleocene and which extends eastward onto Sierran basement.

The western margin of the Del Puerto terrane is the late Cenozoic Tesla-Ortigalita fault (Page, 1981). The eastern boundary with Sierran basement is buried beneath the Upper Cretaceous and early Cenozoic overlap deposits, but is believed from geophysical evidence to be a fault (see Wentworth and others, this volume).

Sixty kilometers south of figure 2 near Llanada (Hopson and others, 1981, p. 438), preliminary paleomagnetic data from the ophiolite and basal Upper Jurassic strata of the Del Puerto terrane suggest that these rocks formed near the equator (Beebe and Luyendyk, 1983). In contrast, similar rocks from the Del Puerto terrane at Del Puerto Canyon (fig. 1) were found to contain a magnetic overprinting similar to that of the present geomagnetic field, suggesting that it was acquired long after deposition and folding (Frei and others, 1983).

Elder Creek terrane

North of the Diablo Range, along the west side of the Sacramento Valley, the Coast Range ophiolite has a more oceanic character; the siliceous volcanic rocks of the Del Puerto terrane are entirely absent. In addition, the uppermost part of the ophiolite consists of a breccia, locally containing well-rounded clasts, made up entirely of pieces of the underlying ophiolite (Bailey and others, 1970). The overlying Jurassic sedimentary rocks consist largely of mudstone, sandstone, and chert-pebble conglomerate. Sedimentary serpentine breccias occur in the Lower Cretaceous strata (Carlson, this volume) and a major unconformity has been identified near the base of the Turonian (McLaughlin and others, 1980; Blake and Jones, 1981, but see Wentworth and others, this volume).

The Coast Range ophiolite and overlying Upper Jurassic and Lower Cretaceous sedimentary strata from this terrane near Stonyford and Paskenta, north of the area of figure 1, were also sampled for paleomagnetic study and, like those at Del Puerto Canyon, were found to have a magnetic overprinting that post-dates both deposition and folding (Frei and others, 1983).

Healdsburg terrane

Near Healdsburg (fig. 1) and to the southeast is a dismembered ophiolite structurally overlain by a sequence of turbidites of Late Jurassic and Early Cretaceous age, which in turn are overlain by a thick pile of Upper Cretaceous conglomerate. Because the Early Cretaceous section is so different from coeval strata to the east in the Elder Creek and Del Puerto terranes (Ingersoll, 1978; Ingersoll and Dickinson, 1981), we describe it in more detail.

At the Dry Creek Dam site 15 km north of Healdsburg, sandstone facies C and B of Mutti and Ricci-Lucci* are interbedded with thin-bedded turbidites (facies E) and concretion-bearing mudstone (facies G). Carbonaceous debris is conspicuous in the thick bedded graywackes. In a few places bedding thickness patterns suggest thickening upward sequences.

Upsection from these probable submarine fan deposits, the bedding style changes abruptly to lenticular beds of alternating sandstone and mudstone lacking sedimentary structures. Grading is restricted to thin (about 1 cm thick) sandstone beds and carbonaceous debris is ubiquitous in the sandstone. This association is not indicative of submarine fan facies and more likely represents a shallow marine, possibly deltaic, environment of deposition.

*Generalization of the Mutti and Ricci-Lucci (1972) classification is used throughout this paper. It can be generalized as follows: Facies A: conglomerate, pebbly sandstone and coarse sandstone, thick to massive bedding, lenticular beds with cut and fill structures, nongraded and graded, Bouma Ta or Tae sequences; facies B: pebbly sandstone, coarse- to medium-grained sandstone, beds 0.4 to 1.0 m thick, beds often amalgamated, pelitic partings often present, broad erosional concave lower surface, diffuse laminations, sole marks, generally nongraded; facies C: fine- to coarse-grained sandstone, beds 0.1 to 1.5 m thick mudstone or siltstone interbeds, generally even and parallel bedding surfaces, sole marks, Bouma Tabcde to Tae sequences, graded; facies D: thinner and finer grained than facies C, sequences, beds laterally persistent. These strata in the past have been called "distal" turbidites; facies E: very fine- to coarse-grained sandstone, mudstone interbeds, beds 3 to 20 cm thick, lenticular bedding, parallel bases and wavy tops of beds, generally nongraded, Bouma Tce of Tde; facies F: chaotic slump deposits, pebbly mudstone, retransported deposits; facies G: hemipelagic pelite, massive bedding.

Up to 3000 m of conglomerate overlies and is locally channelized on top of the shallow marine strata. The conglomerate clasts are principally well-rounded pebbles and cobbles of light-colored rhyotite porphyry and rhyolitic welded ash-flow tuff, and lesser amounts of chert. Buchias of Early Cretaceous (Valanginian) age are found throughout the turbidites and in the conglomerate.

The conglomerate is massive, with lenticular structureless beds of sandstone (fig. 3). Because of its great thickness, the lack of bedding, and the stratigraphic position at the top of an apparent retrogradational sequence, this conglomerate package probably represents non-marine deposition, possibly in an outwash-plain, braided stream setting.

Figure 3 Steeply-dipping, right-facing Lower Cretaceous cobble conglomerate with lenses of coarse sandstone. Healdsburg terrane, near Healdsburg. Beer bottle for scale.

Franciscan terranes

Although the Franciscan Complex overlaps in age with the Great Valley sequence, there are notable differences in lithology, structural state, and metamorphic history that led to their early recognition as separate geologic units. In particular, the Franciscan rocks are distinguished by containing greenstone (altered volcanic rocks) and radiolarian chert and by their degree of stratal disruption, which everywhere varies from broken formation to tectonic melange (Hsu, 1968). Greenstone and radiolarian chert occur in the Great Valley sequence only in the basal part, in what is now considered to be Coast Range ophiolite.

Yolla Bolly terrane

The Yolla Bolly terrane occurs in the San Francisco Bay region as a number of widely distributed fault slices, in contrast to its broad coherent extent along the eastern margin of the northern Coast Ranges (Blake and others, 1981). The distinguishing features of the Yolla Bolly terrane include: graywacke interbedded with laterally persistent radiolarian chert horizons (fig. 4), a metamorphic overprinting manifested by a weak to pronounced penetrative schistosity (textural zone 2 of Blake and others, 1967), and the presence of high P/T metamorphic minerals including lawsonite, jadeite, and glaucophane. In addition to the predominant graywacke-chert lithology, there are local bodies of metamorphosed intrusive rocks (diabase and gabbro) and rare metaserpentinite, such as that on Angel Island in San Francisco Bay (Schlocker, 1974). Radiolarians from the cherts, together with scarce megafossils, suggest a range in age from Late Jurassic (Tithonian) into Early Cretaceous (Valanginian) (Blake and Jones, 1974; Murchey and Jones, this volume). Numerous thrust faults and zones of isoclinal folding tectonically thicken the stratigraphic sequence.

The graywackes are characterized principally by two lithofacies associations. One association includes thick intervals of fine grained facies E, G

Figure 4. Interbedded metagraywacke (mgwy) and metachert (mch) of the Yolla Bolly terrane, Del Puerto Canyon, Diablo Range.

and D that are interbedded with a second facies composed of thinning- and fining-upward megasequences of lithofacies A, B, C, D, E and F. These lithofacies characterize submarine fan deposition in a mid-fan or supra-fan setting that includes transquil, dilute turbidite deposition, probably in channel and interchannel environments.

The second association includes dominantly thick intervals of thick-bedded turbidites, fluxoturbidites or grain flows (facies B and C) interbedded with thin intervals of thin-bedded turbidites (facies D or E). This association is similar to the strata of the Alcatraz and San Bruno Mountain terranes; in the Yolla Bolly terrane, however, numerous chert horizons are interbedded with the graywacke. Most chert units are 20 to 50 m thick and a common sequence involves sandstone passing upward into siliceous shale which grades into pure chert. Sandstone usually overlies the chert, commonly with a thin (50 cm) basal breccia of chert and mudstone.

Within the context of a submarine fan model, the second lithofacies association also indicates a middle or supra-fan setting, though the occurrences of interbedded chert are enigmatic. The accumulations of chert must represent periods of diminished to nonexistent clastic deposition. The chert beds may correspond to areas between places of active deposition in a supra-fan setting, analogous to mud blankets between depositional lobes; alternatively, chert deposition may have occurred more uniformly over the entire fan during intervals of high stands of sea level, events that trap detrital material in inner shelf, estuarine, or bay settings.

Chemically and petrographically, sandstone of the Yolla Bolly terrane is quartzofeldspathic. This composition, together with limited isotopic data (Blake and Jones, 1981), indicates a continental source area. Additional petrographic studies by Jacobson (summarized by Jayko and Blake, this volume) and Seiders (this volume) show that the sandstones and conglomerates of the Yolla Bolly terrane differ markedly from coeval sedimentary rocks of the Great Valley sequence.

Numerous radiometric dates have been obtained on metagraywacke and mineral separates from rocks of the Yolla Bolly terrane. These metamorphic ages range widely from 76-122 m.y. (Suppe and Armstrong, 1972; Lanphere and others, 1978). We believe, however, that a U/Pb age of 92 m.y. determined on metamorphic sphene from the Diablo Range (Mattinson and Echeverria, 1980) reflects the true age of recrystallation because the dates on metagraywacke are contaminated by extraneous radiogenic argon from detrital white mica that is common in the graywackes (see Lanphere and others, 1978, p. 809, for a discussion of this problem).

Pickett Peak(?) terrane

West of Healdsburg and extending to the northwest for many kilometers (fig. 2) is a narrow belt of schistose metagraywacke (textural zone 3 of Blake and others, 1967) that is more deformed and metamorphosed than the rocks of the Yolla Bolly terrane. This belt is tentatively correlated with the Pickett Peak terrane of northern California (Blake and others, 1982). Rocks of comparable grade, but consisting almost entirely of metavolcanic rocks and minor metachert, occur in the vicinity of Ward Creek (Coleman and Lee, 1963). None of these rocks have yielded any fossils but radiometric ages from the rocks at Ward Creek (Lee and others, 1963) suggest an Early Cretaceous metamorphic event (~130 m.y.). A few blocks of similar metagraywacke are also present in melange east of the Maacama fault (McLaughlin and Ohlin, this volume).

The dominant metamorphic mineral assemblage seen in metagraywacke of this area is quartz-lawsonite-glaucophane-white mica-jadeitic pyroxene-sphene ± garnet. This mineral assemblage is compatible with the Type III metabasalt of the Ward Creek area reported by Coleman and Lee (1963).

Burnt Hills terrane

The Burnt Hills terrane crops out in the core of the northern Diablo Range (Cotton, 1972; Crawford, 1975, 1976; Blake, 1981) where it occupies the lowest structural position. Rocks in this terrane consist chiefly of thin-bedded sandstone and mudstone (fig. 5) with local lenticular bodies (channels) of coarse-grained sandstone and pebble conglomerate. Also present are local interbeds of radiolarian chert of Early Cretaceous (Aptian-Albian) age (Murchey and Jones, this volume). Megafossils from the thin-bedded sequence are of Late Cretaceous (Coniacian) age (D. L. Jones, written commun., 1980).

The sandstones of the Burnt Hills terrane are arkoses consisting largely of quartz, feldspar, and biotite derived from a plutonic source (Jacobson, 1978). Metamorphic mineral assemblages in the sandstone range from quartz-albite-pumpellyite through quartz-albite-lawsonite to quartz-lawsonite-jadeitic pyroxene ± glaucophane (Cowan, 1974; Morrell, 1978). The sedimentary facies of the Burnt Hills terrane were analyzed by Telleen (1977) who suggested that they represent a mid-fan environment (similar to many of the Franciscan terranes described herein).

Rio Nido terrane

Along the Russian River near Rio Nido (fig. 1) and in several fault-bounded belts to the north are several deformed but relatively unmetamorphosed, sandstone-rich belts (broken formation) which were originally thought to be part of the Coastal terrane

Figure 5. Thin-bedded turbidites in the Burnt Hills terrane, Burnt Hills, northern Diablo Range. Sandstone beds are about 10 cm thick.

(Blake and others, 1971). Subsequent petrographic studies (M. C. Blake, Jr., and R. H. Wright, unpublished data; Jayko and Blake, this volume) indicated that the rocks here assigned to the Rio Nido terrane lack the abundant K-feldspar characteristic of the Coastal terrane and appear to be older, based on a single Late Cretaceous (Turonian) fossil (Bailey and others, 1964, p. 117, no. 7).

Novato Quarry terrane

This terrane forms a northwest-trending belt between the Hayward fault and the San Andreas fault (fig. 2). It consists of sandstone, mudstone, and some conglomerate containing rare fossils of Late Cretaceous (Campanian) age (Bailey and others, 1964). These strata are isoclinally folded and locally disrupted (broken formation). The sandstone contains prehnite, pumpellyite, and detrital potassium feldspar, an assemblage also found in the Coastal and Alcatraz terranes. The sandstones are arkosic (Jayko and Blake, this volume) and thus are similar to the Burnt Hills terrane.

Good exposures of the Novato Quarry terrane crop out near the east abutment of the San Rafael Bridge at the north end of San Francisco Bay. There, thickening upward cycles about 10 m thick involve lithofacies G, E, C and B in upward order. Elsewhere, thin- and thick-bedded turbidites crop out with no obvious cyclicity in bedding thicknesses.

The inferred depositional environment for strata of the Novato Quarry terrane is the fringe of a submarine supra-fan. The common occurrence of mudstone and thin-bedded turbidites is typical of mid-fan fringe deposits. The thick-bedded units of facies C and B units may represent high concentration sheet flows beyond distributary channels and the upward thickening sequences are commonly associated with prograding lobes or aggrading channels. The Novato Quarry terrane is one of the youngest and most coherent terranes within the Franciscan tectonic collage. To the south, in the Coast Ranges of central California, similar rocks of this age crop out at Point San Luis, Cambria and Pfeiffer Beach (Howell and others, 1977). There the coherent graywacke packages are inferred to lie depositionally on top of Franciscan melange in a trench-slope setting (Smith and others, 1979). A similarly overlapping depositional environment is an appealing interpretation for the Novato Quarry terrane, despite its present fault boundaries.

Nicasio Reservoir terrane

This terrane consists largely of pillow basalt plus minor intrusive diabase-gabbro, tuff, hyaloclastite, and overlying radiolarian chert and graywacke. The radiolarian chert, which is of Early Cretaceous (Valanginian) age, locally overlies the pillow lava depositionally and is, in turn, overlain by lithic graywacke (Wright, this volume).

Chemical and petrologic studies of the volcanic and intrusive rocks (Gluskoter, 1962; Swanson and Schiffman, 1979) suggest that the basaltic rocks are tholeiitic. New chemical analyses (Table 1) have been obtained on samples of basalt and intrusive gabbro taken from drill holes at the Peters Dam Site near Lagunitas (fig. 1). When plotted on the P_2O_5:TiO_2 diagram (fig. 6) of Ridley and others (1974) they suggest that most of the rocks are tholeiitic but that they overlap the fields of ocean ridge and ocean island. In addition, abundant serpentinized ultramafic rocks diapirically intrude the sedimentary rocks in the central part of the terrane (see Wright, this volume, for a discussion of the origin of the serpentinite diapirs).

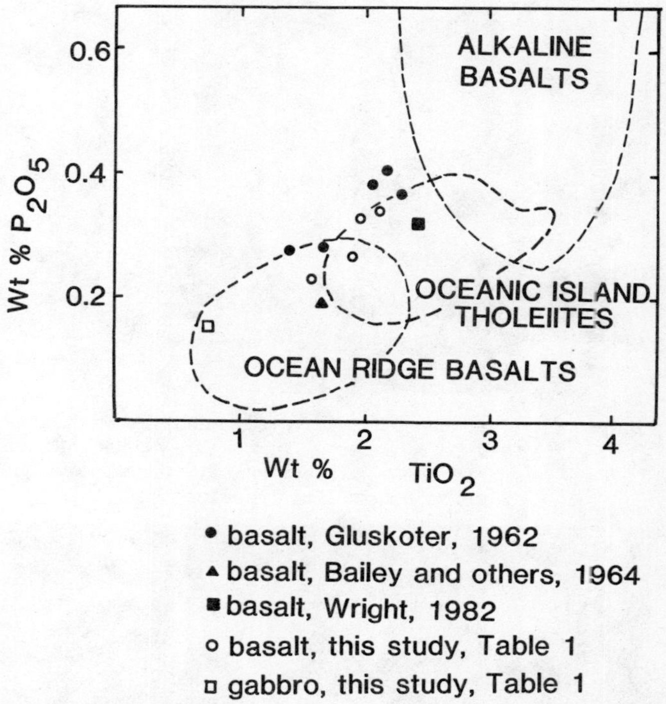

Figure 6. P_2O_5:TiO_2 diagram showing rocks from the Nicasio Reservoir terrane. Compositional fields are from Ridley and others, 1974.

Alcatraz terrane

On Alcatraz and Yerba Buena Islands in San Francisco Bay and nearby (fig. 2) is found another graywacke-rich terrane (broken formation) that was originally part of the Lawson's type Franciscan Formation (Lawson, 1914). Fossils from Alcatraz Island (Armstrong and Gallagher, 1977) indicate a Valanginian age and represent shallow-water environments, although these fossils may be displaced. Elsewhere, this terrane contains lithofacies C and E, with beds displaying Bouma Ta-e and Tc-e turbidite sequences (figs. a, b, c).

Detrital potassium feldspar is present in minor quantities (Wright, 1974) and coexists with metamorphic prehnite and pumpellyite at several localities. Although not obviously highly deformed in most places, there is abundant evidence of isoclinal folding on Russian Hill in San Francisco and, also cores from drill holes under San Francisco Bay, midway between San Francisco and Yerba Buena Island.

Marin Headlands terrane

From Marin Headlands (Fig. 1) south through San Francisco to the Coyote Hills (fig. 2) is the Marin Headlands terrane which is composed of pillow lava, radiolarian chert, and overlying graywacke (Blake and others, 1982; Wahrhaftig, this volume). The basalts from this terrane have not received adequate attention, although a good petrographic description is given by Schlocker (1974) and several chemical analyses are presented by Bailey and others (1964). The cherts, in contrast, have been the subject of several recent studies (see those by Murchey , this volume, and Karl, this volume) which demonstrate that they

Figure 8a. Overturned section of massive to thick-bedded, facies C and B graywacke of the Marin Headlands terrane, north end of Baker's Beach, San Francisco.

Figure 8b. Closeup of 8a showing stacked, thick-bedded fluxoturbidites.

Figure 8c. Closeup of 8a showing facies B fluxoturbidites with carbonaceous debris along diffuse laminations.

Figure 8d. Thin-bedded turbidites, facies E, from the Marin Headlands terrane, south end of Baker's Beach, San Francisco.

The San Bruno Mountain terrane is dominated by lithofacies C, B and E. The occurrences of bundles of facies B sandstone interbedded with the thinner-bedded facies E turbidites characterize supra-fan deposition (Normark, 1972). The one occurrence of lobe deposition is consistent with a generally restricted range of depositional patterns, which is typical of sand-dominated middle fan or supra-fan settings. Paleocurrent trends based on flute casts and imbricated mudstone rip-up clasts indicate northeast to southwest flow, although until the extent of tectonic rotations can be assessed, the actual flow patterns must be appraised in a guarded fashion.

In Marin County, a probable extension of the San Bruno Mountain terrane underlies Bolinas Ridge (fig. 2). Gluskoter (1962, 1969), recognized that the graywacke in this area contained detrital as well as secondary potassium feldspar. Average detrital modes are nearly identical to that of the San Bruno Mountain area (Jayko and Blake, this volume).

Permanente terrane

Between the Pilarcitos and San Andreas faults, and in an offset segment to the southeast (fig. 2), is a terrane that consists largely of basaltic volcanic rocks, includes minor intercalations of Lower Cretaceous (Valanginian) radiolarian chert (Murchey and Jones, this volume) and is overlain by the Calera Limestone of Late Cretaceous age (Sliter, this volume). Also present is graywacke that probably was deposited on the limestone. This graywacke has not been studied petrographically or chemically and no fossils have been found in it.

Paleontological (Sliter, this volume) and paleomagnetic studies (John Tarduno, personal commun., 1984) suggest that the Calera Limestone formed at low latitudes and was carried northward by large-scale plate motions to its present position.

Coastal terrane

The southern terminus of the Coastal terrane of Blake and others, (1982) crops out in the northwest corner of the map area (fig. 2). This body of rock was originally called the Coastal belt (Bailey and others, 1964) but has since been divided into two terranes, the Coastal and Yager terranes (McLaughlin and others, 1983). Within our study area, we identified

Figure 7a. Distrupted strata, broken formation, of the Alcatraz terrane, Tiburon Peninsula.

Figure 7b. Thin-bedded turbidites, facies E/D interbedded with facies B, fluxoturbidites, Alcatraz terrane, Tiburon Peninsula.

record a history of nearly 100 m.y. of oceanic sedimentation far-removed from any continental land mass.

The graywacke unconformably overlying the chert and basalt has yielded ammonites of Albian and Cenomanian age (Bailey and others, 1964). Where exposed, it consists principally of thick-bedded to massive lithofacies B (figs. 8a, b) with interbeds of lithofacies E (fig. 8c)--wavy thin-bedded turbidites. Carbonaceous debris is locally abundant. Woody fragments up to 10 cm in long dimension are randomly dispersed in thick sandstone beds, but more typcally, comminuted carbonaceous debris accentuate diffuse laminations in thick-bedded facies B fluxoturbidites. Grooves and convolutions suggest paleoflow in a N to NW or S to SE direction, although the fact that these rocks may have been rotated as much as 129° in a clockwise direction (Curry and others, this volume) means that paleo-flow may have been W to NW or E to SE.

The rocks of this terrane display a complete range of structural disruption from no obvious deformation in outcrop, as at the north end of Bakers Beach

Figure 7c. Thick turbidite, Bouma Ta-e sequence, Alcatraz terrane, Yerba Buena Island, San Francisco Bay.

in San Francisco (although the entire section is overturned there), to broken formation containing isolated fold hinges, as at the north end of the Golden Gate Bridge, to complete stratal disruption, as occurs in parts of San Francisco (Wahrhaftig, 1984). Sandstone at the Marin Headlands contains abundant laumontite veinlets. Some sandstones mapped as part of this terrane in central San Francisco contain fine-grained pumpellyite but these occur in a poorly exposed area of disrupted rock and could be pieces of the melange matrix tectonically incorporated into the Marin Headlands terrane.

San Bruno Mountain terrane

In San Francisco, another graywacke-rich terrane (broken formation) of unknown age, occurs that is characterized by rather abundant detrital quartz and potassium feldspar. The strata occur in outcrops of coherent graywacke that are locally disrupted by high angle faults and zones of broken formation (fig. 9a); where the graywacke is undisturbed, thin to thick-bedded turbidites predominate (fig. 9b). Outcrops are limited to road and quarry cuts where sections are less than 20 meters thick.

Exposures on San Bruno Mountain (fig. 1) and nearby are principally medium-grained sandstone with bedding thickness ranging from massive (amalgamated turbidite flows) to thin (individual turbidite flows with a relatively thick hemipelagic mudstone component). The sandstone:shale ratio is approximately 8:1, though the nature of the outcrops biases this ratio; unexposed areas of extensive grassy slopes are probably underlain by more mud-rich units. Some outcrop display a crudely developed, thickening-upward cycle (fig. 9c); such cycles are commonly ascribed to prograding lobes. Capping these cycles are mud-rich, thin-bedded turbidites (fig. 9d).

Figure 9a. Thick-bedded fluxoturbidites, facies B, faulted against disrupted bedding of facies B, broken formation (on right). San Bruno Mountain terrane, San Francisco.

Figure 9b. Thin-bedded turbidites, facies D and E, San Bruno Mountain terrane, South San Francisco .

Figure 9c. Thickening-upward sequence, facies C and D overlain by composite beds of facies B. San Bruno Mountain terrane, San Francisco

Figure 9d. Closeup of basal part of figure 9c showing thick-bedded turbidites, Bouma T,b,c,d,e.

Figure 9e. Facies E, thin-bedded turbidites capping the thickening-upward sequence of figure 9c.

the Coastal terrane on the basis of its arkosic sandstone composition, abundant potassium feldspar content and the presence of abundant veins of laumontite. For the most part, the terrane consists of sandstone, shale, and conglomerate, but it also includes minor volcanic rocks and rare limestone (Harbert and others, this volume). Fossils found in the Coastal terrane range in age from Upper Cretaceous to upper Eocene (Blake and Jones, 1974; Bachman, 1978).

Farther north the Coastal terrane has been extensively studied, both from the standpoint of its sedimentary facies and deformational history (Kleist, 1974; O'Day, 1974; Kramer, 1976; Bachman, 1978, 1981; Underwood and Bachman, 1981; Underwood, 1983). There, the Coastal terrane represents a series of coalescing submarine fans (turbidites) deposited in a trench or trench-slope environment. The rocks have been complexly deformed, including early soft-sediment folds and several phases of tectonic folding. The scarce volcanic rocks are considered by some to be interbedded with the clastic sedimentary rocks (Kramer, 1976), but others believe them to be faulted into the sedimentary section (Bachman, 1978; Harbert and others, this volume).

The source of the Coastal terrane is also somewhat controversial. Studies to the north (Kleist, 1974; Kramer, 1976) indicate a predominantly arkosic composition in which potassium feldspar and plagioclase are equally abundant. Rock fragments vary both in abundance and composition and locally include reworked Franciscan(?) detritus in the form of red and green chert, glaucophane grains, blueschist rock fragments, and serpentinite (Kramer, 1976). Petrographic studies from rocks collected in the Bay area (Jayko and Blake, this volume) indicate that there are two distinctive facies present, but this was not recognized in the field. The dominant facies is represented by quartzofeldspathic modes (average QFL = 45:46:9). The other facies is more volcanic with an average QFL = 30:39:31.

Central terrane

All of the previously-described Franciscan terranes in the San Francisco Bay region are tectonically enclosed in a melange matrix defined as the Central terrane (Blake and others, 1982). In addition to the mapped terranes, the Central terrane includes numerous blocks too small too show on figure 2, including blocks of pinkish Laytonville-like limestone (Alvarez and others, 1980; Sliter, this volume). The melange matrix consists of sheared, interbedded fine-grained mudstone, lithic sandstone, and greenish radiolarian tuff. Pillow lava is probably also interlayered in the matrix, but is commonly detached by shearing. The fossils thus far identified in the radiolarian tuffs are all of Late Jurassic or Early Cretaceous age (Jones and Murchey, this volume). Buchias, usually found in limy concretions, range in age from Late Jurassic (Tithonian) to Early Cretaceous (Valanginian). Other fossils recovered from the matrix (largely dinoflagellates) are also restricted to the Late Jurassic and Early Cretaceous (W. R. Evitt, written commun., 1976).

In addition to the penetrative shearing seen in the Central terrane matrix, the sandstones contain abundant metamorphic minerals that include quartz, albite, chlorite, white mica, and pumpellyite. Locally, however, the Central terrane matrix is more metamorphosed, as in the Diablo Range, where the sheared graywacke lenses and blocks contain a pervasive but weak schistosity (Textural zone 2A of Blake and Jones, 1977). This schistosity is accompanied by the growth of lawsonite, aragonite, and locally, by jadeitic pyroxene. Detailed study by Raymond (1973a) in the Diablo Range indicates that the shearing characteristic of the melange is superposed on the schistosity.

The lithology of the Central terrane matrix--mudstone and minor turbidite sandstone, radiolarian tuff, and other volcanic rocks--suggests a deep-water environment of deposition. As pointed out by Blake and Jones (1974), the Central melange matrix is faunally and lithologically similar to the basal strata of the Great Valley sequence. It is possible that the Central terrane was a more distal volcanic- and chert-rich facies of the Great Valley sequence into which the other Franciscan terranes were tectonically emplaced.

Salinian Block

Salinia terrane

West of the San Andreas fault is a large continental fragment known as Salinia or the Salinia terrane (Blake and others, 1982). It consists, in the San Francisco Bay region, of several fault-bounded blocks (not differentiated on fig. 2) that possibly represent separate terranes (Nilsen, 1981). The western block consists of the Upper Cretaceous (Campanian-Maestrichtian) Pigeon Point Formation, possibly underlain by locally exposed siliceous volcanic rocks. The Pigeon Point Formation consists of interbedded sandstone, mudstone, and conglomerate of deep-sea fan facies. East of the block composed of the Pigeon Point block is another block (La Honda) composed of schist, micaceous quartzite, calc-hornfels, and graphitic marble, intruded by middle Cretaceous granitic rocks (Mattinson, 1978). Unconformably overlying the crystalline basement are sedimentary rocks of the Paleocene Locatelli Formation and the Eocene Butano Sandstone. Although many early reconstructions placed Salinia at the southern end of the Sierra Nevada prior to initiation of movement on the San Andreas fault, others have marshalled geologic arguments against such a position (Ross, 1978). Recent paleomagnetic data suggests that the strata of the Pigeon Point Formation may have been deposited as much as 2,500 km south of their present location (Champion and others, 1984).

Point Arena terrane

North of Ft. Ross and west of the San Andreas fault is a sequence of sandstone, shale, and conglomerate of Late Cretaceous to Eocene age. These include two distinctive Cretaceous units, strata of Stewart's Point and strata of Anchor Bay, the former being derived from a granitic source area similar to Salinia, the latter from gabbro (Wentworth, 1968; Ross and others, 1973). Although these rocks lie west of the San Andreas fault, there is no evidence that they are underlain by granitic basement and, in fact, the K-feldspar-rich strata of Stewart's Point are structurally underlain by spilitized basalt, which probably represents oceanic crust.

Preliminary paleomagnetic data (Kanter and others, 1982) suggests that the spilitized basalt and overlying Cretaceous sedimentary rocks of the Point Arena terrane have anomalously low inclinations, similar to those seen in the Salinia terrane.

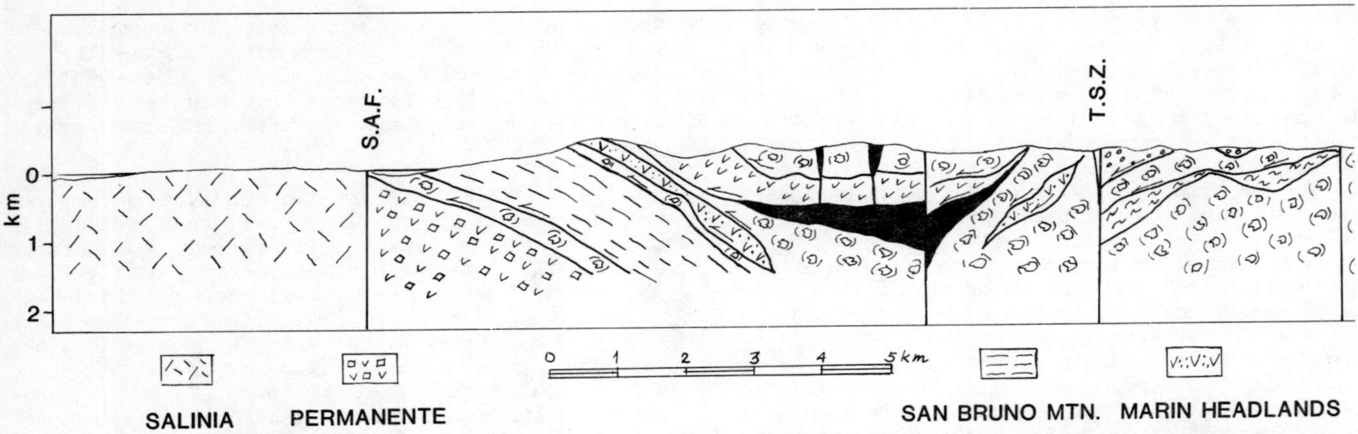

Figure 10. Structural cross-section A-A' (see figure 2 for location).

DISTRIBUTION OF BLUESCHIST KNOCKERS

Within the Central terrane melange (fig. 1), knockers consisting of exotic blueschist, amphibolite, and eclogite occur in several well-defined belts. In San Francisco they are very uncommon being restricted to the melange zones separating the more coherent terranes. There, the blueschists contain the low-grade assemblage glaucophane-lawsonite-sphene, which is superposed on an earlier amphibolite-facies protolith. Farther south, a few similar blueschist knockers found in the vicinity of Redwood City are probably related to the serpentinite body that is present along the northern boundary of the Permanente terrane (Atwater, 1975). Knockers are much more abundant in the New Almaden area (Bailey and Everhart, 1964), where they consist of both glaucophane schist and amphibolite and are clearly concentrated in two shear zones within and along the margin of the Permanénte terrane, associated with serpentinite.

Along the Hayward fault zone, high-grade blueschist knockers occur along an older serpentinite-marked suture between the Franciscan terranes and the Great Valley sequence that was subsequently reoccupied by the Hayward fault. This belt also includes many dismembered slabs of jadeite-bearing metagraywacke (Yolla Bolly(?) terrane).

In the Diablo Range, blueschist knockers are largely confined to the Central terrane melange, with the bulk of the mapped blocks occurring near the contact with the structurally lowest Burnt Hills terrane (Maddock, 1964; Cotton, 1972). A few, however, are found in narrow melange zones within the Yolla Bolly terrane and were considered by Crawford (1975, 1976) to represent olistostromal deposits.

North of San Francisco, high-grade blueschist blocks occur in several belts usually in close proximity to serpentinite and lower-grade metagraywacke, metagreenstone, and metachert of the Yolla Bolly terrane. Perhaps the best known occurrences are on the Tiburon Peninsula, where the blueschist knockers occur near the base of a thrust sheet (nappe) of serpentinite that overlies jadeitized metagraywacke of the Yolla Bolly terrane.

Farther north, many subparallel belts contain both serpentinite and blueschist knockers. Probably these result from both folding and faulting of older suture zones.

In conclusion, the high-grade blueschist knockers of the San Francisco Bay region occur in rather well-defined linear belts. They appear to be spatially related to regionally metamorphosed Franciscan terranes, typically in close proximity to parallel serpentinite belts. In a few areas they appear to be localized near the faulted boundary of the Franciscan Complex and the Great Valley sequence (Coast Range thrust), but in other areas they appear to be related to terrane boundaries within the Franciscan. The metamorphic age of the blocks, where determined, is about 160 m.y. (M. A. Lanphere, oral commun., 1983; Mattinson, 1982) and is thus older than nearly all of the fossils found in the Franciscan terranes. The metamorphic minerals of the blocks formed under P/T conditions that clearly exceed those of any of the regional blueschist-facies rocks of the Yolla Bolly or Pickett Peak terranes. Furthermore, many of the blocks started out as amphibolites or garnet amphibolites that were overprinted by blueschist assemblages. All of these facts suggest that the high-grade blueschists and related rocks formed during a Middle Jurassic event that predated all of the accretionary events in the San Francisco Bay region. The fact that these rocks and spatially related serpentinite are found along several different terrane boundaries extending as far east as the Great Valley (see Carlson, this volume for a description of similar knockers in the Great Valley sequence) suggests that the source of the blocks is a cryptic terrane that underlies all of the exposed terranes.

STRUCTURAL RELATIONSHIPS

Figure 10 is a geologic structure section (see fig. 2 for location of A-A') drawn across southern Marin County where the geology has been mapped at 1:24,000 or better (Gluskoter, 1962; Berkland, 1969, Wright, 1982). As a consequence, there is more detail shown on this section than is contained in the generalized geology on figure 2. The structure consists of a stack of rootless thrust sheets (nappes), generally less than 1 km thick, imbricated within the Central terrane melange matrix. High-angle strike-slip faults and gentle open folds of Neogene age disrupt the imbricate nappe structure. One of the faults shown on figure 10, the Tama Ranchero shear zone (TSZ) of Berkland (1969), appears to mark a major boundary between the more oceanic terranes of the west and the more continental terranes to the east. Also shown on this structure section is the internal imbrication of the Novato Quarry terrane. Such imbricate thrust faults have been described from many terranes (Crawford, 1975, 1976; Blake and Jayko, 1983; Wahrhaftig, this volume; Aalto, this volume). Because they involve both oceanic basement and continental-derived clastic graywacke cover, the thrusting must have occurred during the accretion of the terranes to the continental margin.

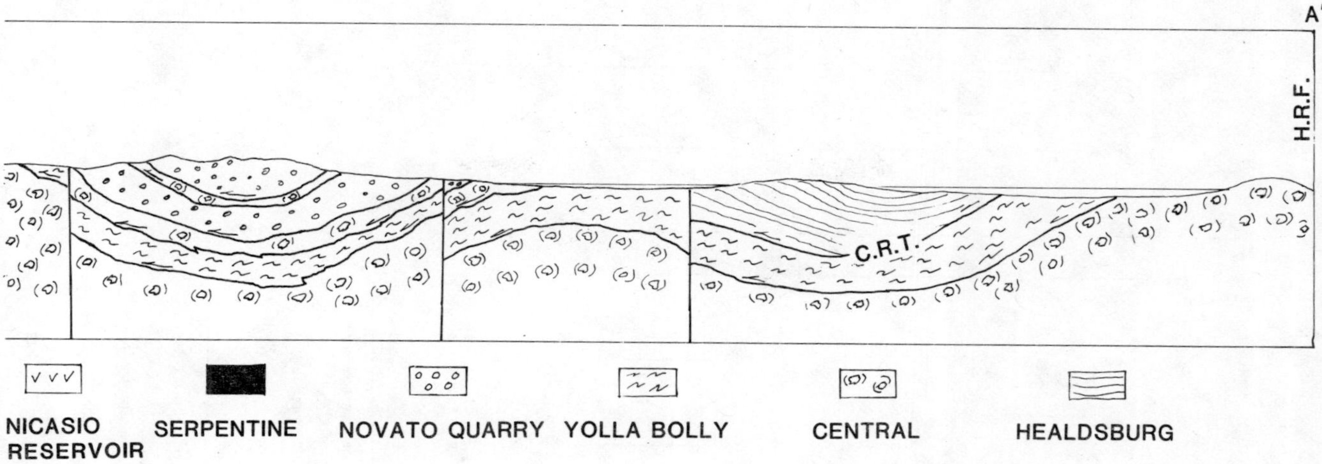

The structural style seen in figure 10 is identical to that seen in other well-studied areas of the Cordillera except that the scale is much smaller in the Bay area. In British Columbia (Monger and others, 1983), similar nappe structures formed during two orogenies, by the accretion of two large allochthonous terranes to the continental margin. A similar mechanism is proposed for the Bay area; however, why the nappes are so thin here is a presently unsolved problem that deserves further attention.

TERRANE ANALYSIS AND TIMING OF ACCRETION

In order to analyze the timing of accretion of the various terranes of the San Francisco Bay region, we have prepared a series of tectonostratigraphic columns including known times of deposition, unconformities, and ages of blueschist metamorphism (fig. 11). Patterns seen in this diagram plus sparse paleomagnetic data suggest the following sequence of events: First, accretion of the Elder Creek, Del Puerto, and Healdsburg terranes to the North American continental margin. The magnitude of movement associated with this event is unknown. The terranes could have formed adjacent to the continental margin or could have been swept in from a considerable distance. It is also possible, based on its Early Cretaceous metamorphic age, that the Pickett Peak terrane was accreted during the same event. The timing of accretion, based on the oldest overlap deposits that cover both Klamath basement rocks and the Elder Creek terrane in northern California, is Early Cretaceous. Farther south, at the latitude of San Francisco, the oldest sedimentary rocks known to overlie the Sierran basement are of Late Cretaceous (Turonian) age. The second event was the accretion and related(?) deformation and high P/T metamorphism of the Yolla Bolly terrane. The age of the youngest strata within the terrane (Valanginian), its U/Pb blueschist metamorphic age of about 92 m.y. (Mattinson and Echeverria, 1980), and the presence of a major disconformity (post-Albian-preTuronian) in both the Elder Creek and Del Puerto terranes (Raymond, 1973b; McLaughlin and others, 1980; Blake and Jones, 1981), suggest accretion along the Coast Range thrust in about middle Cretaceous time (Cenomanian-Turonian boundary). This age, however, would appear to be slightly too old to account for the high P/T metamorphism of the Burnt Hills terrane, which contains fossils as young as Coniacian (88 my). Whether all or some of the oceanic terranes came in at this time is also not clear either. The data of figure 11 suggest that the Marin Headlands terrane approached a continental margin and received its capping of clastic sedimentary rocks during the Cenomanian. The Permanente terrane, on the other hand, because of its equatorial fauna (Sliter, this volume) and similarity to the far-travelled limestone blocks at Laytonville (Alvarez and others, 1980) must have still been en route, in that its youngest rocks were being deposited at low altitudes about 90 m.y. ago.

The third major accretionary event appears to have occurred at about 55-60 m.y. This event is inferred from another major disconformity in the Great Valley sequence sedimentary rocks to the east (Bartow, 1984), and the presence on the San Francisco Peninsula of a lower(?) Eocene overlap assemblage across the Permanente and Central terranes (Brabb and Pampeyan, 1983). Also occurring at about this time was the accretion of Salinia, to the south (Page and Engebretson, 1984; Champion and others, 1984), and possibly the deformation seen in the Campanian rocks of the Novato Quarry terrane.

The fourth and last accretionary event seen in the San Francisco Bay region involved the Coastal terrane. Because some of these rocks are as young as latest Eocene (Bachman, 1978) this event must have occurred sometime after about 40 m.y. ago but prior to initiation of the San Andreas fault system ($\sim$10 m.y. for the latitude of San Francisco, according to Engebretson, 1982).

EPISODIC COLLISION AS A SPECULATIVE MODEL FOR ACCRETION

Sedimentological, structural, and stratigraphic data presented here suggest that the terranes present in the San Francisco Bay region are fragments of oceanic crust, oceanic islands or plateaus, continents, and continental margins and that accretion occurred during collision of larger continental or oceanic elements against North America. Although large plates, with the exception of Salinia, are no longer present, ample evidence in British Columbia and Alaska, suggest that numerous plates were at the paleolatitude of California on their way north (Monger and Irving 1980; Jones and others, 1977; Coney and others, 1980). In addition there could have been many small oceanic islands or rises lying on the oceanic plates that were subducted. Coherent oceanic terranes such as Nicasio Reservoir, Marin Headlands, and Permanente could represent the accreted upper parts of these oceanic elements. The Central terrane would represent tectonically sheared sedimentary and volcanic rocks that presumably formed at the North American continental margin prior to the collision.

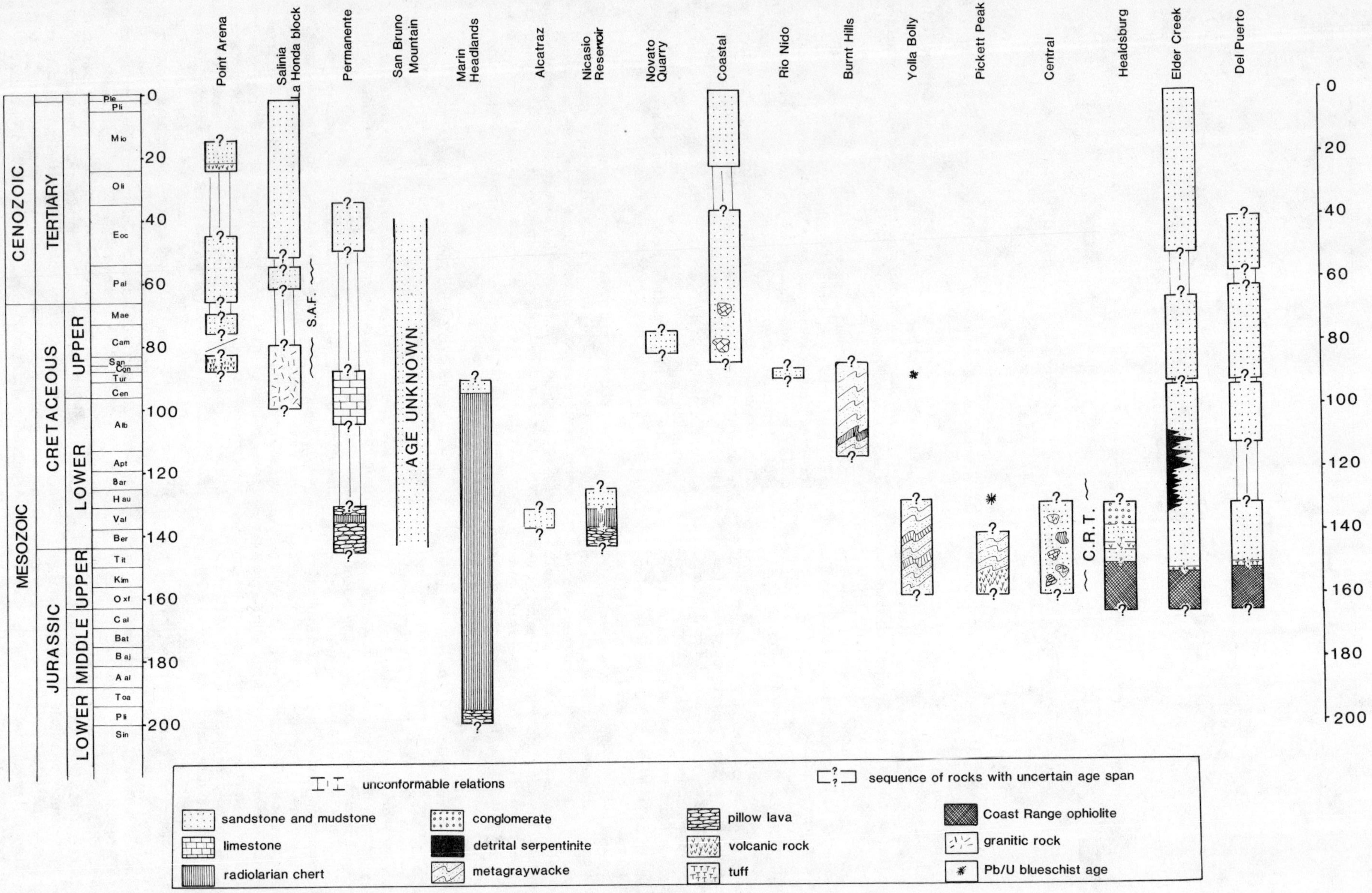

Figure 11. Tectonostratigraphic terrane columns for San Francisco Bay region.

Table 1.--Chemical analyses of basalt and gabbro (DH-9) from Peters Dam site, Nicasio Reservoir terrane, Marin County, California

	DH-9	DH5-211	DH2-141	DH5-98	DH6-136
SiO_2	43.3	46.1	46.5	46.8	42.1
Al_2O_3	21.1	16.2	17.8	16.5	14.6
Fe_2O_3	5.96	9.70	9.86	10.1	10.9
MgO	8.60	6.68	5.95	6.46	5.54
CaO	12.0	11.0	11.2	10.5	11.1
Na_2O	1.85	3.00	2.57	2.74	3.26
K_2O	0.58	0.49	0.44	0.38	0.49
TiO_2	0.76	1.58	1.95	2.10	1.89
P_2O_5	0.14	0.23	0.32	0.34	0.27
MnO	0.09	0.16	0.15	0.14	0.15
L.O.I. (900C)	5.84	4.29	3.15	3.39	9.37
RB ppm	14	10	10	10	10
SR ppm	722	360	325	272	283
ZrO_2	.009	.015	0.019	0.021	0.018
CrO_3	.035	.014	0.021	0.027	0.020
NiO	.030	.010	0.008	0.008	0.009
BaO	.027	.0130	.026	0.019	0.012

Major oxides by X-ray spectroscopy, U.S.G.S. Branch of Analytical Laboratories, Lakewood, Colorado, A. Bartel and J. Taggarat, analysts. Minor elements by quantitative XRF, U.S.G.S. Branch of Analytical Laboratories, Menlo Park, California, L. Espos and B. King, analysts.

The South Pacific region between Australia and Indo-China (Silver and Smith, 1983) is a possible modern analog for an early stage collision setting. In the Molluca Sea region, for example, colliding volcanic arcs are squeezing deep marine sediments into a melange-like fabric (Moore and others, 1981). In nearby areas of Indonesia other microplates of continental and oceanic affinities are posed in a region that will surely be involved in a major collision episode and the entire area is presently being blanketed by continental margin and oceanic pelagic sediments.

ACKNOWLEDGEMENTS

Numerous people contributed data and ideas concerning the tectonic evolution of the San Francisco Bay region. In particular, we would like to acknowledge Brian Atwater, Alan Bartow, Jim Berkland, Earl Brabb, Bill Cotton, Virgil Frizzell, Darrell Herd, Porter Irwin, David Jones, Bob McLaughlin, Bonnie Murchey, Tor Nilsen, Salem Rice, George Schlocker, Bill Sliter, Clyde Wahrhaftig, John Wakabayashi, Carl Wentworth, and above all, Edgar Bailey.

REFERENCES CITED

Alvarez, Walter, Kent, D. V., Premoli Silva, Isabella, Schweickert, R. A., and Larson, Roger, 1980, Franciscan complex limestone deposited at 17° south paleolatitude: Geological Society of America Bulletin, v. 91, p. 476-484.

Armstrong, C. F., and Gallagher, Kathy, 1977, Fossils from the Franciscan assemblage, Alcatraz Island: California Geology, v. 30, p. 134.

Atwater, B. F., 1975, Serpentinite diapir in Woodside and Redwood City, California: Unpublished Ms. C. thesis, Stanford University, Stanford, California, 118 p.

Bachman, S. B., 1978, A Cretaceous and early Tertiary subduction complex, Mendocino Coast, northern California, in Howell, D. G., and McDougall, K. A., eds., Mesozoic Paleogeography of Western United States: Pacific Section, Society of Economic Paleontologists and Mineralogists, Pacific Coast Paleogeography Symposium 2, p. 419-430.

-----,1981, The Coastal Belt of the Franciscan: Youngest phase of northern California subduction, in Leggett, J. K. (ed.), Trench and forearc sedimentation and tectonics in modern and ancient subduction zones: Geological Society of London Special Publication, p. 401-418.

Bailey, E. H., and Everhart, D. L., 1964, Geology and quicksilver deposits of the New Almaden district, Santa Clara County, California: U.S. Geological Survey Professional Paper 360, 206 p.

Bailey, E. H., Irwin, W. P., and Jones, D. L., 1964, Franciscan and related rocks and their significance in the geology of western California: California Division of Mines and Geology Bulletin 183, 177 p.

Bailey, E. H., Blake, M. C., Jr., and Jones, D. C., 1970, On-land Mesozoic ocean crust in California Coast Ranges, in Geological Survey Research 1970: U.S. Geological Survey Professional Paper 700-C, p. C70-C81.

Bartow, J. A., 1984, Cenozoic nonmarine sedimentation in central California, in Ingersoll, R. V., and Ernst, W. G., eds., Cenozoic Basin Development of coastal California, Rubey Volume, no. 6 (in preparation).

Beebe, Ward, and Luyendyk, B. P., 1983, Preliminary paleomagnetic results from the Llanada remnant of the Coast Range ophiolite, San Benito County, California (abs.): Eos, Transactions of the American Geophysical Union), v. 64, no. 45, p. 686.

Berkland, J. O., 1969, Geology of the Novato quadrangle, Marin County, California: Unpublished M.S. thesis, San Jose State University, San Jose California.

Blake, M. C. Jr., 1981, Geologic transect of the northern Diablo Range, California, in Frizzell, Virgil, ed., Upper Mesozoic Franciscan and Great Valley sequence sequence, central Coast Range, California: Annual Pacific Section Field Trip Guidebook, p. 35-43.

Blake, M. C., Jr., Howell, D. G., and Jones, D. L., 1982, Preliminary tectonostratigraphic terrane map of California: U.S. Geological Survey Open-File Report 82-593, 9 p., 3 maps, scale 1:750,000.

Blake, M. C., Jr., Irwin, W. P., and Coleman, R. G., 1967, Upside-down metamorphic zonation, blue-schist facies, along a regional thrust in California and Oregon: U.S. Geological Survey Professional Paper 575C, p. 1-9.

Blake, M. C., Jr., Jayko, A. S., and Howell, D. G., 1981, Geology of a subduction complex in the Franciscan assemblage of northern California: Oceanologica Acta, 1981 N^{o} Sp, 167-172.

Blake, M. C., Jr., and Jayko, A. S., 1983, Geologic map of the Yolla Bolly-Middle Eel Wilderness and adjacent roadless areas: U.S. Geological Survey Miscellaneous Field Studies Map MF-1595-A, scale 1:62,500.

Blake, M. C., Jr., Smith, J. T., Wentworth, C. M., and Wright, R. H., 1971, Preliminary geologic map of western Sonoma County and northernmost Marin County, California: U.S. Geological Survey open-file map, scale 1:62,500.

Blake, M. C., Jr., and Jones, D. L., 1974, Origin of Franciscan melanges in northern California: Society of Economic Paleontologists and Mineralogists, Special Paper no. 19, p. 255-263.

-----,1977, Plate tectonic history of the Yolla Bolly Junction, northern California: Geological Society of America, Fieldtrip Guidebook for 73rd Annual Cordilleran Section Meeting, Sacramento, California.

-----,1981, The Franciscan assemblage and related rocks in northern California: a reinterpretation, in Geodynamic development of California: W. G. Ernst, ed., Prentice-Hall, Inc., p. 307-328.

Brabb, E. E., and Pampeyan, E. H., 1983, Geologic map of San Mateo County, California: U.S. Geological Survey Miscellaneous Investigations Series, Map I-1257A, scale 1:62,500.

Champion, D. E., Howell, D. G., and Gromme, C. S., 1984, Paleomagnetic and geologic data indicating 2500 km of northward displacement for the Salinian and related terranes, California: Journal of Geophysical Research, v. 89, p. 7736-7752.

-----,1963, Glaucophane-bearing metamorphic rock types of the Cazadero area, California: Journal of Petrology, v. 4, pt. 2, p. 260-301.

Coney, P. J., Jones, D. L., and Monger, J. W. H., 1980, Cordilleran suspect terranes: Nature, v. 288, p. 329-333.

Cotton, W. R., 1972, Preliminary geologic map of the Franciscan rocks in the central part of the Diablo Range, Santa Clara and Alameda Counties, California: U.S. Geological Survey Miscellaneous Field Studies Map MF-343, scale 1:62,500.

Cowan, D. S., 1974, Deformation and metamorphism of the Franciscan subduction zone complex northwest of Pacheco Pass, California: Geological Society of America Bulletin, v. 85, p. 1623-1634.

Crawford, K. E., 1975, The geology of the Franciscan tectonic assemblage near Mt. Hamilton, California: unpublished Ph.D. thesis, University of California, Los Angeles, 137 p.

-----,1976, Reconnaissance geologic map of the Eylar Mountain quadrangle, Santa Clara and Alameda Counties, California: U.S. Geological Survey Miscellaneous Field Studies Map MF-764, scale 1:24,000.

Engebretson, D. C., 1982, Relative motions between oceanic and continental plates in the Pacific basin: Ph.D. thesis, Stanford University, Stanford, Calif., 150 p.

Evarts, R. C., 1977, The geology and petrology of the Del Puerto ophiolite, Diablo Range, central California Coast Ranges, in Coleman, R. G., and Irwin, W. P., eds., North American ophiolites: Oregon Department of Geology and Mineral Industries Bulletin 95, p. 121-140.

-----,1978, The Del Puerto ophiolite: structural and petrologic evolution: Unpublished Ph.D. thesis, Stanford University, Stanford, Calif., 409 p.

Frei, L. S., Cox, Allan, and Blake, M. C., Jr., 1983, Paleomagnetic results from the Coast Range ophiolite in northern California and southern Oregon (abs.): Eos, Transactions of the American Geophysical Union), v. 64, no. 45, p. 685.

Gluskoter, H. J., 1962, Geology of a portion of western Marin County, California: Unpublished Ph.D. thesis, University of California, Berkeley, California, 184 p.

Gluskoter, H. J., 1969, Geology of a portion of western Marin County, California: California Division of Mines Map Sheet 11, scale (1:48,000).

Hopson, C. A., Mattinson, J. M., and Pessagno, E. A., Jr., 1981, Coast Range ophiolite, western California, in W. G. Ernst, ed., The Geotectonic Development of California, Rubey Vol. 1,: Prentice-Hall, Englewood Cliffs, N. J., p. 307-328.

Howell, D. G., Vedder, J. G., McLean, H., Joyce, J. M., Clarke, S. H., Jr., and Smith, Greg, 1977, Review of Cretaceous geology, Salinian and Nacimiento blocks, Coast Ranges of central California, in Howell, D. G., and others, eds., Cretaceous geology of the California Coast Ranges, west of the San Andreas fault, Pacific Coast Paleogeography field guide 2, Society of Economic Paleontologists and Mineralogists, Pacific Section, p. 1-46.

Hsu, K. J., 1968, Principles of melanges and their bearing on the Franciscan-Knoxville paradox: Geological Society of America Bulletin, v. 79, p. 1063-1074.

Ingersoll, R. V., 1978, Petrofacies and petrologic evolution of the Late Cretaceous fore-arc basin, northern and central California: Journal of Geology, v. 86, p. 335-353.

Ingersoll, R. V., and Dickinson, W. R., 1981, Great Valley (Group) sequence, Sacramento Valley, California, in Frizzell, Virgil, ed., Upper Mesozoic Franciscan and Great Valley sequence, Central Coast Range, California: Pacific Section, Society of Economic Paleontologists and Mineralogists, Annual meeting field trip guide, p. 1-33.

Irwin, W. P., 1972, Terranes of the western Paleozoic and Triassic belt in the southern Klamath Mountains, California, in Geological Survey Research, 1972: U.S. Geological Survey Professional Paper 800-C, p. C103-C111.

Jacobson, M. I., 1978, Petrologic variations in Franciscan sandstone from the Diablo Range, California: in Howell, D. G., and McDougall, K. A., eds., Mesozoic Paleogeography of Western United States: Pacific Section, Society of Economic Paleontologists and Mineralogists, Pacific Coast Paleogeography Symposium 2, p. 401-417.

Jones, D. L., Silberling, N. J., and Hillhouse, J. W., 1978, Microplate tectonics of Alaska--significance for the Mesozoic history of the Pacific Coast of North America, in Howell, D. G., and McDougall, K. A., eds., Mesozoic Paleogeography of Western United States: Pacific Section, Society of Economic Paleontologists and Mineralogists, Pacific Coast Paleogeography Symposium 2, p. 71-74.

Kanter, Lisa, Engebretson, David, and Cox, Allan, 1982, Paleomagnetism of continental and oceanic terranes of coastal California: Transactions of the American Geophysical Union, v. 63, no. 45, p. 915.

Kleist, J. R., 1974, Deformation by soft-sediment extension in the Coast Belt, Franciscan Complex, Geology, v. 2, p. 501-504.

Kramer, J. C., 1976, Geology and tectonic implications of the Coastal Belt Franciscan, Ft. Bragg-Willits area, northern Coast Ranges, California (Ph.D. thesis): University of California, Davis, 128 p.

Lanphere, M. A., Blake, M. C., Jr., and Irwin, w. P., 1978, Early Cretaceous metamorphic age of the South Fork Mountain Schist in the northern Coast Ranges of California: American Journal of Science, v. 278, p. 798-815.

Lawson, A. C., 1914, Description of the San Francisco District: U.S. Geological Survey Folio 193.

Lee, D. E., Thomas, H. H., Marvin, R. F., and Coleman, R. g., 1963, Isotope ages of glaucophane schists from Cazadero, California: U.S. Geological Survey Professional Paper 475-D, p. 105-107.

Maddock, M. E., 1964, Geology of the Mt. Boardman quadrangle, Santa Clara and Stanislas Counties: California: California Division of Mines and Geology, Map Sheet 3, scale 1:62,500.

Mattinson, J. M., 1978, Age, origin, and thermal histories of some plutonic rocks from the Salinian block of California: Contributions to Mineral Petrology, v. 67, p. 233-245.

Mattinson, J. M., 1981, U-Pb systematics and geochronology of blueschists: preliminary results: Transactions of the American Geophysical Union, v. 62, no. 45,p. 1059.

Mattinson, J. M., and Echeverria, L. M., 1980, Ortigalita Peak gabbro, Franciscan complex: U-Pb date of intrusion and high-pressure-low temperature metamorphism: Geology, v. 8, p. 589-593.

McLaughlin, R. J., Ellen, S. D., Miller, S. G., Lajoie, K. R., and Morrison, S. D., 1983, Terrane boundary relations and tectonostratigraphic framework south Eel River Basin, northwestern California: American Association of Petroleum Geologists Meeting, Program and Abstracts, 58th Anual Meeting, Pacific Sections, A.A.P.G., S.E.G., S.E.P.M., Sacramento, p. 112-113.

McLaughlin, R. J., Ohlin, H. N., Thormahlen, D. J., Miller, J. W., and Carlson, Christine, 1980, Structure of late Mesozoic rocks in the core of the Wilbur Springs antiform, northern Coast Ranges, California: Geological Society of America Abstracts with Programs, v. 12, p. 119.

Monger, J. W. H., and Irving, E., 1980, Northward displacement of north-central British Columbia: Nature, v. 285, no. 5763, p. 289-294.

Monger, J. W. H., and Ross, C. A., 1971, Distribution of fusulinaceans in the western Canadian Cordillera, Canadian Journal of Earth Sciences, v. 8, p. 259-278.

Monger, J. W. H., Price, R. A., and Tempelman-Kluit, D. J., 1982, Tectonic accretion and the origin of the two major metamorphic and plutonic welts in the Canadian Cordillera: Geology, v. 10, p. 70-75.

Moore, G. F., Evans, C. A., Hawkins, J. W., and Kadarisman, 1981, Geology of the Taland Islands, Molucca Sea collision zone, northeast Indonesia: Journal of Structural Geology, v. 3, p. 467-475.

Morrell, R. P., 1978, Geology and mineral paragenesis of Franciscan metagraywacke near Paradise Flat, northwest of Pacheco Pass, California: Unpublished M.S. thesis, Stanford University, Stanford, Calif., 73 p.

Mutti, Emiliano, and Ricci-Lucchi, Franco, 1972, Le turbiditi dell' Appennio settentriorale: introduzione ali' analisi di facies (Turbidites of the northern Appenines: Introduction to facies analysis): Memorie della Societa Geologica Italiana, v. 11, p. 161-199 (Translated by Nilsen, T. H., 1978, International Geological Review, v. 20, no. 2, p. 125-166).

Nilsen, T. H., 1981, Geology of the Santa Cruz Mountains, California, in Frizzell, Virgil, ed., Upper Cretaceous and Paleocene turbidites, central California coast: Society of Economic Paleontologists and Mineralogists, Annual Meeting, Pacific Section Field Trip Guidebook, p. 5-12.

Normark, W. R., 1972, For valleys, channels, and depositional lobes on modern submarine fans; characters for recognition of sandy turbidite environments: American Association of Petroleum Geologists Bulletin, v. 62, no. 6, p. 912-931.

O'Day, M. S., 1974, The structure and petrology of the Mesozoic and Cenozoic rocks of the Franciscan Complex, Leggett-Piercy area, northern California Coast Ranges: Ph.D. thesis, Davis, California, University of California, 152 p.

Page, B. M., 1981, The southern Coast Ranges, in Ernst, W. G., ed., The Geotectonic Development of California: Prentice Hall, New Jersey, p. 329-417.

Page, B. M., and Engebretson, D. C., 1984, Correlation between the geologic record and computed plate motions for central California: Tectonics, v. 3, p. 133-155.

Raymond, L. A., 1973a, Franciscan geology of the Mt. Oso area, California: Unpublished Ph.D. theseis, Davis, California, University of California, 185 p.

-----,1973b, Tesla-Ortigalita fault, Coast Range thrust fault, and Franciscan metamorphism, northeastern Diablo Range, California: Geological Society of America Bulletin, v. 84, p. 3547-3562.

Ridley, W. I., Rhodes, J. M., Reis, A. M., Jakes, P., Shih, C., and Pass, M. N., 1974, Basalts from Leg 6 of the deep-sea drilling project: Journal of Petrology, v. 15, p. 140-159.

Ross, D. C., 1978, The Salinian block--a Mesozoic granitic orphan in the California Coast Ranges in Howell, D. G., and McDougall, K. A., eds., Mesozoic Paleogeography of Western United States: Pacific Section, Society of Economic Paleontologists and Mineralogists, Pacific Coast Paleogeography Symposium 2, p. 509-522.

Ross, D. C., Wentworth, C. M., and McKee, E. H., 1973, Cretaceous mafic conglomerate near Gualala offset 350 miles by San Andreas fault from oceanic crustal source near Eagle Rest Peak, California: Journal of Research, U.S. Geological Survey, v. 1, p. 45-52.

Schlocker, Julius, 1974, Geology of the San Francisco North Quadrangle, California: U.S. Geological Survey Professional Paper 782, 109 p.

Silver, E. A., and Smith, R. B., 1983, Comparison of terrane accretion in modern Southeast Asia and the Mesozoic North American Cordillera: Geology, v. 11, p. 198-202.

Smith, G. W., Howell, D. G., and Ingersoll, R. V., 1979, Late Cretaceous trench-slope basins of central California: Geology, v. 7, p. 303-306.

Suppe, John, and Armstrong, R. L., 1972, Potassium-argon dating of Franciscan metamorphic rocks, American Journal of Science, v. 272, p. 217-233.

Swanson, S. E., and Schiffman, P., 1979, Textural evolution and metamorphism of pillow basalts from the Franciscan Complex, western Marin County, California: Contributions to Mineralogy and Petrology, v. 69, no. 3, p. 291-299.

Telleen, K. E., 1977, Paleocurrents in part of the Franciscan Complex, California: Geology, v. 5, p. 49-51.

Underwood, M. B., 1983, Depositional setting of the Paleogene Yager Formation, northern Coast Ranges of California, in Larue, D. K., and Steel, R. J., eds., Cenozoic Marine Sedimentation: Pacific Section, S.E.P.M. Symposium Volume, Pacific Margin, U.S.A., p. 81-101.

Wentworth, C. M., 1968, Upper Cretaceous and lower Tertiary strata near Guala, California, and inferred large right slip on San Andreas fault: Stanford University Publications in Geological Sciences, v. 11, p. 130-143.

Wright, R. H., 1974, Map showing the distribution of potassium feldspar and fossils in Mesozoic rocks of Marin and San Francisco Counties and parts of Alameda, Contra Costa, and Sonoma Counties, California: U.S. Geological Survey Miscellaneous Field Studies Map MF-573, scale 1:250,000.

-----,1982, Geology of western Marin County, California: Unpublished Ph.D. dissertation, University of California, Santa Cruz, California, 204 p.

AGE AND SIGNIFICANCE OF CHERT IN THE FRANCISCAN COMPLEX IN THE SAN FRANCISCO BAY REGION

Benita L. Murchey
U.S. Geological Survey
Menlo Park, California 94025

David L. Jones
U. S. Geological Survey
Menlo Park, California 94025

ABSTRACT

Radiolarian chert, an important component of many Franciscan terranes in the San Francisco Bay area, is the diagenetic product of biogenic silica deposition over a wide region of the Mesozoic Pacific Ocean. Faunal and lithostratigraphic analyses of chert-bearing Franciscan terranes indicate that depositional sites varied from temperate(?) to tropical localities and from continental margin to open ocean settings. Tectonic processes accompanied by northward translation juxtaposed these pieces of the paleo-Pacific Ocean.

INTRODUCTION

Radiolarian chert is an important component of many Franciscan terranes in the San Francisco Bay area, including the Marin Headlands, Nicasio Reservoir, Permanente, Central, Burnt Hills, and Yolla Bolly terranes (Fig. 1, terrane map of the San Francisco Bay region) (Blake and others, this volume). The radiolarian chert in these terranes occurs in a wide variety of sedimentary associations (Murchey and others, 1983). In this study, we compare and correlate radiolarian faunas in the different terranes, analyze their depositional associations, and postulate their original paleogeographic settings.

MARIN HEADLANDS RADIOLARIAN ASSEMBLAGES

A locally continuous chert sequence in the Marin Headlands on the north side of the Golden Gate, contains seven radiolarian assemblages (MH-1 to MH-7) that range from Pliensbachian to late Albian or early Cenomanian age (Fig. 1, localities 1-29) (Murchey, this volume). We have used these assemblages as references for correlating radiolarian assemblages of the Franciscan Complex from other localities throughout the San Francisco Bay region (Tables 1 and 2). Important faunal elements from these seven assemblages are shown in Table 1.

MARIN HEADLANDS TERRANE

In the San Francisco Bay area, the Marin Headlands terrane (Fig. 1) is characterized by a stratigraphic sequence of Pliensbachian or older pillow basalt depositionally overlain by eighty-two meters of chert that ranges from Pliensbachian to upper Albian or early Cenomanian age. These rocks are, in turn, depositionally overlain by thickly bedded lithic arkose containing the Albian ammonite *Douvilleiceras* cf. *D. mammillatum* (Schlotheim) in San Francisco and with the Cenomanian ammonite *Mantelliceras* sp. in the Marin Headlands (Schlocker and others, 1954; Hertlein, 1956). In the San Francisco Bay region, the Marin Headlands terrane trends southeastward from the Marin Headlands on the north side of the Golden Gate, through the Twin Peaks

Table 1. Radiolarian assemblages at the map localities shown on Figure 1, the terrane map of the San Francisco Bay region.

Map number	Radiolarian assemblages described in this report
Marin Headlands terrane:	
1-29	MH-1, MH-2, MH-3, MH-4, MH-5, MH-6, MH-7 (Marin Headlands)(Murchey, this volume)
48	MH-1 (Twin Peaks)
49	MH-1 (Coyote Hills)
51	MH-2, MH-3, MH-4, MH-5, Mh-7 (The Geysers-Clear Lake)
Nicasio terrane	
37	Nicasio assemblage (correlates with MH-5)
40	Nicasio assemblage (correlates with MH-5)
Central terrane (blocks in melange)	
30	MH-4
31	MH-3
32	MH-1
33	MH-4
34	MH-3
35	MH-5
36	MH-1
38	MH-1
39	MH-5
41	MH-3, MH-4, MH-5
42	MH-2
43	MH-1, MH-3,MH-4, MH-6
44	MH-1, MH-2, MH-3, MH-5, MH-7
45	MH-1
46	MH-1, MH-2, MH-3, MH-4, MH-6
Central terrane (melange association)	
47	Late Jurassic assemblage
Central terrane (limestone block near Laytonville	
50	Cenomanian (slightly younger(?) than MH-7 but many species in common)
Permanente terrane	
52	Lower Cretaceous (Black Mountain assemblage, correlative with MH-5)
53	Turonian or Coniacian (Kennedy Road assemblage)
Burnt Hills terrane	
54	Early Cretaceous or possibly earliest Late Cretaceous
Yolla Bolly terrane	
localities not figured	
	Late Jurassic and Early Cretaceous assemblages

in Blake, M.C. Jr., ed., 1984, **Franciscan Geology of Northern California:** Pacific Section S.E.P.M., Vol. 43, p. 23-30

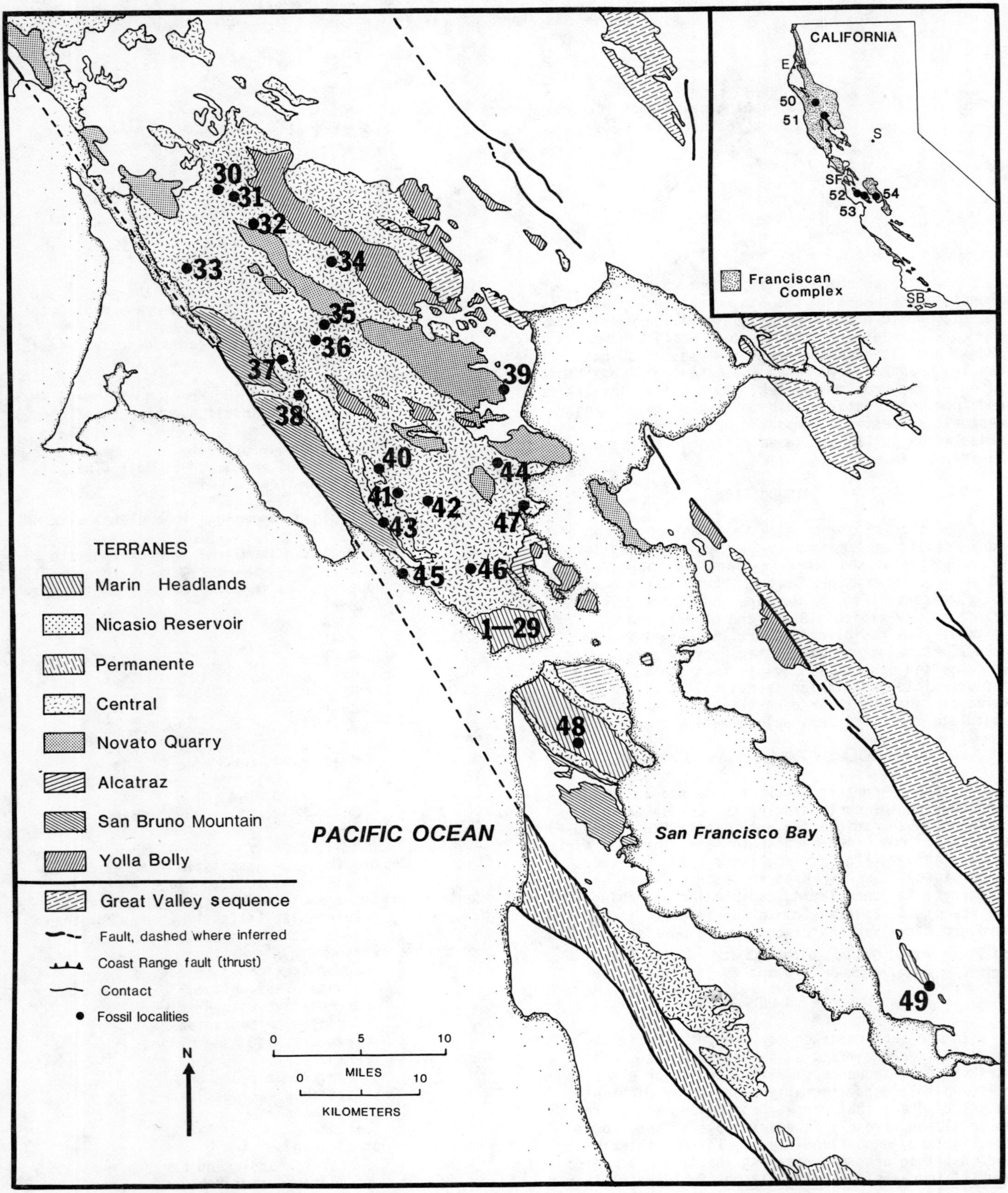

Figure 1. Map showing distribution of Franciscan terranes within the San Francisco Bay area, northern California. Numbered dots indicate radiolarian samples discussed in the text.

area of San Francisco, and across the bay to the Coyote Hills. MH-1 radiolarian assemblages have been recovered from chert near the chert-basalt contact in both the Twin Peaks area (Fig. 1, locality 48) and the Coyote Hills (Fig. 1, locality 49). One hundred kilometers to the north in The Geysers-Clear Lake area (Fig. 1, locality 51), is a similar stratigraphic sequence (McLaughlin and Pessagno, 1978) that Blake and others (this volume) consider part of the Marin Headlands terrane. The biostratigraphy of the 67 m The Geysers-Clear Lake chert sequence was previously described by Pessagno (1977a, 1977b). Pessagno assigned the basal ribbon chert interval lying between the chert-basalt contact to 10.6 m above the contact a Late Jurassic or older age, and assigned overlying chert to late Kimmeridgian and (or) early Tithonian (Pessagno's radiolarian subzone (2A)), Valanginian (Pessagno's subzone 5C), and early Cenomanian (Pessagno's zone 10) ages. In our recollection of The Geysers sequence we identified five radiolarian assemblages which are virtually identical to radiolarian assemblages MH-2, MH-3, MH-4, MH-5, and MH-7 in the Marin Headlands (Tables 1 and 2). The Middle Jurassic MH-2 equivalent assemblage occurs near the base of the Geysers chert sequence. Middle to Late(?) MH-3 and MH-4 assemblage equivalents are those faunas Pessagno assigned to Late Jurassic zone 2A (see Murchey, this volume, for a discussion of age discrepancies). The Cretaceous MH-5 and MH-7 assemblage equivalents are those faunas Pessagno assigned to radiolarian zones 5C and 10 respectively. The Geysers sequence radiolarian assemblages are so similar, both qualitatively (at the specific level) and quantitatively (as determined by population counts), to those in the Marin Headlands, that we believe that the sequences now exposed in the Marin Headlands and The Geysers area must have a shared paleogeographic history. The two sequences are further tied together by similarities in lithostratigraphy. They both possess a basal manganiferous zone of Pliensbachian and (or) Toarcian age in the San Francisco Bay area and Bajocian and older(?) age at The Geysers, a middle zone of interbedded red and green chert and shale of Jurassic age, a recrystallized zone separating Jurassic faunas from Early Cretaceous Valanginian faunas, and a thin Cretaceous interval. In both the Marin Headlands and The Geysers sequences, the Cretaceous interval includes a thin zone (less than 1 meter) near the chert-sandstone contact in which late Albian or early Cenomanian radiolarians (MH-7 assemblage) are infilled with chlorite and are crushed. We consider the absence of MH-1 assemblage in The Geysers sequence and some geochemical differences between the Marin Headlands and The Geysers sequences (Karl, this volume) to be less significant than the astonishing similarity between coeval radiolarian assemblages. We agree, therefore, with Blake and others (this volume) that The Geysers-Clear Lake sequence should be considered part of the Marin Headlands terrane.

The chert sequence in the Marin Headlands terrane is a classic example of the ophiolitic chert association described by Murchey and others (1983) in which pillow basalt is overlain by a thick (greater than 50 m) chert sequence devoid of interbedded clastic rocks. Because clastic depostit were not laid down in the Marin Headlands terrane sedimentary sequence for up to one hundred million years following basaltic volcanism, a mid-oceanic origin for the basalt is probable. Deposition of only biogenic siliceous sediments for one hundred million years requires a stable depositional environment with high biologic productivity--the most obvious environmental candidate is the equatorial high productivity zone. The radiolarian assemblages of the Marin Headlands terrane support sedimentological arguments for a low latitude origin; the assemblages have more affinity with those from lower latitudes than those in the Coast Range ophiolite and Great Valley sequence of California (Murchey, this volume). Using interpretations of chert geochemistry, Karl (this volume) favors a low latitude oceanic depositional site north of the Mesozoic equator for the Marin Headlands sequence but a more continent-proximal depositional site for The Geysers sequence. We feel that Karl's data do not rule out a common regional origin for the two sequences, although the original area encompassing the Marin Headlands terrane prior to tectonic dismemberment may have been large. In our paleogeographic reconstruction (Fig. 2), we have situated the Marin Headlands terrane in the central Pacific at and south of the equator during Pliensbachian time and in the eastern Pacific (continental margin of the Americas) south of present position during early Cenomanian time.

NICASIO RESERVOIR TERRANE

In the Nicasio Reservoir terrane (Fig. 1), tuffaceous manganese-stained maroon chert and shale (10-15 m), cut by jasper dikes and depositionally overlain by basalt breccia and ash, is depositionally(?) sandwiched between hundreds of meters of pillow basalt (alternating pillow basalt/chert association of Murchey and others, 1983). The radiolarian faunas of this terrane are Early Cretaceous, Valanginian to Hauterivian or early Barremian. They are are similar but not identical to those of the MH-5 assemblage in the Marin Headlands (Tables 1 and 2). Like MH-5, the Nicasio Reservoir assemblage contains *Archaeodictyomitra apiaria* (Rust), *Acanthocircus dicranocanthos* (Squinabol), *Thanarla conica* (Aliev), *Pantanellium corriganensis* Pessagno, *Archaeodictyomitra simplex* Pessagno, *Pseudodictyomitra* sp. A, and *Zifondium* sp. The Nicasio Reservoir assemblage lacks *Mirifusus* sp. and *Cecrops septemporata* (Parona), which are common in MH-5, but contains *Xitus* sp. and *Ristola* sp. aff. *R. boesii*, which are absent(?) in MH-5. Abundant, well-preserved cryptocephalic and cryptothoracic nassellarians characterize the Nicasio Reservoir assemblage; abundant poorly preserved spheroidal radiolarians that are probably cryptocephalic and cryptothoracic nassellarians characterize the MH-5 assemblage.

Paleomagnetism studies of diabase and basalt indicate that the Nicasio Reservoir terrane originated at a paleolatitude of $30° \pm 7°$ N (Gromme, this volume). Its Early Cretaceous proximity to North America is uncertain. The faunal similarities and differences between the Nicasio Reservoir assemblage and MH-5 assemblage suggest that the Nicasio Reservoir terrane and the Marin Headlands terrane were in regional proximity but were not contiguous. In our paleogeographic reconstruction (Fig. 2), we placed the Nicasio Reservoir terrane to the south of present position but north of the equator during Early Cretaceous time. Because we were uncertain of its proximity to the continent at this time, we delineated a wide west trending belt from a region somewhat proximal to the continental margin to middle of the open ocean as a possible paleogeographic site for the Nicasio Reservoir terrane during the Early Cretaceous.

TABLE 2 Correlation of Radiolarian Assemblages in Franciscan Complex Terranes

FOSSIL AGE \ TERRANE			MARIN HEADLANDS	NICASIO RESERVOIR	PERMANENTE	CENTRAL: Laytonville limestone	CENTRAL: S.F. Bay area chert blocks	CENTRAL: melange	YOLLA BOLLY	BURNT HILLS
CRETACEOUS	Late	Coniacian to Turonian			▒					
CRETACEOUS	Late / Early	Cenomanian to l. Albian	MH-7 Holocryptocanium barbui Dumitrica, H. japonicum Nakaseko and Nishimura, H. geysersensis Pessagno, Novixitus mclaughlani Pessagno, Novixitus sp. A, Thanarla veneta (Squinabol), Dictyomitra sp. A, and Pseudodictryomitra spp. including rare P. pseudomacrocephala (Squinabol)			▒	▒			▒
CRETACEOUS	Early	Albian(?) to Haut.–Barr.	MH-6 Zifondium sp., Archaeodictyomitra sp., Acanthocircus trizonalis(?) (Rust), and poorly preserved cryptocephalic nassellarians				▒			▒
CRETACEOUS	Early	Haut.–Barr. to Valanginian	MH-5 Archaeodictyomitra apiaria (Rust), Acanthocircus dicranocanthos (Squinabol), Cecrops septemporata (Parona), Mirifusus sp. (M. mediodilatatus (Rust)(?), Pantanellium corriganensis Pessagno, Thanarla conica (Aliev), Pseudodictyomitra sp. A, Archaeodictyomitra simplex Pessagno, Zifondium sp., Podobursa sp., and probable cryptocephalic nassellarians	▒	▒		▒		▒	▒
CRETACEOUS / JURASSIC	Early / Late	Valangininian to Oxf.-Kimm.(?)						▒		
JURASSIC	Late / Middle to Late(?)	Oxf. Kimm.(?) to Bathonian	MH-4 Ristola turpicula Pessagno and Whalen, Tricolocapsa(?) sp. C, Dictyomitrella(?) kamoensis Mizutani and Kido, Ristola sp. A, Acanthocircus suboblongus (Yao), Hsuum sp. C Group, Tricolocapsa sp. cf. T. plicarum Yao, Napora sp. D, and Ristola sp. B				▒			
JURASSIC	Middle to Late(?)	Oxf.-Kimm.(?) to Bajocian	MH-3 Unuma sp. A, U. sp. B, Tricolocapsa sp. cf. T. plicarum Yao, Dictyomitrella(?) kamoensis Mizutani and Kido, Ristola sp. A, R. sp. B, Acanthocircus suboblongus (Yao), Hsuum sp. C Group, Stichocapsa(?) sp. B, Napora sp. D, Trillus(?) sp. cf. T. seiderse, Pantanellium ultrasincerum Pessagno and Blome, Gorgansium sp., Pachyoncus sp., and rare Mirifusus sp. cf. M. guadalupensis Pessagno				▒			
JURASSIC	Middle	Bajocian	MH-2 Parvincingula sp. A, Wrangellium sp. A, Acanthocircus hexagonus (Yao), Hsuum sp. B, Emiluvia sp., Perispyridium sp., Trillus elkhornensis Pessagno and Blome, Napora sp. B, Napora sp. C, Riedelius sp. A, Bipedis(?) sp. A, Pantanellium sp. aff. P. buntonense, Lupherium sp. and Foremanina sp.				▒			
JURASSIC	Early	Toarcian to Pliensbachian	MH-1 Canutus rockfishensis Pessagno and Whalen, Canoptum anulatum Pessagno and Whalen, Canoptum poissoni Pessagno, Hsuum(?) sp. A, Napora sp. A, Katroma sp. aff. K. neagui Pessagno and Poisson, Pantanellium sp. A, Orbiculiforma mclaughlani Pessagno, O. sp. cf. O. multifora Pessagno and Poisson, Eucyrtidium(?) sp. A, E.(?) sp. B, Bagotum sp. aff. B. modestum Pessagno and Whalen, Stichocapsa sp. A, Jacus sp. A, Dumitricaella(?) sp. aff. D(?) cucurbitina De Wever, and Trillus elkhornensis Pessagno and Blome				▒			

PERMANENTE TERRANE

The Permanente terrane (Fig. 1) is distinguished from other Franciscan terranes by large blocks of Cretaceous limestone that range in age from Albian (or older) to Turonian (Sliter, this volume). Graywacke, basalt, and minor chert also characterize this terrane. Basaltic rocks appear to have formed as seamounts (Garrison and Bailey, 1967) or ridges (Wachs and Hein, 19) upon which the limestone was deposited. Paleomagnetic studies of limestone in the Permanente Quarry indicate a paleolatitude at 21° ± 6° North for Albian time (Courtillot and others, 1983). Both megafossils (Blake and Jones, 1974) and foraminifers (Sliter, this volume) have low latitude affinities. On the basis of lithology and fauna, Sliter (this volume) believes that the limestone of the Permanente terrane was deposited within the Cretaceous equatorial productivity zone and that the underlying volcanic edifice formed in the southern hemisphere. Several chert blocks in graywacke-shale matrix that is structurally interleaved with limestone contain Early Cretaceous radiolarians correlative with the MH-5 assemblage and with the Nicasio Reservoir assemblage (Fig. 1, locality 52; Tables 1 and 2). The radiolarians include Thanarla conica (Aliev), Archaeodictyomitra sliteri Pessagno, and Pseudodictyomitra sp. At a small outcrop along Kennedy Road (Fig. 1, locality 53), two meters of siliceous argillite and thinly bedded chert (less than 2.5 cm) depositionally overlying Turonian limestone (Sliter, this volume) contains the radiolarian Dictyomitra koslovae Foreman of late Turonian or Coniacian age. This is the youngest radiolarian fauna known from Franciscan bedded chert.

LIMESTONE AT LAYTONVILLE

North of the San Francisco Bay region (Fig. 1, locality 50), middle Cretaceous pelagic limestone blocks in melange of the Franciscan Complex near Laytonville, have yielded paleomagnetic inclinations of 17° ± 7° south latitude (Alvarez and others, 1980). Sliter (this volume) believes that the sedimentology and foraminifer faunal affinities of the limestone blocks near Laytonville indicate their deposition within the equatorial high productivity zone. Siliceous interbeds within one block (LL-2 of Alvarez and others, 1980) contain a radiolarian assemblage very similar to MH-7 assemblage in the Marin Headlands and in The Geysers sequence. Radiolarians include Novixitus weyli Schmidt-Effing, Thanarla elegantissima (Cita), Thanarla conica (Aliev), and Pseudodictyomitra pseudomacrocephala (Squinabol), all of which are present but not common in MH-7. In addition, all but T. conica are reported from Costa Rica (Schmidt-Effing, 1980). The Laytonville samples also contain foraminifers characteristic of the Rotalipora cushmani foraminifer zone of middle to late Cenomanian age (Alvarez and others, 1980). The faunal similarity between the Laytonville radiolarian assemblage and the slightly older (?) MH-7 assemblage suggests that the Marin Headlands was also near- equatorial during Cenomanian time. Significantly, Cenomanian radiolarian faunas from Costa Rica contain many species in common with both the Laytonville radiolarian assemblage and with the MH-7 radiolarian assemblage (Schmidt-Effing, 1980). In our paleogeographic reconstruction (Fig. 2) we conservatively place the Marin Headlands in the northern hemisphere near the equator during earliest Cenomanian time. The Marin Headlands terrane was adjacent to a landmass at this time because MH-7 zone chert is gradational with overlying clastic rocks. Conversely, the limestone blocks near Laytonville lack interbedded clastics and, for this reason, we place the limestone farther from the Americas in our reconstruction (Fig. 2).

CENTRAL TERRANE

The Central terrane (Fig. 1) is typified by melange with a predominantly graywacke and shale matrix. Within the melange, radiolarian chert occurs as part of the melange matrix and also as tectonically incorporated blocks. The former occurrence is a sedimentary association of chert, basalt, tuff, shale, and graywacke (melange chert association of Murchey and others, 1982).

The melange chert association represents deposition in a marine environment (as indicated by the radiolarian chert) with active volcanism (indicated by basalt and tuff) near a continental margin (indicated by continentally derived graywacke and shale) (Murchey and others, 1983). Melange association chert near San Quentin (Fig. 1, locality 47) has a Late Jurassic (probably Tithonian) radiolarian assemblage that differs from any Marin Headlands assemblage. The locality 47 assemblage contains the genera Mirifusus sp. (M. baileyi Pessagno (?)), Parvicingula sp., Emiluvia sp., Hsuum sp., and Tripocyclia sp. and appears to be correlative with faunal zones missing in the Marin Headlands terrane (between MH-4 and MH-5). A recent study by Pessagno and Blome (in press) proposes that Parvicingula is characteristic of Boreal latitudes while Ristola is characteristic of tropical latitudes. In the poorly preserved locality 47 fauna Parvicingula spp. are definitely present but Ristola spp. have not been positively identified, possibly due to poor preservation. Therefore, the possibility of deposition at temperate latitudes, although not definitively proven by this assemblage, should be seriously considered. Future comparative studies of better preserved faunas will provide more definite clues as to the paleolatitude at which the matrix of the Central terrane melange was deposited. Faunal comparison with Coast Range ophiolite-Great Valley sequence faunas may substantiate or refute the argument for a depositional relationship between the clastic rocks of the melange and the clastics of the Great Valley sequence. In general, age relationships and clastic rock compositions support a depositional relationship between the matrix of the Central terrane and the Great Valley sequence (Blake and others, this volume; Seiders and Blome, this volume; Jayko and Blake, this volume).

Chert blocks which have been tectonically incorporated into the Central terrane melange vary in length from a few centimeters to a kilometer or more. From the Marin Headlands northward for forty kilometers, chert blocks in the Central terrane have the same faunal assemblages and the same stratigraphic range (MH-1 to MH-7) as the Marin Headlands terrane (Fig. 1, localities 30-36, 38-39, 41-46; Tables 1 and 2). Missing Late Jurassic and earliest Early Cretaceous faunas between the MH-4 and MH-5 assemblage zones in the Marin Headlands terrane are also missing in these blocks. We interpret these blocks to be dismembered pieces of the Marin Headlands terrane which were tectonically incorporated into the older Central terrane melange matrix during post-Cenomanian time.

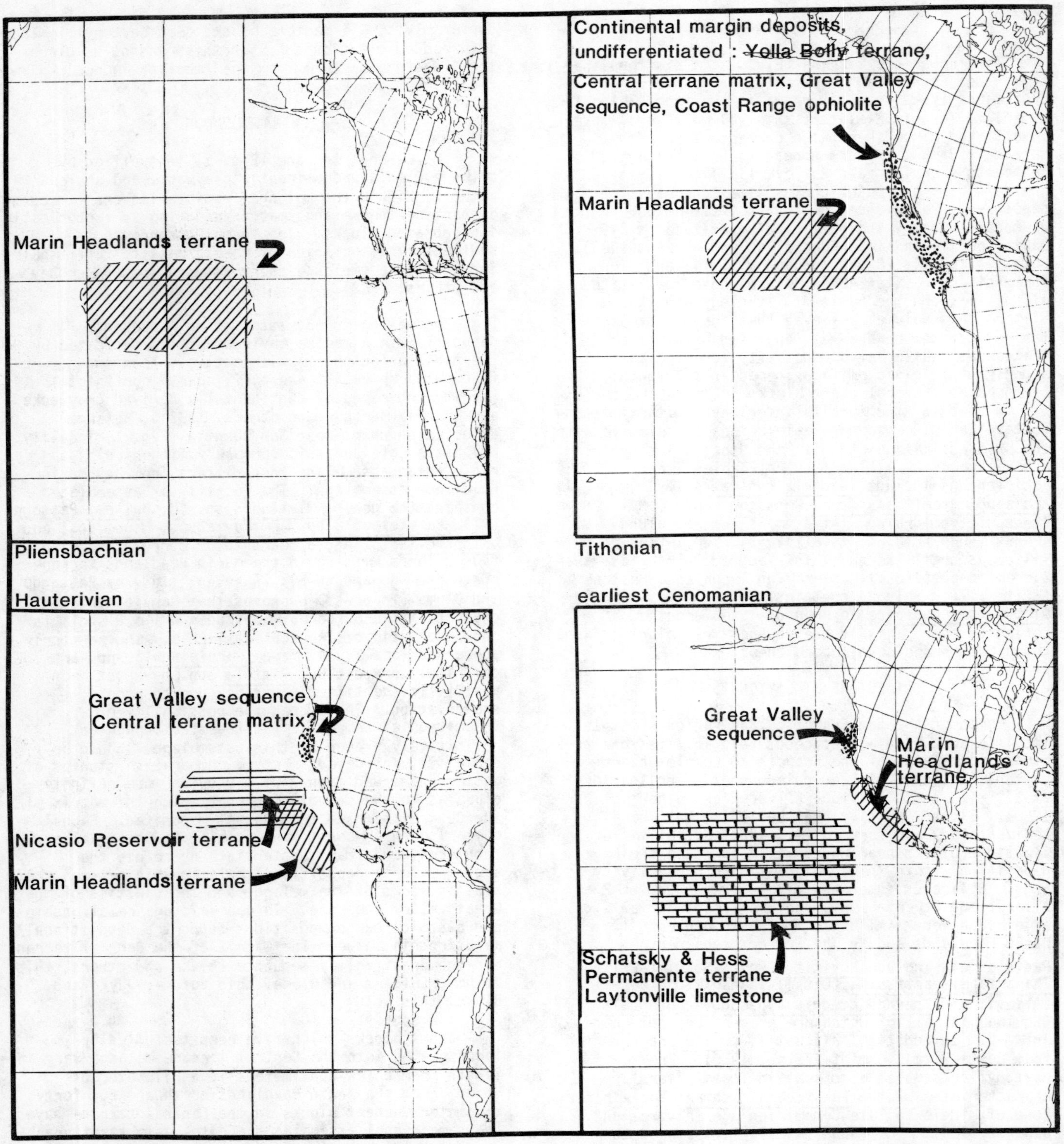

Figure 2. Jurassic and Cretaceous paleogeography of selected Franciscan Complex terranes with radiolarian chert. The Franciscan terranes include Marin Headlands, Nicasio Reservoir, Permanente, Central (matrix as well as limestone blocks near Laytonville), and Yolla Bolly. The paleogeographic positions of the Great Valley sequence, Coast Range ophiolite, and Schatsky and Hess rises are shown for comparison. Reconstructions are based upon paleomagnetic data, lithologic associations, and fossil affinities (radiolarians, foraminifers, and megafossils). Late Jurassic paleogeographic sites of continental margin sequences (Yolla Bolly terrane, Central terrane (matrix), Coast Range ophiolite, and Great Valley sequence) relative to one another are unknown; therefore, we have not differentiated them. The size of the patterned areas does not reflect present terrane size. The base maps are from A. G. Smith and J. C. Briden, 1977, Mesozoic and Cenozoic Paleocontinental Maps: Cambridge University Press, 63 p.

BURNT HILLS TERRANE

The Burnt Hills terrane contains well bedded arkose with interbedded chert up to fifty meters thick (Blake and others, this volume). The depositional interlayering of sandstone with siliceous biogenic sediments (clastic/chert association of Murchey and others, 1983) indicates deep water deposition (probably greater than 500 meters) along a continental margin. One chert sample from this terrane (Fig. 1, locality 54) contains a Cretaceous (late Valanginian? to Albian or early Cenomanian) assemblage with deformed and recrystallized specimens of *Zifondium* sp., *Thanarla* sp. cf. *T. elegantissima* (Cita), and *Archaeodictyomitra* sp. The radiolarian fauna is not well enough preserved to make affinity comparisons. Molluscan fossils from the arkosic sands are Coniacian (Jones, unpublished data). We have not included the Burnt Hills terrane in our paleogeographic reconstructions (Fig. 2) for lack of paleolatitudinal indicators. However, the sedimentary associations indicate that the Burnt Hills terrane formed along a continental margin during the same Cretaceous interval when the thick clastic deposits of the Great Valley sequence were being laid down along North America essentially in place and when thick bedded lithic arkose (Albian and Cenomanian) was being deposited in the Marin Headlands terrane as it approached a continental margin to the south.

YOLLA BOLLY TERRANE

In the Yolla Bolly terrane, bedded radiolarian chert is depositionally interlayered between continentally derived quartzofeldspathic sandstone (clastic/chert association of Murchey and others, 1983). Some individual chert bodies can be traced for kilometers. Like the Burnt Hills terrane, the Yolla Bolly terrane lithologic association indicates deep-water deposition along a continental margin. *Buchia* sp. in sandstone of the Yolla Bolly terrane are Late Jurassic and Early Cretaceous in age (Blake and Jones, 1974). An ichthyosaur snout in chert near the Ladd-Buckeye mines is Tithonian (Camp, 1942). Yolla Bolly radiolarians are recrystallized and deformed due to regional metamorphism of the entire Yolla Bolly belt. Thus, radiolarian preservation is poor, and the dating of samples is difficult. Despite these problems, numerous samples outside of the San Francisco Bay region are dated as Late Jurassic based on the presence of abundant *Mirifusus* sp. cf. *M. baileyi* Pessagno and *Praeconocaryomma magnimamma* Pessagno and Early Cretaceous based on the presence of *Thanarla* sp. (authors' unpublished data). The Late Jurassic radiolarian association is common in the Coast Range ophiolite (Pessago, 1977a) but absent in the Marin Headlands terrane--an absence possibly related to missing Late Jurassic and earliest Early Cretaceous faunas between MH-4 and MH-5 assemblage zones, to differences in Late Jurassic paleogeography, or both.

The regionally extensive Yolla Bolly terrane (Blake and others, this volume) formed along a continental margin during a period (Late Jurassic and Early Cretaceous) when the autochthonous Great Valley sequence prograded over the Coast Range ophiolite along the continental margin of paleo-California, when the melange chert association (interbedded basalt, chert, and graywacke) of the Central terrane was deposited along a continental margin, possibly at temperate latitudes, and when the Marin Headlands and Nicasio Reservoir terranes lay in the open ocean to the south of present position.

CONCLUSIONS

Radiolarian faunas from chert and sedimentary associations in chert-bearing terranes of the Franciscan Complex in the San Francisco Bay region indicate that depositional sites may have been scattered throughout a wide region of the Mesozoic Pacific Ocean from temperate(?) to tropical waters and from continental margin to open ocean settings (Fig. 2, paleogeographic reconstruction). Tectonic processes have subsequently juxtaposed all these pieces of the paleo-Pacific.

In modern oceans, biogenic siliceous sediments are deposited at low latitudes beneath the equatorial high productivity zone, at very high latitudes characterized by high biologic productivity, and in continental margin basins associated with upwelling. Two recent papers (Jenkyns and Winterer, 1982; Hein and Karl, 1983) suggest that most Mesozoic radiolarian chert is the diagenetic product of biogenic silica deposition in marginal basins. However, in the Jurassic and Early Cretaceous Pacific Ocean, local upwelling in marginal basins was not the only process reponsible for silica deposition. We are only now beginning to recognize the extent to which a wide westward flowing equatorial current in the Mesozoic Pacific (Kennett, 1982) was responsible for deposition of radiolarian-rich sediments even in the open ocean. Many of these low latitude deposits were subsequently accreted to the land masses surrounding the Pacific Ocean. The accretionary processes were commonly accompanied by northward translation. We estimate that most of the radiolarian chert of the Franciscan Complex has been transported northward from equatorial to subequatorial latitudes.

ACKNOWLEDGEMENTS

We especially thank Roy Couzin for collecting many of the samples used in this report. We also thank W. Sliter and C. Blome for review and comments, M. C. Blake and C. Wahrhaftig for helpful discussions and some samples, and P. Jefferies and M. Liniecki for technical assistance.

REFERENCES CITED

Alvarez, W., Kent, D. V., Premoli-Silva, I., Schweikert, R. A., 1980, Franciscan Complex limestone deposited at 17° south latitude: Geological Society of America Bulletin, v. 91, p. 476-484.

Blake, M. C., Jr. and Jones, D. L., 1974, Origin of Franciscan melanges in northern California *in* Dott, R. H., Jr. and Shaver, R. H., eds., *Modern* and Ancient Geosynclinal Sedimentation, Society of Economic Paleontologists and Mineralogists Special Publication No. 19, p. 345-357.

Camp, C. L., 1942, Ichthyosaur rostra from central California: Journal of Paleontology, v. 6, p. 362-371.

Courtillot V., Feinberg, H., McWilliams, M., and Cox, A., 1983, Franciscan Complex limestone deposited at 21° North: EOS, Transactions, American Geophysical Union, v. 64, p. 686.

Garrison, R. E. and Bailey, E. H., 1967, Electron microscopy of limestones in the Franciscan Formation of California: U.S. Geological Survey Professional Paper 575-B, p. B94-B100.

Hein, J. R. and Karl, S. M., 1983, Comparisons between open-ocean and continental margin chert sequences, in Iijima, A., Hein, J. R., and siever, R., eds., Siliceous Deposits in the Pacific Region, Developments in Sedimentology v. 36, Elsevier, Amsterdam, p. 25-43.

Hertlein, L. G., 1956, Cretaceous ammonite of Franciscan group, Marin County, California: American Association Petroleum Geologists Bull., v. 40, p. 1985-1988.

Jenkyns, H. C. and Winterer, E. L., 1982, Palaeoceanography of Mesozoic ribbon radiolarites: Earth and Planetary Science Letters, v. 60, p. 351-375.

Kennett, J., 1982, Marine Geology, Prentice-Hall, Inc., Englewood Cliffs, N. J., 813 p.

McLaughlin, R. J. and Pessagno, E. A., Jr., 1978, Significance of age relations of rocks above and below Upper Jurassic ophiolite in The Geysers-Clear Lake area California: U.S. Geological Survey Journal of Research, v. 6, p. 715-726.

Murchey, B. L., Jones, D. L., and Holdsworth, B. K., 1983, Distribution, age, and depositional environments of radiolarian chert in western North America, in Iijima, A., Hein, J. R., and Siever, R., eds., Siliceous Deposits in the Pacific Region, Developments in Sedimentology v. 36, Elsevier, New York, p. 109-126.

Pessagno, E. A., Jr., 1977a, Upper Jurassic Radiolaria and radiolarian biostratigraphy of the California Coast Ranges: Micropaleontology, v. 23, p. 56-113, pl. 1-12.

Pessagno, E. A., Jr., 1977b, Lower Cretaceous radiolarian biostratigraphy of the Great Valley sequence and Franciscan Complex, California Coast Ranges: Cushman Foundation Foram. Research Spec. Publ. 15, p. 1-86, pl. 1-12.

Pessagno and Blome, in press, Tectonogenesis of Mesozoic rocks in the Blue Mountains province of eastern Oregon and western Idaho: U.S. Professional Paper.

Schlocker, J., Bonilla, M. G., and Imlay, R. W., 1954, Ammonite indicates Cretaceous age for part of Franciscan group in San Francisco Bay area, California: Am. Assoc. Petroleum Geollgists Bull., v. 38, p. 2372-2381.

Schmidt-Effing, R., 1980, Radiolarien der Mittel-Kreide aus dem Santa Elena-Massiv von Costa Rica: Neues Jahrb. Geol. Pal. Abh., v. 160, p. 241-257, 33 figs.

Wachs, and Hein, J. R., 1975, Franciscan limestones and their environment of deposition: Geology, v. 3, p. 29-33.

STRUCTURE OF THE MARIN HEADLANDS BLOCK, CALIFORNIA: A PROGRESS REPORT

by Clyde Wahrhaftig
United States Geological Survey, Mail Stop 917,
345 Middlefield Road, Menlo Park CA 94025
and Department of Geology and Geophysics, University of California, Berkeley

ABSTRACT

The Marin Headlands block, exposed for 8 km E-W and 4 km N-S, is a coherent Franciscan terrane composed of approximately ten S to SW-dipping stacked slices of Mesozoic ocean floor. Individual slices are 300-500 m thick; some fray laterally into fragments in a melange matrix; others pinch out between fault branches. No slice can be shown to extend the length of the block.

The block rests on melange of Marin County on a thrust contact that is near-vertical on the west and gently south-dipping on the east. The block is probably separated by a profound dislocation beneath the Golden Gate from the Twin Peaks block of San Francisco, which it closely resembles in age and lithology.

Each slice consists of one or more of the following components, whose thicknesses vary widely:- 0-300 m (av. 150 m) of pillow basalt, overlain by 3-300 m of radiolarian ribbon chert, overlain by 0-250 m of a thick-bedded turbidite sequence of medium to coarse locally conglomeratic greywacke, with infrequent carbonaceous and argillaceous interbeds. All up-section indicators show the rocks to be right-side up. Basalt of the southernmost slice is intruded by a N-trending vertical diabase dike, possibly of oceanic layer 3.

The stack of slices has been folded into a syncline plunging 40° S 25° W, whose limbs make an angle of 155°-160°. In many places the chert is pervasively folded into tight chevron folds with amplitudes and wavelengths of a few m, also plunging SW. The wide variation in chert thickness can only partly be due to thickening associated with these folds, and in part to submarine erosion prior to deposition of sandstone. Abnormally thick sections of chert must have duplication by faulting.

If the direction of tectonic accumulation was normal to fold axes, the block represents a 60-70-km width of oceanic crust. Obduction of the slices seems unlikely, as most of the turbidite sequence that should have overlain the chert is missing, and could have been squeezed out from between the slices only during underthrusting. The short limbs of minor anticlines in the chert are to the north, so if the subducted oceanic crust was moving northeast relative to North America, the Headlands block has been rotated more than 90° clockwise to its present anomalous attitude.

The history of geologic thought on the headlands is reviewed, and the techniques used in this investigation, which involve the application of plate tectonics, are described.

INTRODUCTION

The Marin Headlands are the steep hills at the southern tip of Marin Peninsula, California, along the north side of the Golden Gate (Figs. 1 and 2). They total 8 km in length in an east-west direction, and are 4 km wide. Their highest point is about 339 m above sea level. Their steep rugged slopes contrast with the more gently rolling terrain to the north, and their high sea cliffs form a magnificant backdrop to the view of the Golden Gate from San Francisco.

The rocks of the Headlands are coherent bodies of pillow basalt presumably erupted along a midocean ridge, thin-bedded red radiolarian chert thought to have accumulated on the basalt as deep-sea ooze, and sandstone with minor shale and conglomerate thought to have been deposited on the radiolarite by turbidity currents from the American continent or a bordering volcanic island arc. This assemblage of coherent basalt, chert, and sandstone, called here the Marin Headlands block[1], rests in thrust contact on melange that makes up the Marin Peninsula to the north. It is separated from the rocks of San Francisco by a profound dislocation.

The rocks along the Golden Gate were the subject of the first geologic observations in California. Later, they were the type locality for the original descriptions of the Franciscan assemblage. Accounts of them through the 19th and 20th centuries mark the circuitous path, with many branches ending in wrong conclusions, to knowledge of the origin of the Franciscan. Reading these persuasively argued errors makes one wonder how one's work will fare in the future. The interpretation of these rocks presented here, including the map itself, depends heavily on the theory of plate tectonics; hence, it seems salutary to review older ideas of their origin to see where their proponents went wrong.

This paper therefore includes (1) a historical review, (2) the methods of the present investigation; and (3) their results.

HISTORICAL REVIEW

Investigation of the rocks around the Golden Gate falls into three periods: (1) reconnaissance surveys and specimen collecting from 1816 to about 1890; (2) systematic mapping from 1890 to about 1950, based on a model of coherent rock units of gradually varying thickness deposited in regular stratigraphic succession, following Lyellian principles; and (3) Since 1950 detailed mapping which gradually disclosed the chaotic nature of the Franciscan. Plate tectonics in the late 1960's provided a mechanism to account for this chaos, and the full implications of this theory have been gradually adopted over the last 20 years.

Early Surveys

The first geologic observations were made during the Russian expedition of Von Kotzebue in 1816. Its botanist, Adelbert von Chamisso, recognized flinty slate (the chert) on the north side of the Golden Gate, serpentine at Fort Point, and chert and sandstone between Fort Point and Point Lobos

---

[1] This block, the similar Twin Peaks block, the Coyote Hills (see Fig. 1), and many smaller blocks to the north, with chert and basalt of similar age-span and lithology, are collectively called the Marin Headlands terrane (Blake and others, 1982; this volume).

in Blake, M.C. Jr., ed., 1984, **Franciscan Geology of Northern California:** Pacific Section S.E.P.M., Vol. 43, p. 31-50

(Chamisso, in Kotzebue, 1821, see also VanderHoof, 1951). Specimens of serpentine and sandstone collected by Eschscholtz, the ship's surgeon, were interpreted by Moritz von Engelhardt (in Kotzebue, 1821, v. 3, p. 337-345,) of the University of Dorpat in terms of Wernerian Neptunism, a geological model as popular in the late 18th century as plate tectonics is today. Engelhardt had been a student of Werner (Schiffner, 1935, cf. Sarjeant, 1980) and thought that the serpentine was the latest of the primary rocks, against which the sandstone had been deposited from the secondary seas (cf. Adams, 1938).

Between Nov. 6 and Dec. 28, 1826, Captain Beechey of the British Navy visited San Francisco, and Edward Belcher, surveyor, and Alex Collie, surgeon, made notes and the first map showing the distribution of rocks along the shore. Belcher (quoted in Richardson et al., 1839, p. 174-176) noted that the rocks around the Golden Gate are chiefly "sandstone, jasper, and serpentine" and that the "jasper" (chert) formed the ridge-crests. Of the "jasper" he said, "It resembles an immense mass of sheets of paper, or bands of list, crumpled and contorted by lateral pressure."

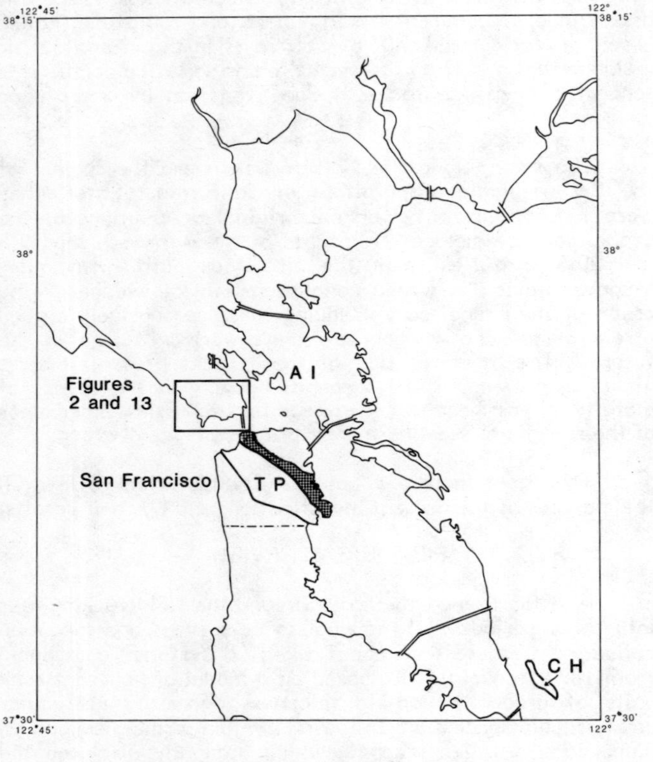

Figure 1. Map of San Francisco Bay showing location of Marin Headlands (Figures 2 and 13) and other fragments of the Marin Headlands terrane:- T P, Twin Peaks block, C H, Coyote Hills. Patterned area is serpentine belt in San Francisco. A I, Angel Island.

The next significant geologic investigations were made by William P. Blake in 1853 and J. S. Newberry in 1855, during the Pacific Railroad Surveys of the U., S. Army (Blake, 1857, Newberry, 1856). Blake's report includes a colored geologic map of the Golden Gate and central part of the bay. He grouped all the sandstone he encountered into his San Francisco Sandstone, and including in it the fossiliferous Merced Formation, he classified it as Tertiary.

He described the chert at Lime Point quite accurately, noting the thin bedding and the local tight folding "resembling the compressed and crumpled leaves of a book." He thought that the chert was metamorphosed sandstone:

"Although the direct transition from the unaltered sandstone and shale to these metamorphic strata has not been observed, there is little doubt they are of the same formation, and that the jaspery condition is due to igneous agencies."

In his discussion of the serpentine of Fort Point, Blake argues that it is an intrusion into the sandstone. Newberry's (1856) comments in general agree with those of Blake.

J. D. Whitney, State Geologist from 1860 to about 1864, and George F. Becker of the U. S. Geological Survey, saw but did not write specifically on the Marin Headlands. Their thoughts on the origin of the rocks are the most widely divergent from those held today. Whitney (1865) thought that the red jaspers (radiolarian chert) were metamorphosed shale and that at least some of the basalt and gabbro and all the serpentine resulted from the metamorphism of sandstone. He seems to have come to these conclusions from a study--and misinterpretation--of contact relations on Mt. Diablo.

Becker (1888) went further and stated (based on petrographic study) that (1) shales were metamorphosed into rocks he called "phthanites"--the radiolarian cherts, and (2) sandstone was metamorphosed to "pseudodiabase" and "pseudodolerite"--the basalt and gabbro of the Franciscan--and finally to serpentine. At one place (p. 123-124) he describes a petrographic thin section of a quartz grain being converted to serpentine, with "long, slender microlites (of serpentine) pierc(ing) the quartz grain...like pins in a cushion." One wonders what mineral he was observing!

Systematic surveys on the Lyellian model

Systematic geologic mapping beginning about 1890 produced three important Geological Survey folios (Fairbanks, 1904; Branner et al., 1909; Lawson, 1915 see also 1895) and two exceptionally fine studies adjacent to the Golden Gate (Ransome, 1893, 1894). The geologic assumptions underlying the gathering and interpretation of field data seem to have been: (1) the Franciscan rests on the presumably older granitic and metamorphic basement exposed in the Santa Lucia and Gabilan Ranges; (2) it is an orderly succession, younger upward, of sedimentary and volcanic rocks; (3) it was deposited essentially where we now find it; and (4) with luck, and enough distinctive marker beds, one could divide it into formations. Fairbanks' folio on the San Luis (30-minute) quadrangle 300 km south of San Francisco (1904) was the most perceptive work of the time, for he recognized the chaotic sheared nature of many Franciscan outcrops, which puzzled him, and which his contemporaries overlooked or ignored. In other work (Fairbanks, 1892), he concluded that serpentine was hydrated intrusions of peridotite and that the Franciscan was overlain unconformably by the Knoxville Formation (which he considered earliest Cretaceous in age) and was therefore wholly Jurassic.

Lawson (1915) divided the Franciscan in the Marin Headlands into five formations of either chert or sandstone. The structural simplicity he deduced contrasts with our present-day understanding of the structure (cf. Fig. 3 with Section E-E', Fig. 10). He interpreted the serpentines as hydrated peridotite intrusions and the bulk of the pillow basalts as intrusions, citing their spheroidal (pillowed) structure as evidence. (Ransome, 1893, had already shown that the pillows resulted from flow, most likely on the surface.) Lawson first concluded (1895) because the cherts were lenticles interspersed with sandstone that they were probably inorganic precipitates from submarine thermal springs; but after Hinde (in Ransome, 1894) reported on the radiolarians, he regarded (1915) the chert as lithified radiolarian ooze.

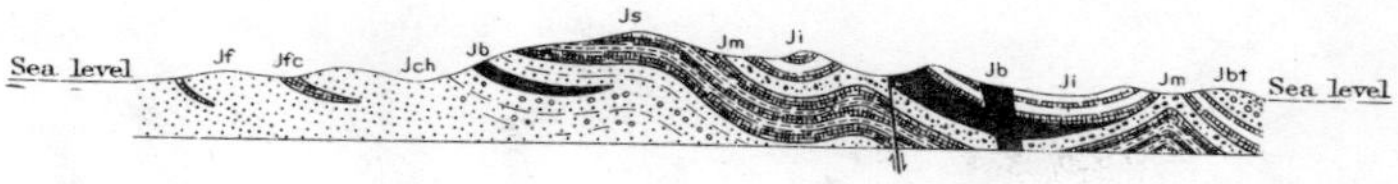

Figure 3. Cross section of Marin Headlands in Lawson (1915). For location, see Figure 2.

Both Lawson and Fairbanks had difficulties with the age of the Franciscan, which was barren of then diagnostic fossils. It appeared to them to be overlain unconformably by the Knoxville Formation (which they considered as earliest Cretaceous in age), but its feldspathic sandstone, they thought, must be derived from granites correlative with those of the Sierra Nevada, which intruded uppermost Jurassic slates, thus requiring an elsewhere unrecorded geologic period for the intrusion of the granite, deposition of the Franciscan, and unconformity on top of the Franciscan.

E. F. Davis (1918a, b) doing some of the most cogently argued, evenly balanced, and thoroughly researched work on the Franciscan sandstone and chert, concluded that the sandstone was most probably deposited in an arid continental environment (based on its immature character and lack of fossils) and that the chert precipitated inorganically from silica-discharging submarine volcanic springs during frequent intervals of shallow submergence of the desert basin of accumulation, because it occurred as many small lenses interbedded (he thought) with the sandstone. The source for the silica-charged waters he thought to be the numerous basaltic and serpentine intrusions into the underlying sandstone.

N. L. Taliaferro (1943) concluded that the Franciscan and apparently overlying Knoxville accumulated in a great geosyncline that came into existence along the present site of the Coast Ranges at the end of the Nevadan orogeny; that they were deposited entirely within Tithonian (late Late Jurassic) time; and that their source was a rugged land-mass containing much granodiorite that lay west of the Coast Ranges. He agreed with Davis on the shallow-water origin of the chert and with Ransome (1894) and others on the contact-metamorphic origin of the schists.

Taliaferro's conclusions were based on examination of a great many outcrops, especially those he thought showed critical relations to overlying rocks (The unconformable contact of the Franciscan with the granitic rocks on which it was supposed to rest was never found). He ignored or never noticed the intensely sheared nature of many exposures. Just before his paper was published, Thalmann (1942) reported Cretaceous foraminifers from the Calera Limestone within the Franciscan, and Schlocker et al. (1954) destroyed his main conclusion with a Cretaceous ammonite from the south shore of the Golden Gate.

E. H. Bailey, W. P. Irwin, and D. L. Jones (1964) summarized the results of 20 years' investigation of the Franciscan, mainly by the U. S. Geological Survey, since Taliaferro's paper. They speculated that the pillow basalts might have erupted as submarine lava jets that broke into discrete blobs which accumulated on the sea-floor while still plastic, moulding themselves to underlying pillows, and that the chert precipitated from deep sea-water which acquired its silica by reaction with the hot lava and precipitated it rapidly as it cooled. They showed that the serpentine had to be emplaced as "cold intrusions" mainly along fault zones. They pointed out the pervasive shearing of much of the Franciscan, and that the schists and associated metmorphic rocks are not the result of contact metamorphism, but must have formed at depths exceeding 20 km, yet at temperatures much lower than normal for those depths, implying rapid burial and equally rapid return to the surface. They also showed that the Franciscan assemblage ranges in age from Middle Jurassic to Late Cretaceous, and is thus contemporaneous with the adjacent strikingly different Great Valley sequence along the east side of the Coast Ranges.

They recognized a number of unexplained problems, such as the source of the serpentine; the nature of the Franciscan basement (which they concluded to be basaltic); the intense and pervasive deformation; the deep burial and rapid unloading of the blueschist-facies minerals; the sandwiching of the Salinian block of apparently contemporaneous granitic basement between two Franciscan areas; and above all, the strange juxtaposition with the coeval but different and much less deformed Great Valley sequence. Clearly, the Franciscan could no longer be interpreted by simple application of the Lyellian model.

The hypotheses they offered to explain these problems included great right-lateral displacement on the San Andreas fault and thrusting of the Great Valley sequence westward over the Franciscan, with intrusion of a sheet of serpentine along the thrust, and stated (p. 164-165) "Or, alternatively, the Franciscan rocks may have been carried eastward beneath the Great Valley sequence by eastward-moving subcrustal currents that plunge under the continental crust." This seems to be the first expression of what is now the accepted origin of the Franciscan.

Using the Plate Tectonic Model

Shortly after the advent of plate tectonics, Warren Hamilton (1969) explicitly explained the Franciscan as a deep-sea accumulation scraped off a subducting slab and welded to the base of the leading edge of the North American Plate; K. J. Hsü's (1971) paper on melanges in the Franciscan made it unnecessary to seek a coherent stratigraphy. Research since then has used as its paradigm the plate tectonic model.

The full implications of the model were slow to be adopted. Perhaps what was needed was the radiolarian biostratigraphy (Pessagno, 1973; Murchey, 1980). Schlocker and his coworkers (1958) mapped the San Francisco North Quadrangle before plate tectonics, but the final report was not published until 1974. In this excellent compendium of the petrology, mineralogy, and engineering geology of the quadrangle, the basalt was still thought to be largely intrusive, the chert to be an inorganic chemical precipitate, and the schist to be in large part metasomatic. The rocks between Sausalito and Pt. Diablo were mapped as a conformable sequence of sandstone and chert into which basalt had been intruded.

Rice (in Rice and Strand, 1971) recognized that most of Marin County was underlain by melange on which the coherent Marin Headlands block was in thrust contact. He did not recognize thrusting within the block and thought that its internal structure involved large folds.

Bedrossian (1974) also showed no internal thrusting, although recognizing that the rocks are a subducted ocean-floor sequence.

The first recognition in print of the complicated structure of the headlands is in the explanation to the map of Blake et al. (1974) which states "In southern Marin County and parts of San Francisco, greenstone occurs as well-bedded pillow lavas and minor intrusive diabase which forms the lowest exposed portion of section, overlain by chert and greywacke and is possibly repeated several times by faulting;". The map does not show these faults.

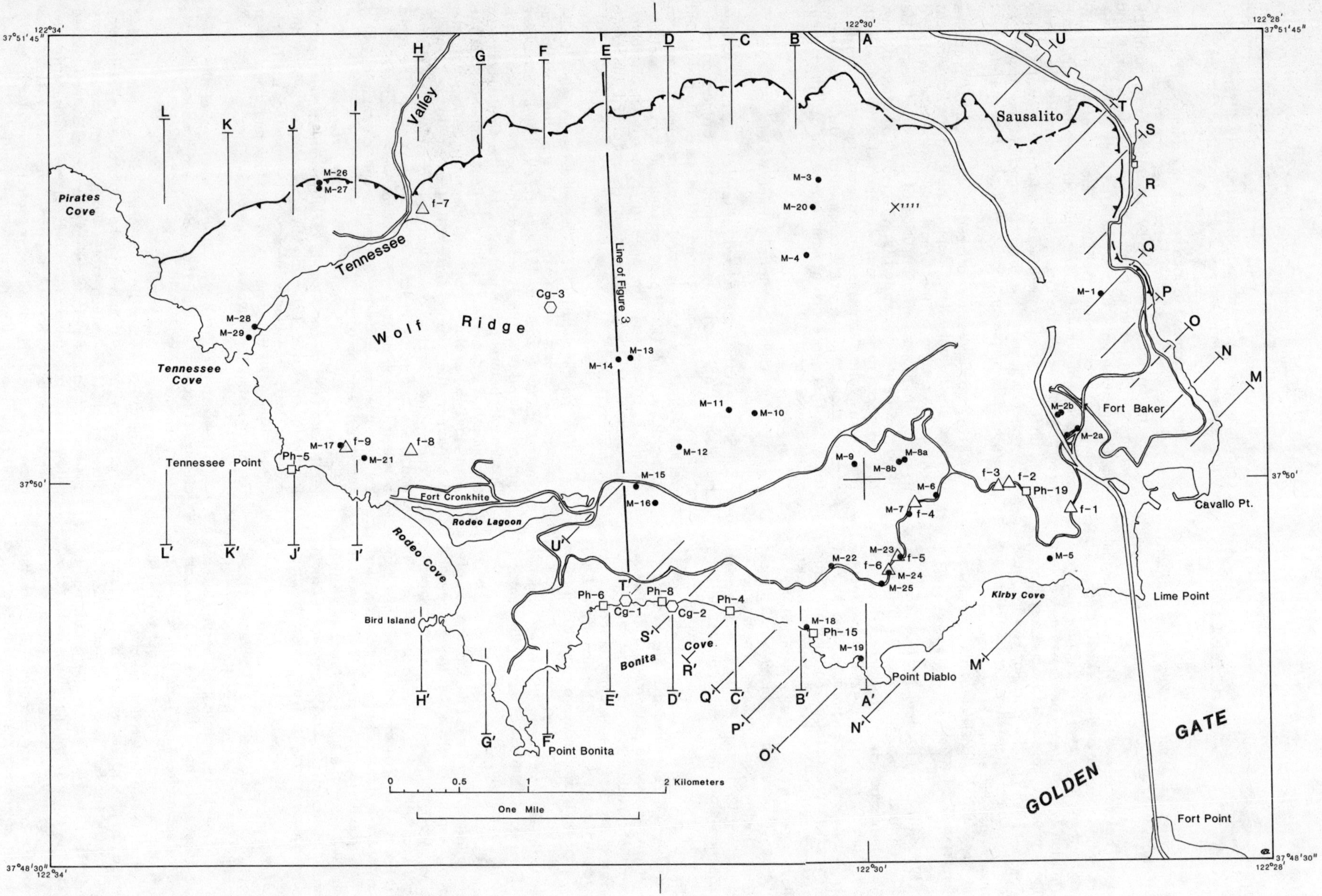

Figure 2. Index map of the Marin Headlands, showing location of sections shown in Figures 3, 10, and 11, locations of conglomerate outcrops (Cg1--Cg3), fossil localities (M-1--M-29), and localities of photographs (Ph-4--Ph-19). Heavy toothed line near north edge of map is the northern boundary fault of the Marin Headlands block. For location see Figure 1.

FIELD WORK FOR THIS INVESTIGATION

In Sept.-Nov. 1974, I was the geologist on a team investigating the natural resources of the Golden Gate National Recreation Area (GGNRA) for the National Park Service, and was impressed with the potential of the GGNRA and especially the Marin Headlands for teaching geology related to plate tectonics (Wahrhaftig and Lehre, 1974). The following year, Cathy Busby-Spera did a senior honors thesis on the eastern part of the headlands under my supervision, and in 1978, 1979, and 1980, I taught Field Geology for 2 weeks each year in the Headlands. I included a trip to the Marin Headlands in A Streetcar to Subduction (Wahrhaftig, 1984). In the last three years, I have spent about 30 days in field work in the Headlands.

Field base maps were (1) ozalid prints of 1:2,400 and 1:4,800 enlargements of Caltrans BATS photographs, taken in 1965, and (2) 1:12,000 enlargements of the Point Bonita and San Francisco North 7.5' U. S. Geological Survey topographic sheets. Nearly all accessible outcrops in the GGNRA west of Fort Baker were examined. Inaccessibility is due to location (1) on steep sea-cliffs or (2) within dense patches of poison-oak (Rhus diversiloba). The geology of the Fort Baker and Sausalito areas has been modified from Schlocker, 1984, Plate 1.

DESCRIPTION OF THE ROCK TYPES

As mentioned, the commonest rocks of the Marin Headlands are basalt, chert (with interbedded red shale), and sandstone (with grey to black shale and rare conglomerate lenses). Melange and broken formation are structural states of the Franciscan consisting largely of tectonically mixed sandstone and shale.

Basalt

Basalt (also called greenstone) outcrops over 20-25 percent of the Marin Headlands. Where exposed in sea-cliffs and deep roadcuts it is dark green to black, hard, and resistant to erosion. Weathering first pales the green color and cuts the basalt with closely spaced curved fractures, and ultimately weathers it to an orange clay-rich soil with fragments of moderately weathered basalt.

Pillow basalt predominates, and where well exposed in the sea-cliffs, the pillows show many of the features described by Moore, 1975, and Moore et al., 1973, that demonstrate their eruption beneath the sea (Figs. 4 and 5). Spectacular exposures are listed in Wahrhaftig (1984) and Ransome (1893).

The pillows commonly have fine-grained to aphanitic rims and somewhat coarser interiors. Locally, they have radial cracks penetrating a few cm inward from the pillow surface.

Rounded pillow tops and downward projecting keels moulded against the tops of older pillows are up-section indicators. Feeder tubes and bifurcation of some of the bulbous lower ends of pillows indicate the direction of flow and probable paleoslope.

Some pillows show an internal layering consisting of 2-5-cm-thick sheets of basalt alternating with thicker bodies of quartz or calcite, or both (Fig. 6). These structures, first pointed out to me by Tanya Atwater, result from partial draining of the pillows after they reached their full size, lowering the level of lava within the pillow. Sea-water immediately invaded the pillow, freezing a solid crust on the lava. Further draining of the lava repeated the process to give several shelves within the pillow. Since the surface of the still-liquid lava within the pillow would have been horizontal, these internal lava shelves coincide with the true horizontal at the time the pillows were formed, and are the only reliable indicators of the paleo-horizontal, in view of the discovery by Fornari et al., 1978, of pillows draping the vertical outer walls of extrusive dikes along the submerged extension of the East Rift Zone of Kilauea.

Interstices between the pillows are filled with soft material consisting mainly of (1) chlorite or nontronite (according to Schlocker, 1974), (2) masses of coarsely crystalline calcite or quartz, or (3) pink pelagic limestone and chert, the last probably deposited on some pillows to a thickness of a few cm before the next pillows were erupted.

Keels and pillow-tops show that in general the bodies of basalt are upright, and within each body younger basalt is to the south. Paleoslopes indicated by feeder tubes and bifurcating lobes, for example, at Tennessee Point, Bonita Cove, and northwest of Lime Point, are also to the south. However, basalt at Tennessee Point pinches out to the east, and that at Lime Point pinches out to the northwest. The pinchout must therefore be due to inclined thrusts beneath the basalt and not to original volcanic topography.

Flood basalt, such as reported by Ballard et al., 1982, along the Galapagos Rift, appears to be a minor component of the basalt at the Marin Headlands. It has been identified at the south end of Tennessee Cove, at the south end of Rodeo Beach, at Point Diablo, and at the north abutment of the Golden Gate Bridge. It may be present in greater abundance on some of the massive and inaccessible sea-cliffs of the headlands. Tuff and agglomerate, apparently of submarine origin, have been recognized at the beach just north of Bird Rock (cf. Ransome, 1893) and at the west end of Kirby cove (Schlocker, 1974).

Petrographic descriptions of the basalt by Ransome and Schlocker stress the presence of plagioclase laths, augite, and glass, together with alteration products such as chlorite and nontronite. Veinlets of albite and pumpelleyite reported by Schlocker establish that metamorphism of the basalt, where it has occurred, is in the sub-greenschist facies, and probably was caused by circulating hot sea-water shortly after eruption.

Chert

Radiolarian chert underlies 50 percent or more of the Headlands, and because of its resistance to weathering, makes up nearly all of the ridge-tops and summits. It occurs predominantly in layers 2-10 cm thick alternating with much thinner layers of dark red shale (hence called ribbon chert.). Individual layers of both chert and shale pinch out, usually abruptly. Locally the chert is intensely crumpled, usually into sharp-crested, straight-sided chevron folds (Fig. 7). The ribbon chert is commonly red, but both chert and shale may be pale green, either along fractures and bedding or throughout, due to reduction of iron.

According to Murchey (1980 and this volume) the 32-m-section exposed on Alexander Avenue at Highway 101 (Localities M-2a and M-2b, Figures 2 and 13) has all the biostratigraphic zones present in the Headlands, rests conformably on basalt, is overlain conformably by dark shale overlain by sandstone, and is undisturbed by minor faulting and folding. She shows that the chert consists essentially of two units of ribbon chert separated by a zone 2 m to more than 6 m thick, of jasper breccia which has been cemented by white quartz veins. The breccia is red, orange, white, or grey. This breccia zone is very resistant to erosion, makes conspicuous craggy outcrops on hilltops, and is the predominant ridge-former in the chert. According to Murchey it may mark an 8-25 m. y. hiatus in chert deposition lasting from some time in the Tithonian to some time in

TABLE I
USGS ACCESSION NUMBERS AND ZONES
OF RADIOLARIAN SAMPLES

Map No.	USGS Access Number(s)	Zone
M-1	4988	MH-1
M-2a	4983-4987, 5100-5110, 5114	MH-1-MH-5
M-2b	5115, 5067, 5064	MH-5-MH-7
M-3	5022	MH-7
M-4	5018	MH-3
M-5	See Murchey, this vol.	MH-1
M-6	5062	MH-3
M-7	5063	MH-5
M-8a	5070	MH-1
M-8b	5071	MH-1
M-9	5074, 5073r	MH-4, 5
M-10	5013	MH-5
M-11	5012	MH-5
M-12	5009	MH-1
M-13	5027	MH-1
M-14	5025	MH-1
M-15	5036	MH-3
M-16	5030	MH-7
M-17	5006	MH-4
M-18	5112, 5113, 5116, 5117	MH-4-MH-7
M-19	4342a, b	MH-2, 3
M-20	4343	MH-2?
M-21	4353	MH-3
M-22	4347	MH-1
M-23	4352	MH-1
M-24	4351	MH-3
M-25	4349 (cryptocephalic nasellarians)	K?
M-26	4518	MH-1
M-27	4515	MH-1
M-28	4524	MH-1
M-29	4522	MH-6

Samples M-1 through M-18 collected by B. L. Murchey, M-19 through M-29 by Clyde Wahrhaftig.

the Valanginian. A few meters of the basal chert has manganese oxide coatings on fractures and partings.

Localities of samples analyzed paleontologically by Murchey (oral communication, 1983, 1984, and this volume) are shown by open circles numbered M-1 through M-29. Table I lists the USGS accession numbers of the collections and their assignment to radiolarian zones. Table II summarizes the age assignments by Murchey (this volume) of these zones. All stratigraphic sections are upright and within each chert body are younger to the south.

Sandstone and shale

Sandstone is mainly lithic arkosic wacke, a microbreccia consisting of angular fragments of plagioclase, quartz, and volcanic rocks, in a dark matrix of presumably the same material. It is resistant and grey in fresh exposures in sea-cliffs, but weathered is soft. buff, and highly fractured. For the most part, sandstone beds are internally structureless, except for a distinct grading from coarse at the base to fine at the top. Locally, there is cross-bedding, most notably in cliffs of Bonita Cove (Fig. 8).

Each sandstone bed grades upward into thinly bedded fine sandstone or shale, or into a sequence of sandstone and siltstone several cm thick marked by abundant carbonaceous fragments parallel to the bedding. Whereas the individual sandstone beds range in thickness from 1--2 m to more than 20 m, the shale interbeds are rarely more than a meter and commonly less than 15 cm thick. The sandstone and shale are

TABLE II
AGES OF RADIOLARIAN ZONES
(From Murchey, this volume)

Zone	Period	Ages
MH-1	Je (<u>ca.</u> 200 Ma)	Pliensbachian-Toarcian
MH-2	Jm	early Bajocian (=Aalenian)-Bathonian
MH-3	Jml	Bajocian-Kimmeridgian(?)
MH-4	Jml	Bathonian-Kimmeridgian(?)
Hiatus (and/or)	Jl-Ke recrystallized	Kimmeridgian--Valanginian
MH-5	Ke	Valanginian-early Barremian
MH-6	Ke	Barremian-Albian(?)[1]
MH-7	Kel (<u>ca.</u> 98 Ma)	late Albian-early Cenomanian

Je, Early Jurassic; Jm, Middle Jurassic, Jl, Late Jurassic; Jml, Middle and Late Jurassic; Ke, Early Cretaceous; Kl, Late Cretaceous; Kel, Early or Late Cretaceous.

[1] Possibly only to Aptian, with a probable hiatus marked by a change in preservation style and thin shale bed (B. L. Murchey, written comm., 1984).

clearly a turbidite sequence. At the Alexander Ave. section 1-3 m of dark shale lies immediately above the chert (Murchey, this volume).

Up-section indicators, such as cross-bedding, graded bedding, and channelling, all indicate that the sandstone is upright and, within each fault slice, is younger to the south.

Conglomerate is rare but has been observed in three places: Locs. Cg-1 and Cg-2, Fig. 2, on the north shore of Bonita Cove, and Cg-3 north of Wolf Ridge. Conglomerate at Cg-1 is matrix-supported with subangular to subrounded boulders, cobbles, and pebbles of red and green chert, grey limestone, sandstone, and greenish altered basalt, of apparently Franciscan provenance. Loc. Cg-2 consists of huge blocks of conglomerate at the base of the sea-cliff, apparently the relict of a landslide that involved sandstone from the fault-slice immediately below that of Locality Cg-1. Here, the conglomerate is well-sorted clast-supported conglomerate consisting of well-rounded pebbles and cobbles of grey to black chert and quartzite of an unknown provenance. The difference between these two conglomerates suggests that the two conglomerates are unrelated and are either of different age or have been juxtaposed from widely different venues.

Conglomerate at Cg-3, near the top of a large sandstone body, consists of large angular blocks, mainly of limestone, floating in a coarse sandstone matrix.

Limestone

Pink, presumably pelagic limestone has been observed at two localities, intimately associated with chert, and deposited directly on top of, or in the interstices between pillows. One locality is at the south end of Rodeo Beach, just east of Bird Rock, where a few cm of pink to purple limestone, with chert nodules, rests on a large nearly flat mound of pillows, and is overlain by pillows. The other locality is in blocks fallen from an inaccesible outcrop in the sea-cliff of Bonita Cove,about 500 m east of Cg-2, Figure 2. Here the limestone is also intimately associated with chert and appears to rest on basalt pillows. The two limestone occurrences suggest that at least part of the Marin Headlands sequence accumulated on bottoms as shallow as the carbonate compensation level.

Figure 4 (top). Large pillow with feeder on right. Locality Ph-4, Figure 2, Bonita Cove.

Figure 5 (bottom, left). Pillow basalt, Locality Ph-5, Tennessee Point, looking northwest.

Figure 6 (bottom, right). Alternating layers of basalt and quartz-calcite mixture within a large pillow, locality Ph-6, Bonita Cove. Layers indicate a tectonic inclination of 40°. Scale is 50 cm long.

Figure 7 (top). Chert with chevron folds, headlands road at locality f-4, Figure 2, looking northwesterly. Scale is 1 m long.

Figure 8 (middle). Cross-bedded coarse arkosic lithic wacke, base of cliff at locality Ph-8, Figure 2, Bonita Cove.

Figure 9 (bottom). Broken formation--sheared shale with phacoids of sandstone--at base of cliff just east of locality Cg-2, Figure 2. Scale is 1 m high.

Diabase

A diabase dike extends from Point Bonita, northward along the cliffs facing the Pacific, to the south end of Rodeo Cove. Ransome (1893) discusses its petrography. The trend of the dike, almost due north, is nearly perpendicular to the regional bedding. It is interpreted here as the only representative in the Marin Headlands of the sheeted dike complex (Coleman, 1977).

Serpentine

Two or three blocks of serpentine a few m across and 10-20 m long occur in a melange belt (see below) at the NW corner of Bonita Cove.

A belt of serpentine crosses San Francisco diagonally to Fort Point at the the southeast corner of Fig. 13 (see Fig. 1). At Fort Point it consists of two gently east-dipping sheets of serpentine at least 50 m thick separated by a sheet of melange about 15-30 m thick. The serpentine sheets consist of rounded blocks of massive bastite-bearing serpentine sheathed in intensely sheared serpentine. The general strike of the serpentine belt in San Francisco is directly toward the Marin Headlands near Point Diablo.

Melange and Broken Formation

Franciscan melange is essentially a chaotic and nearly impervious mixture of intensely sheared sandstone and shale, in which are embedded blocks of exotic lithologies such as basalt, gabbro, chert, limestone, serpentine, schist, and eclogite. A chaotic mixture of intensely sheared sandstone and shale without the exotic blocks is known as broken formation (Fig. 9). Melange underlies all of the Marin Peninsula north of the Headlands, and is separated from the Headlands block by a fault which is near-vertical at its west end (west of the crossing of Tennessee Valley) and dips gently southward beneath the Headlands block east of Tennessee Valley. Topographically, melange areas have broader ridge-crests and gentler slopes than coherent terranes such as the Marin Headlands, with many earthflows. Ground-water is close to the surface, supports patches of phreatophytes such as tules (Scirpus sp.), and emerges as seeps. The exotic

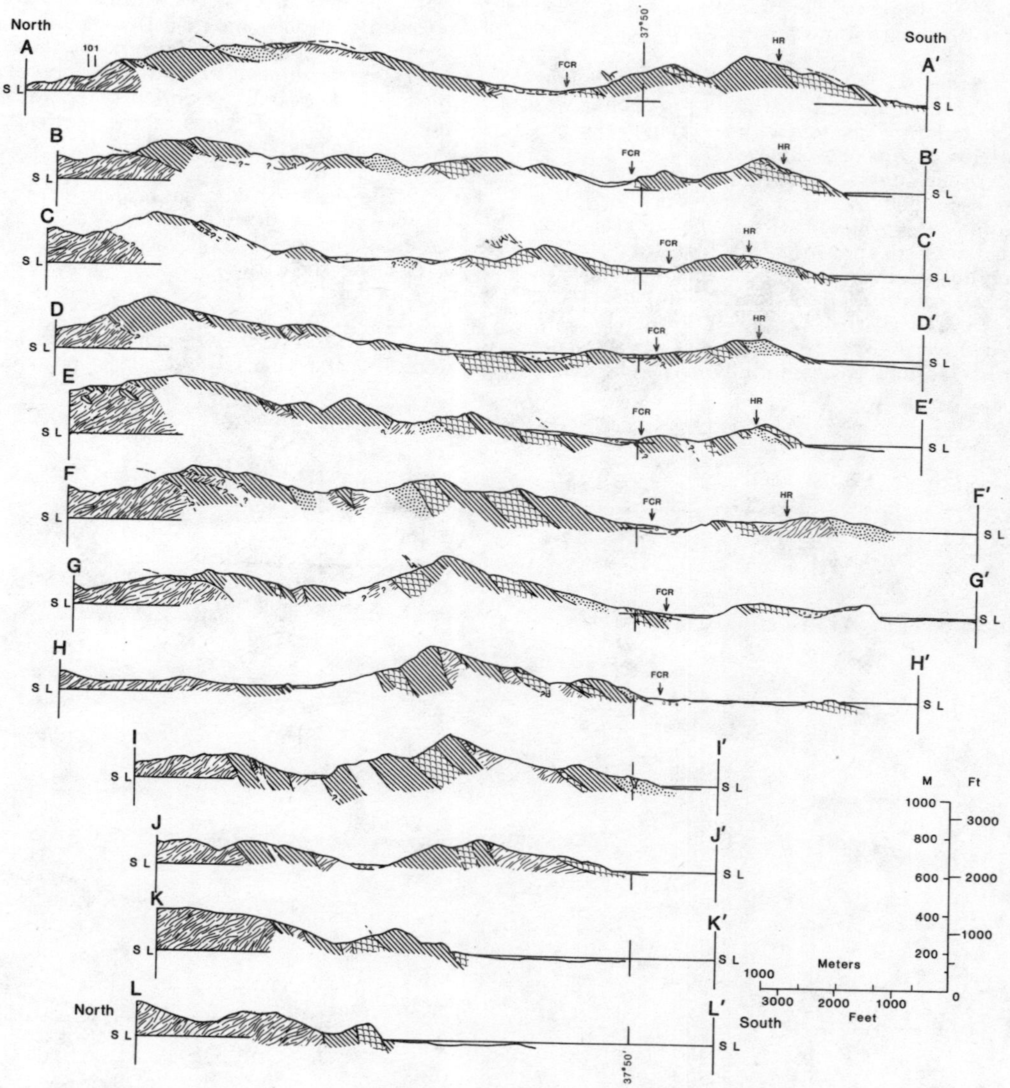

Figure 10. N-S vertical cross-sections of the western two-thirds of the Marin Headlands. Locations are shown on Figure 2. Horizontal scale, vertical scale, and spacing of sea-level lines are all equal. For explanation of patterns see Figure 12 or 13. Note that upright direction of pillow basalt pattern is upward to the right (south).

blocks crop out as picturesque crags. The matrix of sheared sandstone and shale is exposed only in sea-cliffs, as at Pirate's Cove.

Within the Marin Headlands block, melange and broken formation are exposed in sea-cliffs between Tennessee Cove and Tennessee Point, south of Rodeo Lagoon, and along the shores of Bonita Cove and the tiny cove to the east. They are commonly in contact with basalt on the south and chert or sandstone on the north. The melange at the west end of Bonita Cove contains serpentine blocks, and is marked by a huge landslide. The 300-m-thick body of melange or broken formation between Tennessee Point and Tennessee Cove, splits eastward into three narrow belts around two coherent bodies of chert and basalt. Huge blocks of chert in this body are thought to be fragments from the coherent bodies to the east.

CONSTRUCTING THE GEOLOGIC MAP

Exposures along sea-cliffs and in roadcuts are continuous and magnificent. Away from these natural or artificial cliffs, exposures are meager. Only the chert makes conspicuous outcrops, and its detritus commonly mantles other lithologies. Creating a geologic map from these scattered exposures was a circuitous process.

The first step was an outcrop map, showing lithology and structure of every exposure. Lithologic mapping was extended from the outcrops as far as local structures and topographic expression gave unarguable results.

The next steps involved a modification of the down-structure method described by J. Hoover Mackin (1950). Geologic information was projected from the outcrop map onto vertical cross-sections drawn 1,500 feet (ca. 500 m) apart and trending due north in the area west of 122°30' W. (Fig. 10, cf. Fig. 2), and 1,000 feet (ca. 300 m) apart and trending N 45° E in area east of 122°30' W. (Fig. 11). The latter cross-sections were extended southwest as shown in Figure 2. The cross-sections are essentially perpendicular to structural trends.

The geology of each cross-section was then projected onto a line perpendicular to the regional dip. West of 122°30' W. these lines plunge 45° N. East of 122°30' W. they plunge NE.

These linear projections were assembled at appropriate intervals along a reference line to form a projection of the geology onto a plane perpendicular to the average dip. West of 122°30' W. the reference line was the intersection at sea-level with the parallel 37° 50' N. East of 122°30' W. it was their intersection at sea-level with an arbitrary line trending N 45° W (see Fig. 2). The projection plane for the N-S cross-sections strikes due E and dips 45° N. The projection plane for the N 45° E cross-sections strikes N 45° W and dips generally 45° NE. In making this projection I had to eliminate the effect of variations in dip across the strike. The extended outcrop map guided the connecting of contacts between the linear projections.

Joining the geology of the western two-thirds, where the strike of the units is due west, with that of the eastern third, where the units strike northwest, created some problems, especially since the area of the junction just west of hill 1111 has poorly exposed and structurally confusing geology. Some liberties had to be taken with the geology to make a fit. Additional adjustments were made, either to the plunge of the projection lines or to their intersections with the reference lines, to eliminate abrupt variations in the thickness of units or direction of contacts, variations which were more likely fictions of the method than reflections of the true geology.

The resulting diagram (Figure 12) is essentially a down-dip view of the geology of the Marin Headlands, and shows the true thicknesses and variations in thickness of the rock units along the strike, the only direction for which we have reliable data.

The structural interpretation of Figure 12, the next stage in the analysis, depends on the theory of plate tectonics. I assumed that the Franciscan is an ocean-floor

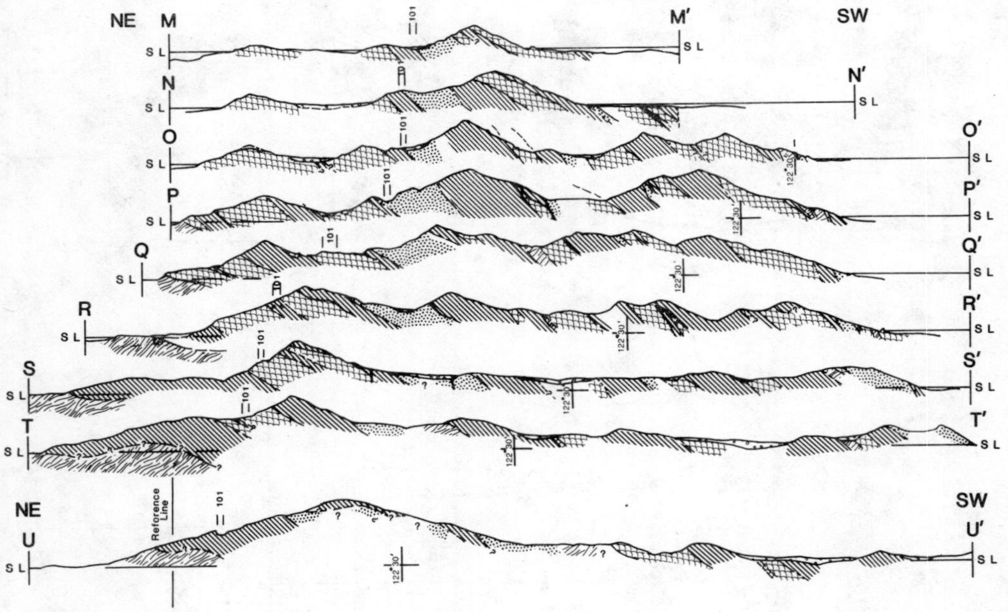

Figure II. N 45° E vertical cross-sections of the eastern third of the Marin Headlands. Locations are shown on Figure 2. Horizontal scale, vertical scale, and spacing of sea-level lines are equal. For explanation of patterns see Fig. 12 or 13.

sequence, born at a spreading axis, transported across the Pacific, and subducted beneath the leading edge of the North American Plate. Hence, all basalt was interpreted to be mid-ocean ridge basalt (MORB), unless the evidence clearly indicates otherwise. The cherts were interpreted to have accumulated on the MORB as radiolarian ooze as it travelled from the spreading axis to the subduction zone, and the clastic rocks to result from floods of terrigenous clastic sediment when the ocean-floor approached the subduction zone--floods that covered the chert and effectively masked any further radiolarian contribution. In this interpretation the basalt would not be deposited on or intruded into chert or sandstone, nor would chert be deposited on sandstone. Any contact with chert above sandstone or with basalt above either chert or sandstone must either be overturned or a fault. All up-section indicators showed the beds to be upright and, within each outcrop or unit, younger to the south. Therefore all such contacts are faults.

Furthermore, isolated exposures or indicators of basalt, sandstone, or melange within areas dominated by chert outcrops and chert rubble on hillsides are interpreted to mark the traces of otherwise unexposed faults. Many thrusts shown in predominantly chert areas on Figure 13, e. g. west of 1111 hill, are based on only one or two such exposures.

Where most of these inferred faults crop out on cliffs or road-cuts, they are marked by narrow belts of melange or broken formation, confirming that they are faults. In only a few places, where there is good biostratigraphic evidence for faulting, is the fault zone difficult to discern. One such place is the intersection of the northeast contact of the basalt body on the east side of Kirby Cove with the Headlands road northwest of the Golden Gate Bridge.

A few small bodies of basalt do not fit this interpretation and may be intrusions into the sandstone. The most accessible of these are two narrow dike-like bodies of basalt that appear to cut the sandstone at a considerable angle to the bedding on the ridge and seacliff just north of Rodeo Cove. The most conspicuous "dike", however, has good pillow structure and a narrow belt of chert along its eastern side, both attributes of MORB.

The information from Figure 12 was then transferred back to the outcrop map to produce Figure 13, the geologic map.

STRUCTURAL INTERPRETATION

Thrust sheets

The panel and map show that the Headlands are made of a series of thrust slices, predominantly of chert with some of its basalt substrate. These slices pinch out laterally or break up at their edges into blocks of chert and basalt in melange, e. g., on the south flank of Wolf Ridge northwest of Fort Cronkhite. Presumably the slices also pinch out or break up in the down-dip direction. No single thrust slice appears to extend the entire length of the Marin Headlands.

The slices are stacked one on the other like overlapping shingles or scales. Chert predominates in most of the slices. The basalt does not exceed 300 m in thickness, and the turbidite sequence overlying the chert is similarly thin. In view of the fact that the basalt layer beneath the ocean floor is commonly 1.4 km thick (Coleman, 1977, p. 11), and that the Franciscan consists predominantly of sandstone and shale, it appears that most of the thickness of uppermost three layers of the sea-floor is missing. It was either scraped off the down-going slab before the slab had reached its present position with respect to the rocks around it, or was subducted to greater depths.

These flakes of chert resting in part on basalt and overlain in part by turbidites are interpreted as fragments of an ocean floor that took about 100 million years to travel from its birthplace thousands of km to the southwest in the Pacific to its interment along the western edge of North America (cf Murchey, this volume). For a while (perhaps a million years) as it was subducted, the sea-floor broke into flakes, each of which successively docked beneath the preceding and now overlying flake as part of the accretionary wedge. Friction against the overlying plate may have slowed the leading edge of the subducted plate and caused the subduction fault to jump downward to a level within the basalt or at the base of the chert, leaving the thereupon detached flake plastered onto the base of the continental plate (Fig. 14-a). This process could have continued until 10 or more such slabs were stacked one beneath the other.

But obduction is also a possible mechanism (Fig. 14-b). In this latter case, the weight of the leading edge would provide the resisting force against being shoved up and onto the continent, and the crust would break into flakes that stacked one on top of the other. The obduction scenario fails to account for the disappearance of the clastic wedge from above much of the chert (of which the present sandstone in the Headlands is a meager relict). In the case of subduction, the sandstone could have docked farther seaward beneath the continental plate; or it could have been squeezed from between the greenstone-chert slabs and carried downward to form melange matrix.

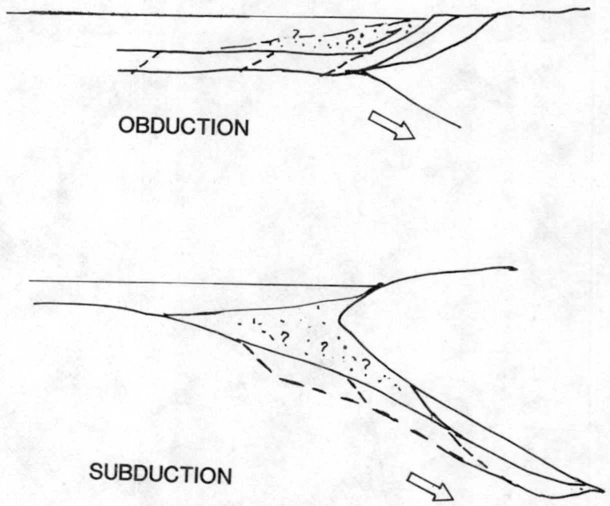

Figure 14. Two mechanisms for the accretion of the Marin Headlands block: (a) subduction; (b) obduction.

Theoretically, one could discriminate between subduction and obduction with a sample of oceanic crust wide enough to have its age range (indicated by radiolarians from the basal chert) cross one or more biostratigraphic zone boundaries. Unfortunately, the 60-km or so width of oceanic crust preserved here could have been formed in less than a million years, at current spreading rates in the Pacific, a time-span much shorter than the average length of a biostratigraphic zone in the Mesozoic.

A particular problem shown in Fig. 12 is the abrupt thickening and thinning of the chert, even where sandwiched depositionally between basalt and sandstone. The Alexander Avenue Section (Locs. M-2a, M-2b on Figs. 2 and 13) is about 80 m thick and has all the biostratigraphic zones present in the Marin Headlands (Murchey, this volume). Thinner sections could have been eroded by turbidity currents. Fig.

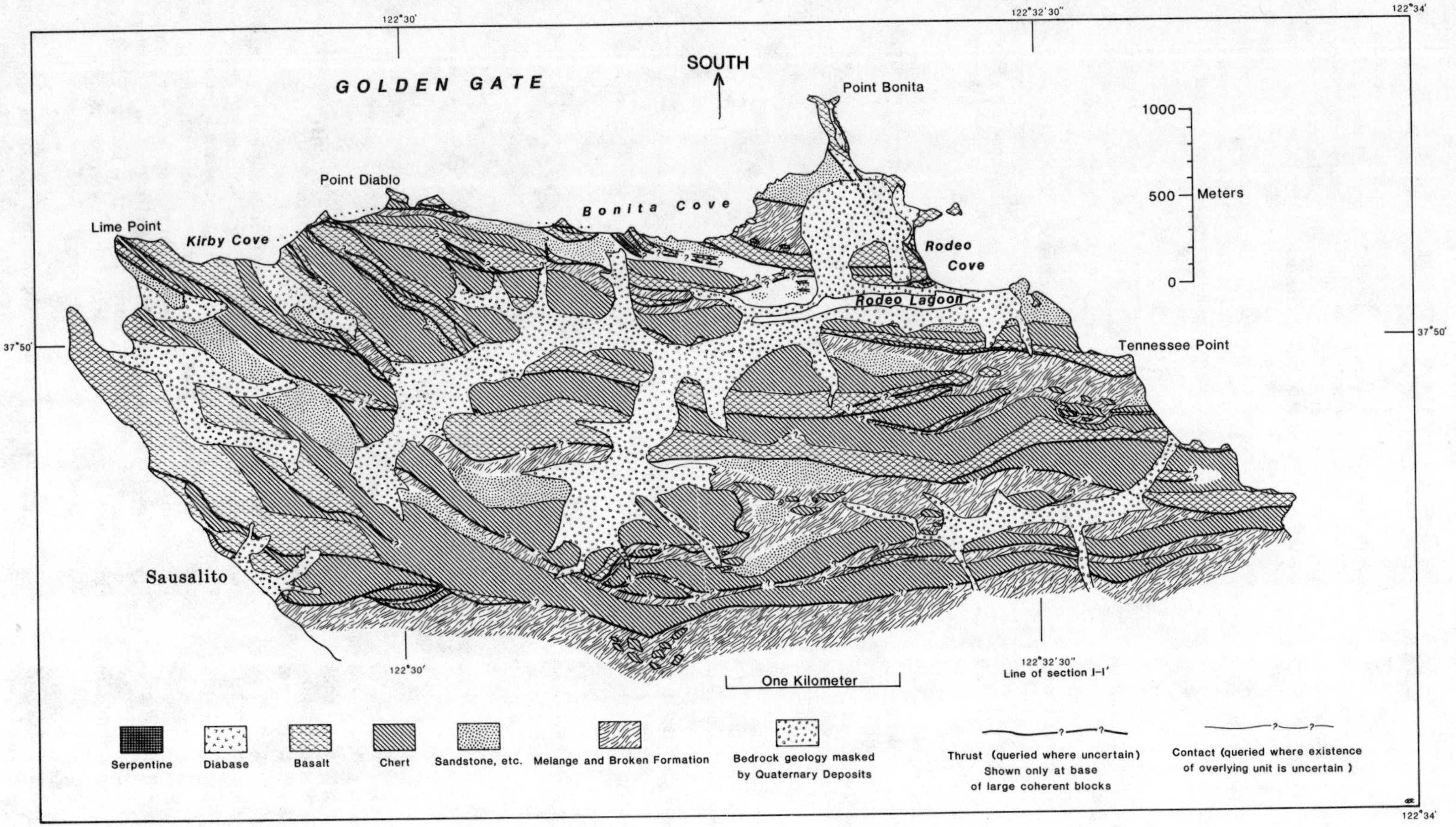

Figure 12. Down-dip projection of the geology of the Marin Headlands. South is at the top. Note that upright direction of pillow basalt pattern is upward. All units shown, except the one titled "Bedrock geology obscured by Quaternary Deposits", are part of the Franciscan assemblage.

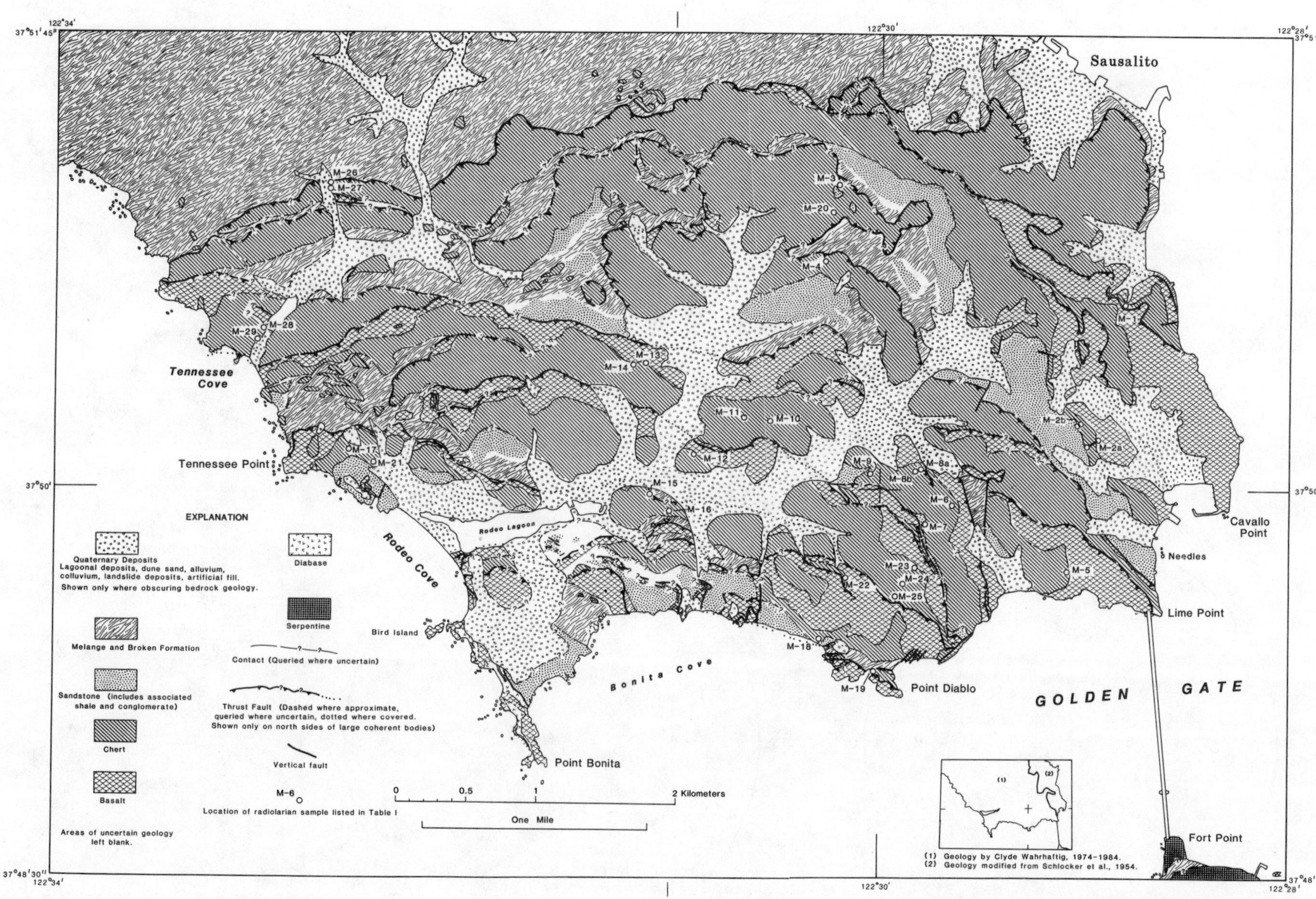

Figure 13. Geologic map of the Marin Headlands. Area shown is same as in Figure 2. See Figure 1 for location. Note that upright direction of pillow basalt pattern is to the south. All units shown, except Quaternary deposits, are part of the Franciscan assemblage.

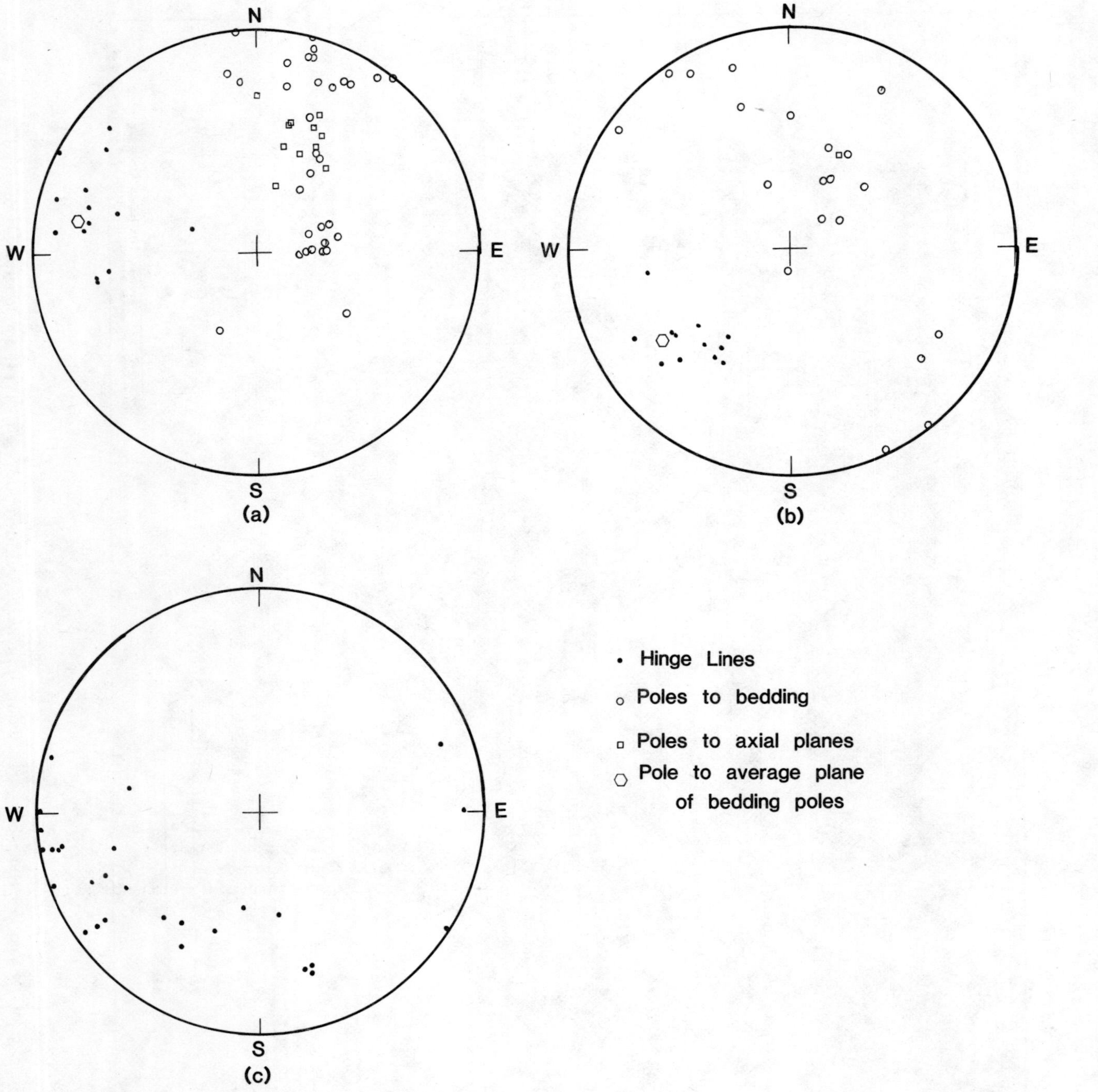

Figure 16. Equal-area projections of minor fold hinge lines, and of poles to bedding and axial planes for localities f-1 and f-4, Figure 2; and of individual hinge lines measured elsewhere in the Marin Headlands; a, projection for f-1; b, projection for f-4; c, other hinge-lines.

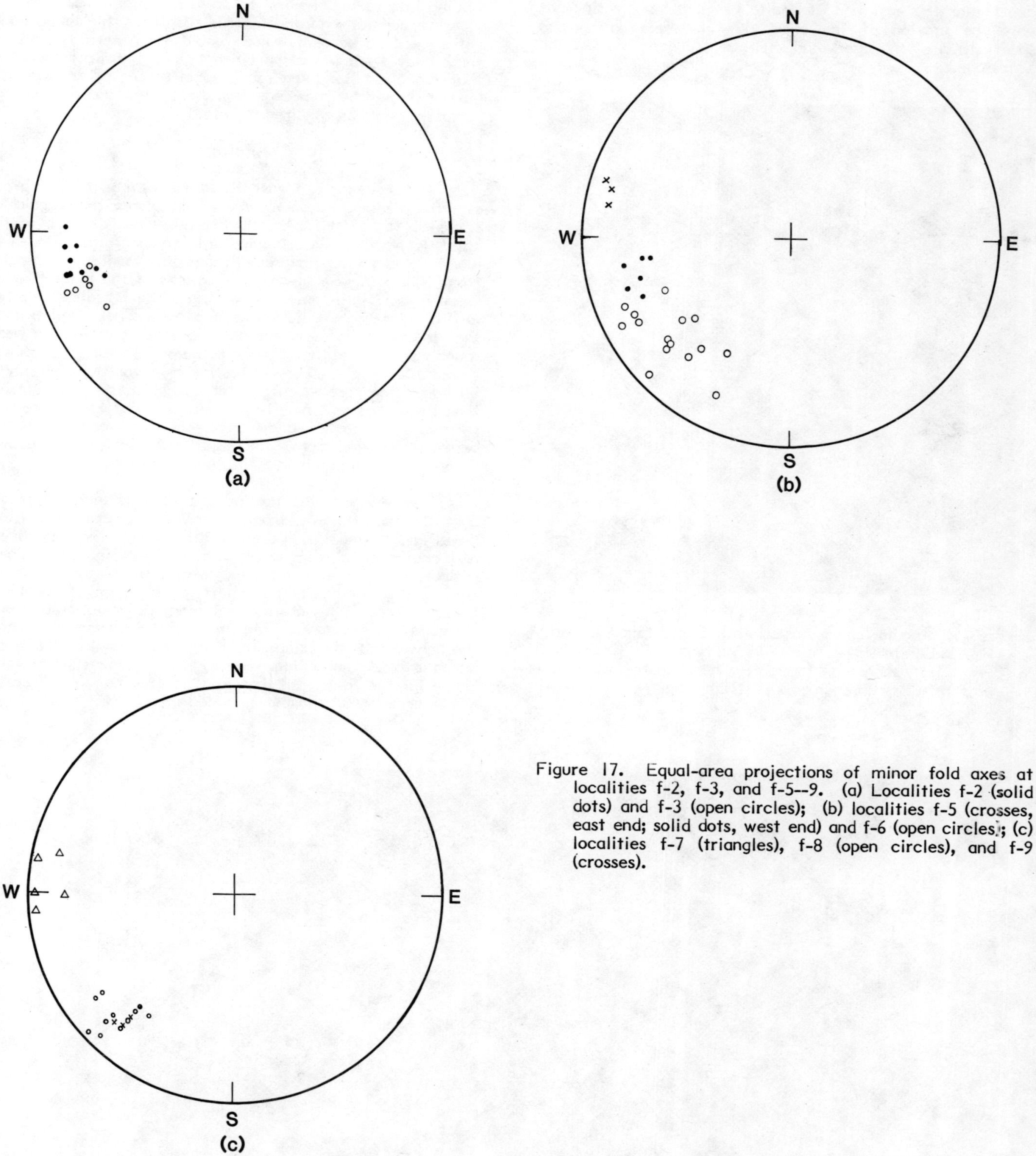

Figure 17. Equal-area projections of minor fold axes at localities f-2, f-3, and f-5--9. (a) Localities f-2 (solid dots) and f-3 (open circles); (b) localities f-5 (crosses, east end; solid dots, west end) and f-6 (open circles); (c) localities f-7 (triangles), f-8 (open circles), and f-9 (crosses).

15 shows angular chert clasts in basal sandstone resting on chert near the east end of Bonita Cove, demonstrating not only the erosion of the chert but also segregation and lithification of the chert and shale before deposition of the sandstone.

Figure 15. Angular chert clasts in basal sandstone. Locality Ph-15, Bonita Cove. View looking northeast; chert is at lower left, sandstone to right. Chert clasts outlined for emphasis. Brunton compass is 15 cm long.

Chert units could be thickened through pervasive folding into minor chevron folds with planar limbs (See Fig. 7). The geometry of the folds is such that thickening could not exceed 50 percent, which would not generate a 200-m-thick body from a section 80 m thick. I suspect therefore that the chert has been repeated in places by thrusts that have not yet been recognized. Detailed biostratigraphic sampling would be required to detect these thrusts.

Folds

The sequence of thrust flakes has been bent into a broad syncline plunging 40° S 25° W, whose planar limbs make an angle of 155°-160°. As indicated previously, much of the chert is pervasively crumpled into minor folds with sharp hinges, planar limbs, and amplitudes and wavelengths ranging from 1 to 3 m. Equal-area projections of the hinge lines of these folds, and of poles to bedding and axial planes for two localities, are shown in Figures 16 and 17. These figures show that the minor folds in the chert plunge southwest to west, roughly parallel to the axis of the major fold but commonly at a shallower angle. They also show that variations in hinge-line directions are much greater within continuous outcrops than they are from one site to another, and furthermore that there appears to be no difference in the average direction between the two limbs of the major fold, which is to expected if the major and minor fold hinges are parallel.

The minor folds die out or change their character abruptly when followed from bed to bed in any given exposure (See Figs. 18 and 19). Likewise they change their amplitude and character along the hinge line (See Fig. 20). These irregularities in the folding, coupled with the difficulty envisaged in folding brittle chert, caused many to conclude that the minor folds are soft-sediment deformation or slump structures. However, fold axes are in general parallel throughout all the slabs, and the minor fold hinges are parallel to the major fold. This regularity suggests a tectonic origin for the folds, and that folds are normal to a direction of compression or transport-induced shear.

Figure 19. Stereoscopic pair of ground photographs of a folded single bedding surface in chert exposed in a road-cut, looking east. Locality Ph-19, north of Kirby Cove.

a

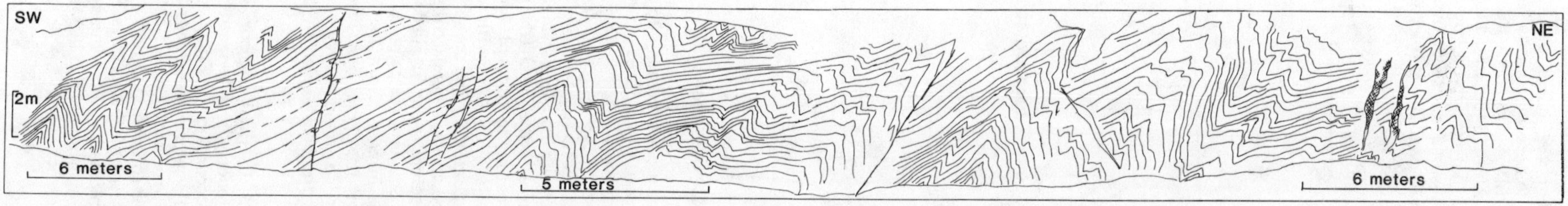

b

Figure 18. Sketches of folds exposed in roadcut outcrops at localities f-1 and f-4, Figure 2. Sketches were compiled from tracings of color Xerox prints of kodachrome slides; a, locality f-1; direction of exposure N 3°--10° E, approximately 35°--40° to left of fold hinge lines; b, locality f-4; direction of exposure N 70°--74° E, approximately 25° to left of fold hinge lines.

Figures 18 and 19 show that the minor folds are decidedly asymmetric. If projected to perpendicularity to the hinge lines, the short limbs are still on the north sides of anticlines, and the axial planes dip south. Except for locality f-3, figure 2, this is the general pattern throughout the headlands. This asymmetry would result from shear induced if the lower plate were moving south or S 20° E relative to the upper.

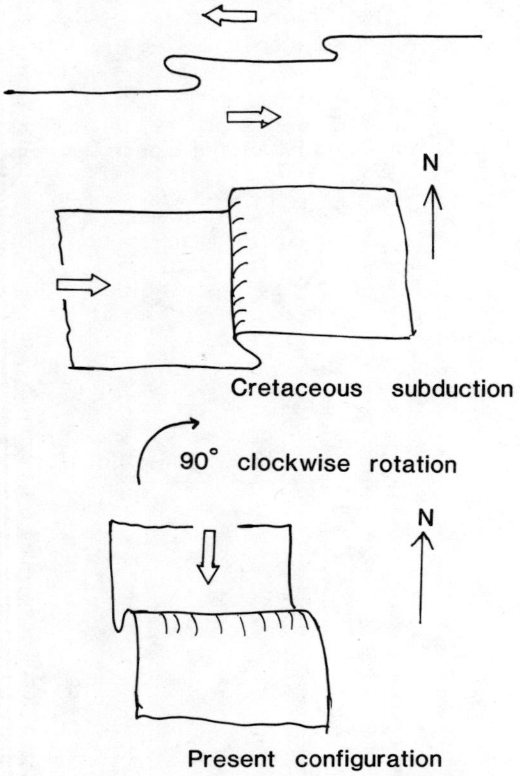

Figure 20. Diagram showing rotation necessary to give present configuration of minor fold axes from a subduction direction to the northeast.

S. D. Ellen (1971) showed that the minor folds are kink-folds that have been rotated by a shear couple in which the upper plate moved northward relative to the lower (the same sense as described above). Attempting to account for the shear by movement on the flanks of shallow folds with wave-lengths of about 50-100 m, he was puzzled by the consistency of the shear direction. As discussed above, a persistent sense of shear is what would be expected under a subduction model. Walter Alvarez and Mary Savina (oral comm., 1976) have found evidence in hinge lines in chert from the Marin Headlands for solution folding of the type described by Alvarez et al., 1976. Thus radiolarians would not necessarily be distorted during the folding of the chert, even if it was hard and brittle.

A direction of south to S 20° E is not an appropriate subduction direction of the Pacific ocean-floor beneath North America 100 m. y. ago which, according to Engebretson, 1982, was northeasterly. But if the Marin Headlands block was rotated 90-130 degrees clockwise about a vertical axis after subduction (Fig. 20), the present configuration of the folds and the anomalous dip directions of the thrust slabs of the Marin Headlands would be explained. According to the paleomagnetic data (Curry et al., this volume) this hypothesis is not improbable and is probably correct.

CONCLUSIONS

The following conclusions are deduced from this study:

The Marin Headlands block is a stacking of about ten thrust slabs that strike east to south-east, dip to the south, are folded into a broad syncline about an axis plunging 40° S 40° W, and are all upright (with internal bedding that is younger upward and to the south).

The block rests on melange that underlies most of the Marin Peninsula. The contact is a thrust fault that is near-vertical on the west, but shallows eastward to dip gently to the south on the east. The block is separated from the rocks of San Francisco, including the structurally, lithologically, and temporally similar Twin Peaks block, by a fault of unknown character beneath the Golden Gate. Hence the entire Marin Headlands block is an allochthon, resting on a surface of detachment at no great depth.

The individual slabs that make up the Marin Headlands block consist predominantly of the chert layer of the late Mesozoic Pacific Ocean floor, with attached remnants of the underlying basalt and overlying turbidite sequence. The bulk of the turbidite sequence has been squeezed out from between the slabs and may be represented by melange matrix elsewhere in the Franciscan assemblage; the remainder of the basalt and underlying diabase dikes and gabbro probably returned to the mantle by subduction.

The direction of subduction indicated by the present configuration of minor folds and the dip of the layers is south to southeast. At the time of presumed emplacement, the Pacific Ocean floor was being subducted in an easterly or northeasterly direction beneath North America. Therefore a 90°+ clockwise rotation of the Marin Headlands block since subduction is indicated.

Confirmation of these hypotheses requires detailed study of many more radiolarian samples and minor fold localities, including statistical analysis of the hinge line data. Hence this is a progress report.

ACKNOWLEDGEMENTS

I am grateful to the National Park Service, especially to Judd Howell, Resource Manager for the GGNRA, for permission to collect rocks in the Marin Headlands, and to the staffs of the Yosemite Institute and the Marin Headlands Hostel for accommodations. Many students and teaching assistants have contributed stimulating questions and ideas, and have discovered critical localities. These include, among others, Cathy Busby-Spera, David Bice, Donna Marron, Martha Miller, Mark Muir, Mary Savina, and Warren Sharp. I thank also many professional colleagues who were on field trips to the Headlands with me for discussions and ideas, especially Walter Alvarez, Tanya Atwater. M. C. Blake, Allan Cox, David Engebretson, H. C. Helgeson, Evan Lietch, and Victor Seiders. Gary Collier helped make some of the field measurements. Michael Cusick gave advice on drafting and reproduction. I am especially grateful to Benita L. Murchey for identifying radiolarian samples that provided the only proof of the speculations on thrusts, and to Foster Curry and his colleagues for the paleomagnetic evidence of rotation.

REFERENCES CITED

Adams, F. D., 1938, The Birth and Development of the Geological Sciences: New York, Dover Publications, 506 p.

Alvarez, Walter, Terry Engelder, and William Lowrie, 1976, Formation of spaced cleavage and folds in brittle limestone by dissolution: Geology, v. 4, p. 698-701.

Bailey, E. H., W. P. Irwin, and D. L. Jones, 1964, Franciscan and Related Rocks and their significance in the geology of western California: California Division of Mines and Geology Bull. 183, 177 p.

Ballard, R. D., T. H. van Andel, and R. T. Holcomb, 1982, The Galapagos Rift at 86° W 5: Variations in volcanism, structure, and hydrothermal activity along a 30-kilometer segment of the Rift Valley: Journal of Geophysical Research, v. 8, No. B2, p.1149-1161.

Becker, G. F., 1888, Geology of the Quicksilver Deposits of the Pacific Slope: U. S. Geological Survey Monograph 13, 486 p., atlas.

Bedrossian, T. L., 1974, Geology of the Marin Headlands: California Geology, v. 27, No. 4, p. 75-86.

Blake, M.C., Jr., J. A. Bartow, V. A. Frizzell, Jr., J. Schlocker, D. Sorg, C. M. Wentworth, and R. H. Wright, 1974, Preliminary geologic map of Marin and San Francisco Counties and parts of Alameda, Contra Costa and Sonoma Counties, California: U. S. Geological Survey Miscellaneous Field Investions Map MF-574, 2 sheets, 1:62,500.

Blake, M. C., Jr., Jones, D. L., and Howell, D. G., 1982, Preliminary Tectonostratigraphic Terrane Map of California: U. S. Geological Survey Open File Report No. OF-82-593, 1:750,000.

Blake, W. P., 1857, Geological Report, in Reports of Explorations and Surveys to Ascertain the Most Practicable and Economical Route for a Railroad from the Mississippi River to the Pacific Ocean: Senate Ex. Doc. No. 78, 33d Congr., 2nd Session, Vol. V. part 2, p.1-310 (Chapter on San Francisco Bay, p. 145-162).

Branner, J. C., Newsom, J. F., and Arnold, Ralph, 1909, Description of the Santa Cruz (30') Quadrangle: U. S. Geological Survey Folio No. 163.

Coleman, Robert G., 1977, Ophiolites, Ancient Oceanic Lithosphere?: Berlin, Springer-Verlag, 229 p.

Davis, E. F., 1918a, The Franciscan Sandstone: University of California Publications, Bulletin of the Department of Geology, vol. 11, No. 1, p. 1-44.

Davis, E. F., 1918b, The Radiolarian Cherts of the Franciscan Group: University of California Publications, Bulletin of the Department of Geology, v. 11, No. 3, p. 235-432.

Ellen, S. D., 1971, The development of folds in layered chert of the Franciscan Assemblage near San Francisco, California: Ph. D. thesis, Stanford University, 320 p.

Engebretson, D. C., 1982, Relative motion between oceanic and continental plates in the Pacific Basin: Ph. D. thesis, Stanford University, 211 p.

Fairbanks, H. W., 1892. The Pre-Cretaceous age of the metamorphic rocks of the California Coast Ranges: American Geologist, v. 9, p. 153-166.

Fairbanks, H. W., 1904, Description of the San Luis (30') Quadrangle: U. S. Geological Survey Folio No. 101.

Fornari, D. J., A. Malahoff, and B. C. Heezen, 1978, Volcanic structure of the crest of the Puna Ridge, Hawaii: Geophysical implications of submarine volcanic terrain: Geological Society of America Bulletin, v. 89, No. 4, p. 605-616.

Hamilton, Warren, 1969, Mesozoic California and the underflow of the Pacific mantle: Geological Society of America, Bulletin v. 80, No. 12, p. 2409-2429.

Hsu, K J., 1971, Franciscan melanges as a model for eugeosynclinal sedimentation and underthrusting tectonics: Journal of Geophysical Research, v. 76, No. 5, p. 1162-1170.

Kotzebue, Otto von, 1821, A voyage of discovery into the South Seas and Beering's Straits for the purpose of exploring a north-east passage, undertaken in the years 1815-1818, Trans. by H. E. Lloyd: London, Longman, Hurst, Rees, Orme, and Browne, 3 vols.

Lawson, A. C. 1895, Sketch of the Geology of the San Francisco Peninsula: Fifteenth Annual Report of the U. S. Geological Survey, p. 405-476.

Lawson, A. C., 1915, Description of the San Francisco District: U. S. Geol. Survey Folio 193, field edition.

Mackin, J. H., 1950, The down-structure method of viewing geologic maps: Journal of Geology, v. 58, No. 1, p. 55-72.

Moore, J. G., 1975, Mechanism of formation of pillow lava: American Scientist, v. 63, no. 3, p. 269-277.

Moore, J. G., R. L. Phillips,. R. W. Griggs, D. W. Peterson, and D. A. Swanson, 1973, Flow of lava into the sea, 1969-1971, Kileauea Volcano, Hawaii: Geol. Soc. America Bull., v. 82, No. 2, 537-546.

Murchey, B. L., 1980, Significance of chert age determination in the Marin Headlands, California (abstract): Geological Society of America Abstracts with Programs, v. 12, No. 3, p. 144.

Newberry, J. S., 1856, Report upon the Geology of the Route, in U. S. War Dept., Reports of Explorations and Surveys, etc.: U. S. Senate Ex. Doc. No. 78, 33d Congr., 2nd Session, Vol. 6, Part 2, p. 9-68.

Pessagno, E. A., Jr., 1973, Age and geologic significance of radiolarian cherts in the California Coast Ranges: Geology, v. 1, No. 4, p. 153-156.

Ransome, F. L., 1893, The Eruptive Rocks of Point Bonita: University of California Publications, Bulletin of the Department of Geology, v. 1, No. 3, p. 71-114.

Ransome, F. L., 1894, The Geology of Angel Island: University of California Publications, Bulletin of the Department of Geology, v. 1, No. 7, p. 193-240.

Rice, S. J., and Strand, R. G., 1971, Report to accompany Geologic and Slope Stability Maps of the Tennessee Valley, Lucas Valley, and North Coastal Areas, Marin County, California: California Division of Mines and Geology Report, 16 p.

Richardson, J., N. A. Vigors, G. T. Lay, E. T. Bennett, Richard Owen, J. E. Grey, W. Buckland, and G. B. Sowerby, 1839, The Zoology of Captain Beechey's Voyage (1825-28): London, Henry G. Bohn.

Sarjeant, William A. S., 1980, Geologists and the history of Geology, 5 vols.: New York, Arno Press.

Schiffner, C., 1935, Aus dem Leben alter Freiberger Bergstudenten: Freiberg, Ernst Mauckisch; xv + 375 p.

Schlocker, Julius, 1974, Geology of the San Francisco North Quadrangle, California: U. S. Geological Survey Professional Paper 782, 109 p.

Schlocker, Julius, Bonilla, M. G., and Imlay, R. W., 1954, Ammonite indicates Cretaceous age for part of Franciscan group in San Francisco Bay Area, California: American Association of Petroleum Geologists Bulletin v. 38, p. 2372-2381.

Schlocker, J., M. G. Bonilla, and D. H. Radbruch, 1958, Geology of the San Francisco North Quadrangle, California: U. S. Geological Survey Miscellaneous Geological Investigations Map I-272, 1:24,000.

Taliaferro, N. L. , 1943, Franciscan-Knoxville Problem: American Association of Petroleum Geologists Bulletin, v. 27, No. 2, p. 109-219.

Thalmann, H. E., 1942, Globotruncana in the Franciscan limestone, Santa Clara County, California (abstr.): Geological Society of America Bulletin, v. 53, p. 1838.

VanderHoof, V. L., 1951, History of Geologic Investigation in the Bay Region: California Division of Mines Bulletin 154, p. 109-116.

Wahrhaftig, Clyde, 1984, A Streetcar to Subduction and other Plate Tectonic Trips by Public Transport in San Francisco, revised ed.: Washington, American Geophysical Union, 72 p.

Wahrhaftig, Clyde, and Lehre, Andre K., 1974, Geologic and Hydrologic Study of the Golden Gate National Recreation Area: Manuscript report to the National Park Service, Fort Mason, San Francisco, 246 p., 37 figs., 3 plates, 19 photographs.

Whitney, J. D., 1865, Geology, Vol. I, Report of Progress and Synopsis of the Field-work, from 1860 to 1864, Geological Survey of California: Sacramento, Legislature of California. 498 p.

Chert and pillow basalt exposed in the cliff north of Tennessee Cove, Marin Headlands, from the top of the point just south of the cove. The hilltop above the cliff is about 115 m above sea level.Muir Beach and Inverness Ridge in the distance.

Pillow basalt on shore platform, point at east end of Bonita Cove. Rectangular board is about 0.25 m wide, 0.35 m long. For location see Wahrhaftig, this volume, Figures 2 or 13.

BIOSTRATIGRAPHY AND LITHOSTRATIGRAPHY OF CHERT IN THE FRANCISCAN COMPLEX, MARIN HEADLANDS, CALIFORNIA

Benita Murchey
U. S. Geological Survey
Menlo Park, California 94025

ABSTRACT

Upper Pliensbachian to upper Albian or lower Cenomanian radiolarian chert of the Franciscan Complex in the Marin Headlands, California, forms a significant part of a structurally complex but locally coherent stratigraphic sequence that includes underlying pillow basalt and overlying lithic arkose. This study documents seven radiolarian assemblages, including sixty-three species, and four lithostratigraphic units within the eighty-two meter chert interval. Both the biostratigraphic and lithologic units are recognized in successive imbricated thrust slices within the Marin Headlands and, thus, are useful in unravelling the structure of this allochthonous block.

The pelagic sediments of the Marin Headlands were deposited under conditions of moderately high biologic productivity below the calcium carbonate compensation depth in an oxygenated environment. Sedimentary processes and silica diagenesis created or enhanced variations in the clay to silica ratios, resulting in rhythmically alternating chert and shale beds. Basal Lower Jurassic sediments are iron and manganese rich with lenticular jasper beds; these sediments may have been affected by hydrothermal fluids. A leached, extremely siliceous, and locally brecciated horizon lies marks a major unconformity between Middle or lower Upper Jurassic strata and Lower Cretaceous (Valanginian) strata.

Radiolarian faunal affinities suggest a low latitude origin for at least the Middle Jurassic to Cretaceous faunas. A near equatorial origin in the mid-Pacific (perhaps on a spreading ridge) and subsequent low latitude trajectory with a probable northward component toward the Americas is postulated for the Pliensbachian to late Albian or early Cenomanian pelagic depositional sites.

INTRODUCTION

Bedded radiolarian chert forms a small but significant component of the Franciscan Complex of California. Some of the most well known and best exposed chert is in the Marin Headlands area north of the San Francisco (Fig. 1). This study entails a systematic examination of the biostratigraphy and lithostratigraphy of the chert sequence from measured sections within the Marin Headlands area. Only one other systematic biostratigraphic examination of any substantially thick Franciscan chert sequence is documented (Pessagno, 1977a, 1977b, a single chert section in the Geysers-Clear Lake area 100 km north). For lithostratigraphic and comparative studies of Franciscan chert, internal structural complexities make biostratigraphic control a necessity.

GEOLOGIC SETTING

The Marin Headlands, which form the north shore of the Golden Gate to San Francisco Bay (Fig. 1), are part of the Marin Headlands tectonostratigraphic terrane (Blake and others, 1982; Blake and others, this volume). Imbricated Franciscan rocks form a gently curving southwest-dipping syncline reflected by a series of ridges and valleys. The dominant Franciscan lithologies in the Marin Headlands are basalt, radiolarian chert, and lithic arkose. Where depositional contacts occur, Lower Jurassic radiolarian chert overlies pillow basalt and clastic rocks overlie Cretaceous radiolarian chert. Chert forms resistant ridges while sandstone and basalt underlie saddles and valleys. The best outcrops are in roadcuts and seacliffs. Unconsolidated Pleistocene and Holocene sediment covers lower slopes and fills valleys.

The four kilometer by one kilometer Marin Headlands is fault-bounded. Its coherent rocks structurally overlie Franciscan melange to the north along a south-dipping thrust fault. The San Andreas fault lies offshore to the west. Submarine faults lie beneath San Francisco Bay to the east because the nearest Franciscan outcrops on Angel Island and the Tiburon Peninsula are age correlative but lithologically dissimilar. To the south, the Marin Headlands terrane apparently extends across the mouth of San Francisco Bay to the Twin Peaks area. However, the Twin Peaks sequence is offset and dips to the east rather than the west.

Before the development of a technique for extracting radiolarians from chert (Pessagno and Newport, 1972), previous workers (Schlocker and others, 1958; Schlocker, 1974; Bedroissian, 1974) lacked any means with which to correlate rock units. The imbricated structure in the Marin Headlands was, therefore, unrecognized. The radiolarian faunas documented in this report prove the imbricated nature of the stratigraphic units. Wahrhaftig has now mapped the geologic structure of the Marin Headlands in great detail (this volume and Fig. 1, geologic and locality map).

Thrusting and folding have determined the apparent thickness of units throughout the area. Within the chert unit, thrusting has apparently repeated or cut out sections in some slabs and intense folding has increased the apparent thickness in many sections.

Along Alexander Avenue (Fig. 1, localities M-2a-M-2b), where the section is relatively undeformed, the chert sequence is approximately 82 m thick. The radiolarian faunas in this chert sequence range from Early Jurassic (Pliensbachian) near the depositional contact with underlying basalt to middle Cretaceous (late Albian or early Cenomanian) at the depositional contact with overlying clastic rocks. All other

in Blake, M.C. Jr., ed., 1984, **Franciscan Geology of Northern California:** Pacific Section S.E.P.M., Vol. 43, p. 51-70

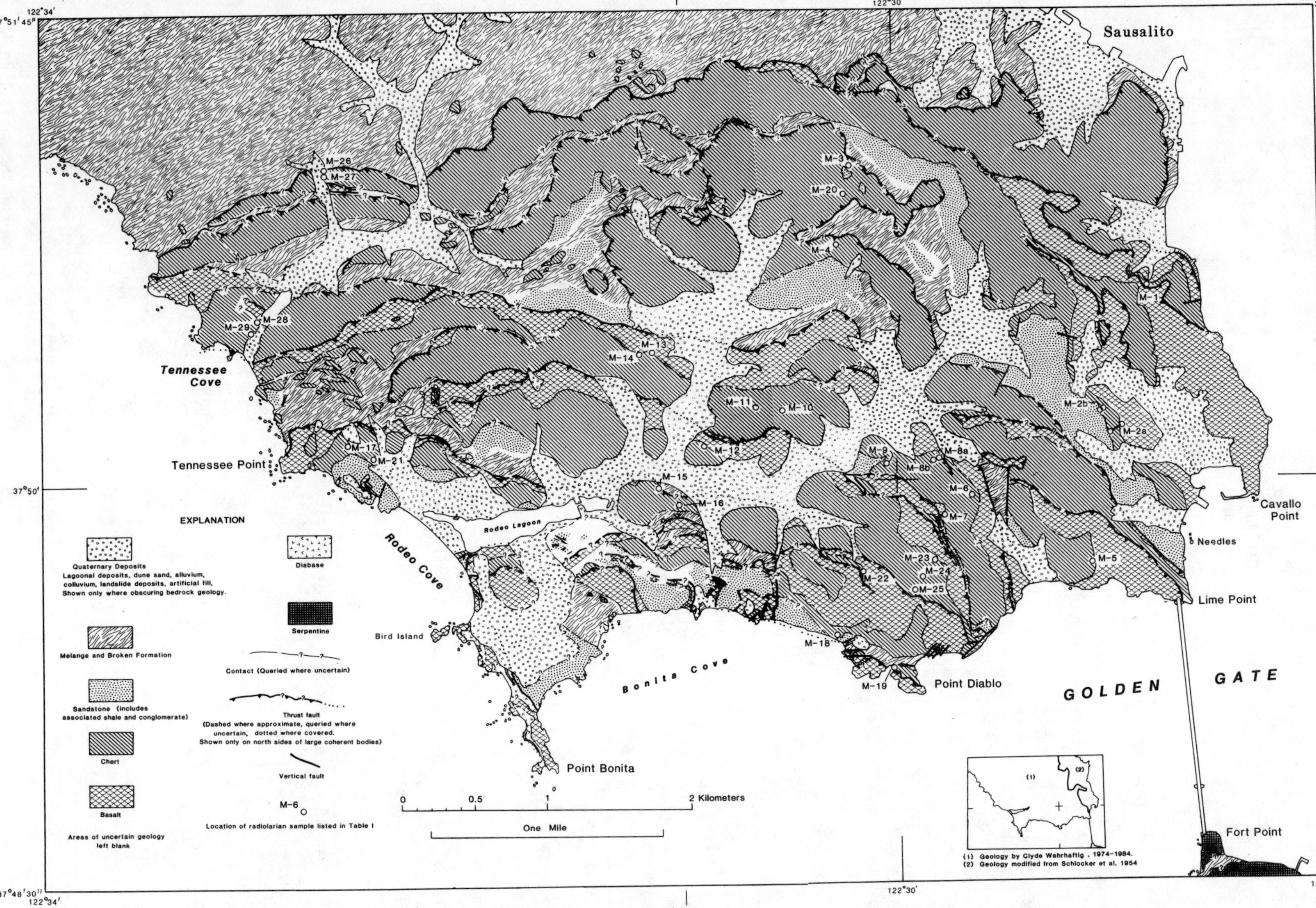

Figure 1. Geologic map showing distribution of Mesozoic rocks of the Franciscan Complex, Marin Headlands, California. Radiolarian fossil localities are denoted by small circles; radiolarian taxa in each sample are shown in Table 1.

radiolarian samples in the Marin Headlands fall within this stratigraphic range. The Alexander Avenue section was also studied for geochemistry (Karl, this volume) and for paleomagnetism (Curry and others, this volume).

BIOSTRATIGRAPHY OF THE MARIN HEADLANDS

The chert sequence in the Marin Headlands ranges in age from Pliensbachian to late Albian or early Cenomanian. Seven sequential radiolarian assemblages (Marin Headlands assemblages MH-1 to MH-7) are described in this report. Their ages have been determined by correlation with independently dated samples from North America, Europe, Japan, and the Atlantic, Pacific, and Indian Oceans. Important taxa are figured in Plates 1 and 2. Table 1 shows occurrences of these taxa in individual samples. Sample localities are shown in Fig. 1, the geologic map of the Marin Headlands. Table 2 summarizes the ranges of important taxa in the Marin Headlands assemblages.

At map locality M-2a-M-2b (Fig. 1) the chert sequence has depositional contacts with underlying basalt and with overlying clastic rocks. Six assemblages (MH-1, MH-3, MH-4, MH-5, MH-6, and MH-7) are present within this interval. A seventh assemblage, MH-2, is not exposed at this locality, probably because of a covered interval between assemblages MH-1 and MH-3. All samples collected within the Marin Headlands are correlative with these seven biostratigraphic assemblages.

Assemblage MH-1

In the Marin Headlands the base of MH-1 is the basalt-chert contact. This assemblage is characterized by Canutus rockfishensis Pessagno and Whalen (Plate 1, Fig. 30), Canoptum anulatum Pessagno and Poisson (Plate 1, Fig. 32), Canoptum poissoni Pessagno (Plate 1, Fig. 31), Hsuum(?) sp. A (Plate 1, Fig. 33), Napora sp. A (Plate 2, Fig 31), Katroma sp. aff. K. neagui Pessagno and Poisson (emend. DeWever) (Plate 1, Fig. 28), Pantanellium sp. A (Plate 2, Fig. 24), Orbiculiforma mclaughlani Pessagno (not figured), O. sp. cf. O. multifora Pessgno and Poisson (not figured), Praeconocaryomma spp. (not figured) and "Cenosphaera" (not figured). Taxa first appearing in the upper part of the stratigraphic sequence containtaining this assemblage include Nassellaria sp. A (Plate 2, Fig. 27), Eucyrtidium(?) sp. A (Plate 25, Fig. 25), Bagotum sp. aff. B. modestum Pessagno and Whalen (Plate 1, Fig. 29), Stichocapsa sp. A (Plate 1, Fig. 27), Eucyrtidium(?) sp. B (Plate 2, Fig. 26), Jacus sp. A (Plate 2, Fig 28), Dumitricaella(?) sp. aff. D(?) cucurbitina DeWever (Plate 2, Fig. 30), and Trillus elkhornensis Pessagno and Blome (Plate 2, Fig 23). Low diversity (ten to twelve genera) near the basalt-chert contact is, at least in part, a function of poor preservation. Diversity increases upsection within the MH-1 interval to more than twenty genera per sample.

Based on the known ranges of these taxa, assemblage MH-1 is Pliensbachian (probably late) and (or) Toarcian. Assemblage MH-1 is similar to Liassic faunas in southwestern Turkey (Pessago and Poisson, 1981; DeWever,1982), part(?) of the Parahsuum simplum Assemblage in Japan (Yao and others, 1982) which overlies Triassic rocks, and Pliensbachian faunas in eastern Oregon (Pessagno and Blome, 1980; Pessagno and Whalen, 1982).

Assemblage MH-2

At Alexander Avenue, locality M-2a, the stratigraphic interval between MH-1 and MH-3 is covered. A transitional assemblage has been recovered from a structurally complex outcrop on the east side of Bonita Cove, map locality M-19. This assemblage marks the first appearance of Parvicingula (P. sp. A, Plate 1, Fig. 22), common to abundant Wrangellium (W. sp. A, Plate 1, Fig. 21), Acanthocircus species with "ridged" rings (A. hexagonus (Yao), Plate 2, Fig. 18), well-defined Hsuum (H. sp. B, Plate 1, Fig. 23), Emiluvia sp. (not figured), and Perispyridium sp. (not figured). Few genera and fewer species carry over from MH-1; Trillus elkhornensis is a notable exception. Species first appearing in MH-2 assemblage include Napora sp. B (Plate 2, Fig. 20), Napora sp. C (Plate 2, Fig. 19), Riedelius sp. A (Plate 1, Fig. 26), Bipedis(?) sp. A (Plate 1, Figs. 24, 25), Pantanellium sp. aff. P. buntonense (Plate 2, Fig. 22), and Foremanina sp. A (Plate 1, Fig. 14). The one sample that was counted contains 30% nassellarians (.63-1.80 mm sieve size fraction), most of which are simple conical forms. Unlike older or younger Marin Headlands assemblages, several nassellarian genera, including Parvicingula, Wrangellium, Hsuum, and Lupherium, have species with well-developed horns.

This assemblage seems to be correlative with Bajocian strata of the Snowshoe Formation in Oregon (Pessagno and Blome, 1980; Pessago and Blome, 1982; Pessagno and Whalen, 1982) even though Wrangellium species are reported only in Early Jurassic assemblages (Pessagno and Whalen, 1982). I consider this assemblage to be Middle Jurassic, probably Bajocian (U.S. Geological Survey does not recognize Aalenian, but considers such strata early Bajocian).

Assemblage MH-3

Assemblage MH-3 has several genera but few species in common with MH-2. Taxa which appear first in MH-3 incude Unuma (U. sp. A, U. sp. B, Plate 1, Figs. 19, 20), Tricolocapsa (T. sp. cf. T. plicarum Yao, Plate 2, Fig. 11; T. sp. A, Plate 2. Fig. 10; T. sp. B, Plate 2, Fig. 9), Dictyomitrella(?) kamoensis Mizutani and Kido (Plate 1, Fig. 13), Ristola spp. A and B (Plate 1, Figs. 12, 10), Acanthocircus suboblongus (Yao) (Plate 2, Fig. 12), Hsuum sp C Group (Plate 1, Fig. 17), Stichocapsa(?) sp. B (Plate 1, Fig. 16), Napora sp. D (Plate 2, Fig. 13), Trillus(?) sp. cf. T. seiderse (Plate 2, Fig. 15), Pantanellium ultrasincerum Pessagno and Blome (Plate 2, Fig. 14), Gorgansium sp. (Plate 2, Fig. 17), Pachyoncus sp. (Plate 2, Fig. 16), and rare Mirifusus (M. sp. cf. M. guadalupensis Pessagno, Plate 1, Fig. 15).

Nassellarians constitute thirty to sixty-five percent of the .63 to 1.80 mm sieve size population fraction, the majority being small theoperids with enlarged rounded terminal chambers or closed tapered distal portions such as Tricolocapsa and Unuma species. Conical nassellarians lack horns (e.g., Ristola species) or have very small horns (e.g., some Hsuum species). Pantanellids form as much as two percent of the .63 to 1.80 mm population).

MH-3 contains rare specimens of the genus Mirifusus which is a marker for the base of the Upper Jurassic Zone 2 in the California Coast Ranges (Pessagno, 1977a), yet it is correlative with the Japanese Unuma echinatus assemblage (Yao and others, 1982) which is considered to be Callovian or older.

Yao and others, (1982) report Mirifusus sp. in the Unuma echinatus assemblage. Baumgartner (in press) also reports Mirifusus sp. in apparently correlative Bathonian or early Callovian European material. Disagreements concerning the megafossil control of Mirifusus sp. bearing strata and, hence, the true range of the genus are unresolved. The North American ranges of Trillus spp., P. ultrasincerum, Pachyoncus spp., and Gorgansium spp. (Pessagno and Blome, 1980) favor a Middle Jurassic, possibly Bajocian, age for MH-3. The possible range of assemblage MH-3 is considered to be Middle Jurassic (Bajocian to Callovian) and possibly lower Upper Jurassic (Oxfordian?, Kimmeridgian?).

Assemblage MH-4

The appearance of Ristola turpicula Pessagno and Whalen (Plate 1, Fig. 9) characterizes assemblage MH-4. In addition, Tricolocapsa(?) sp. C (Plate 1, fig. 8) is apparently confined to this assemblage. MH-4 taxa which also occur in MH-3 include Dictyomitrella(?) kamoensis, Ristola sp. A, Acanthocircus suboblongus, Hsuum sp. C Gp. (Plate 1, Fig. 11), Tricolocapsa sp. cf. T. plicarum, Napora sp. D, and Ristola sp. B (Plate 1, Fig. 10).

R. turpicula alone comprises two to fifteen percent of the total .63 to 1.80 mm population. In contrast to MH-3, nassellarians with rounded or tapered distal portions such as Tricolocapsa, form less than ten percent of the total .63 to 1.80 mm radiolarian population, although nassellarians in general form ten to forty percent of this population fraction. Many Hsuum species have horns, which are lacking or very small on Hsuum species in MH-3.

MH-4 appears to be correlative with Dictyomitrella(?) kamoensis-Pantanellium foveatum Assemblage (Mizutani and Kido, 1983) and Lithocampe(?) nudata Assemblage (Yao and others, 1982) in Japan, both of which apparently overlie Unuma echinatus Assemblage. In the Snowshoe Formation of Oregon, Ristola turpicula first appears in upper Bathonian strata (Pessagno and Whalen, 1982). No Mirifusus were recovered in this assemblage in the Marin Headlands but Mirifusus sp. occurs with R. turpicula elsewhere in Franciscan chert near The Geysers (Pessagno, 1977a). MH-4 is upper Middle Jurassic (Bathonian and (or) Callovian) and (or) lower Upper Jurassic (Oxfordian?, Kimmeridgian?)

Stratigraphic Gap between MH-4 and MH-5

Early Cretaceous MH-5 assemblage is separated from Middle or lower Upper Jurassic MH-4 by a one to ten meter recrystallized and commonly brecciated interval from which radiolarians cannot be extracted. Several Late Jurassic and Early Cretaceous radiolarian assemblages or zones which can be predicted to occur between these two assemblages are missing from the sequence. This biostratigraphic gap represents fifteen to thirty million years, a duration indicating nondeposition or erosion within or above the thin (to ten meters) recrystallized interval. The biostratigraphic gap between MH-4 and MH-5 is characteristic of many allochthonous blocks within the Franciscan (Murchey and others, this volume) and it was apparently caused by widespread regional events in the Mesozoic Pacific Ocean.

Assemblage MH-5

Taxa characterizing assemblage MH-5 include Archaeodictyomitra apiaria (Rust) (Plate 1, Fig. 3), Acanthocircus dicranocanthos (Squinabol) (Plate 2, Fig. 7), Cecrops septemporata (Parona) (Plate 2, Fig. 8), Mirifusus mediodilatatus (Rust) (?) (Plate 1, Fig. 2), Pantanellium corriganensis Pessagno (not figured), Thanarla conica (Aliev) (Plate 1, Fig. 4), Pseudodictyomitra sp. A (Plate 1, Fig. 7), Archaeodictyomitra simplex Pessagno (Plate 1, Fig. 5), Zifondium sp. (Plate 1, Fig. 6), and Podobursa sp. (Plate 1, Fig. 1). Probable cryptocephalic nassellarians are too poorly preserved to be identified. MH-5 is correlative with Valanginian to Hauterivian or early Barremian radiolarian zones in circum-Pacific onland sequences as well as with Deep Sea Drilling Project zonation schemes. The range of MH-5 is Valanginian to Hauterivian or early Barremian.

Assemblage MH-6

Assemblage MH-6 is a poorly preserved, low diversity assemblage characterized by Zifondium sp., Archaeodictyomitra sp., Acanthocircus trizonalis (Rust), and poorly preserved cryptocephalic nassellarians. Its range is herein considered to be Hauterivian(?) to Albian(?) although correlation with DSDP material (Schaaf,1981) suggests a more restricted late Barremian or Aptian range from the Crolanium pythia zone (postdates last occurrence of Acanthocircus dicranocanthus, Archaeodictyomitra apiaria, Cecrops septemporata) to the middle of the Acaeniotyle umbilicata zone (last occurrence of Acanthocircus trizonalis and Pantanellium spp.). A sharp change in the preservation of radiolarians between the MH-6 assemblage (poorly preserved and recrystallized) and the MH-7 assemblage (beautifully preserved, broken) suggests a probable hiatus that could be as long as six(?) million years. At map locality M-18, this contact is marked by a 15 cm red siliceous shale interval.

Assemblage MH-7

Radiolarians characterizing assemblage MH-7 include Holocryptocanium barbui Dumitrica (Plate 2, Fig. 6), H. japonicum Nakaseko and Nishimura (not figured), H. geysersensis Pessagno (not figured), Novixitus mclaughlani Pessagno (Plate 2, Fig. 3), Novixitus sp. A (Plate 2, Fig. 2), Thanarla veneta (Squinabol) (Plate 2, Fig. 1), Dictyomitra sp. A (Plate 2, Fig. 4, Pseudodictyomitra spp., Pseudodictyomitra pseudomacrocephala (Squinabol) (rare, not figured), and Stichomitra(?) sp. A (Plate 2, Fig. 5). Holocryptocanium species dominate the assemblage. Two samples contain abundant massive tri-bladed, grooved spumellarian spines. Holothurian(?) sclera are also common in this assemblage. The probable range of this assemblage is late Albian to early Cenomanian. A Cenomanian ammonite, Mantelliceras sp., has been recovered from float from the overlying clastic unit (Hertlein, 1956).

Faunal Affinities

Jurassic and Cretaceous faunas from the Marin Headlands are quite similar to faunas from the Mediterranean region, the Himalayas, Japan, the northwestern Pacific, and other Franciscan rocks from low paleolatitudes (as determined by paleomagnetism). Paleolatitudes for faunas from the Mediterranean and Himalayas were near equatorial. Japanese faunas are probably displaced from the south and northwestern Pacific faunas have been backtracked southward. Pessagno and Blome (in press) believe that Middle and Late Jurassic faunas with abundant and diverse Ristola and pantanellids are low latitude assemblages while faunas with abundant and diverse Parvicingula but few pantanellids are higher latitude assemblages. In the Marin Headlands, Middle Jurassic MH-2 assemblage contains abundant Parvicingula species and relatively abundant pantanellid species. MH-3 and MH-4 contain few or no Parvicingula species but abundant Ristola and pantanellid species. According to Pessagno and Blome's model and assuming a symmetrical arrangement of faunal realms both north and south of the equator, the depositional site of the Marin Headlands may have originated south of the equator and crossed into the equatorial region (0° + 15°) during Middle Jurassic. Assemblages MH-3 and MH-4 are very different from the Jurassic faunas described by Pessagno in the California Coast Range ophiolite and basal Great Valley sequence. Either MH-3 and MH-4 represent very different environments from the latter or they are simply older or both.

The Cretaceous Great Valley sequence faunas described by Pessagno (1977b) are clearly temperate continental margin assemblages. The Marin Headlands Cretaceous assemblages (MH-5, MH-6, MH-7) are more like correlative low latitude and probable low latitude faunas from the Mediterranean area (Dumitrica, 1975), the Himalayas, Indonesia, Costa Rica (Schmidt-Effing, 1980), Japan (Nakaseko and others, 1979; Taketani, 1982), and northward displaced sites in the northwestern Pacific (Foreman, 1973; Schaaf, 1982) than like the Great Valley sequence assemblages as described. Furthermore, the Cretaceous Marin Headlands assemblages can be directly correlated with other Franciscan radiolarian assemblages for which there is paleomagnetic data indicating displacement from the south.

Plate 1. Scanning electron micrographs of Jurassic and Cretaceous Radiolaria from the Franciscan Complex, Marin Headlands, California.

Figures 1-7: assemblage MH-5

1. Podobursa sp. A (USGS MR 5063). Scale bar=125 um
2. Mirifusus sp. (M. mediodilatatus?) (USGS MR 5063). Scale bar=167um
3. Archaeodictyomitra apiaria (USGS MR 5063). Scale bar=96um.
4. Thanarla conica (USGS MR 5073). Scale bar=56um.
5. Archaeodictyomitra simplex (USGS MR 5063). Scale bar=125um.
6. Zifondium sp. A (USGS MR 5063). Scale bar=96um.
7. Pseudodictyomitra sp. A (USGS MR 5073). Scale bar=300um.

Figures 8-11: assemblage MH-4

8. Tricolocapsa sp. C (USGS MR 5110). Scale bar=50um.
9. Ristola turpicula (USGS MR 5110). Scale bar=83um.
10. Ristola sp. B (USGS MR 5110). Scale bar=83um.
11. Hsuum sp. C (USGS MR 5110). Scale bar=104um.

Figures 12-20: assemblage MH-3

12. Ristola sp. A (USGS MR 5102). Scale bar=83um.
13. Dictyomitrella(?) kamoensis (USGS MR 5103). Scale bar=83um.
14. Foremanina sp. A (USGS MR 5018). Scale bar=96um.
15. Mirifusus sp. A (USGS MR 5103). Scale bar=152um.
16. Stichocapsa(?) sp. B (USGS MR 5108). Scale bar=114um.
17. Hsuum sp. C (USGS MR 4982). Scale bar=96um.
18. Hsuum sp. (USGS MR 5104). Scale bar=75um.
19. Unuma sp. A (USGS MR 5104). Scale bar=50um.
20. Unuma sp. B (USGS MR 5108). Scale bar=60um.

Figures 21-26: assemblage MH-2

21. Wrangellium sp. A (USGS MR 4342A). Scale bar=69um.
22. Parvicingula sp. A (USGS MR 4342A). Scale bar=75um.
23. Hsuum B (USGS MR 4342A). Scale bar=75um.
24, 25. Bipedis(?) sp. A (USGS MR 4342A). Scale bar=69um.
26. Riedelius sp. A (USGS MR 4342A). Scale bar=69um.

Figures 27-33: assemblage MH-1

27. Stichocapsa sp. A (USGS MR 5101). Scale bar=50um.
28. Katroma sp. aff. K. neagui (USGS MR 5101). Scale bar=69um
29. Bagotum sp. aff. B. modestum (USGS MR 5101). Scale bar=50um.
30. Canutus rockfishensis (USGS MR 5101). Scale bar=104um.
31. Canoptum poissoni (USGS MR 5101). Scale bar=83um.
32. Canoptum anulatum (USGS MR 5101). Scale bar=104um.
33. Hsuum(?) sp. A (USGS MR 5101). Scale bar=64um.

Plate 2. Scanning electron micrographs of Jurassic and Cretaceous Radiolaria from the Franciscan Complex, Marin Headlands, California.

Figures 1-6: assemblage MH-7

1. Thanarla veneta (USGS MR 5064). Scale bar=69um.
2. Novixitus sp. A (USGS MR 5064). Scale bar=104um.
3. Novixitus mclaughlani (USGS MR 5064). Scale bar=104um.
4. Dictyomitra sp. A (USGS MR 5064). Scale bar=125um.
5. Stichomitra(?) sp. A (USGS MR 5064). Scale bar=69um.
6. Holocryptocanium barbui (USGS MR 5064).

Figures 7-8: assemblage MH-5

7. Acanthocircus dicranocanthos (USGS MR 5073r). Scale bar=139um.
8. Cecrops septemporata (USGS MR 5063). Scale bar=64um.

Figures 9-17: assemblage MH-3

9. Tricolocapsa sp. B (USGS MR 5108). Scale bar=83um.
10. Tricolocapsa sp. A (USGS MR 5104). Scale bar=50um.
11. Tricolocapsa plicarum (USGS MR 5018). Scale bar=64um.
12. Acanthocircus suboblongus (USGS MR 5103). Scale bar=83um.
13. Napora sp. D (USGS MR 5108). Scale bar=64um.
14. Pantanellium ultrasincerum (USGS MR 5018). Scale bar=64um.
15. Trillus(?) sp. cf. T. seiderse (USGS MR 5108). Scale bar=75um.
16. Pachyoncus sp. (USGS MR 5104). Scale bar=45um.
17. Gorgansium sp. (USGS MR 5018). Scale bar=60um.

Figures 18-22: assemblage MH-2

18. Acanthocircus hexagonus (USGS MR 4342A). Scale bar=69um.
19. Napora sp. C (USGS MR 4342A). Scale bar=96um.
20. Napora sp. B (USGS MR 4342A). Scale bar=100um.
21. Pantanellium sp. (USGS MR 4342A). Scale bar=104um.
22. Pantanellium sp. aff. P. buntonense (USGS MR 4342A). Scale bar=83um.

Figures 23-31: assemblage MH-1

23. Trillus elkhornensis (USGS MR 5101). Scale bar=96um.
24. Pantanellium sp. A (USGS MR 5101). Scale bar=104um.
25. Eucyrtidium(?) sp. A (USGS MR 5101). Scale bar=60um.
26. Eucyrtidium(?) sp. B (USGS MR 5101). Scale bar=69um.
27. Nassellaria sp. A (USGS MR 5101). Scale bar=64um.
28. Jacus sp. B (USGS MR 5101). Scale bar=60um.
29. Jacus sp. A (USGS MR 5101). Scale bar=69um.
30. Dumitricaella(?) sp. aff. D(?). cucurbitina (USGS MR 5101). Scale bar=50.
31. Napora sp. A (USGS MR 5101). Scale bar=69um.

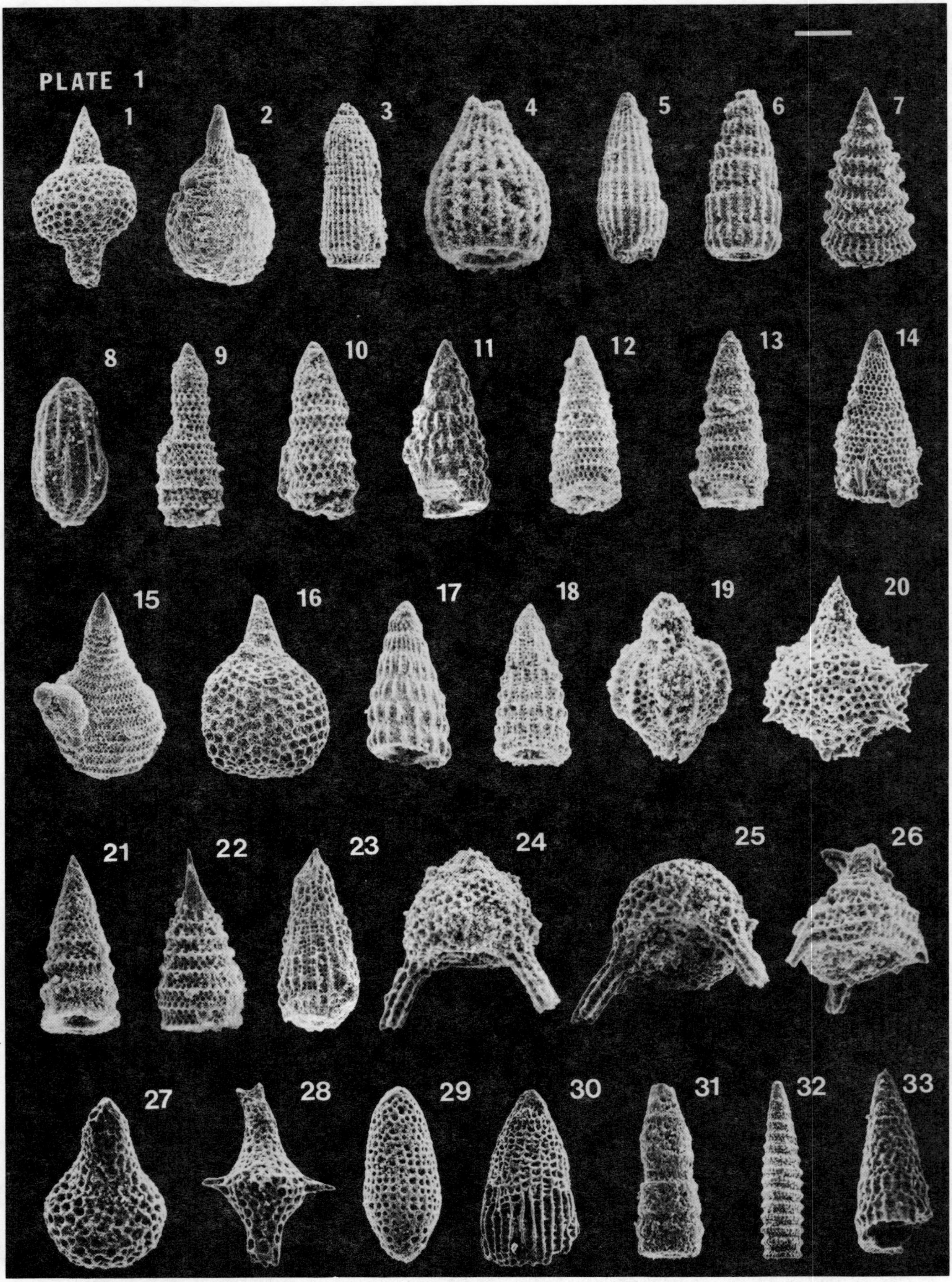
PLATE 1
1
2
3
4
5
6
7
8
9
10
11
12
13
14
15
16
17
18
19
20
21
22
23
24
25
26
27
28
29
30
31
32
33

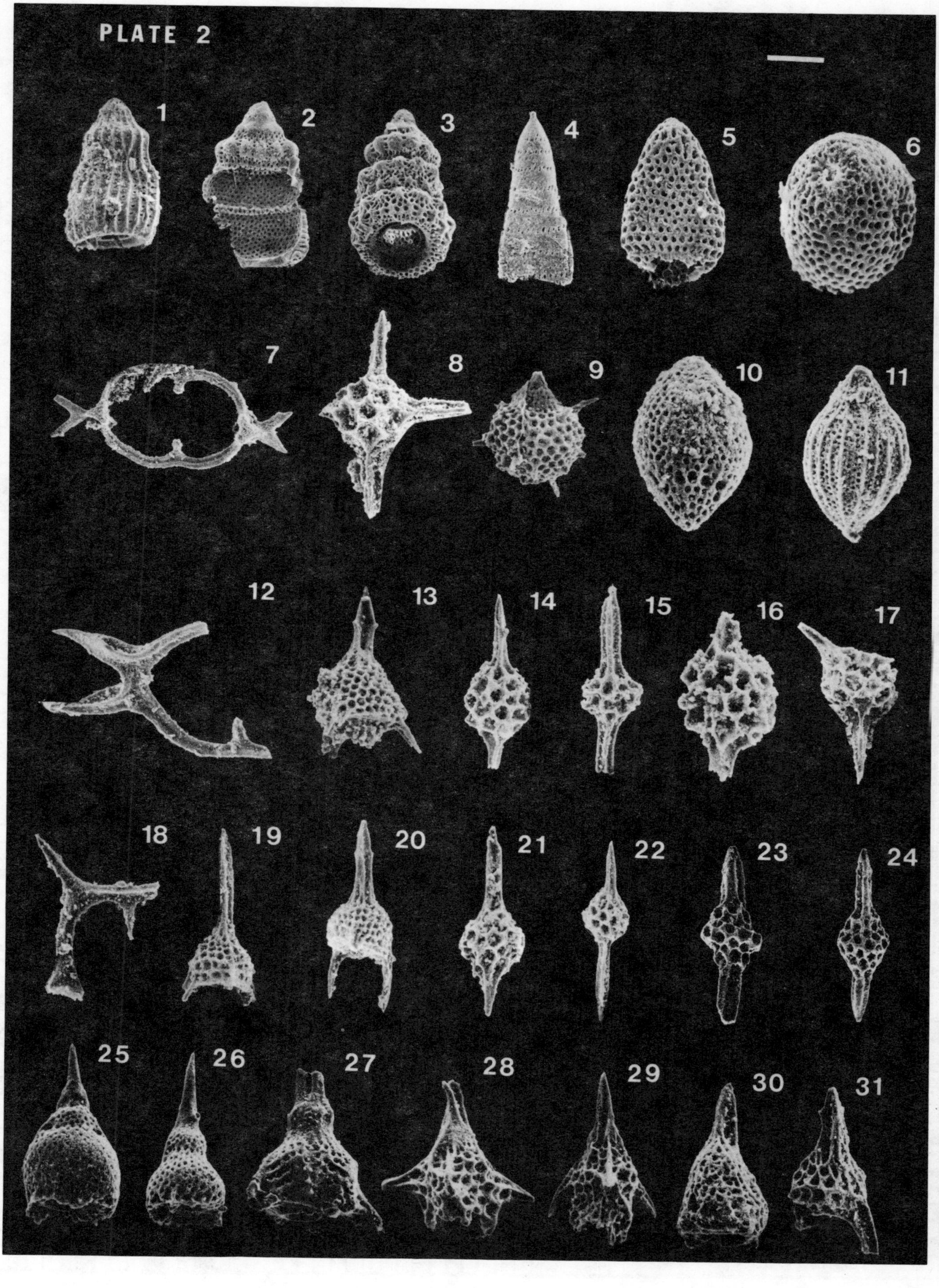
PLATE 2
1
2
3
4
5
6
7
8
9
10
11
12
13
14
15
16
17
18
19
20
21
22
23
24
25
26
27
28
29
30
31

Table 1. Occurrence of Selected Radiolaria in Samples from the Marin Headlands

Radiolaria / Map numbers	M-1	M-2aA	M-2aB	M-2aC	M-2aD	M-2aE	M-2aF	M-2aG	M-2aH	M-2aI	M-2aJ	M-2aK	M-2aL	M-2aM	M-2aN	M-2aO	M-2aP	M-2aQ	M-2aQ	M-2bR	M-2bS	M-2bT	M-3	M-4	M-5A	M-5B	M-5C	M-5D	M-5E	M-5F	M-5G	M-5H	M-5I
U.S.G.S. Accession	4988	4983	4984	4985	4986	4987	5100	5101	5102	5103	5104	5105	5106	5107	5108	5109	5110	5114	5115	5066	5067	5064	5022	5018	4971	4972	4973	4974	4975	4976	4977	4978	4978
Assemblages	MH-1	MH-1	MH-1	MH-1	MH-1	MH-1	MH-1	MH-1	MH-3	MH-3	MH-3	MH-3	MH-3	MH-3	MH-3	MH-3	MH-4	MH-5	MH-5	MH-6	MH-7	MH-7	MH-7	MH-3	MH-1	MH-1	MH-1	MH-1	MH-1	MH-1	MH-1	MH-1	MH-1
1 Acanthocircus dicranocanthos																			■														
2 Acanthocircus hexagonus																																	
3 Acanthocircus suboblongus									■	■	■				■									■									
4 Archaeodictyomitra apiaria																		■	■														
5 Archaeodictyomitra simplex																				■													
6 Bagotum sp. aff. B. modestum	■				■	■	■	■																					■				
7 Bipedis(?) sp. A																																	
8 Canoptum anulatum					■		■	■																				■					
9 Canoptum poissoni			■			■		■																			■						
10 Canutus rockfishensis	■	■		■	■	■	■	■																				■					
11 Cecrops septemporata																																	
12 Dictyomitra sp. A																						■	■										
13 Dictyomitrella(?) kamoensis									■	■	■																						
14 D.(?) sp. aff. D(?) cucurbitina							■	■																									
15 Eucyrtidium(?) sp. A								■																									
16 Eucyrtidium(?) sp. B						■	■	■																									
17 Foremanina sp. A										■	■	■	■	■	■									■									
18 Gorgansium sp.									■		■													■									
19 Holocryptocanium barbui																						■	■										
20 Hsuum(?) sp. A				■			■	■																									
21 Hsuum sp. B																																	
22 Hsuum sp. C									■	■	■	■	■	■	■	■	■																
23 Jacus sp. A					■			■																					■				
24 Jacus sp. B								■																									
25 Katroma sp. aff. K. neagui				■	■	■	■	■																	■			■	■	■	■	■	■
26 Mirifusus mediodilatatus (?)																		■	■														
27 Mirifusus sp. A										■																							
28 Napora sp. A							■	■																	■	■							
29 Napora sp. B																																	
30 Napora sp. C																																	
31 Napora sp. D									■	■					■		■							■									
32 Nassellaria sp. A						■	■	■																									
33 Novixitus mclaughlini																						■	■										
34 Novixitus sp. A																						■											
35 Pachyoncus sp.											■	■		■										■									
36 Pantanellium sp. aff. P. buntonense																																	
37 Pantanellium ultrasincerum														■		■								■									
38 Pantanellium sp. A	■		■	■	■	■	■	■																									
39 Parvicingula sp. A																																	
40 Podobursa sp. A																			■														
41 Pseudodictyomitra sp. A																		■	■	■													
42 Riedelius sp. A						■																											
43 Ristola turpicula																	■																
44 Ristola sp. A									■	■	■	■	■	■	■	■																	
45 Ristola sp. B																■																	
46 Stichocapsa sp. A		■	■		■	■	■	■																									
47 Stichocapsa(?) sp. B															■		■							■									
48 Stichomitra(?) sp. A																																	
49 Thanarla conica																																	
50 Thanarla veneta																				■			■										
51 Tricolocapsa sp. A										■	■													■									
52 Tricolocapsa sp. B										■	■	■	■	■	■	■								■									
53 Tricolocapsa sp. C.															■																		
54 Tricolocapsa sp. cf. T. plicarum									■	■	■	■	■	■	■	■								■									
55 Trillus elkhornensis							■	■																									
56 Trillus(?) sp. cf. T. seiderse															■																		
57 Unuma sp. A									■	■	■													■									
58 Unuma sp. B										■	■													■									
59 Wrangellium sp. A																																	
60 Zifondium sp.																				■													

Table 1 - continued

	M-5J 4979 MH-1	M-5K 4980 MH-1	M-5L 4982 MH-1	M-5M 5084 MH-1	M-5N 5085 MH-1	M-5O 5086 MH-1	M-5P 5087 MH-1	M-5Q 5088 MH-1	M-5R 5089 MH-1	M-5S 5090 MH-1	M-5T 5091 MH-1	M-5U 5092 MH-1	M-5V 5093 MH-1	M-5W 5094 MH-1	M-5X 5095 MH-1	M-5Y 5096 MH-1	M-5Z 5097 MH-1	M-5AA 5098 MH-1	M-5BB 5099 MH-1	M-6 5062 MH-3	M-7 5063 MH-5	M-8a 5070 MH-1	M-8b 5071 MH-1	M-9 5074 MH-4	M-9fit 5073r MH-5	M-10 5013 MH-5	M-11 5012 MH-5	M-12 5009 MH-1	M-13 5027 MH-1	M-14 5025 MH-1	M-15 5036 MH-3	M-16 5030 MH-7	M-17 5006 MH-4	M-18A 5112 MH-4	M-18B 5116 MH-5	M-18C 5117 MH-6	M-18D 5112 MH-7	M-18E 5113 MH-4	M-19A 4342A MH-2	M-19B 4342B MH-3	M-20 4343 MH-2?	M-21 4353 MH-3	M-22 4347 MH-1	M-23 4352 MH-1	M-24 4351 MH-3	M-25 4349 Cret?	M-26 4518 MH-1	M-27 4515 MH-1	M-28 4524 MH-1	M-29 4522 MH-6
1																					■				■	■	■								■															
2																																							■											
3																															■																			
4																					■				■	■	■																							
5																					■				■																									
6						■	■				■		■		■		■		■																															
7																																							■											
8		■	■	■																																													■	
9											■				■				■				■						■																					
10		■	■	■	■		■		■	■	■		■	■	■	■																											■	■			■		■	
11																					■				■																									
12																																																		
13					■	■				■																																								
14						■				■																																								
15					■	■				■							■																																	
16										■	■																																							
17																				■				■							■								■	■					■					
18																																																		
19																																■					■									cf				cf
20										■																																								
21																																							■		■									
22																																																		
23																							■																											
24																																																		
25	■	■		■	■		■	■		■	■											■	■						■																					
26																					■				■																									
27																																																		
28	■	■				■				■																																								
29																																							■											
30																																							■											
31																								■									■																	
32			■		■	■	■				■				■																																			
33																																																		
34																																																		
35																				■																														
36																																							■											
37																																																		
38																																																		
39																																							■		■									
40																					■				■		■																							
41																					■				■		■									■														
42																																							■											
43																								■									■	■				■												
44																																																		
45																																																		
46																																															■	■	■	
47																																																		
48				■	■		■	■	■	■	■	■		■	■							■							■														■	■						
49																					■				■		■								■															
50																																																		
51																																																		
52																															■									■										
53																								■									■																	
54																																								■		■								
55							■		■	■																																								
56																																																		
57																																																		
58																																																		
59																																																		
60																					■															■														■

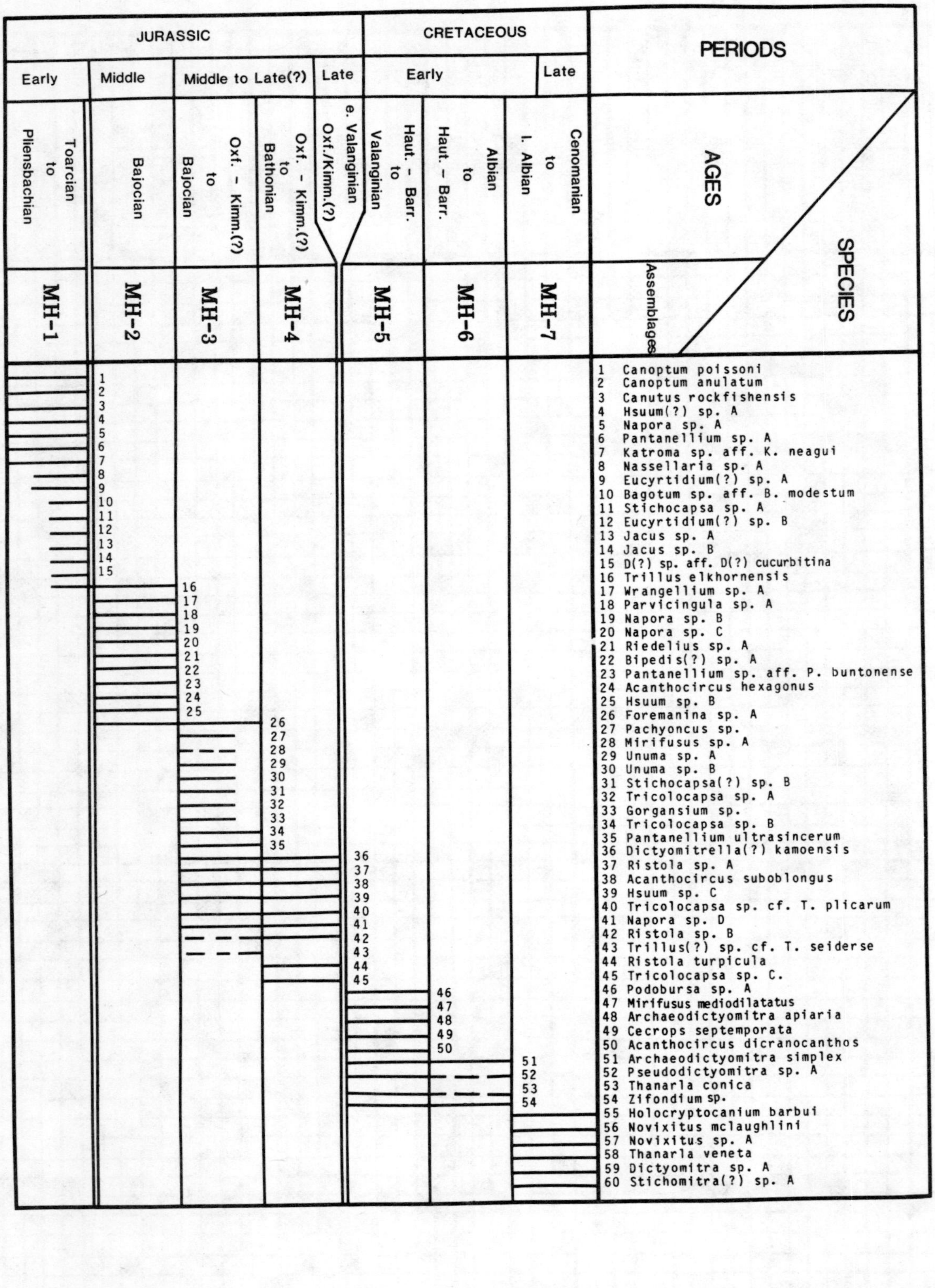

Table 2. Assemblage ranges for sixty radiolarian taxa in the Marin Headlands, California. Radiolarians are figured in Plates 1 and 2.

Table 1. Occurrence of selected Radiolaria in samples from the Marin Headlands. Species 1-60 are figured in Plates 1 and 2. U.S. Geological Survey accession numbers denote individual samples curated by the U.S. Geological Survey, Branch of Paleontology and Stratigraphy, in Menlo Park, California.

Acanthocircus dicranocanthos, abundant in the MH-5 assemblage and present in Nicasio Reservoir assemblage, may be a good low latitude indicator (Pessagno and Blome, in press). Assemblage MH-7 is similar to Cenomanian radiolarians from limestone near Laytonville which had a paleolatitude of 17° + 7° south of the equator (Alvarez and others, 1980; Murchey and Jones, this volume). Paleolatitudinal or water temperature differences seem a likely explanation for the difference between Great Valley sequence faunas and Cretaceous Marin Headlands assemblages. The probability seems high that the Marin Headlands Cretaceous faunas are tropical or subtropical assemblages. If so, the depositionally overlying clastic unit was also derived from the south. Curry and others, this volume, did not find anomalous paleolatitudinal values for the Marin Headlands chert sequence but, because the polarity of the entire chert sequence is normal, they have suggested that the sequence was remagnetized long after deposition.

LITHOSTRATIGRAPHIC UNITS

Descriptive studies of Franciscan chert by Davis (1918) and Bailey and others (1964) are now classics. Techniques for radiolarian extraction from chert that were subsequently developed (Pessagno and Newport, 1972) provide the means for comparative megascopic and microscopic stratigraphic analyses of chert. The eighty-two meter radiolarian chert sequence in the Marin Headlands depositionally overlies basalt and underlies lithic arkose. Four sequential lithostratigraphic units within this chert sequence can be correlated between thrust sheets. The lithologic characteristics of the units, designated L-1, L-2, L-3, and L-4, are summarized in Figure 2, which shows the correlation of these lithologic units units between six sections. Radiolarian biostratigraphy indicates that lithostratigraphic zone boundaries are essentially synchronous within the Marin Headlands area. These units must have been regionally extensive (40 km minimum). Karl (this volume) has subdivided the chert sequence at Alexander Avenue (Fig. 1, locality M-2a-M-2b) into even smaller units which are not as laterally extensive.

Lithostratigraphic unit L-1

The basal chert unit is characterized by maroon chert and shale covered by a silvery gray manganese oxide coating. A clear depositional contact between chert and underlying pillow basalt is exposed in many localities including localities M-1, M-2a, M-8, and M-12. Two lithologic assemblages characterize L-1: (A) ribbon chert--rhythmically alternating chert and siliceous shale and (B) lenticular jasper beds.

Ribbon chert and associated shale in L-1 is maroon or brick-colored with a surficial manganese oxide stain. Radiolarians extracted from L-1 are coated with hematitic clay. Chert beds in L-1 are thicker (100 mm or more) than those in overlying lithostratigraphic units (see measured sections Fig. 2). Cryptocrystalline quartz (<.01mm) and hematite-stained, partially oriented illite are the primary constituents of the chert matrix. The chert tends to have a dull luster and may lack chonchoidal fracture. Much of the "chert" within this interval approaches a siliceous mudstone. Radiolarians are abundant but L-1 chert is matrix-supported, not grain-supported. Preservation is poor and most radiolarians are preserved as microcrystalline quartz or chalcedony casts of external molds.

Thick (.2 m to 3 m) lenticular chert beds occur at two stratigraphic intervals within lithostratigraphic unit L-1: (1) at the chert-basalt contact and (2) 10-18 m above the chert-basalt contact. The lower horizon is as much as six meters thick while the upper horizon is rarely more than two meters thick. These saucer-shaped lenses vary lithologically from dark maroon homogeneous siliceous mudstones to brightly colored red and (or) tan veined, brecciated, microcrystalline jasper (most common lithology). Radiolarians are never well preserved and are generally present only as ghosts. The lenticular beds commonly show "slump" structures. Sections through the L-1 lithostratigraphic unit may contain one, two, or no lenticular jasper bed horizons. The top of the manganese-stained zone lies approximately one meter above the upper jasper bed horizon wherever it is present.

L-1 varies in thickness from approximately ten to sixteen meters. Thinner L-1 units lack lenticular jasper beds and have thinner chert-shale couplets than thicker L-1 intervals. This relationship between bed thickness and total unit thickness suggests that thicker L-1 intervals were deposited in topographic depressions or they were more resistant to compaction because of more rapid lithification.

All of L-1 lies within radiolarian assemblage MH-1. A distinct increase in radiolarian diversity within the MH-1 assemblage zone occurs directly above (within .5 m) the L-1/L-2 lithostratigraphic boundary.

Lithostratigraphic unit L-2

L-2 consists of red and (or) green ribbon chert (rhythmically alternating chert and shale). The base of L-2 is the first chert above the manganese stain zone. The upper part of L-2 grades into microcrystalline chert without shale partings. At Alexander Avenue (locality M-2a), relatively undeformed L-2 is 52 m thick, but folding and faulting have obscured depositional thicknesses elsewhere.

The chert beds (and chert-plus-shale couplets) are thinner, on average, than those in the L-1 ribbon chert. Some beds are quite thin (chert 20 to 40 mm, shale 20 to 40 mm) and nodular. Shale interbeds are nearly as thick as chert beds in parts of the L-2 lithostratigraphic unit, particularly within the MH-3 assemblage zone. Shale to chert ratios approaching 1:1 are not characteristic of the other lithostratigraphic units in the Marin Headlands pelagic sequence. Chert beds in L-2 are generally more indurated than chert beds in L-1 (with the exception of some of the L-1 "jasper" beds) and chonchoidal fracture is common. On average, chert in L-2 is less indurated, less vitreous, and less deeply leached than in overlying lithostratigraphic units L-3 and L-4.

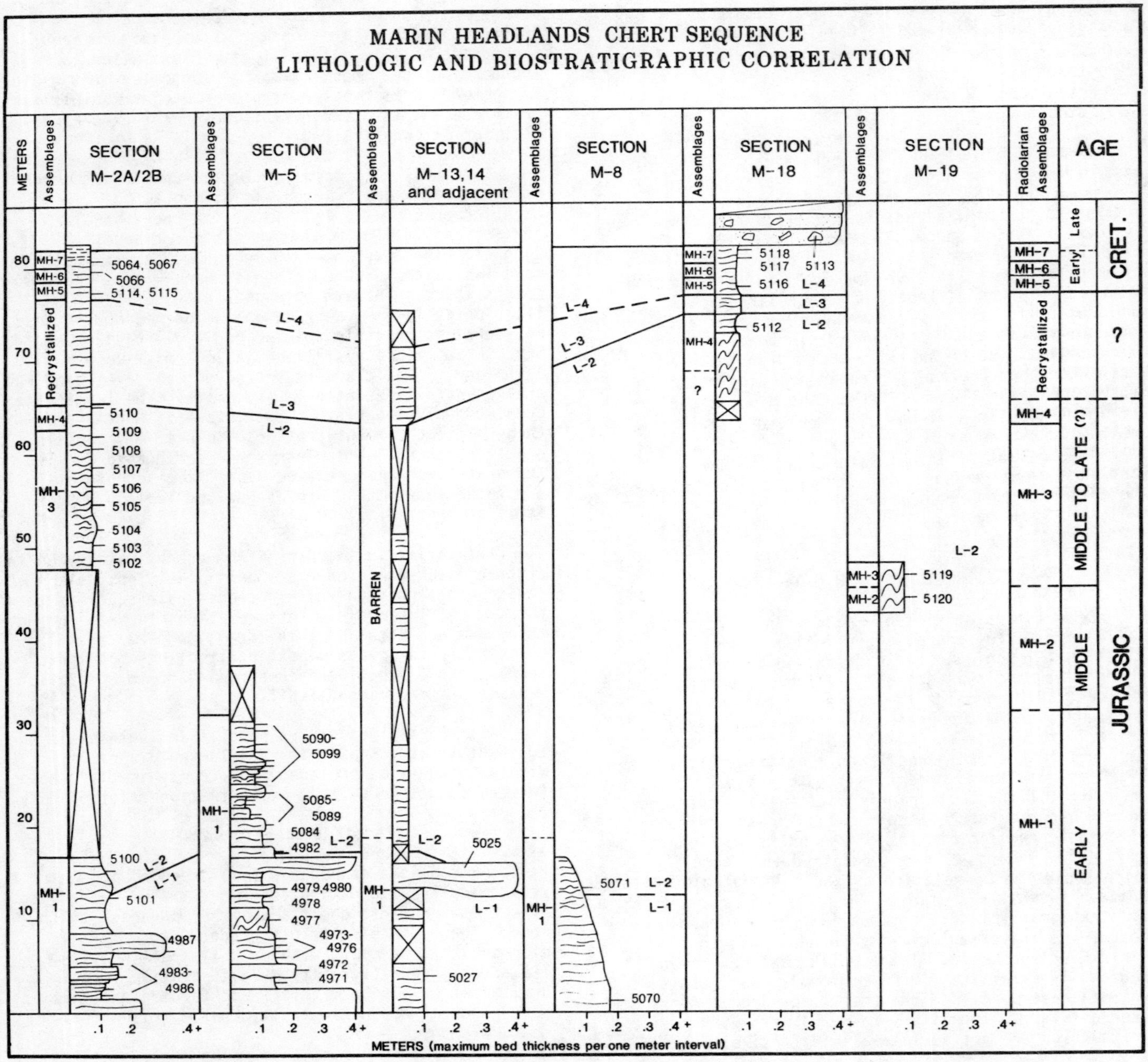

Figure 2. Lithostratigraphic and biostratigraphic correlation between measured sections in the chert sequence of the Marin Headlands, California. Section numbers correspond to map localities on Figure 1. L-1 to L-4 are lithostratigraphic units described in this manuscript. Width of measured sections indicates maximum bed thickness within a one meter interval. MH-1 to MH-7 are radiolarian biostratigraphic units. Four digit numbers to the right of the sections are U.S. Geological Survey accession numbers for radiolarian faunas.

Thin sections of the chert show less opaque material than ribbon chert in L-1. In addition, silica in the matrix of L-2 chert is commonly microcrystalline quartz (approximately .01 mm) as well as cryptocrystalline (less than .01 mm) , whereas L-1 ribbon chert has less microcrystalline quartz in the matrix. Radiolarians abundances in L-2 are highly variable from being absent altogether to com more than fifty percent of the rock in a grain supported fabric. They are quite well preserved in some strata.

Assemblages MH-1(part), MH-2, MH-3 and MH-4 occur within L-2.

Lithostratigraphic unit L-3

The resistant ridge-forming L-3 unit is formed from light green thinly bedded chert (less than 50 mm thick) with localized brightly colored "jasper" breccia. L-3 has no interbedded shale. It is the most indurated, most siliceous unit within the Marin Headlands pelagic sequence. With the exception of the brightly colored jasper breccias, the entire unit has been pervasively leached. The pale green bedded chert is dominantly microcrystalline mosaic quartz and non-oriented clay. The matrix of the jasper breccia is dominantly microcrystalline subequant quartz and iron oxides or oxyhydroxides. The space between breccia fragments contain geopetal fill of silicified radiolarian-bearing mud indicative of synsedimentary brecciation. Radiolarians in this interval are very recrystallized. No datable faunas have been extracted except from the very top of the unit which contains Early Cretaceous MH-5 faunas. Faunas from L-2 near the transition to L-3 contain late Middle Jurassic or early Late Jurassic MH-4 faunas. The faunal break between MH-4 and MH-5 represents an interval lasting at least fifteen million years and possibly as long as thirty million years but the maximum thickness of L-3 is little more than ten meters. The minumum thickness is less than two meters. The immense time interval represented by this ten meter sequence strongly suggests that a hiatus or an erosional event is recorded within or at the top of the L-3 unit. The jasper breccias, the lack of shale, the coarse crystallinity, and the pervasive leaching of this unit resulted from intensive diagenetic processes associated with this sedimentary event.

Lithostratigraphic unit L-4

L-4 consists of red and green ribbon cherts beds characteristically less than fifty mm thick interbedded with very thin shales. The well indurated, very siliceous chert beds are commonly amalgamations of several thin beds. L-4 chert beds have rather consistent lateral thicknesses and lack the lumpy or irregular surfaces common in L-2. Beds are striped with red centers and deeply leached green margins. Interbedded shale is red or green. Matrix silica in the chert beds is dominantly crypto-crystalline and microcrystalline mosaic quartz with some chalcedony (including zebraic) and microchalcedony.

L-4 is less than six meters thick. The contact between L-3 and L-4 is conformable and generally gradational. The upper boundary of L-4 is the contact between the chert sequence and overlying clastic rocks. L-4 contains assemblages MH-5, MH-6, and MH-7. MH-5 and MH-6 radiolarians are preserved as casts and by replacement. The MH-6 faunas are particularly poorly preserved and recrystallized. At localities M-2b and M-18, the MH-6 and MH-7 assemblages are separated by a red shale horizon approximately 15 mm thick. Only the upper 50 cm of L-4 contains assemblage MH-7 faunas; this interval is green siliceous mudstone interbedded with shale. The MH-7 faunas are very well-preserved by volume for volume replacement. Some of the MH-7 radiolarians exhibit pre-consolidation breakage not present in the underlying faunas. The sharp contrast in preservation style between MH-6 and MH-7 and the thinness of the sedimentary interval lying between MH-5 (Valanginian to Barremian(?)) and MH-7 (latest Albian or earliest Cenomanian) is suggestive of another hiatus within the pelagic sequence.

Chert-Sandstone Contact

The contact between the chert unit and the overlying clastic unit is depositional. At locality M-2b, chert grades into calcareous black shale within an interval of a few centimeters. The green chert at the contact contains very well-preserved MH-7 radiolarians which are internally filled with chlorite and an acicular green mineral (pumpellyite?). This chert is laminated with high concentrations of iron oxide parallel to bedding. Iron-rich, manganese-poor micronodules are concentrated along some laminae. These micronodules, studded with radiolarians, have displaced surrounding clay. No microfossils were found in an examination of the overlying calcareous black shale by Kris McDougall. At Bonita Cove (locality M-18), a sharp erosional contact separates MH-7 bedded chert from intraclast-bearing medium-grained lithic arkose. These large angular rip-up clasts (250 mm long) of red and green radiolarian chert are evidence for Cenomanian scouring within the basin. One clast contains Cretaceous radiolarians and another clast contains MH-4 Jurassic radiolarians.

At locality M-3, medium-grained sandstone overlies red and green MH-7 chert. The contact is conformable but not clearly depositional. At locality M-7, massive sandstone overlies thinly bedded MH-5 (Lower Cretaceous) chert. The contact is sharp and the sand adjacent to the contact is somewhat sheared. Some of the folding in the chert resembles slump structures. This particular contact may be a bedding plane fault or may be a depositional contact. If it is a depositional contact, MH-6 and MH-7 assemblages are missing due to erosion or nor-deposition.

Discussion

The following analysis of Marin Headlands chert includes separate discussions of bedded radiolarian chert (with or without shale partings), lenticular jasper beds in the L-1 unit, and jasper breccia in the L-3 unit.

Bedded Radiolarian Chert

Bedded chert occurs as rhythmically alternating chert and shale (ribbon chert) or as thin chert beds with no shale partings (bedded chert in the L-3 interval). Both types of bedded strata may have shared similar depositional histories but slightly different diagenetic histories.

The alternating chert and shale of ribbon cherts have sharp contacts but they differ essentially only in their quartz:clay ratios. Chert beds contain zero to fifty percent or more radiolarian fossils while radiolarians are sparse in shale. Both chert and shale are colored by iron oxides and hydroxides.

Differing interpretations of observed data support two different hypothetical origins for bedding: sedimentary (Hein and Karl, 1983) versus diagenetic (Davis, 1918; Karl, this volume). Ratio grading of radiolarian to clay content, a general upsection decrease in bed thickness, wispy laminae, and lateral bedding characteristics can each be used to support either interpretation.

Both vertical and lateral grading is present in Marin Headlands chert beds. In twenty to thirty percent of etched chert samples the proportion of radiolarians to clay is vertically graded. The radiolarians are not size graded. Long axes of radiolarians are preferentially oriented parallel to bedding but do not exhibit directional orientation. Ratio grading is most common in chert beds greater than 50 mm thick, the majority of which are found in the L-1 and L-2 units. A few beds are reverse ratio graded. Beds rarely have more than one graded interval. Bedding controlled primary structures such as grading are strong evidence for a sedimentary origin for bedding. A diagenetic origin would require the lithification of each bed to proceed in an upward direction for normal grading or downward direction for reverse grading, thereby preserving more radiolarians in the first-lithified sediments. Sano (1983) observed symmetrical mineralogical changes about the center of chert beds in Japan. Rare examples of analogous symmetrical ratio grading occur in the Marin Headlands and must be diagenetically formed. Lateral ratio grading is particularly characteristic of knobby beds and irregular beds in the L-2 interval. The knobs commonly contain more than fifty percent radiolarians (grain-supported) but laterally, within a few centimeters, the percentage of radiolarians can drop to near zero. The knobs were preferentially lithified very early in the diagenetic histories of the sediments and were more resistant to compaction than their lateral equivalents. The lateral variations in radiolarian abundance and silica content in the knobby beds show the degree to which the mobility of biogenic silica during early diagenesis can affect the distribution and concentration of radiolarian microfossils.

Within the chert sequence, bedding thickness decreases upsection although no over-all change in grain size occurs in the fine-grained sequence. The upward decrease in bed thickness can be interpreted either as a decrease in the rate of sedimentation through time (sedimentary model) or as a diagenetic phenomenon related to diagenetic maturity at the time of Cenomanian clastic loading (diagenetic model). In the first interpretation, sedimentation rates could decrease as topographic lows were in-filled by turbidites or contourites and sediments spread out over wider areas, forming thinner beds. A decrease in biologic productivity through time could also decrease the sedimentation rate. Cretaceous hiatuses in the sequence support this possibility although unrecognized hiatuses may also occur in older parts of the sequence. In support of a diagenetic interpretation, compaction features such as knobby beds, bedding parallel microstylolites, and crushed radiolarians (MH-7 assemblage) and diagenetic features such as high silica content, pervasive leaching, and recrystallization are more common in the upper parts of the pelagic sequence. The thinning upward phenomenon is quite likely to be a function of more than one process.

Many chert beds have poorly developed wispy laminae approximately parallel to bedding. Petrographically, these laminae are distinguishable by differences in crystallinity, color, or radiolarian abundance. Laminae may reflect original compositional differences, current features, diagenetic processes, or all three. Stylolites, bedding parallel micro-faults, and leaching are common along laminae and support a diagenetic origin. Davis (1918) illustrates the lenticular character of many laminae, presumably of diagenetic origin. Assuming the converse, a sedimentary origin, the wispy nature of the laminae could result from moderate bioturbation, with homogeneous beds being the product of more complete bioturbation. Well-developed varve-like laminae, commonly interpreted as indicators of an anoxic environment, do not characterize Marin Headlands chert.

Within less deformed parts of the section, individual beds are commonly traceable the length of an outcrop (to 10 m). However, within a given package of ribbon chert, some chert beds may pinch-out. They are replaced by bifurcating shale beds such that the total thickness of the chert and shale package remains constant. The first observation supports the sedimentary bedding hypothesis, the second supports the diagenetic bedding hypothesis. Where folding intensifies, boudinage and pinchouts of beds are common and they obscure pretectonic bedding characteristics of the sedimentary sequence.

Diagenetic processes have obscured the original composition of the pelagic sediments in the Marin Headlands. Clay and radiolarians were major components, but the original carbonate content is not known. Current winnowing, bioturbation, and productivity fluctuations all must have affected the distribution of radiolarians and clay in the original sediments. Chert in the Marin Headlands was formed by the transformation of biogenic opal (opal A) from radiolarians to quartz via an intermediate opal-CT phase (see Karl, this volume, for a discussion). The mobilization of Si during the solution step between opal A dissolution and opal-CT precipitation permits the silicification of the sediment interstices. During early diagenesis, radiolarian-rich laminae are permeable and have a very high silica diagenetic potential. Variations in the radiolarian to clay ratio, therefore, could create microenvironments in which pore fluids are either oversaturated or undersaturated with respect to opal-CT or quartz. The opal-A to opal-CT transition has potential to enhance slight variations in the radiolarian to clay ratio. Diagenesis may proceed by the precipitation of opal-CT in nodules or in laminae. As opal-CT precipitation continues, nodules eventually coalesce (Jenkyns and Winterer, 1982). Laminae increase in thickness and coalesce. Eventually this process would form beds with shale partings.

Petrographic examinations of many samples did not produced any evidence for trapped micrite. Only two silicified foraminifers were recovered from among thousands of radiolarians. This negative evidence suggests that the original carbonate content was low. The somewhat lumpy and irregular bedding surfaces in the L-2 lithostratigraphic interval may be evidence for carbonate dissolution contemporaneous with compaction. Elsewhere, very irregular knobby beds are characteristic in transition zones between bedded chert and pelagic limestone (Bosellini and Winterer, 1975). Such beds may indicate impingement of the calcite compensation depth on the sediment-water interface. Diagenetic siderite rhombs in L-2 formed penecontemporaneously with early dissolution of opal A. Most of these rhombs were subsequently replaced by silica during a period of silica mobilization and precipitation. Tiny chalcedony-filled birds-eyes in the L-4 lithostratigraphic unit suggest the possibility of syndepositional sulfates in the original sediment.

The characteristic bedding of the radiolarian chert, whether it is alternating chert and shale (L-1, L-2, L-4) or bedded chert with hairline partings (L-3), is a function of both depositional as well as diagenetic processes. Fine grained sediments containing radiolarians, clay, and an unknown but probably small amount of carbonate were deposited in a pelagic environment. Some of these sediments were deposited as turbidites or contourites. Current winnowing and bioturbation influenced the distribution of radiolarians and clay. Biological productivity rates probably affected the radiolarian to clay ratio in a general way, but there is no good evidence in these sediments that production periodicity produced alternating chert and shale beds. Silica diagenesis enhanced variations in the radiolarian to clay ratios and overprinted original sedimentary characteristics.

Lenticular Beds in L-1

Saucer shaped chert lenses (to 300 cm thick) in the L-1 lithostratigraphic unit generally occur as a package of two or more beds, but they can also occur as isolated beds. Lithologically these beds vary from silica-rich bright red or tan chert with chonchoidal fracture (jasper) to silica-"poor" dark maroon manganiferous siliceous mudstones.

The quartz component of the brightly colored jasper at the silica-rich end of the spectrum is either cryptocrystalline quartz (less than .01 mm crystals) or microcrystalline equant quartz (approximately .01 mm crystals). Textures when viewed in thin section are either dense homogeneous, pelletal (iron-oxide stained "pellets" surrounded by clear quartz crystals), or curdled. Iron-colored impurities are commonly concentrated outside the rims of the quartz crystals. Spherical zones with high iron content may be radiolarian ghosts. In the jaspers, the red color is caused by hematite, the ochre by limonite. The silica-rich beds contain internal "dessication" structures including arcuate veins, radiating veins, and internal jigsaw breccias, none of which extend to the bed margins. The veins and breccia interstices are filled with bladed and equant megaquartz and (or) length fast or zebraic chalcedony. They are never filled with iron-stained radiolarian chert.

At the silica-poor end of the spectrum is manganese stained siliceous mudstone, composed primarily of fairly homogeneous cryptocrystalline mosaic quartz and oriented clay. In thin section and in hand specimen, the mudstone has a dirty appearance and a dull luster. The mudstone rarely has early diagenetic veins and never has internal breccia. Radiolarians are rare, but they are occasionally preserved in small laminae a few millimeters thick. The radiolarian to matrix ratio in these laminae is approximately 1 to 4. The radiolarian assemblages in these laminae are similar to those in associated L-1 ribbon chert, an indication that any preferential radiolarian dissolution affecting the mudstone lenses also occured in associated ribbon chert.

The radiolarian to matrix ratio in these laminae is approximately 1 to 4. The radiolarian assemblages in these laminae are similar to those in associated L-1 ribbon chert, an indication that any preferential radiolarian dissolution affecting the mudstone lenses also occured in associated ribbon chert.

Gradational changes from silica-rich brightly colored jasper to silica-"poor" manganiferous siliceous mudstone to "normal" radiolarian chert suggest that diagenetic history plays a more important role than depositional history in the formation of these lenses.

Preliminary oxygen isotope values from SiO_2 in four samples collected in a measured section at locality M-2a (Fig. 2) show dO^{18} depletion in a jasper bed near the basalt-chert contact (dO^{18}SMOW=16.30), somewhat less depletion in an overlying manganiferous ribbon chert in the L-1 unit (dO^{18}SMOW=20.59, six meters above basalt), and normal values in radiolarian chert from the L-2 unit (dO^{18}SMOW=26.38, 16m above basalt; dO^{18}SMOW=26.48, 50 m above basalt)(J. Drotleff, written comm., 1983, preliminary results). If one assumes a submarine origin for these beds, then hydrothermal fluids may account for the oxygen isotope depletion. An alternative explanation that I consider less likely is that these rocks were influenced by fresh water during early diagenesis.

The best explanation for the origin of the thick jasper lenses is a model in which they are analogous to similar and stratigraphically equivalent strata-bound jasper bodies of hydrothermal origin such as those associated with manganese deposits elsewhere in the Franciscan (Snyder, 1978; Crerar and others, 1982). In this model, jasper beds and associated L-1 ribbon beds were originally similar pelagic deposits. The jasper beds were lithified very rapidly by hydrothermal fluids; the associated ribbon beds were lithified less rapidly than the jasper beds but more rapidly than overlying ribbon chert in the L-2 interval. These different rates of lithification resulted in varying degrees of compaction, and, therefore, pronounced differences in bed thicknesses between jasper beds, ribbon beds in L-1, and ribbon beds in L-2. Differing rates of lithification and influence by hydrothermal fluids also account for different crystallinity and textures of quartz in these beds and for differences in radiolarian preservation.

The duration of hydrothermal activity after formation of oceanic crust is five to ten million years (Hart and Staudigel, 1978). This figure is consistent with the radiolarian faunal data which suggests that the L-1 interval was deposited over a period that did not exceed 12 million years (Pliensbachian and Toarcian) and was probably considerably shorter in duration.

Stratiform Jasper Breccia in the L-3 Interval

Locally the upper part of the L-3 lithostratigraphic unit is a stratiform jasper breccia. The brightly colored ochre, red, or rare green matrix fragments are composed of clear microcrystalline quartz crystals (subequant, .01 mm) surrounded by iron oxide or iron oxyhydroxide. The matrix commonly has a curdled "cottage cheese" appearance. Radiolarians in the matrix fragments are rare.

Although this breccia is very similar in appearance to the lenticular jasper breccia in the L-1 unit, some significant differences between them exist. First, this type of breccia is not confined within a single bed. In fact, brecciation may have destroyed original bedding. Second, the space between breccia fragments commonly contains geopetal fill of maroon radiolarian bearing sediments in addition to zoned megaquartz, bladed and equant, and (or) chalcedony. Third, this breccia is not associated with the manganese stain zone or with basalt.

Similar breccia horizons occur in the Paleozoic Caballos Novaculite of Texas (McBride and Folk, 1977) where these horizons contain fossil tree trunks. Folk interprets them as the product of diagenetic alteration of subaerial paleosols that once contained evaporites and McBride interprets them as the product of diagenetic alteration of evaporite beds deposited in deep water between radiolarian ooze. In the Marin Headlands, overlying ribbon chert in the L-4 lithostratigraphic unit has tiny (1 mm) birds-eye fenestrae infilled with chalcedony. In at least one case, the chalcedony is zebraic. Zebraic chalcedony is associated with sulfates (McBride and Folk, 1977; Keene, 1983). While these fenestrae may be evidence for sulfate replacement, sulfates, including barite and anhydrite, are known in deep marine deposits and do not require a subaerial genesis.

Karl (this volume) has proposed that the leached zone of L-3 is related to hydrothermal activity. While this model is certainly a possibility, one must then account for a source for the hydrothermal fluids. Because of the great difference in age between this lithostratigraphic unit and the underlying basalt, any hydrothermal system affecting the L-3 interval is not the hypothetical spreading center that formed the basal basalts. This interpretation also fails to account for the major stratigraphic break between MH-4 and MH-5 assemblages, a break that appears to be widespread in Franciscan chert blocks (Murchey and Jones, this volume).

I believe the enigmatic origin of the stratiform breccia in L-3 is directly related to the recrystallized zone and apparent hiatus between MH-4 assemblages and MH-5 assemblages. Subaerial exposure (during the Berriasian-Valanginian regression for instance) such as the model proposed by Folk for the Caballos Novaculite jaspers accounts for some observed features but does not explain the absence of shallow marine deposits overlying the stratiform breccia, intraformational clastic units within the chert sequence, or the apparent absence of erosional surfaces cutting down through lower strata.

A purely submarine hiatus might cause either the formation of a hardground or a partially lithified interval. In the first scenario, a hardground formed which became "jasperized" and brecciated. As new sediments were deposited over the hardground, some filtered down into the cracks which were subsequently cemented with silica. In the second scenario, a hiatus or condensed sequence in L-3 unit resulted in partially lithified L-3 sediments and poorly lithified L-4 sediments at the time of Cenomanian loading by clastics. Loading created increased temperatures, hydrostatic pressures, and silica mobilization. Differences in viscosity between the two litho-stratigraphic units caused the forceful injection of fluid and some sediment into the partially indurated L-3 unit, thereby brecciating it. This particular model has the advantage of paralleling the hydrothermal model for formation of the jasper beds in L-1 without requiring a hydrothermal source. Steinitz (1970) postulates a similar mechanism for the formation of chert dikes with internal breccias in non-pelagic Senonian chert beds in Israel.

CONCLUSIONS

Only a small portion of the Jurassic-Cretaceous Pacific Ocean Basin has not been subducted or accreted. The record of this ancient ocean is preserved in the northwestern part of the modern Pacific and in accreted fragments such as the Marin Headlands. Productivity, the calcite compensation depth, and plate motion are the parameters which controlled the sedimentary history of this ocean.

Oceanography controls the deposition of biogenic silica. Although dissolved silica may have various primary sources, it is redeposited as biogenic silica only in regions with high productivity and nutrient-rich surface waters (Kennett, 1982, p. 473-483). The Marin Headlands pelagic sequence records high siliceous biogenic productivity during an interval of approximately one hundred million years, indicating a stable environment which endured for a long period of time. Assuming that high productivity areas in the Jurassic-Early Cretaceous paleoPacific are analagous to those in the modern Pacific, the Marin Headlands pelagic sequence formed near the equator, at high latitudes, or in a coastal upwelling zone. No evidence whatsoever, either paleomagnetic or faunal, supports a high latitude origin for any Franciscan rocks. However, both faunal and paleomagnetic data indicate that many of the "oceanic" pieces of the Franciscan assemblage formed near the equator (Murchey and Jones, this volume; Sliter, this volume; Gromme, this volume). Although the uppermost part of the Marin Headlands pelagic sequence certainly formed near a continental margin, an equatorial high productivity zone is a more stable depositional environment than a marginal basin over a long period of time.

The radiolarian faunas of the Marin Headlands show more affinity to correlative low-latitude faunas from Europe, Deep-Sea Drilling Project sites, Japan, and Costa Rica than to faunas from the Coast Range Ophiolite and Great Valley sequence of California. The Mesozoic Pacific Ocean probably had very low temperature gradients between latitudes, and the zone of tropical faunas may well have been rather wide. A fairly wide tropical zone would permit some northward plate movement which would remain undetectable. Nevertheless, my tentative interpretation of the faunal data is that the Marin Headlands first came within the sphere of continental influence, as recorded by the influx of Cenomanian clastic deposits, quite far south of present position.

Virtually no calcareous fossils have been preserved in the Marin Headlands pelagic sequence, even in silicified form. The assumption, therefore, that the sediment-water interface was below the CCD (calcite compensation depth), or at least below the FCD (foraminifer compensation depth) seems valid. Because the Marin Headlands is not in place, the local parameters affecting its CCD history remain speculative. Very little is known about the history of the calcite compensation depth prior to the Late Cretaceous. Because carbonate producing plankton did not become important sediment builders until Cretaceous time, the Jurassic CCD in the Pacific may have been relatively shallow in a wide variety of environments. The dual distribution of both carbonate and siliceous biogenic sediments over large oceanic areas did not occur until Cretaceous time (Kennett, 1982). If estimates of a 3500 m or greater calcite compensation depth in the Cretaceous Pacific are correct (Van Andel and others, 1975), then the Cretaceous sediment-water interface of the Marin Headlands was quite deep or lay within a geographic region with a locally elevated CCD during that period.

Hein and Karl (1983) interpret the depositional environment of the Marin Headlands chert sequence as anoxic but the overall evidence fails to support that interpretation. Black organic-rich shale or chert, phosphorite, or glauconite do not occur within the chert sequence. The Marin Headlands chert is not well-laminated. At best, some chert beds have streaky laminae.. Many beds are homogeneous and some show evidence of bioturbation. The original color of the sediments is red, with the possible exception of part of the L-3 lithostratigraphic unit. Pyrite is uncommon in the acid insoluble residue and radiolarians are rarely preserved by pyrite. Non-phosphatic iron-oxyhydroxide nodules developed at or near the sediment-water interface in some of the youngest sediments. Holothurians(?) apparently lived in the MH-7 environment. Throughout the depositional history of the Marin Headlands chert sequence, the sediment-water interface of this block was apparently below or outside the equatorial and continental margin plume of oxygen-depleted water even during postulated oceanic anoxic events.

Although tectonic models generally place the starting position of Franciscan basalt and chert sequences in the oceanic province to the west of present position (Dickinson, 1981), two recent papers on sedimentary environments of bedded chert favor deposition in continental margin basins.

Jenkyns and Winterer (1982) propose that "most and possibly all Mesozoic ribbon cherts" formed in relatively small basins. Their hypothesis is based upon several observations:
(1) Ribbon chert had not been recovered from mid-ocean basins by deep-sea drilling. (2) Tethyan cherts are laterally and vertically associated with pelagic shelf-sea sediments of "less than oceanic" character. (3) The absence of significant cerium anomalies in many cherts suggests continental influence. (4) In small basins, productivity is typically heightened and the CCD is elevated, favoring the accumulation of clay and biogenic silica. Modern analogues in which biogenic silica is accumulating today are the Ventura and Santa Barbara basins, the Gulf of California (their Tethyan analogue), and arc-related basins such as the Sea of Okhostsk, and the Bering and Japan Seas (their California analogues).

Hein and Karl (1983) use some of the same arguments to reach similar conclusions. They specifically cite Franciscan chert in the Marin Headlands as one example of pelagic deposition in a continental margin basin.

Counter arguments to the points listed in these two manuscripts are applicable to an analysis of the depositional environment of the Marin Headlands sequence. (1) Alternating radiolarites and mudstones of Albian-Cenomanian age recently have been found in the western Atlantic on the Bermuda Rise in Hole 386 of Leg 43 (Tucholke, Vogt and others, 1979)). (2) Although Marin Headlands is overlain by clastics derived from a continental source, no evidence indicates continental influence prior to Cretaceous. (3) Jenkyns and Winterer do report the presence of significant cerium anomalies in Franciscan chert from unspecified localities. (4) Productivity is also heightened at very low and very high latitudes as a result of large scale oceanic circulation patterns. In addition, the Jurassic CCD may have been quite shallow all over the Pacific.

A near continent origin for the Cretaceous portion of the Marin Headlands pelagic sequence is indicated by its gradational boundary with overlying clastic rocks. However, a small-basin, near-continent origin for the Jurassic portion of the Marin Headlands sequence, while not impossible, has formidable tectonic and sedimentological arguments against it.

Continental margin basins in areas of active plate margins are generally short-lived before crust is subducted or obducted, the basin is uplifted, or the basin is infilled with clastic detritus. With the exception of the Bering sea which contains trapped oceanic crust, the oldest marginal basins in the Pacific are typically floored by basement no older than 25-35 million years old. In contrast, the volcanic basement of the Marin Headlands sequence was generated approximately one hundred million years before the pelagic sediments were buried by clastic sediments. The Marin Headlands sequence contains very little volcanic detritus or clastic detritus as determined by geochemistry (Karl, this volume) and by petrographic analysis. The one hundred million year interval represented by the Marin Headlands chert sequence was a period of rapid plate convergence and orogenesis throughout the circumPacific region. Yet, widespread Cordilleran compressional events during Middle Jurassic time and later are not reflected in the Marin Headlands pelagic sequence. If the Marin Headlands were adjacent to a continental land mass, one might expect to see an influx of terrigenous clastics during at least one regression in the one hundred million year sequence. The absence of carbonate or clastic deposition during periods of low sea levels and Cordilleran mountain building tends to support a deep basin, oceanic origin for the Marin Headlands sequence. For these reasons, a continental margin basin origin for the basalt and basal pelagic sediments of the Marin Headlands sequence seems very unlikely. Therefore, one must reconsider the older open ocean model for the Marin Headlands sequence.

The abundant radiolarians throughout the pelagic sequence can best be explained by a trajectory within the equatorial high productivity bulge. Such a trajectory would be generally consistent with some of the paleomagnetic data from other Franciscan pelagic sequences (Alvarez and others, 1980; Gromme, this volume). The few studies of Marin Headlands basalt (Bailey and others, 1964) suggest a composition more characteristic of a mid-ocean ridge than a seamount. Marin Headlands, therefore, is one of the Franciscan sequences most likely to be genuine deep ocean floor that has been accreted to the continent. In lieu of further geochemical studies of the basalt compositions within this terrane, a plateau or seamount environment cannot be ruled out.

The unstable opaline skeletons of buried radiolarians provided almost all the silica which eventually cemented these rocks although hydrothermal fluids may have donated a small portion of silica to basal sediments. Silica mobilization during diagenesis magnified any pre-existing variations in the silica to clay ratio in the sediments. Eventually, the combination of sedimentary processes plus diagenetic processes produced bedded chert, with and without interbedded shale.

During Cenomanian time, the pelagic sediments were loaded by an influx of thick-bedded clastics, a process which must have compacted and dewatered unconsolidated or partially consolidated sediments, especially those in the younger parts of the sequence. The concurrent rapid increase in hydrostatic pressure and temperature would increase the dissolution rate of unstable forms of silica. Therefore, the Cenomanian is a logical time for major mobilization of silica rich fluids, particularly in the younger parts of the stratigraphic sequence. Fluid mobilization was probably accompanied by the leaching of L-2, L-3, and L-4 lithostratigraphic units. Some leaching in the Marin Headlands predates the formation of brittle veins. Compaction was accompanied by the formation of pressure solution stylolites parallel to bedding, by the formation of irregular bedding in L-2, and possibly by brecciation in L-3. Isaacs and others (1983) find a direct correlation between degree of compaction of siliceous sediments (Miocene Monterey Formation of California) and diagenetic maturity. Increasing compaction and maturity reduces porosity, permeability and the diagenetic potential of silica.

The Marin Headlands terrane was detached from its basement, imbricated, and emplaced over melange during the Late Cretaceous and (or) early Tertiary, processes possibly beginning as early as Cenomanian time. Although Cenomanian loading probably increased the silica maturation rate of the Marin Headlands pelagic strata, significant silica diagenetic potential may have still existed at the time of detachment. Folding and thrusting of these sediments would have caused further silica mobilization of any diagenetically "immature" strata. Apparent "soft sediment" deformation, brittle deformation, pressure solution stylolites in fold noses, and bedding-plane microfaults developed during this period. The pinching-out diagenetic effect used as supporting evidence for a diagenetic origin for ribbon beds is greatest in folded sections. Silica mobilization during this period of folding enhanced this diagenetic characteristic.

The general northward motion of oceanic plates of the Pacific Ocean relative to North America during the late Mesozoic must have created zones of oblique convergence along plate boundaries. One probable effect of the oblique convergence apparently required by the paleomagnetic data is right lateral translation concurrent with and following collision. Assuming that the Marin Headlands terrane did indeed approach the Americas to the south of its present position, it was moved northward during Cenomanian and (or) post Cenomanian time by right lateral translation to its present position.

In summary, the Franciscan chert sequence in the Marin Headlands north of San Francisco, records a pelagic sequence of approximately 100 million years duration (one of the longest stratigraphic intervals for a single chert sequence in the world). Seven radiolarian assemblages ranging in age from Pliensbachian to late Albian or early Cenomanian can be discriminated. Stratigraphic gaps in recovered faunas occur between the Jurassic and Lower Cretaceous strata and possibly between Lower and Upper Cretaceous strata. The seven radiolarian assemblages and four lithostratigraphic units within the chert sequence can be correlated between thrust sheets within the Marin Headlands.

The pelagic sediments were deposited under conditions of fairly high biologic productivity in an oxygenated or at leastly mildly oxygenated environment. The Marin Headlands probably originated at a spreading center in the near equatorial mid-Pacific during Pliensbachian time and followed a low latitude trajectory (probably with a northward component) through most of its depositional history. Submarine volcanism, represented by pillow basalt, basalt flows and breccia, ended by Early Jurassic (Pliensbachian or earlier) time. For approximately 100 million years, radiolarian-rich siliceous sediment accumulated above the basalt. During latest Albian or early Cenomanian time, the Marin Headlands came within the realm of influence of an emerged continent and received an influx of clastic sediments. Biogenic silica from radiolarians was the principal silica source for the chert deposits. Sedimentary processes created variations in chert to clay ratios but silica diagenesis increased compositional differences. Characteristic alternating chert and shale beds resulted from these processes. Leached crystalline zones are associated with hiatuses or very slow deposition. Such zones contain little or no interbedded shale. Jaspers and jasper breccias occur at three horizons and probably formed by at least two different processes. The lower two jasper horizons are probably related to hydrothermal processes. The upper jasper breccia horizon appears to be related to an hiatus. Detachment of the Marin Headlands sequence from its basement and emplacement into its present position are processes that began during Cenomanian or Turonian time.

ACKNOWLEDGEMENTS

I thank D. L. Jones, C. Carlson, and R. L. Folk for reviews and comments, M. C. Blake, C. Wahrhaftig, and P. O. Baumgartner for discussions of the geology of the area, and P. Jefferies, R. Couzin, and L. Kohnitz for technical assistance.

REFERENCES CITED

Alvarez, W., Kent, D. V., Premoli-Silva, I., Schweikert, R. A., 1980, Franciscan Complex limestone deposited at 17° south latitude: Geological Society of America Bulletin v. 91, p. 476-484.

Bailey, E. H., Irwin, W. P., and Jones, D. L., 1964, Franciscan and related rocks, and their significance in the geology of Western California: California Division of Mines and Geology Bulletin 183, p. 1-177.

Baumgartner, P. O. in press, Comparison of unitary associations and probabilistic ranking and scaling as applied to Mesozoic Radiolaria: Computers and Geosciences, Proceedings of the IGCP Project 148 meeting, 1982.

Bosellini, A. and Winterer, E. L., Pelagic limestone and radiolarite of the Tethyan Mesozoic: a genetic model: Geology v. 3, P. 279-282.

Bedroissian, T. L., 1974, Geology of the Marin Headlands: California Geology v. 19, p. 75-86.

Blake, M. C., Jr., Howell, D. G., and Jones, D. L., Preliminary tectonostratigraphic terrane map of California: U.S. Geological Survey Open-File Report 82-593.

Crerar, D. A., Namson, J., So Chyi, M., Williams, L., Feigenson, M. D., 1982, Manganiferous cherts of the Franciscan Assemblage: I. General geology, ancient and modern analogues, and implications for hydrothermal convection at oceanic spreading centers: Economic Geology, v. 77, p. 519-540.

Davis, E. F., 1918, The radiolarian cherts of the Franciscan group: California univ. Dept. Geology Bull., v. 11, no. 3, p. 235-432.

De Wever, P., 1982, Nassellaria (radiolaires Polycystines) du Lias de Turquie: Revue de Micropaleontologie v. 24, p. 189-232.

Dickinson, W. R., 1981, Plate tectonics and the continental margin of California in Ernst, W. G., ed., The Geotectonic Development of California, Rubey Vol. I, Prentice-Hall, Inc., Englewood Cliffs, N. J., p. 1-28.

Dumitrica, P., 1975, Cenomanian Radiolaria at Podui Dimbovitei: Micropaleontological guide to the Mesozoic and Tertiary of the Roumanian Carpathian (14th European Micropaleontology Colloquium), p. 87-89.

Hart, S. R. and Staudigel, H., 1978, Oceanic crust: Age of hydrothermal alteration: Geophysical Research Letters, v. 5, p. 1009-1012.

Hein, J. R. and Karl, S. M., 1983, Comparisons between open-ocean and continental margin chert sequences, in Iijima, A, Hein, J. R., and Siever, R., eds., Siliceous Deposits in the Pacific Region, Developments in Sedimentology v. 36, Elsevier, New York, p. 25-43.

Hertlein, L. G., 1956, Cretaceous ammonite of Franciscan group, Marin County, California: American Association of Petroleum Geologists Bulletin, v. 40, p. 1985-1988.

Isaacs, C. M., Pisciotto, K. A., and Garrison, R. E., 1983, Facies and diagenesis of the Miocene Monterey Formation, California: a summary, in Iijima, A., Hein, J.R., and Siever, R., eds., Siliceous Deposits in the Pacific Region, Developments in Sedimentology, v. 36, Elsevier, p. 247-282.

Jenkyns, H. C. and Winterer, E. L., 1982, Palaeoceanography of Mesozoic ribbon radiolarites: Earth and Planetary Science Letters v. 60, p. 351-375.

Keene, J. B., 1983, Chalcedonic quartz and occurrence of quartzine (length- slow chalcedony) in pelagic sediments: Sedimentology 30, p. 449-454.

Kennett, J., 1982, Marine Geology, Prentice-Hall, Inc., Englewood Cliffs, N. J., 813 p.

Lawson, A. C., 1914, Description of the San Francisco district; Tamalpais, San Francisco, Concord, San Mateo, and Hayward quadrangles: U.S. Geological Survey Geologic Atlas, Folio 193.

McBride, E. F. and Folk, R. L., 1977, The Caballos Novaculite revisted: part II: chert and shale members and synthesis: Journal of Sedimentary Petrology, v. 47, p. 1261-1286.

Mizutani, S. and Kido, S., 1983, Radiolarians in Middle Jurassic siliceous shale from Kamiaso, Gifu Prefecture, Central Japan: Transactions of Proceedings Palaeontogical Society of Japan, N. S., No. 132, p. 253-262, pl. 51-53.

Nakaseko, K., Nishimura, A. and Sugano, K., 1979, Radiolaria (mainly Cretaceous) of the Shimanto Belt: News of Osaka Micropaleontologists, Special Paper, no. 2, p. 1-49, 8 pls.

Pessagno, E. A., Jr., 1977a, Upper Jurassic Radiolaria and radiolarian biostratigraphy of the California Coast Ranges: Micropaleontology, v. 23, p. 56-113, pl. 1-12.

Pessagno, E. A., Jr., 1977b, Lower Cretaceous radiolarian biostratigraphy of the Great Valley Sequence and Franciscan Complex, California Coast Ranges: Cushman Foundantion Foram. Res. Spec. Publ. 15, p. 1-86, pl. 1-12.

Pessagno, E. A., Jr. and Blome, C. D., 1980, Upper Triassic and Jurassic Pantanelliinae from California, Oregon, and British Columbia: Micropaleontology v. 26, p. 225-273, pl 1-11.

Pessagno, E. A., Jr. and Blome, C. D., 1982, Bizarre Nassellariina (Radiolaria) from the Middle and Upper Jurassic of North America:

Pessagno, E. A., Jr. and Blome, C. D., in press, Tectonogenesis of Mesozoic rocks in the Blue Mountains province of eastern Oregon and western Idaho.

Pessagno, E. A., Jr. and Newport, R. L., 1972, A technique for extracting Radiolaria from radiolarian cherts: Micropaleontology v. 18, p. 231-234, pl. 1

Pessagno, E. A., Jr. and Poisson, A., 1981, Lower Jurassic Radiolaria from the Gumuslu Allochthon of Southwestern Turkey (Taurides Occidentales): Bulletin of the Mineral Research and Exploration Institute of Turkey, v. 92, p. 47-69, 15 pl.

Pessagno, E. A., Jr. and Whalen, P., 1982, Lower and Middle Jurassic Radiolaria (multicyrtid Nasellariina) from California, east-central Oregon, and the Queen Charlotte Islands, B. C.: Micropaleontology v. 28, p. 111-169, pl. 1-13.

Sano, H., 1983, Bedded cherts associated with greenstones in the Sawadani and Shimantogawa Groups, Southwest Japan, in Iijima, A, Hein, J. R., and Siever, R., Siliceous Deposits in the Pacific Region, Devlopments in Sedimentology v. 36, Elsevier, New York, p. 427-440.

Schaaf, A., 1981, Late Early Cretaceous Radiolaria from Deep Sea Drilling Project Leg 62: in Thiede, J., Vallier, T. L., and others, Initial Reports of the Deep Sea Drilling Project, v. 62, p. 419-470, 27 pl., U.S. Government Printing Office, Washington.

Schlocker, J., 1974, geology of the San Francisco North quadrangle, California: U.S. Geological Survey Professional Paper 782, 109p.

Schlocker, J., Bonilla, M. G., and Radbruch, D. H., 1958, Geology of the San Francisco North quadrangle, California: U.S. Geological Survey Map 1-272.

Schmidt-Effing, R., 1980, Radiolarien der Mittel-Kreide aus dem Santa Elena- Massiv von Costa Rica: Neues Jahrb. Geol. Pal. Abh., v. 160, p. 241-257, 33 figs.

Snyder, W., 1978, Manganese deposited by submarine hotsprings in chert- greenstone complexes, western United States: Geology v. 6, p. 741-744.

Steinitz, G., 1970, Chert 'dike' structures in Senonian chert beds, southern Negev, Israel: Journal of Sedimentary Petrology, v. 40, p. 1241-1254.

Taketani, Y., 1982, Cretaceous Radiolarian Biostratigraphy of the Urakawa and Obira Areas, Hokkaido: Tohoku University, Science Rep., 2nd ser. (Geol.), v. 52, p. 1-76, pl. 1-13.

van Andel,Tj. H., Heath, G. R., and Moore, T. C., Jr., 1975, Cenozoic history and paleoceanography of the central equatorial Pacific Ocean: Geological Society of America Memoir 143.

Yao, A., Matsuoka, A., and Nakatani, T., 1982, Triassic and Jurassic radiolarian assemblages in Southwest Japan: News of Osaka Micropaleontologists, special volume, p. 27-43, pl. 1-4.

SEDIMENTOLOGIC, DIAGENETIC, AND GEOCHEMICAL ANALYSIS OF UPPER MESOZOIC RIBBON CHERTS FROM THE FRANCISCAN ASSEMBLAGE AT THE MARIN HEADLANDS, CALIFORNIA

by

S. M. Karl[1]

Department of Geology
Stanford University
Stanford, CA 94305

ABSTRACT

A pilot study of measured and well-dated chert sections from the Franciscan assemblage of California coordinated geochemical, sedimentologic, and petrologic evidence illustrating that distinct chemical patterns characterize cherts with different lithologic associations. Patterns of cherts from the Marin Headlands, California, reflect a change up-section from a site of pelagic deposition to a site within terrigenous influence. Multivariate statistial analysis of 268 Upper Mesozoic and Tertiary DSDP and northeastern Pacific rim chert samples determined that the same elements proven effective for discriminating sediment source components and depositional provinces for Recent sediments are also effective for ancient cherts in spite of diagenetic and paleo-oceanographic processes which distinguish them from unlithified seafloor sediments. Important elements which define unique ratios for identifying chert sediment sources include Al, Fe, Mn, Ni, Ba, Cu, K, Ti, Zr, Pb, V, Y, Mg, and B. Chemical patterns of DSDP cherts, backtracked to sites of deposition, compared favorably with modeled provincial chemical patterns. Element ratios plotted on Fe-Ba-Ni, Cu-Ba, and Fe-Mg-K discriminant diagrams designate samples in distinct sediment source fields. When applied to Franciscan cherts, this method delineates sediment sources and depositional provinces consistent with observed local geology and emphasizes differences in the provenance of cherts and associated lithologies. Chemical parameters combined with other geologic data suggest that some Franciscan cherts were initially deposited near, but north of, the equator. Other Franciscan cherts were deposited in marginal basins associated with volcanic arcs. Differences in northern and southern Pacific depositional provinces in Late Mesozoic time may explain observed differences between DSDP and orogenic belt cherts.

INTRODUCTION

Detailed sedimentologic and geochemical analysis of an 81-meter measured section of Lower Jurassic to middle Cretaceous radiolarian ribbon chert at Alexander Avenue on the Marin Headlands, central California, formed the basis of a pilot study to test the feasibility of characterizing cherts chemically and of delineating depositional environments.

Reddish-brown, radiolarian ribbon chert represents approximately three percent of the rocks assigned to the Franciscan assemblage in the San Francisco area, according to Schlocker (1974). Associated lithologies include graywacke (80%), shale (10%), mafic volcanic rocks (6%), and occasional limestone, conglomerate, serpentine, and blueschist (1%).

[1]Now at U.S. Geological Survey, 4200 University Drive, Anchorage, AK 99508-4667

Depositional environments for the radiolarites which formed the ribbon chert can be inferred from lithologic associations. The full perspective on the paleoenvironment, however, is elusive due to the extreme deformation of these rocks. Most apparent stratigraphic sections are tectonically dislocated, having been broken into blocks and juxtaposed by faults lubricated by the less competent rocks such as shale and serpentine. This has complicated environmental reconstruction considerably.

Traditionally, another complicating factor has been the problem of the origin of ribbon cherts. Nowhere in the deep sea have such cherts been recognized in cores obtained by the Deep Sea Drilling Project (Winterer and Jenkyns, 1979; Hein and Karl, 1983). Although it is generally accepted that this missing evidence may well be an artifact of the drilling process (chert is very brittle and difficult to core), the fact remains that there is no place geologists can yet point to as an environment where red radiolarian ribbon cherts are forming.

A few outcrops of ribbon chert on the Marin Headlands have been examined in detail. In particular there is a fairly continuous section at Alexander Avenue on the north side of the Golden Gate, 5 km from central San Francisco, (fig. 1) that has been dated by B. Murchey (this volume)[2] and congruently measured, sampled and geochemically analyzed by myself. This section consists of approximately 81 m of ribbon chert, composed of interlaminated red radiolarian chert and red shale, resting depositionally on pillowed greenstone, and conformably overlain by olive-gray shale and medium-grained graywacke sandstone (figs. 2, 3). At the road level, part of the section is covered, but it appears to be reasonably intact from greenstone to graywacke on the ridge above. The 81 m section represents approximately 100 my of fine grained siliceous sedimentation, followed by a fairly abrupt change in depositional regime to shale and sandstone. This study involves an attempt to characterize geochemically the depositional environment of the chert, and to discern whether a geochemical pattern parallels the sedimentary changes in the section. In this chapter I will discuss the evolution of the environment in which these radiolarites were deposited, and were subsequently transformed to chert, and consider the ensuing tectonic implications.

ALEXANDER AVENUE SECTION DESCRIPTION

From the underlying pillowed greenstone to the overlying graywacke, the section at Alexander Avenue consists of 81 m of red radiolarian chert with interlaminated red shale (Fig. 2). The Jurassic base

in Blake, M.C. Jr., ed., 1984, **Franciscan Geology of Northern California:** Pacific Section S.E.P.M., Vol. 43, p. 71-88

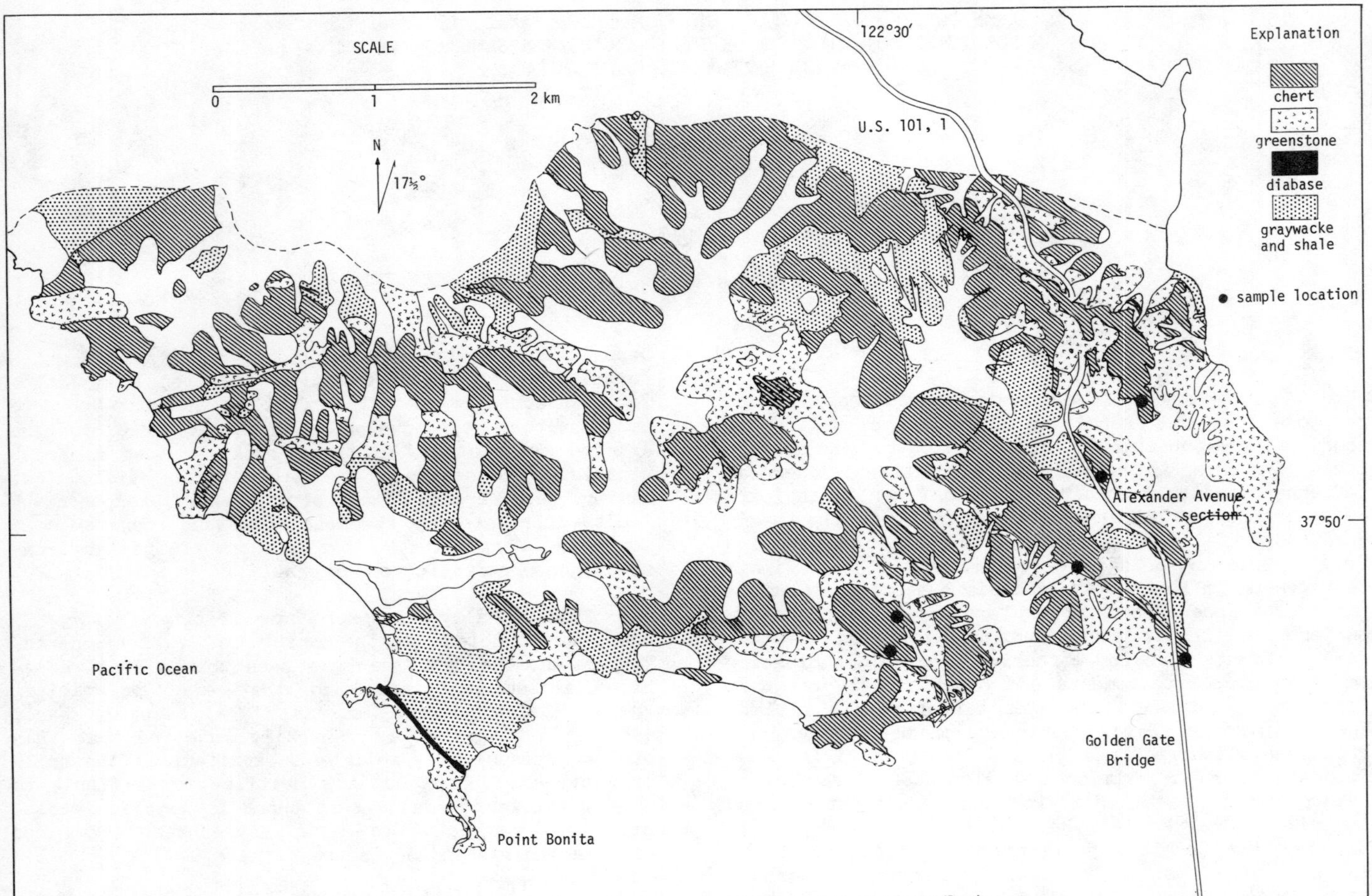

Fig. 1 General geology of the Marin Headlands area. Modified from Bedrossian (1974).

of the section is separated by a covered portion from Cretaceous cherts. The Lower Cretaceous cherts (Zone 5c of Pessagno, 1977) are separated from middle Cretaceous cherts (Zone 10A) at the top of the section by 10 m of quartzose recrystallized ribbon chert.

Chert:shale ratios and bed thicknesses are plotted in figure 4. Generally, with the exception of the recrystallized zones at 12-13 m and 66-76 m, the chert bed thickness and the chert:shale ratio decreases upsection. Chert and shale layers are both lenticular, extending for a few meters, or at most tens of meters before pinching out. Based on composition and sedimentation patterns, the observed section can be broken into 10 parts.

The lowest part of the section consists of 1.5 m of intensely pinched-and-swelled chert layers depositionally overlying pillowed greenstone. The chert is reddish-brown and locally transected by mm-scale quartz veinlets perpendicular to the layering. These veinlets rarely extend beyond a single layer. The shaley layers pinch and swell complementary to the chert layers, having flowed in compensation to the deformation in the chert layers. Chert and shale are both locally stained silvery black by manganese oxides; the shales generally appear blacker because the stain tends to coat partings and fractures. The chert:shale ratio for this portion of the section is approximately 18:1. Average chert bed thickness is 14.5 cm; average shale bed thickness is 0.8 cm.

The second part of the section, 1.5-11.5 m above basement, consists of slightly more than 10 m of chert and shale very similar to the first part except that chert beds are thinner and less deformed, and the chert:shale ratio decreases to approximately 8.3:1 (figs. 5a and b). Average chert bed thickness is 5.1 cm and shale partings average 0.6 cm. There is one contorted layer at 4.4 m, apparently slumped, bounded by parallel layers (fig. 2). Several layers with silty bases which "grade" up into finer grained aphanitic chert may represent reworking or local redeposition. There are three such layers immediately above the contorted layer at 4.4 m.

The third part of the section is just less than a meter thick and consists of several layers of bleached, goethitic, Cu- and Fe-stained, recrystallized quartzose chert. The chert:shale ratio is 2.7:1. Several relatively thick shaley layers alternate with the most recrystallized quartz layers. Average chert bed thickness is 6.2 cm; shale layers average 2.3 cm in thickness.

Above this bleached, Fe-stained zone, from 12.5 to 14.2 m above basement, is 1.7 m of MnO_2-stained beds. This observation is environmentally significant and will be discussed later. The chert:shale ratio for this fourth section is 2.2:1 and average layer thicknesses are 5.0 cm for chert and 2.3 cm for shale.

The fifth section, 14 to 57 m above basement, is divided into two parts by 35 m of covered area. The two parts are very similar in character, and thus

Fig. 2: Sample location on measured chert section at Alexander Avenue, Marin Headlands.

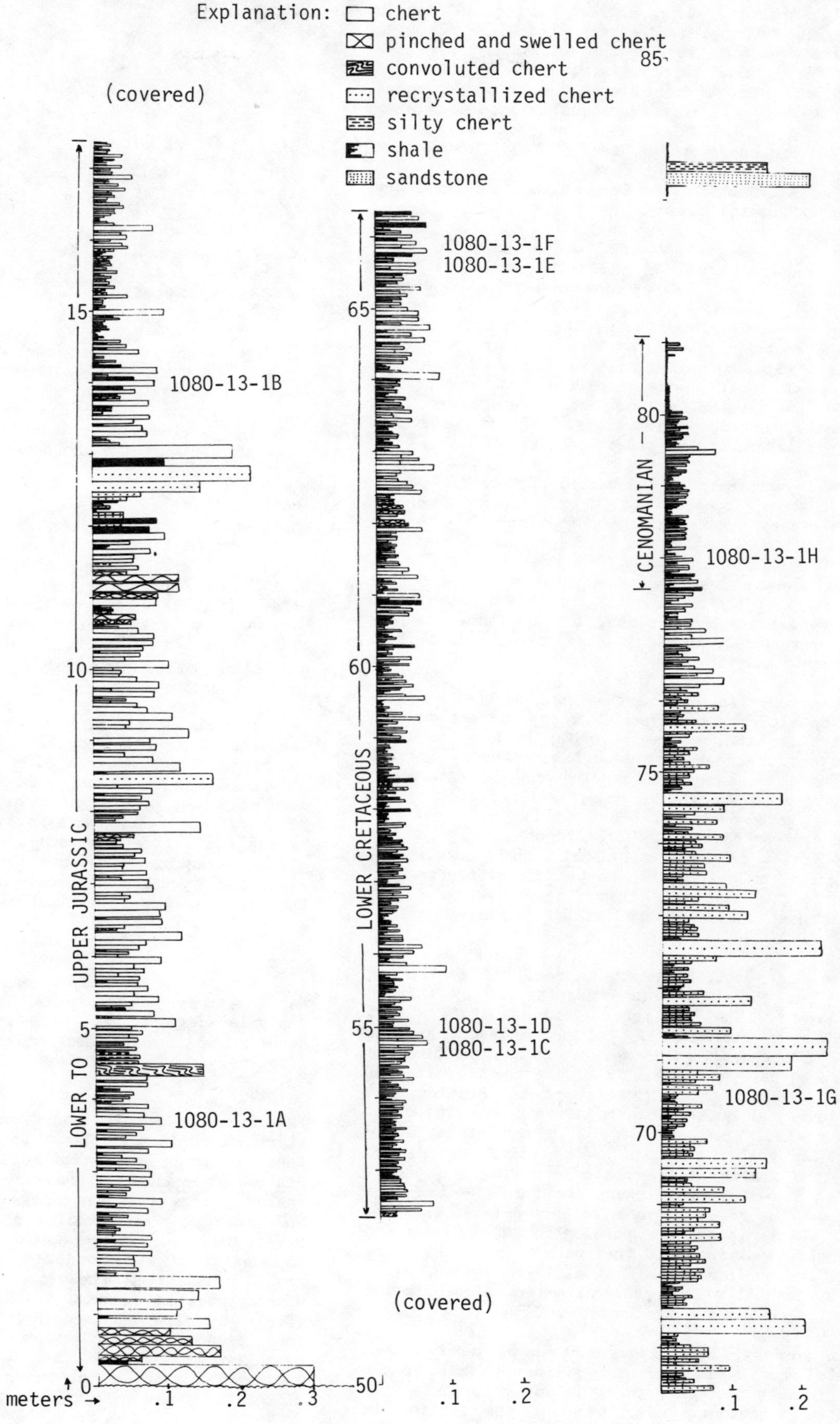

probably represent a single depositional regime. For this section, the chert:shale ratio is slightly higher than sections one through four, but the beds are noticeably thinner. Average chert layer thickness is 1.94 cm and average shale layer thickness is 0.46 cm. From the base to the top of section five the average shale layer thickness increases from 0.31 to 0.54 cm and average chert layer thickness decreases from 2 to 1.9 cm. The chert:shale ratio decreases from 6.4:1 to 3.6:1. The cherts and shales are reddish-brown and conspicuously free of MnO_2 stains after a gradational transition from section four. The cherts in this section tend to be silty, and there are occasional "graded" layers.

The sixth section, from 57 to 65.4 m above basement, is characterized by a still lower chert:shale ratio of 2.3:1, and thicker shaley layers, averaging 1.0 cm in thickness, with chert layer thickness averaging 2.2 cm. This section is also characterized by "mammae", knobs of chert that extend upwards or downwards into adjacent shaley layers. These knobs appear to represent local prelithification readjustments in response to dewatering, compaction processes, and differential overburden stresses. In this section also chert layers may be slightly silty.

The seventh part of the section, 65 to 75 m above the greenstone, consists of 10 m of pale green to buff, bleached, recrystallized, quartzose chert with stylolitic black partings. The chert:shale ratio is 36:1. Chert beds average 3.7 cm in thickness, shale partings average less than a millimeter. The stylolitic nature of the partings is probably a response to the recrystallization process, because they are intensely convoluted, and sedimentary structures tend to be planar or wavy.

The recrystallized part of the section is succeeded by alternating layers of more and less quartzose chert for the next meter and a half of section. Some layers are green, some buff, some red. Several layers have green or buff margins and red interiors, suggesting that the entire recrystallized section may initially have been red chert, and that some diagenetic process reduced the iron in the margins of the beds. Shaley layers are typically red, but may be buff, green, or gray. The chert:shale ratio for this section (eight) is 2.5:1. Chert layers average 2.2 cm in thickness; shaley layers average 0.9 cm in thickness.

The ninth and uppermost part of the section is almost 3 m thick, and is characterized by thinner chert layers. The chert:shale ratio is 12:1. Chert bed thickness averages 1.3 cm; shale layers average 0.1 cm. The chert in this section is also mottled and varicolored buff, green, and red like the transitional section just below. At the top of the section, the chert layers gradually thin to several mm. The chert layers are all green or buff, with dark shaley partings. The shaley partings become silty upward, and grade into grayish olive shale overlying the chert. Forty cm above the last chert are three 1- and 2-cm layers of fine grained graywacke intercalated with the shale. These layers are parallel to the cherts and undeformed, implying that the contact is indeed gradational and conformable. A final 14-cm lens of silty chert at least 6 m long, several meters higher in the section, overlies a 20-cm sandstone layer (fig. 2).

In summary, slightly more than 81 m of interlaminated red chert and red siliceous shale occur depositionally between greenstone and gray shale with graywacke sandstone. The base of the section is characterized by thicker chert layers, moderate chert:shale ratios, and MnO_2 staining. Upward, chert layers become thinner. The middle part of the section appears very red and iron rich, and the top is mottled to buff in color. The color change implies a loss of oxygen in either the sedimentary or the diagenetic environment. The trend of chert:shale ratios and bed thicknesses is a decrease up-section in biogenic deposition relative to red clay, the default sediment in ocean basins.

Fig. 3: Lower Jurassic portion of Alexander Avenue chert section, Marin Headlands. Pillowed greenstone underlying the section is on the right side of the photograph. Lower Cretaceous portion of the section is to the left of the photograph.

The characteristics of the chert noted in the Alexander Avenue section are apparently stratigraphically controlled. Large outcrops of chert are common on the Marin Headlands, though the full section is generally not complete. Dating of Radiolaria by Murchey (this volume) has revealed that cherts of the same age show the same bedding characteristics. The oldest Jurassic cherts are typically manganese-oxide stained and younger Jurassic and Cretaceous cherts are typically thinner bedded and more clay-rich.

PETROGRAPHY

A total of 18 samples were collected from well-dated outcrops on the Marin Headlands for geochemical analysis (figs. 1, 2). On Alexander Avenue, samples were collected from sections 2, 4, 5, 6, 7, and 9 (fig. 4).

In thin section, as well as in outcrop, certain features appear to be characteristic of certain stratigraphic horizons. Cherts from section 2 and equivalent levels at other outcrops, (less than 4 m from pillowed greenstone,) are very dark brownish-red; the shaley layers between them are commonly black with MnO_2 stains. Thin sections show sparse Radiolaria in a very dark, almost opaque matrix. The matrix appears to consist of amorphous clots of the dark material, probably hematite. These clots coat grain surfaces of microcrystalline quartz. The quartz grains in the matrix are finer grained than the quartz grains which form pseudomorphs after radiolarian tests. Both nassellarian (conical) and spumellarian (spherical) radiolarian tests of different sizes are evenly distributed in the matrix. Test rims are composed of

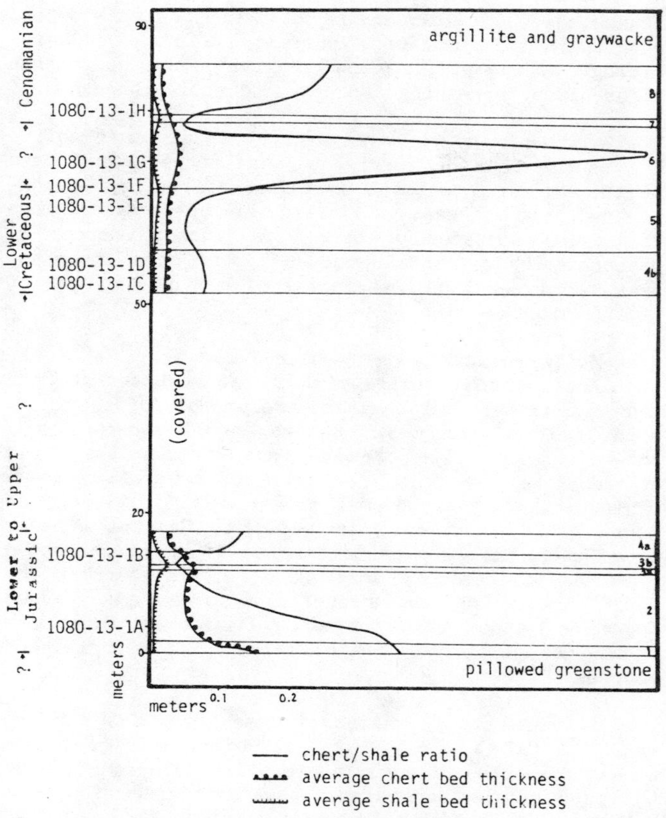

Fig. 4: Chert/shale ratio and average bed thickness at measured section, Alexander Avenue, Marin Headlands.

microgranular quartz, while their interiors are filled with either microgranular quartz or chalcedony. The chalcedony occurs as a single radiating cluster of fibers, with the node attached to the test inner wall. Occasional Radiolaria are filled with murky hematite and rarely with a green clay mineral. In some sections Radiolaria are flattened, but they are not broken. In the sections with flattened Radiolaria, there is also an anastamosing planar fabric in the matrix. This fabric is displaced around the Radiolaria, and thus apparently postdates test recrystallization.

Fine-grained crystals of hematite and magnetite are disseminated in the matrix of some thin sections. Chlorite, prehnite, and feldspar are confined to quartz veinlets. The oldest veinlets are consistently those which are perpendicular to bedding. There are commonly at least three generations of quartz veinlets. XRD mineralogy determinations of veinless chert by Lisa Morgenson, Marla Wilson, and James R. Hein of the U.S. Geological Survey include quartz and hematite and/or pyrite. No calcite was identified in thin-section or XRD analysis.

Samples from section 3 contain more abundant Radiolaria than the samples from lower in the section, but the matrix and mineralogy are the same. Samples from section 4 are 14 or more meters above basement. The sample from Alexander Avenue is in the MnO_2--stained chert more than a meter above the recrystallized zone. The Radiolaria are more abundant than lower in the sections, but still "float" in a murky hematitic matrix. Locally, the murky material is brecciated in a siliceous matrix. Reddish or brownish bands form discontinuous, mm-thin color laminations in the chert. These laminations may extend laterally for a few centimeters and appear to be independent of bedding. In thin section, the color changes are gradational and due to the relative abundance of murky hematitic material. Radiolarian distribution is independent of the banding; tests often straddle color banding contacts.

Accessory minerals include finely disseminated pyrite, hematite, magnetite, and green clay. Chlorite and pumpellyite are associated with some quartz veins. XRD analyses identified quartz, hematite and/or pyrite, and mixed-layer clays (smectites) for these rocks and for all the rest of the Marin cherts.

Samples from section 5 are more than 50 m above basement. This part of the section is characterized by thinner layers, common pinching and swelling of layers, nodular "mammae", some silty chert layers, some "graded" silty chert layers, lower chert:shale ratios, and the lack of MnO_2 stains. In thin section, sponge spicules and siliceous test fragments accompany well-preserved whole radiolarian tests. Some tests are filled with matrix material, but most contain either granular microcrystalline quartz or chalcedony. The matrix is considerably less hematitic and murky than the Upper Jurassic cherts. The matrix contains abundant fine-grained greenish platy minerals which are probably the smectites identified by x-ray analysis. The matrix is also more recrystallized and the radiolaria generally more poorly preserved than lower in the section. This is probably due to diagenetic processes, because studies by other workers (Heath and Moberly, 1971; Lancelot, 1973; Keene, 1976; Kastner and others, 1977) have demonstrated that chert is produced by a solution-redeposition process in which opaline tests of microfossils are the primary source of silica. Tests that recrystallize to quartz earlier in the process escape total annihilation, but clay minerals can retard the diagenetic reaction rate (Kastner and others, 1977; Isaacs, 1979), and tests are less likely to be well-preserved in clayey sediments. Accessory minerals in section 5 include chlorite, pumpellyite, prehnite, stilpnomelane, and epidote-group minerals. Along with some quartz, these commonly form veins, but may occur as independent neomorphic crystals in the chert and siliceous shale as well.

Samples from section 6 are 15 or more meters above samples from section 5. Thin sections are very similar to those from section 5. Laminations within chert layers are defined by proportions of Radiolaria and radiolarian debris to matrix, and by relative amounts of hematite in the matrix. Spicules and Radiolaria in many states of preservation are not size-sorted, and are very unevenly distributed. The matrix is streaky and inhomogeneous. Green platy minerals are common. Hematitic material tends to coat grain boundaries and varies locally in concentration, suggesting mobilization by diagenetic processes. Some quartz veins cut and offset radiolarian tests. Radiolarian tests that are not filled with granular quartz or chalcedony are filled with clay minerals rather than hematitic material. The same metamorphic minerals noted in section 5 are present but very rare.

The sample from section 7 is 70.5 m above the greenstone. This section is composed of intensely recrystallized quartzose chert, commonly with stylolitic partings between layers. A thin section from a red and green mottled layer shows that the color change is gradational and patchy. Radiolarian tests often straddle color boundaries. The reddish areas are hematitic and murky; the greenish areas are composed of finely interlocking quartz grains generally without associated clay minerals. More coarsely crystalline Radiolaria are scattered unevenly through the matrix.

The sample from the Cretaceous part of the section, 78 m above basement, is internally laminated and slightly recrystallized. Clay minerals are rare. Laminations are distinguished by relative abundances of hematitic material and Radiolaria. Faint micro-cross laminations are defined by radiolarian distributions and may be a sedimentary feature. Distribution of hematitic material does not conform to these micro-cross laminations, and thus color banding is probably the result of a different (diagenetic) process. Radiolaria straddle color banding contacts, as is the case with other samples lower in the section.

GEOCHEMICAL DATA

Methods of Analysis

From each sample, 150 grams of clean, non-veined, non-stained, non-weathered chert were crushed in a tungsten carbide mill to 200 mesh. The powder was homogenized and split. One split was analyzed for 10 major elements by x-ray fluorescence analysis, and Fe_2O_3/FeO was determined by wet-chemical methods. The other split was analyzed by quantitative, direct-reader emission spectroscopy for 33 elements, including the 10 major elements. Analyses were performed by the USGS Branch of Analytical Laboratories. Results of the two methods of analysis generally agree within 5 percent. Analyses of replicate samples agree within a few ppm, which is less than analytical error.

Trends in Element Abundance

General trends in the raw element composition data (Table 1) do not show striking changes up-section, but the results are geologically reasonable in terms of the observed characteristics of the section and the samples analyzed.

Silica shows no systematic behavior. All samples contain greater than 90 percent silica. The sample highest in silica is from the highly recrystallized, quartzose part of the section from 75 to 75 m above the greenstone. The sample lowest in silica is of Early Jurassic age and in a metal-enriched zone above a thin recrystallized zone. Total iron generally decreases up-section. FeO is present only in negligible amounts, and shows no systematic pattern. Alumina tends to increase up-section. Magnesium is richest in the middle of the section. Calcium, sodium, and potassium are scarce and variable. Titanium generally increases up-section. Phosphorus is below the detection limit. Manganese decreases up-section.

Only those minor elements found to be useful in statistical analysis and interpretation are listed in Table 1. Of these, Ba is relatively abundant in the middle of the section. Cr and Cu tend to parallel Ba in abundance. Ni is quite variable. Y increases up-section. Zr is variable, but lowest at the base of the section.

Aluminum to Metal Ratio

Important sedimentary sources in the Pacific Ocean basin include terrigenous detritus, biogenic debris, hydrothermal precipitates, and authigenic precipitates (Heath and Dymond, 1977; Leinen and Stakes, 1979). Which component dominates will depend on basin configurations, current patterns, climate, biologic productivity, and proximity to topographic highs, lows, spreading centers, and active trenches.

Although the observed geology indicates a definite change from no terrigenous influence at the base of the section, to total dominance at the top (demonstrated by increasing Al and Ti up-section), the relative importance of other sedimentary sources is obscure. Hydrothermal sources appear more important near the greenstone, as indicated by the decrease of Fe and Mn up-section.

Most previous workers have found that Al can be attributed almost entirely to a detrital source, and that aluminum to metal ratios are useful in discriminating sediment sources. Bostrom and Peterson (1969) showed that a value of less than 0.4 for the ratio Al/(Al + Fe + Mn) is indicative of transition metal enrichment by hydrothermal processes. Steinberg and Marin (1978) applied this ratio to Mesozoic cherts from Greece as well as to samples of chert collected from various Mesozoic orogenic belts elsewhere in the world, including the Franciscan assemblage. They found that the ratio increased for depositional environments inferred to be closer to terrigenous sources.

In this study, values of less than 0.39 for the ratio Al/(Al + Fe + Mn) are found only for the Jurassic samples directly above pillowed greenstone at the base of the Marin section. The values for this ratio are plotted in Figure 6 against TiO_2, chosen because it is also considered to have a detrital source (Bostrom and others, 1971; Debrabant and others, 1980; McMurtry and others, 1981; Baltuck, 1982). Thus TiO_2 is in fact dominantly of detrital origin, as confirmed by its linear relationship with the Al ratio. (fig. 11). The other samples appearing on this plot are Franciscan and Great Valley sequence cherts shown for the sake of comparison. The Geysers cherts are considered to be analogous to Alexander Avenue (McLaughlin and Pessagno, 1978; Murchey, 1980; McLaughlin and Ohlin, this volume; Murchey, this volume). The Stonyford cherts rest on pillowed high-Ti greenstone considered to represent a seamount (Hopson and others, 1980) of probable Franciscan affinity (MacPherson, 1983). Three samples of chert from the base of the Great Valley sequence, one of which is from the Geyser Peak section of the Coast Range ophiolite (McLaughlin), are also plotted.

Several conclusions can be drawn from figure 6. The most obvious is that TiO_2 values are low in the samples that were deposited on the high-Ti Stonyford basalt. Samples with high TiO_2 values also tend to have high values for the Al/(Al + Fe + Mn) ratio. These samples are at the top of the Geyser Peak section, and in the Great Valley sequence, all directly beneath a pronounced transition to sandstone which indicate the presence of a nearby source of detrital sediment. Transition metal enrichment is characteristic only of Marin and Stonyford basal cherts which show no evidence of terrigenous detritus in outcrop, in thin section, or in XRD analyses.

Element Partitioning

Heath and Dymond (1977), and Dymond (1981) combined equations based on compositions of sediments from closely sampled single source environments on the Nazca plate into a single equation for partitioning elements among the various sources. Leinen (1981)

Fig. 5a: Transition from thicker, MnO2-stained chert beds to thinner, MnO_2-free beds in part 2 of the Alexander Avenue section, Marin Headlands. Hammer handle is 33 cm long.

Fig. 5b: Example of thinning upward cycles, including "graded" chert beds with silty bases at Alexander Avenue section, Marin Headlands. Bed tops are toward left of photograph. Scale is a hand lens on a shoestring.

Table 1. Geochemical Analyses of Marin Cherts

Sample	Age	Description	Si02%	Al203%	Fet03%*	Fe0%	Mg0%	Ca0%	Na_2O%	K_2O%	TiO_2%	P_2O_5%	Mn0%	Bappm	Crppm	Cuppm	Nippm	Yppm	Zrppm
1080-13-1H	middle Cretaceous	earthy dk. red chert	95.4	1.16	1.03	0.08	0.20	0.07	<0.15	0.32	0.04	0.06	0.07	710	<10	17	24	17	42
1080-13-1G	Early Creta- middle Cretaceous	gray and red banded chert	97.8	0.73	0.49	0.10	0.12	0.07	<0.15	0.20	0.03	<0.05	0.03	460	<10	16	8.0	<10	28
1080-13-6A	Early Cretaceous	brown chert	92.8	1.97	1.67	0.09	0.36	<0.02	<0.15	0.47	0.13	<0.05	0.03	2500	15	51	9.7	<10	[illegible]9
1080-13-5A	Early Cretaceous	brownish red chert	95.8	1.14	0.68	0.08	0.19	0.14	<0.15	0.32	0.04	0.06	0.08	450	<10	49	21	15	39
1080-13-5B	Early Cretaceous	brown chert	92.1	1.89	1.61	0.10	0.47	0.02	0.16	0.47	0.11	<0.05	0.02	3000	14	90	13	14	53
1080-13-5R	Early Cretaceous	brown chert	93.3	1.88	1.64	0.13	0.43	0.03	<0.15	0.53	0.12	<0.05	0.02	2600	14	110	12	15	55
1080-13-1E	Early Cretaceous	earthy red chert	94.8	1.60	1.13	0.18	0.39	0.03	<0.15	0.44	0.08	<0.05	<0.02	880	12	12	6.8	<10	41
1080-13-1C	Early Cretaceous	brown chert	96.7	0.93	0.56	0.18	0.29	0.05	<0.15	0.18	0.06	<0.05	0.05	1900	11	36	15	<10	32
1080-13-1B	Early Jurassic	brick red chert	90.5	1.92	3.48	<0.01	0.54	<0.02	<0.15	0.85	0.11	<0.05	0.69	300	11	84	12	<10	63
1080-13-3A	Late Jurassic(?)	brown chert	92.3	1.88	2.09	0.22	0.56	0.06	<0.15	0.53	0.11	<0.05	0.09	1000	14	36	12	10	60
1080-13-2	Late Jurassic(?)	red-brown earthy chert	96.0	0.73	2.39	0.06	0.22	<0.02	<0.15	0.29	0.03	<0.05	0.12	250	<10	28	8.8	13	39
1080-13-4	Late Jurassic(?)	red-brown chert	95.8	0.35	1.91	0.08	0.14	0.08	<0.15	0.16	<0.02	<0.05	0.16	230	<10	19	12	10	27
1080-13-1A	Early Jurassic	earthy dk. red chert	95.3	0.28	2.77	0.13	<0.10	<0.02	<0.15	0.08	<0.02	<0.05	0.15	290	<10	20	19	11	32
1080-13-6B	Early Cretaceous	red-brown sili- ceous shale	67.5	13.3	7.30	0.70	1.74	0.03	<0.15	3.91	0.76	<0.05	0.04	740	91	180	38	30	210
1080-13-1F	Early Cretaceous	brown siliceous shale	59.6	16.2	9.02	0.86	2.46	0.25	0.63	5.01	0.85	0.08	0.06	870	79	88	41	27	270
1080-13-1D	Early Cretaceous	brown siliceous shale	64.7	12.3	9.00	0.61	2.33	0.21	0.31	4.06	0.84	0.06	0.09	830	68	160	63	27	250
1080-13-3B	Late Jurassic(?)	red-brown sili- ceous shale	65.6	11.2	10.7	1.50	2.86	0.38	0.90	3.95	0.80	0.18	0.27	1300	60	160	65	35	280

* Fet03%: total iron as Fe+3

applied this method to DSDP Leg 63 sediments and found that the coefficients used in Dymond's (1981) equation required slight adjustment for total closure for analyses of sediments from individual areas in the Pacific basin. The equation is as follows (Dymond, 1981):

$$E_T = (E/Al)_D D_{Al} + (E/Fe)_H H_{Fe} + (E/Si)_B B_{Si} + (E/Ni)_A A_{Ni} + (E/Ba)_R R_{Ba}$$

where: ET = total content of an element in a carbonate- and salt-free sample
D = detrital component of alumina for the sample
H = hydrothermal component of iron
B = biogenic component of silicon
A = authigenic component of nickel
R = dissolution residue component of barium
$(E/Al,Fe,Si,Ni,Ba)_{D,H,B,A,R}$ = the elemental ratio for the element in a pure detrital, hydrothermal, biogenic, authigenic, or dissolution residue component, respectively. The adjusted coefficients used for this study are listed in Table II.

The validity of the partitioning equation was tested by factor analysis on the same Nazca data (Heath and Dymond, 1977; Leinen and Stakes, 1979; Dymond, 1981; Leinen and Pisias, 1982) because the multivariate statistical analysis was thought to be an objective method of estimating elemental interrelationships. The partitioning equation was found to provide geologically reasonable results in good agreement with the results of factor analysis.

Leinen (1981) arrived at two important conclusions from application of the Nazca plate-derived equation to Leg 63 samples west of Baja California: 1) "simple sedimentary components with uniform compositions are adequate to describe most of the variability in deep-sea sediments deposited within a single sedimentary province, and 2) the composition of these sedimentary components within the northeast Pacific basin has not changed much in the last 50 million years."

These two conclusions justify the assumptions necessary to apply Dymond's partitioning equation to the Marin Headlands chert section. The necessary assumptions made for this study are that 1) during the 100 my represented by the Marin cherts the sedimentary source components in this region of the Pacific maintained a uniform composition, and 2) the sedimentary source components contributing to this basin can be discriminated and isolated by simple elemental relationships, such as were determined for the Nazca and Leg 63 sediments. Results of many DSDP studies (Keene, 1976; Lancelot, 1973; Heath, 1974; van Andel and others, 1975; Heath and Dymond, 1977; Leinen and Stakes, 1979) suggest that it is geologically reasonable to assume that the same sedimentary sources were contributing to the Pacific basin in Mesozoic time as are currently in operation. Therefore we can apply the equation without adding or deleting source components.

The coefficients themselves involve more intricate justification. Dymond's (1981) coefficients are based on actual end-members identified within the Nazca depositional province. For the northeast Pacific depositional province Leinen (1981) found it necessary to adjust several of the coefficients. For the Mesozoic cherts we do not have many clues regarding lateral facies equivalents and end-member sources contributing to the province in which they were deposited. In fact, we are not even sure where they were deposited.

How far did these cherts travel? Whether the Marin radiolarites were deposited on the Kula, the Farallon, or the Pacific plate, paleo-pole reconstructions indicate that they would have been transported in a generally northerly, northwesterly, or northeasterly direction since middle Mesozoic time (Larson and Pitman, 1972; Stone, 1977). Paleomagnetic determinations (Curry and others, this volume) on Cenomanian siltstone in the Marin Headlands suggest a minimum northward displacement of about 7°, but paleomagnetic determinations from chert in the Marin Headlands do not pass a fold test or show polarity reversals. Murchey (this volume) believes the Marin Headlands radiolarian faunas have greater affinities to low latitude faunas than to higher latitude faunas such as those in the Great Valley sequence. Thus there is a strong possibility that the Marin cherts were deposited south of their present position relative to the North American craton.

Evidence (Tj. H. van Andel and J. G. Sclater, unpublished data) indicates that middle to late Mesozoic sediments in the equatorial zone were rich in calcareous ooze. However, in the Marin sections, limestone is strikingly lacking. If we constrain these radiolarites to a depositional site beneath an area of high siliceous biogenic accumulation, the Marin radiolarites could initially have been deposited north or south of the equator, in an area characterized by siliceous deposits, (due either to primary deposition or depth-dependent dissolution of $CaCO_3$,) such as west of Baja California or on the northern part of the Nazca plate. If they were deposited south of the equator, the site could already have subsided beneath the CCD when it crossed the equator. If the CCD was elevated with respect to its present position, it would serve to enhance the lack of carbonate accumulation.

Because Hopson and others (1980) suggest slow-spreading rates for Franciscan oceanic crust, because translation vectors probably included eastward or westward components rather than pure northward motion (Stone, 1977), because calcium carbonate is entirely lacking from the section, and because siliceous sedimentary depositional features are not mirrored or repeated up-section, a site north of the equator in the northeastern Pacific depositional province is favored for initial deposition of the Marin radiolarites.

Leinen's (1981) Tertiary northeastern Pacific province coefficients (Table II) approximate average Pacific clay (Scott, 1981) and average hydrothermal sediment (Leinen, 1981). The partitioning model was also found to be effectively insensitive to changes in the amount of elements in biogenic debris (which would be altered considerably during diagenesis), (Leinen, 1981). Therefore the coefficients are reasonable and adequate to a first approximation for application to upper Mesozoic cherts that probably originated in the same province.

Dymond's (1981) partitioning equation solves 8 equations simultaneously for 5 variables through linear programming (Simmons, 1972, and Narula and Wellington, 1977, in Dymond, 1981). Since the program for this solution was not available to the author at the time of this writing, an earlier version, outlined

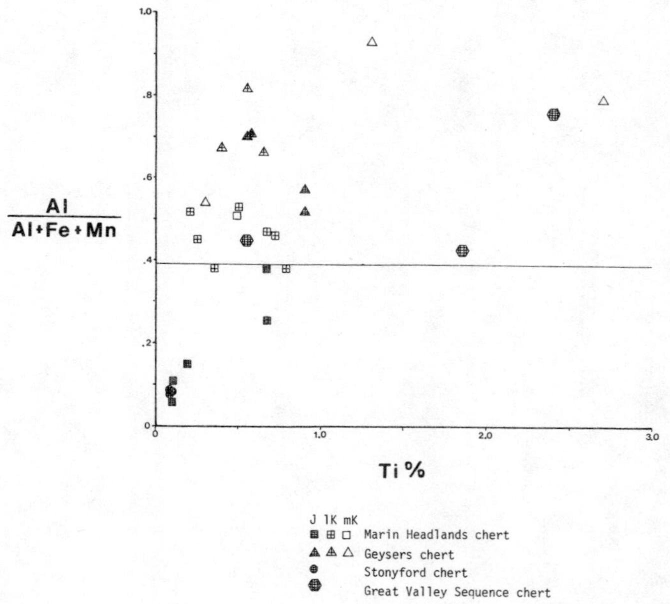

Fig. 6: Al to metal ratio plotted against Ti. Horizontal line at 0.391 distinguishes purely pelagic deposits from those influenced by terrigenous detrital material (Bostrom and Peterson, 1969). All Jurassic, and some Cretaceous samples, from Marin Headlands and Stonyford plot in the purely pelagic realm.

by Heath and Dymond (1977) was employed to partition the sediment source components for the Marin cherts. Dymond's 1981 version adds considerable precision to the results of the 1977 version, but given the assumptions and approximations discussed above, does not significantly alter or improve the conclusions from application of the 1977 method. Obviously, geochemical partitioning of Marin cherts can only be a first approximation of the Mesozoic sedimentary component relationships, regardless of the precision of the solution to the partitioning equation.

The results of partitioning the geochemical data for the Marin cherts are plotted on figures 7 through 15. The partitioning method is successful in clarifying trends through multiple element relationships in each sample that are not obvious in plots of single element variation from the raw data of Table 1. Figure 7 shows the amount of error introduced by changing the composition of the sediment-source end member from which the component coefficient is derived. The coefficient for average Pacific clay (Scott, 1981) and a combination of felsic and intermediate detritus (Leinen, 1981) and the coefficient for mafic detritus (computed from average EPR basalt) were used to determine the hydrothermal Fe component for the Marin chert samples. The discrepancy can be envisioned as defining error bars for the composition of systematically decreasing hydrothermal iron up-section. Waning influence of a hydrothermal sediment source for Fe can be expected from the observed geology, but the influx of Fe-bearing detrital material towards the middle and upper part of the section compensates for this effect and thus both sedimentary patterns are obscured in the raw data. The results of partitioning detrital and hydrothermal iron agree well with the results of applying Bostrom and Peterson's (1969) aluminum to metal ratio to the Marin cherts.

Figure 8 is a triangular plot of the raw data for K, Mg, and total Fe. This plot shows an increase up-section in K, and a slight increase in Mg, relative to total Fe, up-section for the Marin cherts. If Leinen's (1981) conclusion that the detrital component probably maintains a uniform composition over a 50 my timespan is correct, and K is assumed to be effectively purely detrital (residing in eolian plus fluvial illites), then the plot suggests more than one source for Fe. The same conclusion can be drawn for the Geysers chert data. If the same plot is recast with detrital rather than total Fe and Mg (fig. 9), the trend is reversed. Detrital iron shows a slight increase relative to potassium up-section for both the Marin and Geysers cherts. This can be explained if eolian illite is held constant through the time represented and a detrital component from terrigenous fluvial systems rich in chlorite (arc derived, intermediate detritus as represented by overlying graywackes) is gradually introduced above the base of the chert section. This hypothesis is supported by the absence of terrigenous detritus observed at the base of the Marin and Geysers chert sections.

Titanium has been used by geochemists dealing with altered and metamorphosed rocks as a stable, immobile element which reflects original rock compositions (Pearce and Cann, 1973; Garcia, 1978; Hellman and others, 1979). Because Ti is not significant directly above Ti-rich basalts at Stonyford, it was assumed that the TiO_2 analyzed for the cherts in this study reflected a detrital component other than the substrate. Since Ti has not been considered significant in hydrothermal, biogenic, authigenic, or residual sediments (Bostrom and others, 1971; not analyzed by Dymond, 1981), possibly due to the low solubility of Ti with respect to the fluids from which hydrothermal deposits precipitate (Bostrom and others, 1971), it follows that the Ti measured must be attributable to a terrigenous (including volcanic arcs) source component, as is confirmed by the linear relationship of TiO_2 with detrital Fe (fig. 10) and Al (fig. 11).

Titanium, plotted against two other elements generally considered to be representative of original lithologies through alteration and low grade metamorphism, Y, and Zr (Pearce and Cann, 1973; Hellman and others, 1979), is often used to discriminate between tectonic provinces for basalt. The contribution of the basalt substrate in the form of detrital material in the cherts was tested on a ternary plot designed for basalts by Pearce and Cann (1973). This plot for the cherts (fig. 12) shows an increase in Ti relative to Zr up-section at Marin, and an increase in Y relative to to Ti up-section at Geysers. As already mentioned, Stonyford cherts are remarkably lower in Ti than any of the other cherts. The plot shows some things about the composition of the cherts, but nothing about the underlying basalt. Points fall far

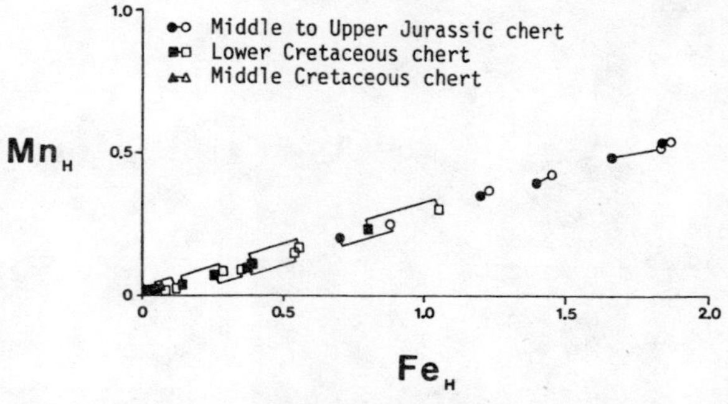

Fig. 7: Plot showing decrease in hydrothermal Fe and Mn components with decreasing age for Marin Headlands chert. Plot also compares the effect of using average Pacific pelagic clay value (open symbols) to using average EPR basalt value (closed symbols) for the Fe_D partitioning coefficient.

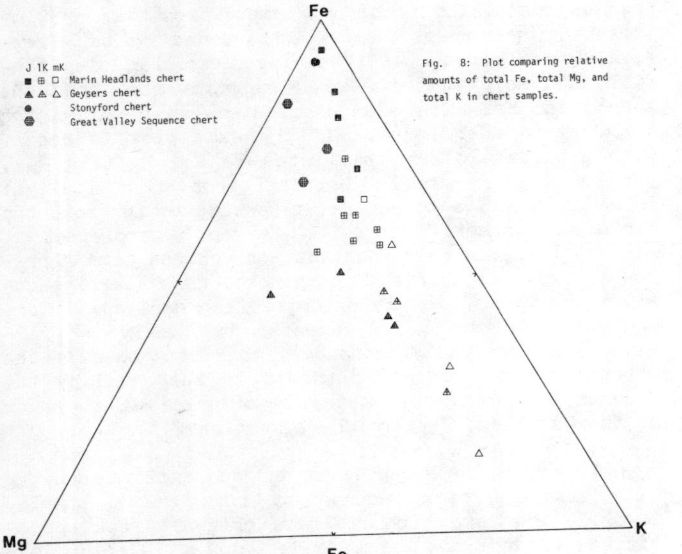

Fig. 8: Plot comparing relative amounts of total Fe, total Mg, and total K in chert samples.

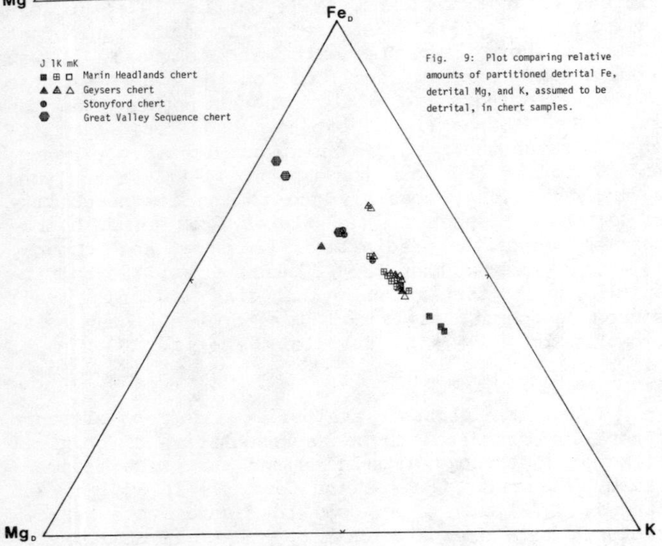

Fig. 9: Plot comparing relative amounts of partitioned detrital Fe, detrital Mg, and K, assumed to be detrital, in chert samples.

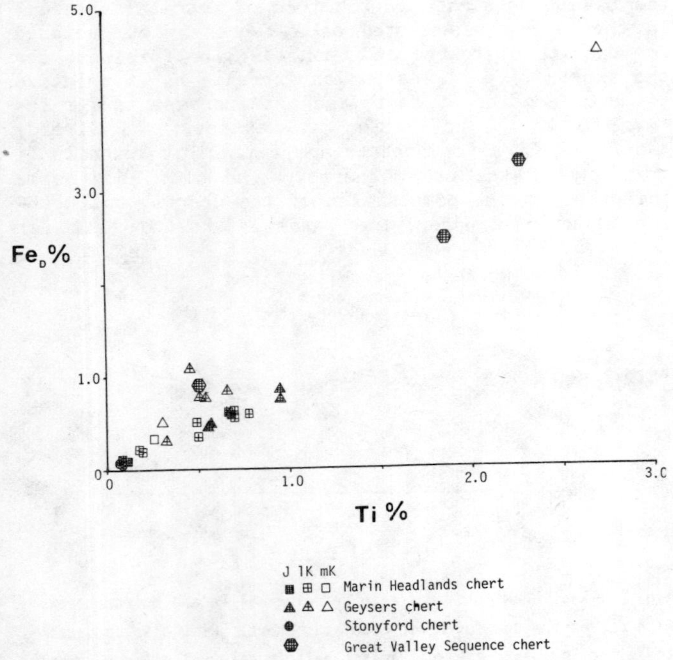

Fig. 10: Detrital Fe plotted against Ti. Linear relationship suggests a similar source for these elements.

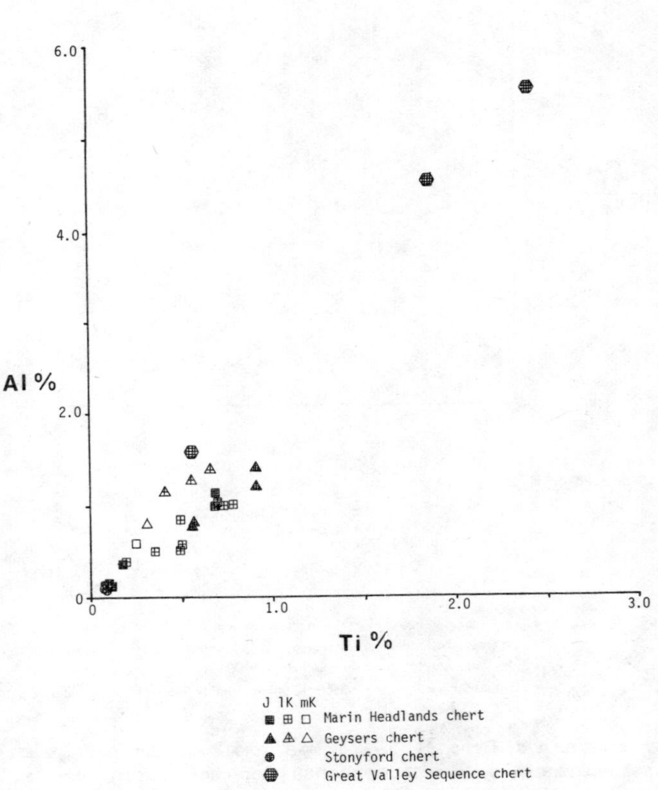

Fig. 11: Al plotted against Ti. Linear relationship suggests a similar source for these elements.

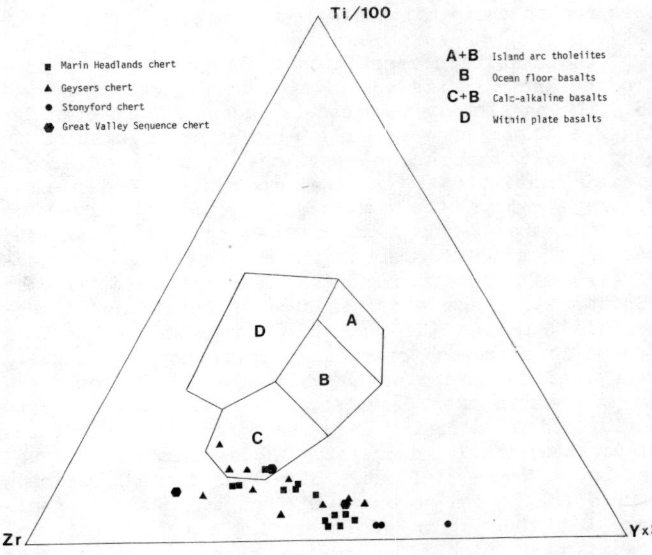

Fig. 12: Plot of Ti/100, Zr, and Y x 3, after Pearce and Cann, (1973). Changes in relative proportions of Ti, Zr, and Y demonstrate that these elements have more than one source for these chert samples.

Table II: Coefficient Array for Linear Programming

Element	Detrital X/Al	Hydrothermal X/Fe	Biogenic X/Si	Authigenic X/Ni	Dissolution Residue X/Ba
Al	1.000000	0.006000	0.005000	1.400000	0.500000
Si	2.600000	0.130000	1.000000	3.700000	0.000000
Mn	0.016000	0.290000	0.000023	30.000000	0.007000
Fe	0.600000	1.000000	0.001000	14.000000	0.350000
Ni	0.001500	0.000900	0.000040	1.000000	0.006500
Cu	0.000700	0.004200	0.000050	0.460000	0.016000
Zn	0.001500	0.001900	0.000080	0.093000	0.004000
Ba	0.010000	0.005000	0.000500	0.186000	1.000000

Weighting: Si=1, Al=300, Mn=300, Fe=300, Ba=300, Ni=1000, Cu=1000, Zn=1000

outside of Pearce and Cann's (1973) basalt provinces. Different sources apparently exist for the three elements relative to each other at either section, but the substrate is apparently the least important source.

Figures 13, 14, and 15 illustrate some insights provided by the partitioning analysis and also some drawbacks of the 1977 method relative to the 1981 version. In spite of the drawbacks, consistent patterns can be recognized and qualitative conclusions can be drawn.

In the 1977 partitioning method, detrital elemental components are subtracted from the total elemental composition first. Subsequently, hydrothermal components are subtracted, next authigenic components, and finally the biogenic components. No dissolution residual component is accounted for. The original equation (Heath and Dymond, 1977) appears as follows, with variables defined as in equation (1):

$$E_T = (E/Al)_D D_{Al} + (E/Fe)_H H_{Fe} + (E/Ni)_A A_{Ni} + EB \quad (2)$$

Any amount of an element not partitioned into the first 3 components was assumed to be biogenic. The partitioning order is supported by objective factor analysis, which assigns relative weights to the components (Heath and Dymond, 1977; Leinen and Stakes, 1979).

Figure 13 shows non-detrital Mn plotted against hydrothermal Fe. The line represents the ideal relationship in which all the Mn which is not detrital is hydrothermal. Thus samples plotting above this line can be interpreted in 3 ways: 1) there is a certain amount of authigenic ± biogenic Mn in the sample, 2) the hydrothermal component of Mn in these samples is enriched relative to average compositions of hydrothermal precipitates, or 3) the partitioning coefficients used are inappropriate for these samples. Samples below the line could be depleted in Mn, but since one of these samples (1A) was collected from a Mn-rich part of the Marin section, it seems more reasonable to conclude that the partitioning coefficients are inexact. Hydrothermal Fe may be overestimated due to lack of partitioning into the hydrogenous, biogenic, and dissolution residue components. The difference between the 1981 and the 1977 partitioning methods accounts for this inadequacy

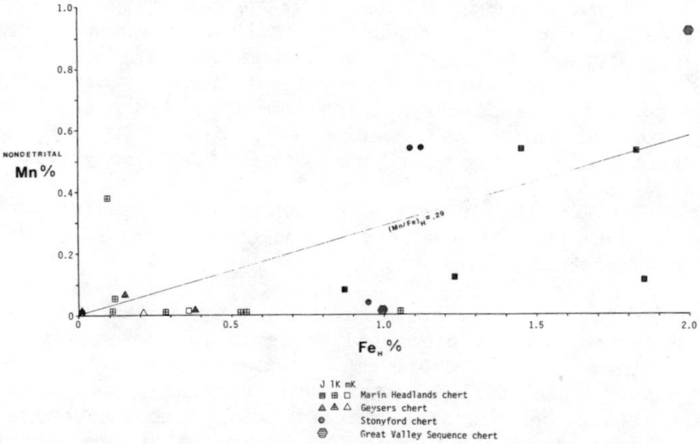

Fig. 13: Scatter plot of nondetrital Mn against nondetrital Fe (hydrothermal Fe) according to the Heath and Dymond, (1977), model. Scatter above the ideal hydrothermal sediment ratio (0.29) may be attributed to a hydrogenous sediment source component. Scatter below the hydrothermal ratio may be a consequence of overestimation of hydrothermal Fe due to lack of Fe partitioning into hydrogenous, biogenic, and residual components.

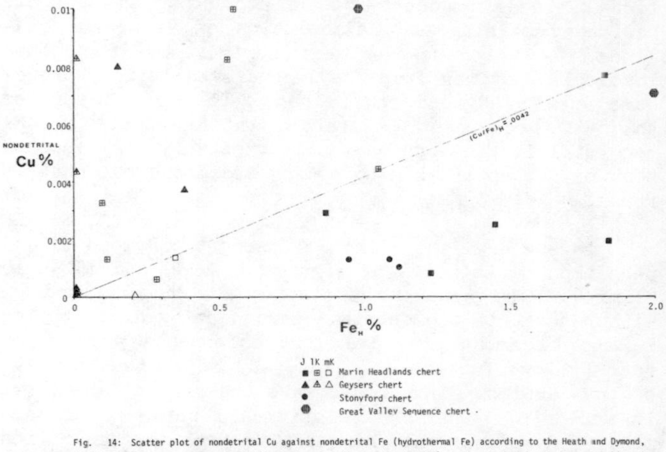

Fig. 14: Scatter plot of nondetrital Cu against nondetrital Fe (hydrothermal Fe) according to the Heath and Dymond, (1977), model. Scatter above the ideal hydrothermal sediment ratio (0.0042) may be attributed to hydrogenous, biogenic, and dissolution residue sediment source components. Scatter below the hydrothermal ratio may be a consequence of overestimation of hydrothermal Fe due to lack of Fe partitioning into hydrogenous, biogenic, and residual components.

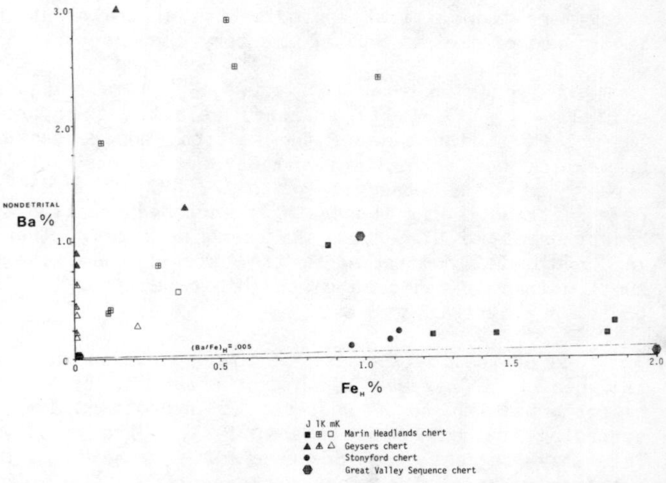

Fig. 15: Scatter plot of nondetrital Ba against nondetrital Fe (hydrothermal Fe) according to the Heath and Dymond, (1977), model. Scatter above the ideal hydrothermal sediment ratio (0.005) may be attributed to hydrogenous, biogenic, and dissolution residue source components. The low ratio partitions most Ba into hydrogenous, biogenic, and residual source components.

without changing the relationship between samples.

There are 3 important improvements in the 1981 version: 1) The equations used to solve the partitioning equation are solved sequentially by the 1977 method, and simultaneously by the 1981 method, which allows for a more objective distribution of any element relative to the other elements. 2) In the 1977 method, all Al, Mn, and Fe, were assumed to be either detrital or hydrothermal. This assumption was based on the effective absence of Fe or Al in biogenic debris, and the derivation of authigenic Al and Fe from hydorthermal or detrital elements rather than adsorbtion from sea water. Although Leinen and Stakes (1979) found that Cu, Zn, Ni, Al, Fe, and Mn were incorporated in biogenic silica, it is also known that the recrystallization process expels impurities during chert formation (Keene, 1976; Kastner and others, 1977). By the 1977 method, the expelled impurities in Marin cherts would be assigned to the authigenic component. Furthermore, it does not seem geologically reasonable to expect that bottom sediments would acquire authigenic metals such as Ni from pore fluids or sea water and not acquire Fe, Al, and Mn. The 1981 solution allows for the partitioning of Fe, Al, and Mn among 5 rather than 2 components,thus partitioning less Mn into the detrital component and lowering hydrothermal Fe values. As a result, all of the samples in Figure 13 would plot on or above the line. Finally, 3) the 1981 method adds the dissolution residue component. which accounts for biogenic residue remaining upon dissolution of $CaCO_3$ and opaline SiO_2 prior to sediment burial, and for other insolubles derived from the leachate analytical procedure used by Dymond and Leinen. It could be argued that this component is trapped in the cherts by the recrystallization process. Alternatively, this component may be applicable to modern sediments but not old rocks.

In summary, the effect of using the Dymond (1981) partitioning solution of figures 13, 14, and 15 would eliminate the apparent depletion of Mn, Cu, and Ba with respect to the detrital component by distributing Fe and Al among more than two components and consequently lowering the Al_D, Fe_D, and Fe_H values for all of the samples. This would make the plot cleaner, but it wouldn't change the relationship between the samples. Thus important changes with time in the section are reiterated in these plots. Due to the necessary initial assumptions discussed above, it is questionable whether the improved plots would still be anything more than a first approximation of the relative importance of the various sedimentary components.

The trends observed in figures 13, 14, and 15 are as follows: 1) Mn is enriched relative to hydrothermal Fe at the base of the section, and decreases up-section for both Marin and Geysers cherts. This agrees with the observed geology. 2) Cu shows a reverse trend, being increasingly enriched relative to hydrothermal Fe higher in the chert sections. There is substantially more Cu in the base of the Geysers section than in the base on the Marin section. 3) Ba plots very similarly to Cu.

One of the implications of these plots is that Mn is present in excess of hydrothermal Mn (3:1 for hydrothermal Mn with respect to hydrothermal Fe, according to Dymond and others, 1973; Leinen, 1981). This may suggest the presence of hydrogenous, or authigenic, Mn at the base of the section. The color change from red to green chert from the centers to the margins of beds, as well as up-section, suggests that the diagenetic (rather than the sedimentary) environment became more reducing with time. There may have been loss of manganese out of the upper cherts, but there is no diagenetic mechanism to transfer Mn from the upper to lower sediments. Thus it seems more reasonable to conclude that there is a hydrogenous component of Mn in addition to hydrothermal Mn in the Marin chert sections. Heath and Dymond (1981) and Lyle (1981) demonstrate that the abundance of hydrogenous Mn in Nazca plate sediments decreases with distance from the spreading ridge because of increasing dilution by terrigenous sediment. This observation, coupled with reductive remobilization, would adequately explain the decrease in Mn up-section at the Marin Headlands.

After the partitioning exercise, authigenic Ni is present only in the Cenomanian sample. In contrast, partitioning suggests that there is apparently a significant amount of authigenic, biogenic, and possibly dissolution-residual Cu and Ba in the Cretaceous part of the section. The upper part of the section is distinctly more clay rich and thinner bedded than the lower part of the section, implying that sedimentation rates were lower, "ambient" clay sedimentation was more important relative to biogenic sedimentation, and sediments were exposed on the sea floor for a longer period of time for the younger Marin radiolarites. These factors would allow the accumulation of a larger fraction of transition metals through precipitation from sea water, and through dissolution of biogenic material. Enrichment in Ba coupled with lower sedimentation rates is in accord with Dymond's (1981) determination that the solution residue component is dominated by Ba.

DEPOSITIONAL ENVIRONMENTS

Siliceous Deposition

The depositional environments for these cherts, based on observed geology in conjunction with various methods of interpreting the geochemical data, can be summarized as follows:

1. The sediments studied represent the sedimentary province in which they were deposited rather than the substrate on which they were deposited because there is little or no contamination of the sediments with basaltic debris.

2. Sedimentary sources contributing to the Pacific basin were approximately constant through Late Mesozoic time. These sources include terrigenous detritus, biogenic debris, hydrothermal and authigenic precipitates, and probably a residual component from dissolution of biogenic opal and carbonate.

3. The composition of a single sedimentary component can be defined in terms of a few characteristic elements, and remains effectively uniform for tens and possibly more than a hundred million years.

4. The relative importance of sedimentary sources for a specific depositional province may change through time.

5. The Jurassic Marin, Geysers, Geyser Peak and Stonyford cherts deposited directly on greenstone are dominated by a hydrothermal sedimentary source component. The influence of this component wanes up-section, and detrital and authigenic components gain increasing relative importance. Detrital components are more important at the Geysers chert section than at Marin or Stonyford. Authigenic components are more

significant for the Marin cherts than elsewhere.

A possible analogue for the Marin cherts may be derived from observations by Heath and Moberly (1971) and Lancelot (1973) concerning the distribution of Eocene and Mesozoic cherts in the north Pacific, from both drilling results and seismic interpretations, suggesting that these siliceous deposits accumulated under a zone of northern equatorial high productivity and were subsequently transported northward into the realm of red pelagic clays. Eocene cherts in the Central Pacific Basin are seismically characterized by numerous discontinuous reflectors, inferred by Heath and Moberly (1971) to represent layers covering a few to a few thousand square meters. Drilled sections include 2-5 cm thick chert layers alternating with porcellanite composed of silicified montmorillonitic and clinoptilolitic claystone, and with one to several meter thicknesses of apparently unaltered siliceous ooze. Heath and Moberly (1971) suggest that these deposits may ultimately be similar to bedded cherts known on land. In addition, the lack of drilling results for ribbon-like cherts may be an artifact of the drilling process (Pimm and others, 1971; Heath and Moberly, 1971): circulation fluid pressure changes required to drill cherts as opposed to clays produce cores full of "chert biscuits" while washing out softer beds.

Although few of the drilled Pacific cherts directly overly basalt, this may be related to a combination of factors including elevation of sea level and the CCD, coincidence of a spreading center with the marginal equatorial zones of productivity, climatic effects on oceanic circulation, and drilling difficulties. The possibility that the Marin radiolarites may have originated on or near a spreading center passing under the area of high biogenic productivity somewhere north of the equator, to the southwest of the North American continent, and transported northward and eastward into the realm of terrigenous depositional influence, as suggested by lithologic, geochemical, and paleontologic data, cannot be excluded (G. R. Heath, personal communication, December 1981). Two important sedimentologic features, the lack of an sediment equatorial bulge (van Andel and others, 1975,) and the lack of mirrored depositional patterns up section, support this hypothesis.

Diagenetic Processes

As suggested by Lancelot (1973), the composition and abundance of clay minerals is a critical factor in the diagenetic transformation of siliceous ooze to chert. Abundant disseminated Fe-rich and Mg-poor clays will aggregate early in diagenesis to block the migration of both seawater and dissolved silica and preserve thin "ribbons." Sparsely disseminated clays would require longer to aggregate, and allow the formation of thicker cristobalitic layers. Ultimately, the cristobalite would transform to quartz, and the layer would become chert. Thus diagenetic processes may provide a way to form ribbon cherts.

The complete segregation of quartz and clay at site 471 (Leinen, 1981) suggests that the diagenetic processes as proposed by Kastner and others (1977) are also capable of producing segregation layers through mobilization of silica. In most cases the process is probably not carried to completion because sufficient quartz may precipitate or clays may aggregate to form impermeable layers, inhibiting further mobilization of opaline silica. If silica migrates and accumulates, segregated from other cations and impurities, authigenic mixed layer clays (smectites) could form as described by Heath and Dymond (1977). Newly formed authigenic smectites, together with other disseminated clays might aggregate until they formed a layer which would block permeability and allow cristobalite precipitation. More likely, clay is a residual accumulation product as silica dissolves and migrates. According to the results of Kastner and others (1977), the pelagic clays and authigenic smectites would compete for available Mg^{++} and (OH)- ions, and diagenetic processes would be arrested when these ions became depleted. Cristobalitic silica would precipitate as a matrix for yet-undissolved radiolarians. Fe-Mn oxy-hydroxides incorporated in the original sediment (Lyle, 1981) could play several roles. They could be incorporated into cherts with clays and smectites, they could coat radiolarians and protect them from dissolution, and they could accumulate at grain boundaries in the cristobalitic matrix. Evidence for all three modes of occurrence was observed in thin-section. Thus, the Fe-Mn oxides and hydroxides may play a subordinate role in the diagenetic transformation of siliceous ooze to ribbon chert.

The implication of this diagenetic segregation process in terms of the Marin cherts would be that the decrease in chert layer thickness, as well as the decrease in chert/shale ratio up-section is due to an increase up-section of clay relative to biogenic silica. This observation supports the hypothesis of the Marin section being transported tectonically from an oceanic zone of high biologic productivity to an oceanic zone of lower biologic productivity but dominated by pelagic clays.

Sedimentary Processes

It is important to be able to discriminate between depositional and diagenetic features. Depositional features potentially preservable in siliceous oozes include: 1) laminations or interbedding of different sediment types, 2) erosional features (including surface solution unconformities), 3) layer thicknesses, and 4) bioturbation.

Rhythmic alternation of sediment types is often considered to be characteristic of turbidite deposition, though not conclusive evidence for turbidites. Pelagic turbidite deposition has been invoked for siliceous and calcareous material ponded in low areas near submarine highs (van Andel and Komar, 1969; Heath and Moberly, 1971; Nisbet and Price, 1974; Barrett, 1981; Kelts and Arthur, 1981). In the Gulf of California (sites 477, 478, 481, Leg 64; Curray, Moore, and others, in Kelts and Arthur, 1981) low density siliceous turbidites redeposit diatoms from bathyal depths due to tectonic activity on unstable slopes. Such beds are generally not graded due to lack of sufficient partical size range, though occasionally the base is siltier than the top because diatoms settle more slowly than clay or silt. Some beds have basal benthic foraminifera hash, and uppermost pure diatom ooze. Beds range from 30 to 80 cm in thickness. Pyrite and organic matter are abundant, and burrowing is minimal due to low oxygen conditions on the basin floor, and to very high sedimentation rates. Calculated accumulation rates are in excess of 1,500 m/my.

Convincing evidence for pelagic turbidite deposition includes textural grading, lamination, cross-lamination, and convolute layers, typical of Bouma sequences but on a finer scale, in addition to alternation of sediment types. Such features have been described for sequences at DSDP sites 386 and 387 (Tucholke and Vogt, 1979) near the Bermuda Rise, and by Torrini and Mattioli (in Speed and Larue, 1982) for Eocene radiolarites on Barbados. Nisbet and Price (1974), and Price (1976) distinguish between layers with traction features deposited by currents, and featureless "current-free" layers for Mesozoic ribbon cherts on Cyprus, and conclude that some layers are probably turbidite deposits. Barrett (1981) found similar relationships for Mesozoic Ligurian cherts in northern Italy. Microfossil tests in these siliceous turbidites may be filled or empty, individual or aggregated in fecal pellets, whole or fragmented (Kelts and Arthur, 1981). Combined sedimentary evidence shows that ribbon cherts may form by purely sedimentologic processes, namely alternation of sediment types due to turbidite deposition.

Some layers, and some thin sections from the Marin cherts appear to be similar to some of these published descriptions of pelagic turbidites. Very rare grading from silt was observed, but no traction features. One slumped layer was noted, and unrecognized unconformities may be present, perhaps obscured by pinching and swelling and other diagenetic features. Layer thicknesses are generally in the 2-10 cm range for Marin outcrops, which corresponds to unlithified deposits such as the Guaymas Basin, where layer thicknesses are 30 to 80 cm. Keene (1976) calculated that compaction and lithification and diagenesis of siliceous oozes could result in a 75 to 90 percent volume loss. A factor of 4 or 5 is generally preferred (G. R. Heath, personal communication, 1981), which would translate to an 8- to 50-cm range for initial layer thicknesses of typical ribbon radiolarite. Hayes and others (1972) suggested that sediment layer thicknesses can be related to spreading rate, or plate/ridge orientations with respect to high biologic productivity zones. If this is true, it may partially explain the overall decrease in layer thickness upsection for the Marin cherts, assuming they were transported out of a high productivity zone by plate motions. Diagenetic processes also influence layer thickness and will be discussed shortly.

No unequivocal evidence for bioturbation was recognized in the Marin chert sections visited by the author. Davis (1918), however, described "concretions" from a quarry in North Berkeley, and included a photograph of these features, which may in fact be relict burrows. Evidence to support the burrow hypothesis include: 1) deflection and/or termination of MnO_2 banding in chert by the concretions, 2) a thin layer of red shale or occasionally MnO_2 may separate the concretion from the matrix, 3) lighter color of the concretion relative to the matrix, 4) internal streaky character of the concretion, 5) ellipsoidal, irregular (with rounded corners), or elongate shape of the concretions, 6) fewer radiolarians in the concretions than the matrix, and 7) less iron oxide in the concretion than the matrix (Davis, 1918). The concretions may be a simple product of diagenetic processes, as suggested by Davis (1918), but they are also compatible with published descriptions of burrows in siliceous deposits (Keene, 1975; Hein and others, 1981).

RIBBON CHERT FORMATION

The layering in the Marin ribbon cherts is most likely produced by a combination of sedimentary, diagenetic, and compaction processes.

Diagenetic features characteristic of siliceous deposits include: 1) pinching and swelling of layers, 2) streaky and discontinuous laminations, 3) stylolites, 4) nodules, concretions, and mammae, 5) cavities, brecciation and recementation, 6) quartz veinlets perpendicular to layering and confined to single layers, 7) color mottling, 8) changing proportions of microfossils to matrix, and 9) the presence of Fe/Mn colloids or micronodules. Sedimentary features include: 1) continuous laminations of different sediment types, 2) cross-lamination, 3) erosional features, and 4) bioturbation.

Layering in the Marin cherts is predominantly lenticular. Most lenses are a few to tens of meters in extent, but some lenses may be in trains of boudins on a centimeter scale. Chert lenses occasionally pinch out entirely from up to 20 cm thicknesses over the space of a few centimeters. When this occurs, red clays with fluxion textures fill in gaps such that overlying and underlying layers maintain parallelism with the rest of the section. Knobs and concretions are generally a few cm in diameter, and are compensated by immediately adjacent layers of chert and shale. Conversely, shaley laminations pinch out entirely within chert layers. These irregularly shaped features are not depositional structures and are considered to be diagenetic. As diagenetic features they imply differential plastic flowage between the two rheologically distinct components within the deposit: siliceous ooze and clays. This plastic flow may also explain the mm-thin streaky color laminations and the lack of burrow features. Parallel boundaries above and below some of the more deformed layers suggest that the flowage takes place prior to lithification, and may have been a response to differential compaction. These preserved diagenetic features may represent a combination of differential lithification rates, controlled by solution-redeposition processes and differential compaction, affected by differing degrees of lithification.

Recemented breccia, preservation of delicate Radiolaria in original spherical form, and single layer quartz veinlets observed in Marin cherts imply that some chert formed prior to compaction of the sediments, as suggested by Lancelot (1973). However, flattened Radiolaria were observed locally, and it is probable that the chertification process is a continuum which may allow the production of an array of plastic to brittle deformational features.

In conclusion, sedimentologic and diagenetic evidence suggest that Marin cherts represent siliceous sediments which were deposited on basaltic ocean crust beneath a region of high biogenic productivity. Subsequently these deposits were overlain by more clay-rich sediments derived from an oceanic region of low biologic productivity. Terrigenous clays gradually increased late in the sedimentary history, until graywacke and shale ultimately dominated the sedimentary regime, indicating proximity to a land mass. Diagenetic processes modified sedimentary structures and textures, enhancing alternation of sediment types, and resulting in ribbon chert formation.

The gradual nature of sedimentologic and compositional changes with time suggest passive transport of these deposits from a spreading center under a realm of predominantly pelagic deposition toward a convergent continental margin similar to Cook's (1972) model. At the continental margin, these sediments, already lithified, were broken and deformed by faulting, and incorporated into an accretionary package, which is the Franciscan assemblage.

IMPLICATIONS FOR THE FRANCISCAN ASSEMBLAGE

Numerous factors influence the depositional environment of siliceous sediments. Cherts are common in the Franciscan assemblage and occur in several distinctly different lithologic associations, indicating that more than one environment of deposition is represented. It is beyond the scope of this paper to consider the various categories of Franciscan cherts. This study has centered on just one category: cherts overlying pillowed greenstone and overlain by graywacke and gray shale. Such cherts are distinguished from chert lenses in greenstone and cherts interbedded with sandstone. Therefore, conclusions drawn about the depositional environment of the Marin cherts should be applied only with extreme caution to other types of Franciscan chert.

Sedimentologic, geochemical, and paleontologic evidence discussed above suggest that Marin cherts are allochthonous with respect to the North American continental margin, and possibly with respect to the other rocks in the Franciscan assemblage. This is not a new idea and has been widely accepted by workers familiar with California geology.

Paleomagnetic results indicate that the siltstone at the top of Marin Headlands section came from at least 7° of latitude south of its present location (Curry, this volume). It is likely that the chert came from even farther south. The results of this study support the paleomagnetic conclusions and provide some constraints on the latitudinal and longitudinal origin for the Marin cherts.

The complete absence of calcium carbonate from the composite section and comparison with other sediments from DSDP cores (Jenkyns, 1978) suggest that the oceanic crust underlying the Marin cherts may always have been north of the equatorial zone of high calcareous production. Abundant siliceous microfossils imply an oceanic zone of high productivity such as the northern equatorial zone of upwelling or zones of upwelling along continental margins. Paleogeographic reconstructions for Late Jurassic through Cretaceous time (Kennett, 1982) indicate that although circulation patterns differed from present circulation patterns, there was a Late Mesozoic equatorial current bounded by zones of upwelling.

An open oceanic site of deposition is preferred over one adjacent to the continental margin for two reasons: 1) the lack of terrigenous detritus through most of the section, and 2) the highly oxidized nature of the sediments. Under reducing conditions iron and manganese are both quite mobile (Dymond, and others, 1980; Lyle, 1981) and both elements are abundant in the Marin deposits. Continental margin siliceous deposits such as the Monterey shale or the diatomites in the Gulf of California, topographically barred from terrigenous material and chemically barred from calcareous material and oxygen, are characterized by gray or green, often carbonaceous and reduced, sediments (Ingle, 1973; 1981). Red, oxidized sediments such as the Marin radiolarites are not characteristic of deposition beneath continental margin zones of upwelling and high biogenic productivity (Ingle, 1973; 1981).

It is more difficult to evaluate the longitudinal distance between the Jurassic continental margin and the site of Marin pelagic sedimentation. A topographic barrier, such as an abandoned spreading ridge or a seismic ridge similar to those within the Panama Basin (Heath and others, 1974), or a starved trench such as the North Chile Trench, could allow pelagic sedimentation very close to the continental margin. In the absence of topographic barriers, typical estimates for the distance of terrigenous influence range around 900 to 1,000 km (Auboin, von Huene, and others, 1981; Rosato and Kulm, 1981). Spreading rates and vectors could add more distance between a hypothetical depositional site and the continental margin.

Finally, a comparison of the Marin cherts with other Franciscan and Great Valley Sequence Mesozoic cherts resting on greenstone and overlain by graywacke shows some important differences that relate to depositional environments. Specifically, terrigenous sedimentary components are more pronounced in the Great Valley Sequence chert sections than for Geysers cherts or Marin cherts. Geysers cherts, on the other hand, have a tuffaceous component absent from the Marin cherts. Thus they were probably deposited either closer to an arc or to a continental margin than the Marin sediments, or they were deposited on the opposite (continentward) side of a significant topographic barrier. These Franciscan cherts all lack calcium carbonate and have approximately similar chemical characteristics, suggesting that they all originated in roughly the same (northeast Pacific) depositional basin. The important conclusion is that although they may be time equivalents of one another, facies relationships cannot be demonstrated, and thus they are probably <u>not</u> spatially related sections.

CONCLUSIONS

Important conclusions supported by this study are the following:
1) Cherts possess a geochemical signature characteristic of a combination of their sedimentary and diagenetic environments. 2) Geochemical patterns in marine pelagic deposits represent a combination of five sedimentary sources as determined by Dymond (1981): detrital, hydrothermal, hydrogenous, biogenic, and solution residue. The relative proportions of these factors can be used to infer characteristics of ancient as well as modern sedimentary environments. 3) Sedimentologic, diagenetic, and geochemical parameters can be used to infer depositional environment characteristics of siliceous deposits. 4) Ribbon cherts form from both sedimentary and diagenetic processes. Most ribbon cherts similar to Marin cherts probably form through diagenetic modifications of original sedimentary features. The primary sedimentary process is probably redeposition of siliceous ooze from topographic highs to lows by means of dilute turbidity currents. The most important factors include the composition and abundance of clay minerals, calcium carbonate relative to siliceous ooze, the alkalinity of the pore water, and heat flow. 5) Marin cherts probably originated as siliceous pelagic deposits near a slow-spreading mid-ocean ridge in the northeast Pacific basin. These deposits were tectonically transported from their initial pelagic site of deposition toward an arc or continental margin. They were deformed and accreted subsequent to lithification. 6) Marin cherts probably were not spatially related to their time equivalent sections at The Geysers, Stonyford, or in the Great Valley Sequence.

REFERENCES CITED

Aubouin, J., von Huene, R., and others, 1981, A summary of Deep Sea Drilling Project Leg 67: Shipboard results from the Mid-America trench transect off Guatemala: Oceanologica Acta, 26th International Geological Congress, Proceedings, Geology of continental margins symposium, Paris, 1980, p. 225-232.

Baltuck, M., 1982, Provenance and distribution of Tethyan pelagic and hemipelagic siliceous sediments, Pindos Mountains, Greece: Sedimentary Geology, v. 31, p. 63-88.

Barrett, T. J., 1982, Stratigraphy and sedimentology of Jurassic bedded chert overlying ophiolites in the north Appennines, Italy: Sedimentology, v. 29, p. 353-373.

Beccaluva, L., Macciotta, G., Savelli, C., Serri, G., and Zeda, O., 1980, Geochemistry and K/Ar ages of volcanics dredged in the Philippine Sea (Mariana, Yap, and Palau Trenches and Parece Vella Basin), in Hayes, D. E., ed., The tectonic and geologic evolution of Southeast Asian seas and islands: American Geophysical Union, Monograph 23, p. 247.

Bostrom, K., Farquharson, B., and Eyl, W., 1971, Submarine Hot Springs as a source of active ridge sediments: Chemical Geology, v. 10, p. 189-203.

Bostrom, K., and Peterson, N. N. A., 1969, Origin of aluminum-poor ferromanganoan sediments in areas of high heat flow on the East Pacific Rise: Marine Geology, v. 7, p. 427-447.

Cook, H., 1972, Stratigraphy and sedimentation, in Hays, J. D., and others, Unit Reports, Vol. 9: Washington, D.C., U.S. Government Printing Office, p. 933-943.

Crouch, J., 1979, Neogene tectonic evolution of the California Continental Borderland and Western Transverse Range: Geological Society of America Bulletin, v. 90, pt. I, p. 338-345.

Davis, E. F., 1918, The Radiolarian cherts of the Franciscan Group: University of California Publications, Bulletin of Department of Geology, v. 11, no. 3, p. 235-432.

Debrabant, J. F., and Maillot, H., 1980, Mathematical treatment of geochemical data, Deep Sea Drilling Project sites 415 and 416, in Lancelot, Y., Winterer, E. L., and others, Unit Reports, DSDP 50: Washington, D.C., U.S. Government Printing Office, p. 695-703.

Dymond, J., 1981, Geochemistry of Nazca plate surface sediments: an evaluation of hydrothermal, biogenic, detrital, and hydrogenous sources: Geological Society of America Memoir 154, p. 133-173.

Dymond, J., Corliss, J. B., Cobler, R., Muratti, C. M., Chou, C., and Conard, R., 1980, Composition and origin of sediments recovered by deep drilling of sediment mounds, Galapagos spreading center, in Rosendahl, B. R., Hekinian, R., and others, Initial Reports, DSDP, Vol. 54: Washington, D.C., U.S. Government Printing Office, p. 377-385.

Edmond, J. M., Measures, C., McDuff, R. E., Chan, L. H., Collier, R., Grant, B., Gordon, L. I., and Corliss, J. B., 1980, Ridge crest hydrothermal activity and the balances of the major and minor elements in the ocean; the Galapagos data: Earth and Planetary Science Letters, v. 46, p. 1-18.

Fujioka, K., 1980, Conglomerates of volcanic rocks of the Deep Sea Drilling Project, site 439, in Initial Reports of DSDP, vol. 57: Washington, D.A., U.S. Government Printing Office, p. 1075-1082.

Garcia, Michael O., 1978, Criteria for the identification of Ancient volcanic arcs: Earth Science Reviews, v. 14, p. 147-165.

Garrison, Robert E., 1974, Radiolarian cherts, pelagic limestone, and igneous rocks in eugeosynclinal assemblages, in Hsu, K. J., and Jenkyns, H. C., Pelagic sediments on land and under the sea: International Association of Sedimentology, Special Publication 1, p. 367-399.

Hayes, J. D., Cook, H., Jenkyns, G., Orr, W., Goll, R., Cook, F., Milow, D., and Fuller, J., 1972, An interpretation of the geologic history of the eastern equatorial Pacific from the drilling results of Glomar Challenger, Leg 9, in Hays, J. D., and others, Initial Reports, DSDP, Vol. 9: Washington, D.C., U.S. Government Printing Office, p. 909-931.

Heath, G. R., 1974, Dissolved silica and deep-sea sediments, in Hay, W. W., Studies in paleo-oceanography: Society of Economic Paleontologists and Mineralogists, Special Publication 20, p. 77-93.

Heath, G. R., and Dymond, Jack, 1977, Genesis and transformation of metalliferous sediments from the East Pacific Rise, Bauer Deep, and Central Basin, northwest Nazca plate: Geological Society of America Bulletin, v. 88, p. 723-733.

Heath, G. R., and Moberly, R., 1971, Cherts from the western Pacific, Leg 7, Deep Sea Drilling Project, in Winterer, E. L., Riedel, W. R., and others, Initial Reports, DSDP, Vol. 7: Washington, D.C., U.S. Government Printing Office, p. 991-1007.

Heath, G. R., Moore, T. C., and Roberts, G. S., 1974, Mineralogy of surface sediments from the Panama Basin, eastern equatorial Pacific: Journal of Geology, v. 82, p. 145-160.

Hein, J. R., and Karl, S. M., 1983, Comparisons between open-ocean and continental margin chert sequences, in Iijima, A., Hein, J. R., and Seiver, R., eds., Siliceous deposits of the Pacific Region: Developments in Sedimentology: v. 36, p. 25-43, Elsevier (Amsterdam).

Hellman, P. L., Smith, R. E., and Henderson, P., 1979, The mobility of rare earth elements: evidence and implications from selected terranes affected by burial metamorphism: Contributions to Mineralogy and Petrology, v. 71, p. 23-44.

Hopson, C. A., Mattinson, J. M., and Pessagno, E. A., 1980, Coast range ophiolite, western California, in Ernst, W. G., ed., Geotectonic development of California: Englewood Cliffs, N.J., Prentice-Hall, p. 232-248.

Hussong, D. M., and Uyeda, S., 1981, Tectonics in the Mariana Arc: resultsof recent studies, including DSDP Leg 60, in Proceedings, 26th International Geological Congress, Geology of Continental Margins Symposium, Paris, July 1980: Oceanol. Acta, (SP), p. 203-212.

Ingle, J. C., Jr., 1973, Summary comments on Neogene biostratigraphy, physical stratigraphy, and paleoceanography in the marginal northeastern Pacific Ocean, in Initial Reports, DSDP Leg 18: Washington, D.C., U.S. Government Printing Office, p. 949-960.

_______1981, Origin of Neogene diatomites around the north Pacific rim, in The Monterey Formation and related siliceous rocks of California: Society of Economic Paleontologists and Mineralogists, Pacific Section, p. 159-179.

Isaacs, C. M., 1979, Lateral diagenesis in Monterey Shale, Santa Barbara Coast, California: American Association of Petroleum Geologists Bulletin, v. 63, no. 3, p. 473.

Jenkyns, H. C., 1978, Pelagic environments, in Reading, H. G., ed., Sedimentary environments and facies: Oxford, U.K., Blackwell Scientific Publications, p. 314-371.

Jones, J. B., and Segnit, E. R., 1971, The nature of opal. I. Nomenclature and constituent phases:

Journal of Geological Society of Australia, v. 18, p. 57-68.

Kastner, M., Keene, J. B., and Gieskes, J. M., 1977, Diagenesis of siliceous oozes; I, Chemical controls on the rate of opal-A to opal-CT transformations--an experimental study: Geochim. et Cosmochim. Acta, v. 41, p. 1041-1051.

Keene, J. B., 1975, Cherts and porcellanites from the North Pacific, Leg 32, Deep Sea Drilling Project, in Moberly, R., Larson, R., and others, Initial Reports, DSDP 32: Washington, D.C., U.S. Government Printing Office, p. 429-507.

_______1976, The distribution, mineralogy and petrography of biogenic and authigenic silica from the Pacific Basin: University of California, San Diego, unpublished Ph.D. thesis.

Kelts, K., and Arthur, M. A., 1981, Turbidites after ten years of deep-sea drilling--wringing out the Map? Society of Economic Paleontologists and Mineralogists, Special Publication 32, p. 91-127.

Kennett, J. P., 1982, Marine geology: Englewood Cliff, N. J., Prentice Hall, Inc., 813 p.

Knauth, L. P., and Epstein, S., 1975, Hydrogen and oxygen isotope ratios in silica from the JOIDES Deep Sea Drilling Project: Earth and Planetary Science Letters, v. 25, p. 1-10.

Kolodny, Y., and Epstein, S., 1976, Stable isotope geochemistry of deep sea cherts: Geochim. et Cosmochim. Acta, v. 40, p. 1195-1209.

Larson. R. L., and Pitman, W. C., III, 1973, Worldwide correlation of Mesozoic magnetic anomalies and its implications: Geological Society of America Bulletin, v. 83, p. 3645-3662.

Leinen, Margaret, 1981, Metal-rich basal sediments from northeastern Pacific Deep Sea Drilling Project sites, in Yeats, R. S., Haq, B. U., and others, Initial Reports, DSDP Leg 63: Washington, D. C., U.S. Government Printing Office, p. 667-676.

Leinen, M., and Pisias, N., 1982, An objective technique for determining end-member compositions and for partitioning sediments according to their sources: Geochim. et Cosmochim. Acta, (in press).

Leinen, M., and Stakes, D., 1979, Metal accumulation rates in the central equatorial Pacific during the Cenozoic: Geological Society of America Bulletin, v. 90, p. 357-375

Lyle, M., 1981, Formation and growth of ferromanganese oxides on the Nazca plate: Geological Society of America Memoir 154, p. 269-293.

MacPherson, G J., 1983, The Snow Mountain Volcanic Complex: an on-land seamount in the Franciscan terrain, California: Journal of Geology, v. 91, p. 73-92.

Martin, J. H., and Knauer, G. A., 1973, The elemental composition of plankton: Geochim. et Cosmochim. Acta, v. 37, p. 1639-1654.

McLaughlin, R. J., and Pessagno, E. A., Jr., 1978, Significance of age relationships above and below Upper Jurassic ophiolite in the Geyser- Clear Lake region, California: U.S. Geological Survey Journal of Research, v. 6, no. 6, p. 715-726.

McMurtry, G. M., Veeh, H. H., and Moser, C., 1981, Sediment accumulation rate patterns on the Northwest Nazca plate: Geological Society of America Memoir 154, p. 211-249.

Murchey, Benita, 1980, Significance of chert age determinations in the Marin Headlands: Geological Society of America, Cordilleran Section Meeting, Corvallis, Oregon, Abstracts with Programs, p. 144.

Nisbet, E. G., and Price, I., 1974, Siliceous turbidites: bedded cherts as redeposited ocean ridge-derived sediments, in Hsu, K. J., and Jenkyns, H. C., eds., Pelagic sediments: on land and under the sea: International Association of Sedimentologists, Special Publication 1, p. 351-366.

Pearce, J. A., and Cann, J. R., 1973, Tectonic setting of basic volcanic rocks determined using trace element analyses: Earth and Planetary Science Letters, v. 19, p. 290-300.

Pessagno, E. A., Jr., 1977, Lower Cretaceous radiolarian biostratigraphy of the Great Valley sequence and Franciscan complex, California Coast Ranges: Cushman Foundation for Foraminiferal Research Special Publication 15, 87 p.

Pimm, A. C., Garrison, R. E., and Boyce, R. E., 1971, Sedimentology synthesis: lithology, chemistry and physical properties of sediments in the Northwestern Pacific Ocean, in Fischer, A. G., Heezen, B. C., and others, Initial Reports, DSDP, Vol. 6: Washington, D.C., U.S. Government Printing Office, p. 1131-1252.

Pisciotto, K. A., 1980, Chert and porcelanite from Deep Sea Drilling Project Site 436, Northwest Pacific, in Scientific Party, Initial Reports, DSDP Legs 56, 57, Pt. 2: Washington, D.C., U.S. Government Printing Office, p. 1133-1142.

_______1981, Distribution, thermal histories, isotopic compositions, and reflection characteristics of siliceous rocks recovered by the deep sea drilling project: Society of Economic Paleontologists and Mineralogists, Special Publication 32, p. 129-147.

Price, Ilfryn, 1977, Facies distinction and interpretation of primary cherts in a Mesozoic continental margin succession, Othris, Greece: Sedimentary Geology, v. 19, p. 321-335.

Rosato, V. J., and Kulm, L. D., 1981, Clay mineralogy of the Peru continental margin and adjacent Nazca plate: Implications for provenance, sea level changes, and continental accretion: Geological Society of America Memoir 154, p. 545-568.

Schlocker, Julius, 1974, Geology of the San Francisco North quadrangle, California: U.S. Geological Survey Professional Paper 782, 109 p.

Schweller, W. J., Kulm, L. D., and Prince, R. A., 1981, Tectonics, structure and sedimentary framework of the Peru-Chile Trench: Geological Society of America Memoir 154, p. 323-349.

Scott, Martha R., 1981, Metallogenesis in the Parece Vela marginal basin complex of the Philippine Sea, Deep Sea Drilling Project Leg 59, in Kroenke, L., Scott, R., and others, Initial reports, DSDP, Leg 59: Washington, D.C., U.S. Government Printing Office, p. 649-652.

Steinberg, M., and Mpodozis-Marin, C., 1978, Classification geochemique des radiolarites et des sediments siliceux oceaniques, signification paleo-oceanographique: Oceanologica Acta I, p. 359-367.

Stone, D. B., 1977, Plate tectonics, paleomagnetism, and the tectonic history of the N.E. Pacific: Geophysical Surveys, v. 3, p. 3-37.

Torrini, R., and Mattioli, G., (in prep), in Speed and Larue, 1982, Barbados architecture and implications for accretion: Journal of Geophysical Research (in press).

Tucholke, B. E., and Vogt, P. R., 1979, Western North Atlantic--sedimentary evolution and aspects of tectonic history, in Tucholke, B. E., and Vogt, P. R., Initial Reports, DSDP, Leg 43: Washington, D.C., U.S. Government Printing Office, p. 791-826.

Vail, P. R., and Mitchum, R. M., Jr., 1978, Global cycles of relative changes of sea level from seismic stratigraphy, in Watkins, J. S.,

Montadert, L., and Dickerson, P. W., eds., Geological and geophysical investigations of continental margins: American Association of Petroleum Geologists Memoir 29, p. 469-472.

van Andel, Tj. H., Heath, G. R., and Moore, T. C., Jr., 1975, Cenozoic history and paleoceanography of the central equatorial Pacific: Geological Society of America Memoir 143, p. 1-134.

van Andel, Tj. H., and Komar, Paul D., 1969, Ponded sediments of the Mid-Atlantic Ridge between 22o and 23o North latitude: Geological Society of America Bulletin, v. 80, no. 7, p. 1163-1190.

Vedder, J. G., Crouch, J. K., and Lee-Wong, F., 1981, Comparative study of rocks from Deep Sea Drilling Project holes 467, 468, and 469, and the southern California borderland, _in_ Yeats, R. S., Haq, S. U., and others, Initial Reports, DSDP, Vol. 63: Washington, D.C., U.S. Government Printing Office, p. 907-918.

Winterer, E. L., and Jenkyns, H. C., 1979, Radiolarians, diatoms, and the origin of Mesozoic ribbon chert: Geological Society of America, Abstracts with Programs, v. 11, no. 7, p. 542-543.

Wolery, T. J., and Sleep, N. H., 1976, Hydrothermal circulation and geochemical flux at mid-ocean ridges: Journal of Geology, v. 84, p. 249-275.

PALEOMAGNETISM OF FRANCISCAN ROCKS IN THE MARIN HEADLANDS

Foster B. Curry, Allan Cox, and David C. Engebretson[1]
Department of Geophysics
Stanford University, Stanford, CA 94305

ABSTRACT

The Marin Headlands terrane, a fragment of ocean floor, consists of pillow basalt, overlain by radiolarian chert ranging in age from about 200 Ma to 95 Ma, overlain directly by clastic sedimentary rocks of continental origin. The structure is one of the least disrupted in the entire Franciscan Complex. Paleomagnetic vectors from 8 sites in pillow basalt show a reduction in scatter when corrected for tilt. Chert samples show stable magnetizations under thermal and alternating field demagnetization but a fold test is inconclusive. Results from the basalt and chert are well-grouped after correction for tilt. Paleomagnetic results from siltstone layers in the clastic sedimentary rocks show a significant clustering of directions when corrected for tilt and are well-grouped with the results from the basalt and chert.

When compared with expected cratonal field directions, the siltstone results indicate a poleward displacement of 7.4°±7.7° and clockwise (eastward) rotation of 128.9°±13.5° (at 95% confidence), similar to results from elsewhere in the Franciscan Complex. Directions from the basalt sites show a similar clockwise rotation, but are steeper in inclination than expected for their age of formation. This could arise from incorrect tilt corrections, from southward movement of the terrane since extrusion of the basalt, or from remagnetization of the basalt at a later time when the expected direction was steeper. The chert samples have directions that also indicate a large clockwise rotation, but with no significant difference in inclination from that expected for their age of deposition. These cherts have radiolarian ages within the Jurassic-Cretaceous Mixed Polarity Superchron (165-119 Ma), yet the polarity of the entire section is normal. This raises the possibility that the cherts were magnetized long after deposition.

Magnetization of the Cenomanian siltstone apparently predates both the deformation of the clastic sedimentary layers and the clockwise rotation of the Marin Headlands terrane. Magnetization of the basalt and chert also occurred before the rotation of the terrane but cannot be dated accurately relative to the time of their deformation.

INTRODUCTION

Portions of the Franciscan Complex in northern California consist of pillow basalt and radiolarian chert that formed in a deep ocean environment, overlain by clastic sedimentary rocks of continental origin. We report results of a paleomagnetic study of Franciscan rocks from the Marin Headlands terrane north of San Francisco.

In the first published paleomagnetic study of Franciscan rocks, Gromme and Gluskoter (1965) reported results from spilite and diabase overlain by Valanginian sedimentary rocks in north-west Marin County that are now interpreted as evidence of 19° of poleward displacement and 106° of clockwise rotation (Gromme, this volume). A more recent paleomagnetic study of two limestone blocks of Albian-Cenomanian age incorporated in Franciscan melange near Laytonville, California indicated deposition in the southern-hemisphere followed by more than 56° of poleward motion (Alvarez and others, 1980). Many pre-Tertiary rock units along the western margin of North America have undergone clockwise rotations and poleward translations (Beck, 1980).

In the Marin Headlands terrane, the basal pillow basalts are overlain by 80-meters of radiolarian cherts spanning approximately 100 million years from Pliensbachian to Cenomanian; the cherts are overlain by clastic sedimentary rocks containing Cenomanian radiolaria and megafossils (Murchey, this volume). The objectives of the present study were to test the internal consistency of paleomagnetic results from the three lithologies present and to determine if the Marin Headlands terrane has been translated or rotated.

GEOLOGIC SETTING

The Marin Headlands terrane is one of the best preserved of the many Mesozoic terranes in the Franciscan Complex, with a coherent stratigraphy of pillow basalt, radiolarian chert, and coarse-grained quartzo-

[1]Now at Department of Geology, Western Washington University, Bellingham, WA 98225

in Blake, M.C. Jr., ed., 1984, **Franciscan Geology of Northern California:** Pacific Section S.E.P.M., Vol. 43, p. 89-98

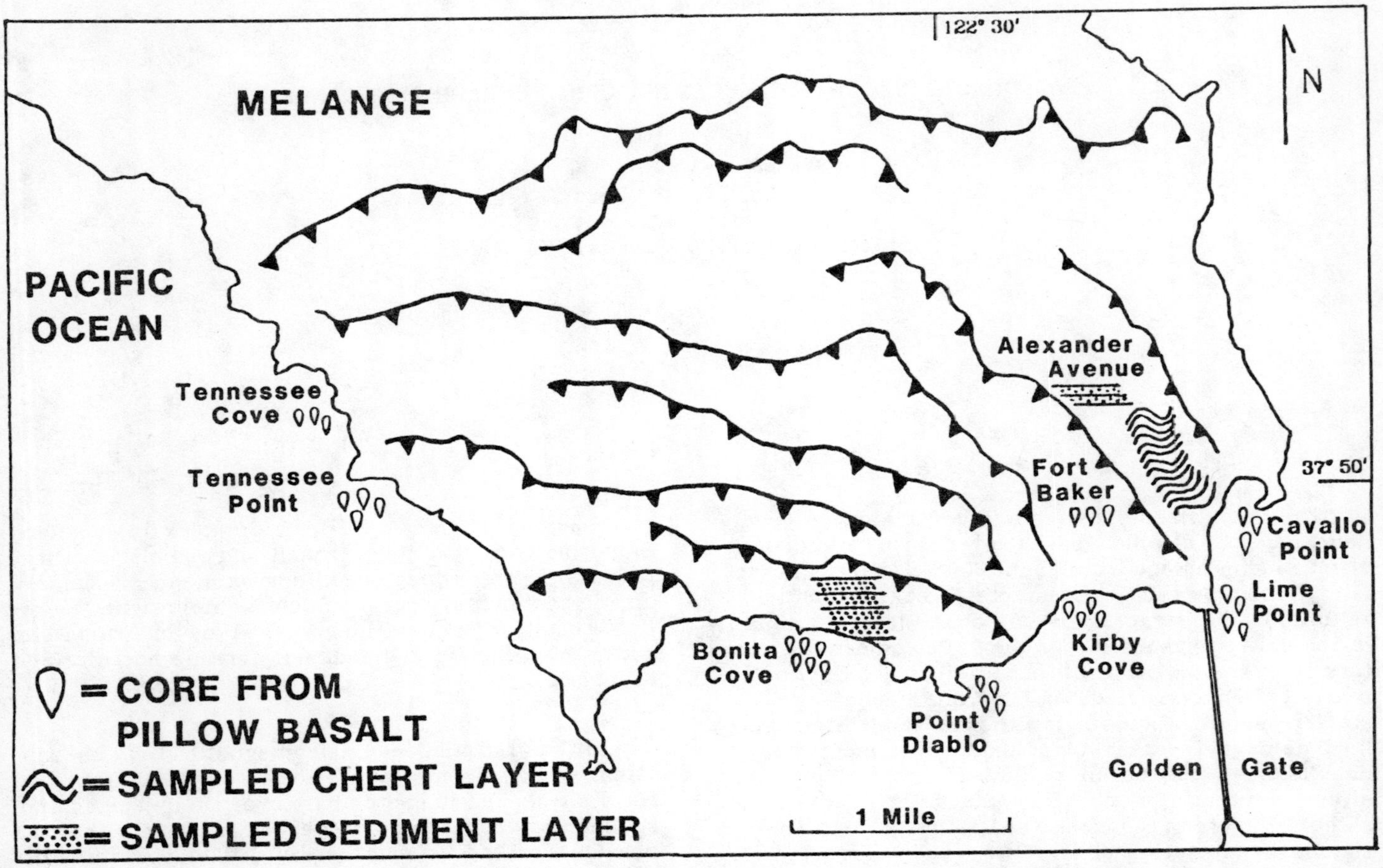

Fig. 1. Paleomagnetic sample locations in the Marin Headlands terrane. Pillow symbols at each basalt site indicate the number of cores drilled that were magnetically stable. Thirteen chert layers were sampled along a road-cut at the Alexander Avenue type-locality. The sediment samples are from sandstone layers at Bonita Cove and two siltstone beds above the Alexander Avenue chert locality.

feldspathic sandstone (Wahrhaftig, this volume; Blake and others, 1982). It stretches from southern Marin County south along the San Francisco Peninsula in fault contact with other Franciscan terranes. These adjacent terranes are composed of melange, blueschist, limestone, sandstone, and mafic rocks (Blake and others, this volume).

Our samples were collected from a 7 square-mile block of the Marin Headlands terrane in southern Marin County (Fig. 1). This block is separated from the continuation of the terrane to the south by a probable fault beneath the Golden Gate. To the north, a south-dipping, E-W striking thrust fault places the block over the melange of the Central Belt of the Franciscan (Blake and others, this volume). The block consists of 8-10 stacked thrust sheets, bent into a broad syncline dipping to the south-west (Fig. 1). Recumbent folds within each thrust sheet are overturned to the north and north-east (Wahrhaftig, this volume).

Fossil dating by Murchey (this volume) and structural mapping by Wahrhaftig (this volume) have shown that the thrust sheets in the Marin Headlands simply repeat the section of pillow basalt, chert, and clastic sedimentary rocks. A detailed sedimentologic and geochemical analysis of the Marin Headlands chert by Karl (this volume) indicates that the cherts were formed mainly by diagenetic processes in a deep ocean basin. Paleomagnetic samples in the present study are from the basalt, chert, and sandstone units.

BASALT INVESTIGATION

The eight basalt sampling sites are located in several of the stacked thrust sheets. Whether each site represents a different eruption is uncertain. At each site from three to six samples were collected over the local outcrop (Fig. 1). Tilt corrections are based on internal pillow structures and the attitude of the nearest contact with the overlying chert.

Individual specimens were thermally demagnetized over 10-12 temperature steps up to 580-650°C in a demagnetizer with a field of less than 2 gammas in the cooling chamber. The intensity and direction of magnetization were measured at each thermal step in a two-axis cryogenic magnetometer with a background noise level of less than 1×10^{-8} emu. A set of alternating field (AF) coils, built into the magnetometer, were used to reduce viscous magnetization acquired after thermal demagnetizations by performing AF demagnetizations at 5 mT. These AF coils were also used to demagnetize duplicate specimens by stepwise treatments up to 80 mT. Several specimens were demagnetized in a tumbler at fields up to 200 mT.

The thermal demagnetization experiments usually revealed the presence of a simple two-component magnetization: a main component displayed over 6-8 temperature steps and a low-temperature present-field overprint. Demagnetizations using the AF method revealed the same two components. Straight lines were fitted to the main thermal components using Kirschvink's algorithm with an error angle of 5 degrees (Kirschvink, 1978). The resulting directions were then averaged together by site and tilt corrections were applied to the site-mean directions.

Applying the tilt corrections brings an cbvious reduction in the scatter of the site-mean directions

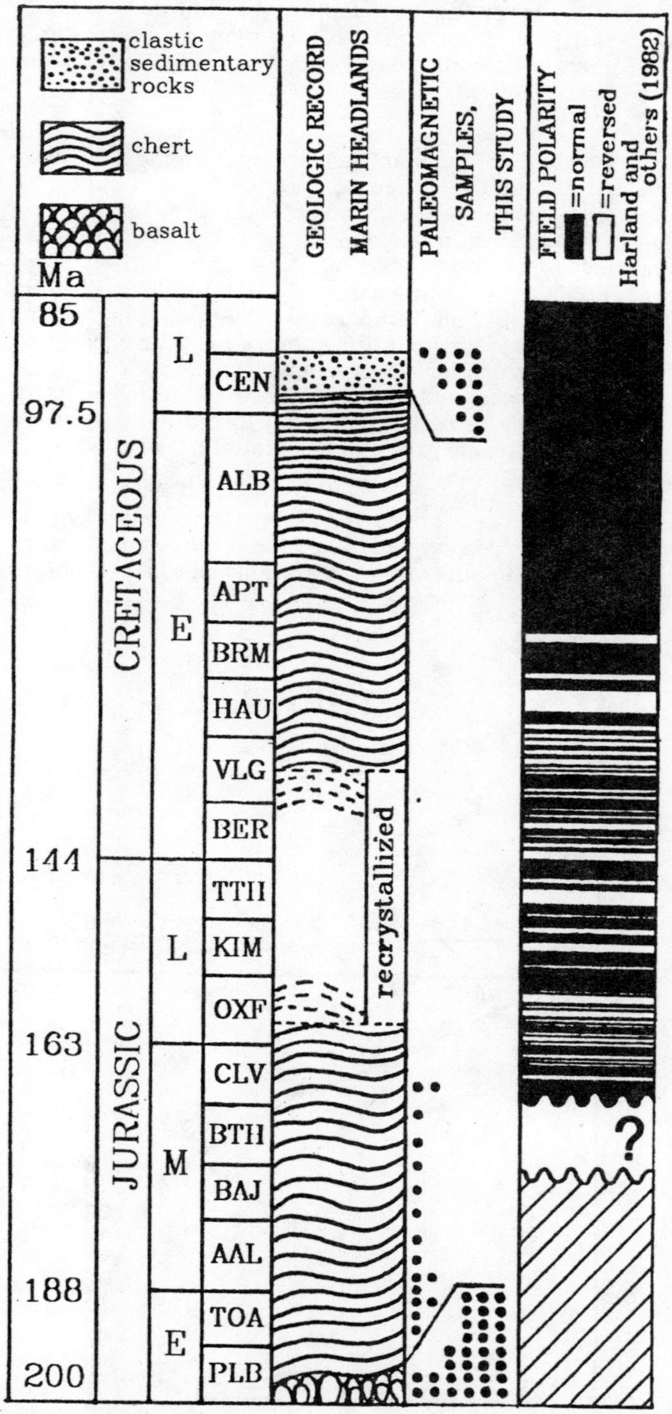

Fig. 2. Stratigraphy and magnetostratigraphy of the Marin Headlands Terrane. Stratigraphic column from fossil dating by Murchey (this volume). Each dot in the central column represents a paleomagnetic sample, with each level of dots representing either a different horizon in the sedimentary units or a different site in the basalt. Polarities of all samples are normal. In column to right, magnetic polarities are normal (black), reversed (white), mixed (diagonal), or un known (?) (Harland and others, 1982).

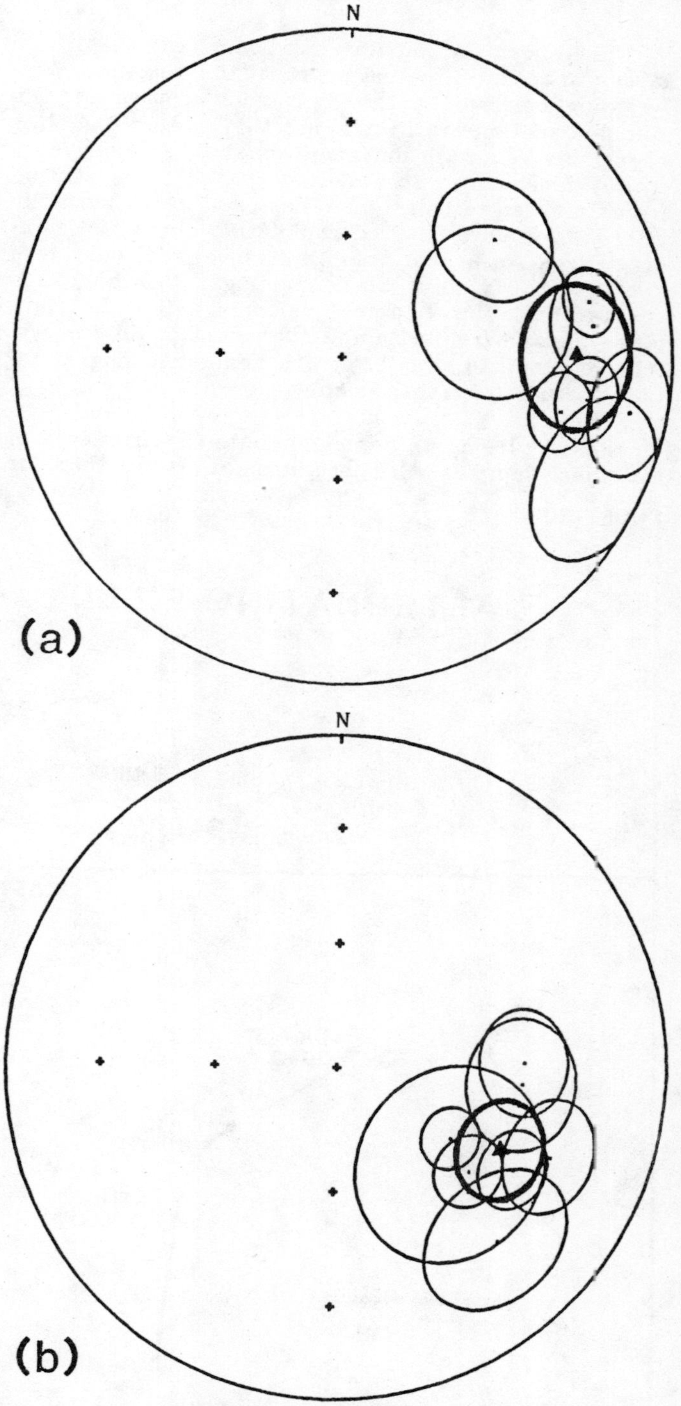

Fig. 3. Paleomagnetic results from basalt before (a) and after (b) correction for tilt. Solid triangle with heavy circle represents group mean results (see Table 1 for statistics). Site means are centered within their circles of 95% confidence. Lower hemisphere equal-area projections.

(Fig. 3, Table 1) that is not, however, significant at the 95% confidence level (McElhinny, 1964). The clustering of directions does suggest the magnetization was acquired before structural deformation, but the test is not definitive because the basalt flows record the different geomagnetic fields that existed at the different times of their respective eruptions.

CHERT INVESTIGATION

The chert section was hand-sampled in thirteen layers, spaced 1-5 meters apart, at the Alexander Avenue type-locality (Fig. 1). The radiolaria present in the sampled section are from zones MH1 and MH2 of Murchey (this volume), and span about 25 million years from Pliensbachian to Callovian (Fig. 2). The section from Callovian to Valanginian has been hydrothermally altered, the normally red radiolarian chert being recrystallized to a pale green-buff, bleached, quartzose chert with stylolitic black partings (Murchey, this volume; Karl, this volume). Paleomagnetic pilot samples collected from within and adjacent to this recrystallized zone had weak or erratic remanent magnetizations; results from these samples are not shown.

Results from the demagnetization experiments of the chert samples were less straightforward than for the basalt, although the techniques of demagnetization and analysis were the same. Thermal demagnetizations often produced two distinct linear components in addition to the present-field overprint. AF demagnetizations generally removed only about half of the remanent magnetization and suggest but do not establish the presence of multiple components (see Fig. 4 for a comparison of results obtained using AF and thermal demagnetization).

A possible primary component appeared consistently over the range 200-500°C. Steps below 200° showed a component aligned with the present geomagnetic field; this low-temperature component may be carried in goethite, a hydrous form of hematite. Steps beyond 500° sometimes revealed an additional component; these high-temperature components had no consistent orientation either before or after correction for tilt. The primary components of cherts from zone MH1 show no significant difference in direction from those of zone MH2 (Table 1), so all 15 samples were treated as a single group in the statistical analysis.

The fold test applied to the primary components of chert magnetization yields an inconclusive result since the scatter in directions is roughly the same before

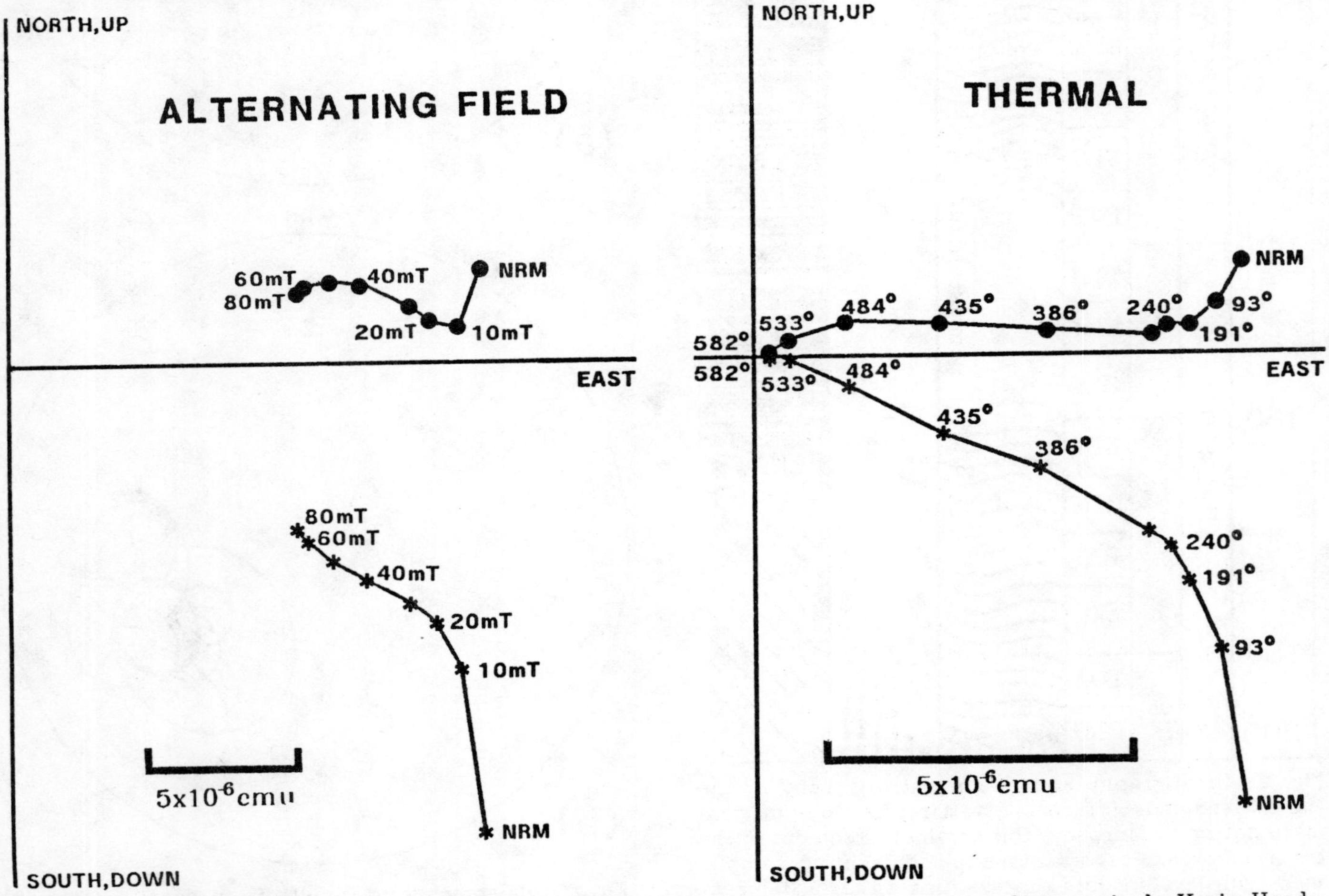

Fig. 4. Alternating-field and thermal demagnetization experiments on two specimens from a single Marin Headlands chert sample. The plots show orthogonal projections of the magnetization vectors into the horizontal plane (circles) and the east-west vertical plane (asterisks) at successive demagnetization steps. A main stable component is directed toward the east and is inclined 30° below the horizontal. A weak overprint is inclined steeply to the north, approximately parallel to the present field.

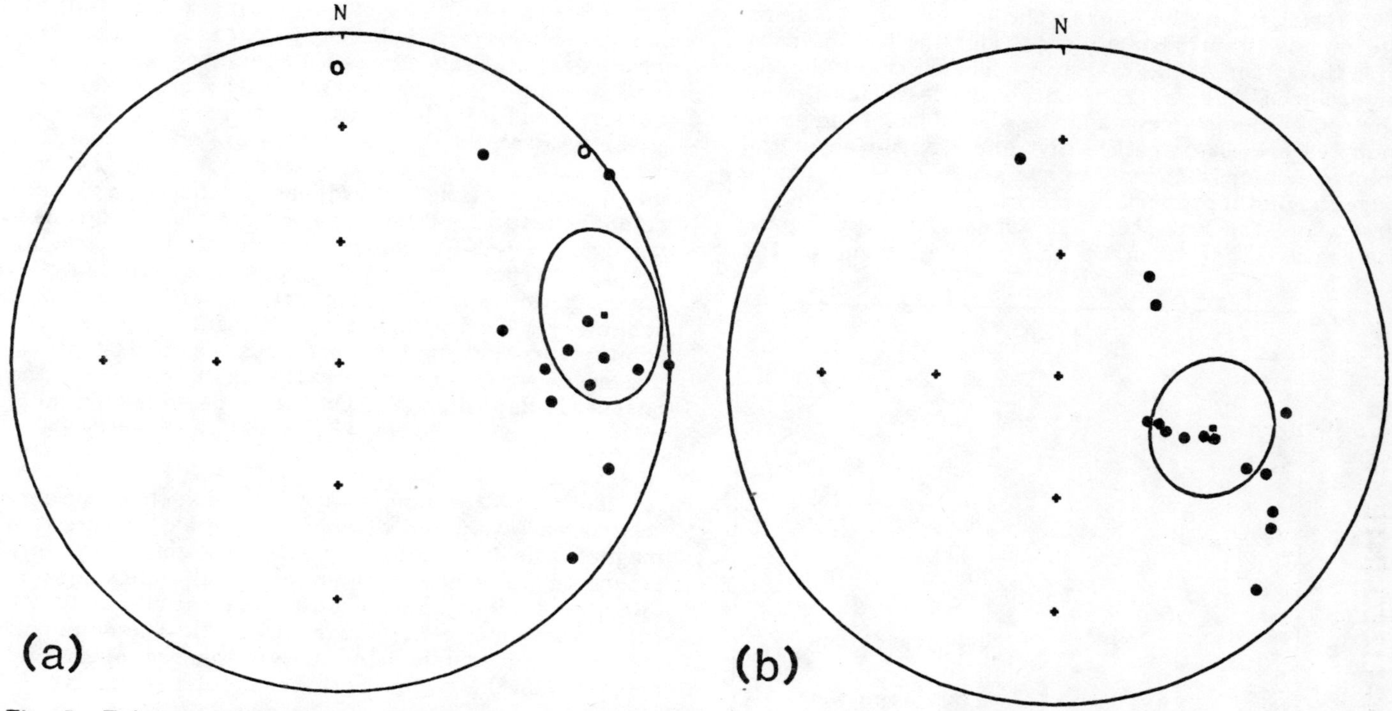

Fig. 5. Paleomagnetic directions from Marin Headlands chert before (a) and after (b) correction for tilt. Each circle is a stable component from one sample. Site means are centered within their circles of 95% confidence. The large amount of scatter present is not significantly affected by the tilt corrections. Open circles, upper hemisphere, solid circles, lower hemisphere of equal-area projection.

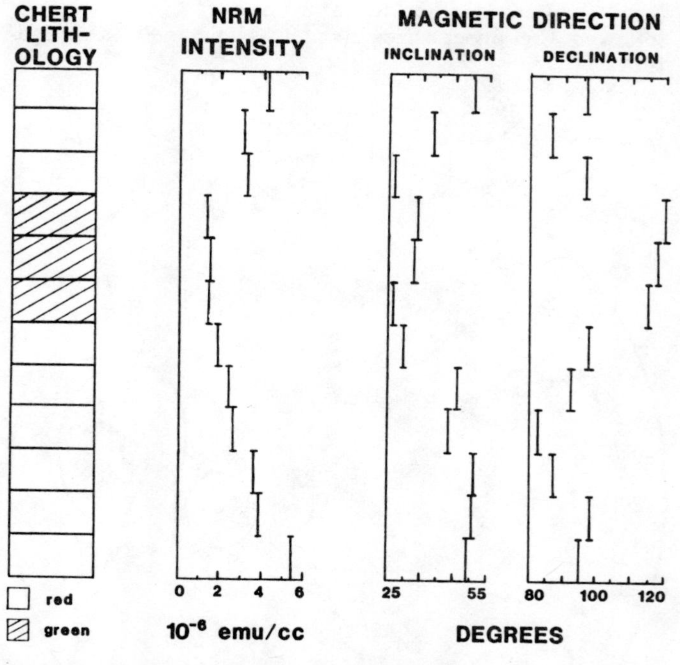

Fig. 6. Paleomagnetic results from 12 mini-specimens 8 mm long cut from one chert core. Magnetic Direction (Inclination and Declination) shows the tilt-corrected primary component resulting from thermal demagnetization up to 650°C. Three green specimens (shaded) have been recrystallized. The recrystallization affected both the intensity and direction of magnetization of the green chert specimens and of adjacent red chert specimens.

and after correcting for tilt (Fig. 5). Because the tilt corrections used for each layer are somewhat similar, the fold test is not definitive. The primary magnetization measured may have been acquired before or after deformation. The apparent streaking in the corrected data may be the result of some unrecognized structural complexity.

According to the time-scale of Harland and others (1982), the polarity of the geomagnetic field was predominantly mixed during the time of deposition of the cherts (Fig. 2). Therefore, reversed polarity might be expected in about half of the samples, yet all samples in this study are of normal polarity. This suggests the possibility of post-depositional magnetization during an interval of normal polarity, such as the Cretaceous Normal Superchron (Aptian to Santonian).

The angular dispersion of the paleomagnetic directions from the chert layers is large, with little or no apparent serial correlation between successive chert layers. To provide some insight into the cause of this dispersion, the top chert layer from zone MH1 was analyzed in more detail by drilling a 4-inch core perpendicular to bedding and slicing it into 12 mini-specimens, each 8 mm long (versus the standard specimen length of 24 mm). Each mini-specimen was measured during thermal demagnetization up to 600°C. The fourth, fifth, and sixth mini-specimens from the top of the chert layer were observed to be recrystallized to greenish chalcedonized quartz with no radiolaria preserved. Similar bands of green chert apparent in other chert layers had been avoided when drilling the standard cores.

Variation in the magnetization among the mini-specimens is surprisingly large (Fig. 6). The intensity of natural remanent magnetization (NRM) of the recrystallized green chert is noticeably less than that of the red mini-specimens and the directions of the green samples are also significantly different, Moreover the mini-specimens adjacent to the green band appear to have magnetic properties intermediate to those of the green and the red chert. It appears that the recrystallization that changed the magnetization of the green chert also changed the magnetization but not the appearance of red chert adjacent to the band of recrystallization. This agrees with the hypothesis of Karl (this volume) that the recrystallized cherts were initially red and were later leached to green by some diagenetic process. These results are also consistent with Baltuck's suggestion that apparently random magnetic directions of radiolarian cherts in Greece could reflect two different remagnetization mechanisms: (1) remobilization of iron during long term diagenetic processes such as pressure solution and stylolitization and (2) complete remagnetization of a section of chert at different times by solution/precipitation of silica phases as plate tectonic processes move and rotate the plate on which the cherts are deposited (Baltuck, 1983, personal communication).

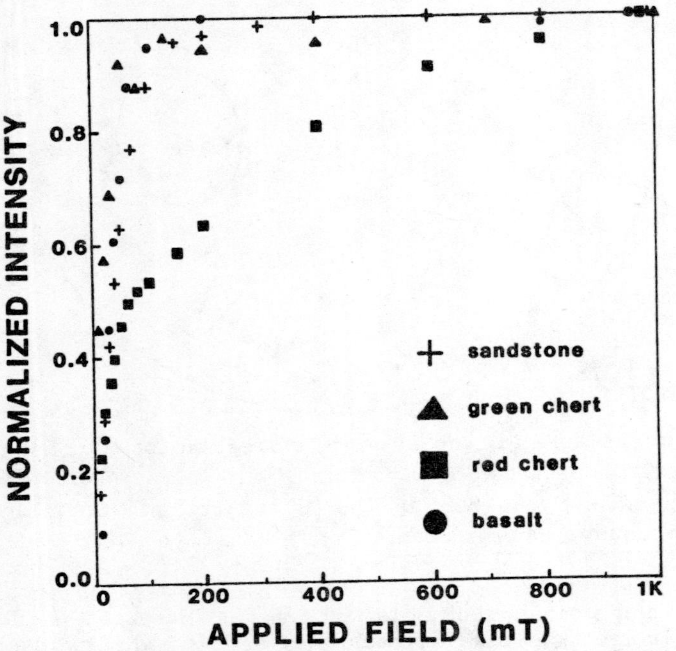

Fig. 7. Isothermal remanent magnetizations of specimens subjected to static magnetic fields.

Isothermal remanent magnetization (IRM) susceptability was evaluated by subjecting rock samples to a magnetic field of increasing intensity until they were saturated. These experiments reveal differences in IRM susceptability among basalt, sandstone, and both red and green chert specimens (Fig. 7). The dominant magnetic mineral is hematite in the red chert and magnetite in the green chert, basalt, and sandstone.

CLASTIC SEDIMENTARY ROCK INVESTIGATION

The presence of sandstone and siltstone at the top of the section indicates the arrival of the terrane at the continental margin in the Cenomanian. One sandstone and two siltstone localities were hand-sampled (Fig. 1). Demagnetization and analysis of the samples followed the procedures used in the basalt investigation.

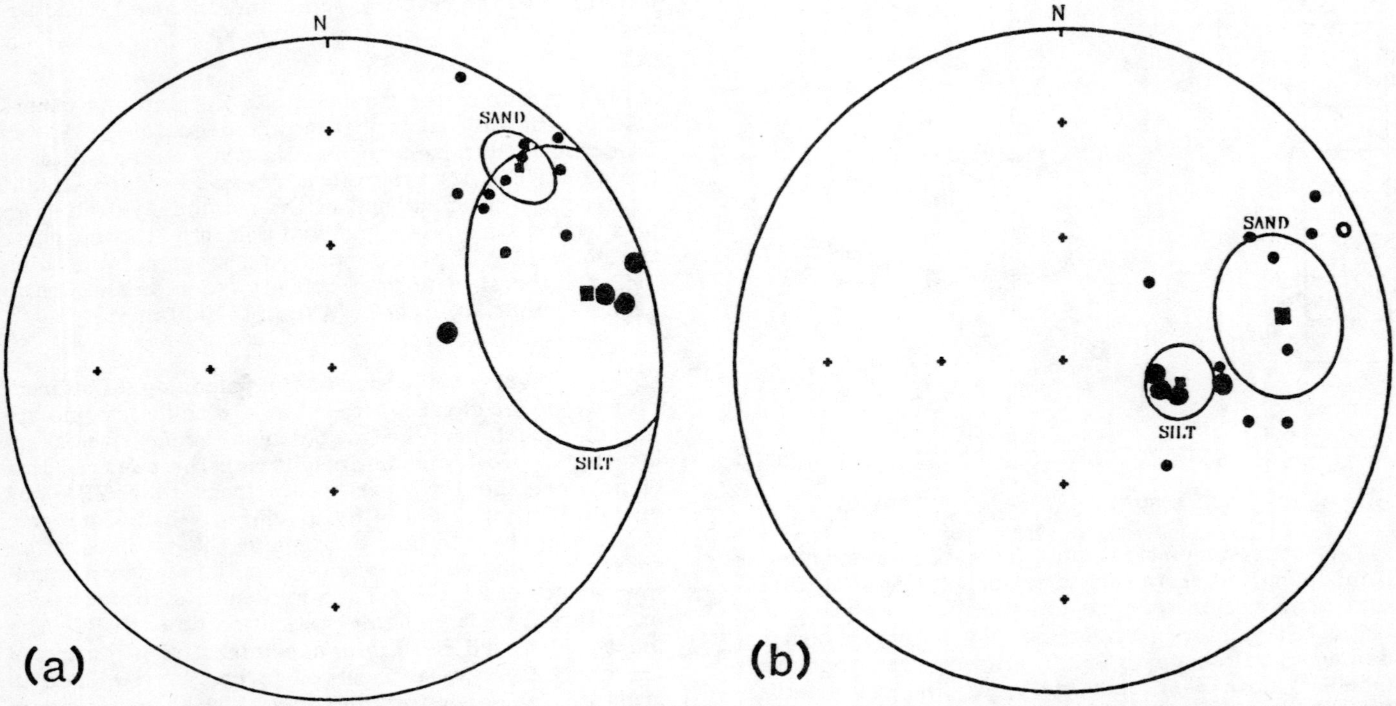

Fig. 8. Paleomagnetic results from clastic sedimentary rocks before (a) and after (b) correcton for tilt. Results are from sandstone (small circles) and siltstone (large circles), with site means (squares) centered within their circles of 95% confidence. Lower-hemisphere equal-area projections.

Thermal demagnetization of the clastic sedimentary rock specimens typically revealed the presence of a main component and a low-temperature overprint. A fold test applied to the main component of the sandstone and siltstone samples produced quite different results (Fig. 8). The sandstone directions group tightly around their mean before correction, while restoring the beds to horizontal causes a scattering of directions that is significant at the 95% confidence level. The siltstone directions instead show a significant reduction in scatter upon correcting for tilt.

The normal polarity of the Cenomanian siltstones is consistent with their having been deposited and magnetized entirely within the Cretaceous Normal Superchron (Fig. 2). The sandstones may not have acquired depositional magnetization because of their larger grain size and may have further suffered chemical remagnetization after folding because of their higher permeability to fluid flow.

COMPARISON WITH CRATONAL DIRECTIONS

In comparing the directions for the Marin Headlands with the expected field directions for cratonal North America we used the pole positions of Irving and Irving (1982) and the statistical techniques of Fisher (1953), Cox (1970, 1980), and Demarest (1983). We find substantial clockwise rotations for all three lithologies in the Marin Headlands (Fig. 9).

Comparison of the observed mean basalt direction with that expected on the basis of paleomagnetic results from cratonal North America indicates a clockwise rotation of 130.5°±11.1° (Fig. 10). The basalt inclinations are 17.0° steeper than expected; this difference is equivalent to a latitudinal displacement of 11.6°±6.8°. This would indicate movement of the block toward the paleo-equator after formation of the basalt, if indeed the directions are taken to represent the average orientation of the geomagnetic field at the site during the Pliensbachian. This anomalous result is not consistent with most tectonic models for the Pacific Basin, which indicate predominantly northward movement since the Jurassic (Atwater, 1970; Engebretson, 1983), nor with the faunal record in the chert sequence, which supports formation at a lower latitude than that of the present site (Murchey, this volume).

Because of uncertainty about the timing of chert magnetization, the entire group of chert directions is compared to the expected cratonal directions for 190, 180, and 170 Ma (Fig. 10). The expected inclination steepens rapidly between 190 and 170 Ma; the observed inclinations fall within this wide band of expected values. The observed declinations shows a clockwise rotation of 120°±20° with respect to the 170 Ma expected direction.

The observed directions from the Cenomanian siltstones are 5.9° shallower and rotated 128.9°±19.0° clockwise compared to the expected cratonal direction at 90 Ma. The difference between the observed and expected inclination indicates a poleward displacement of 7.4°±7.7° (Fig. 10).

For comparison, we plot the expected and observed directions for the Valanginian rocks of the Nicasio terrane (Gromme and Gluskoter, 1965; Gromme, this volume). The observed directions indicate 19.7°±7.8° poleward displacement and 106.4°±9.3° clockwise rotation.

INTERPRETATION

The observed inclinations do not provide conclusive evidence about changes with time of the latitude of the Marin Headlands terrane. The steep inclinations observed in the basalt may reflect their having been remagnetized long after extrusion when the expected directions were steeper. The age of recrystallization and remagnetization of the chert is uncertain as is the question of whether tilting occurred since their magnetization. Results from the siltstones indicate some poleward movement. The range of inclinations in the Marin Headlands are consistent with the results from the Nicasio terrane spilite and diabase (Gromme, this volume) but not with latitudinal shifts as large as those inferred for the Laytonville limestone (Alvarez and others, 1980).

The strongest conclusion that can be drawn from these paleomagnetic results is that the Marin Headlands has rotated nearly 130° since the Cenomanian. The amount of rotation, unlike the change in latitude, is not sensitive to age because the expected declinations change little. Does so large a clockwise rotation make geologic sense?

The main structural features in the Marin Headlands (Figs. 1,11b) are E-W trending thrust faults dipping to the south and asymmetrical folds overturned to the north (Wahrhaftig, this volume). These features, in their current orientation, would imply either subduction to the south or obduction to the north.

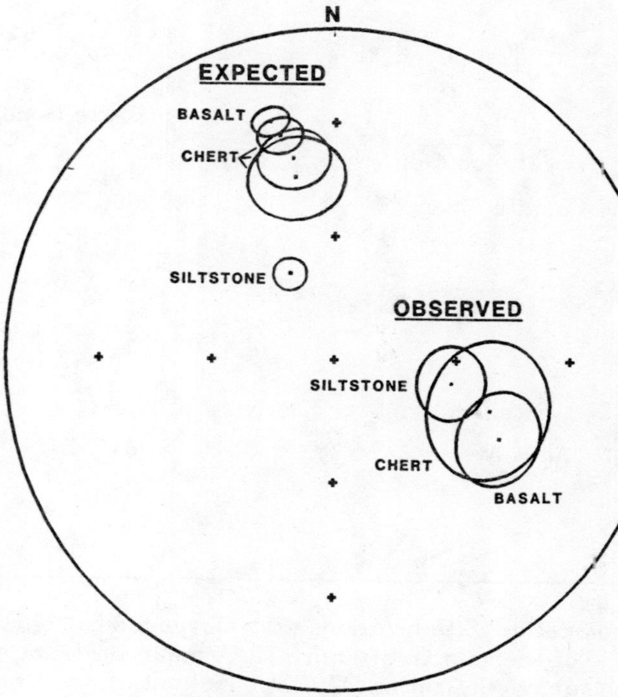

Fig. 9 Combined paleomagnetic results from basalt, chert, and siltstone samples in the Marin Headlands. Lower-hemisphere equal-area projections.

On the other hand, if we use the paleomagnetic data to rotate the Marin Headlands block counterclockwise 120° back to its approximate orientation during magnetization of the rocks (Fig. 11a), the inferred direction of subduction would be toward N60E, nearly perpendicular to the N20W trend of the Cretaceous Sierran magmatic arc (Stern and others, 1981), represented schematically (Fig. 11) by a trench parallel to the magmatic arc.

A counterclockwise restoration of the Marin Headlands based on paleomagnetic data is thus consistent with a simple tectonic model in which the dominant pattern of thrust faulting is associated with subduction perpendicular to the trend of the Sierra Nevada magmatic arc. The clockwise rotation of the faulting pattern and paleomagnetic data is consistent with right-lateral strike-slip faulting along the Pacific/North America plate margin.

Table 1. Site-mean data with group statistics

Marin Headlands Basalt

Site location	N	T	MDT	I_u	D_u	I_c	D_c
Tenn. Cove	3	610	425	50.5	71.2	52.6	128.6
Tenn. Point	4	590	400	41.8	50.1	43.6	93.8
Bonita Cove	6	580	380	25.3	80.9	43.2	87.2
Point Diablo	4	620	390	25.2	75.3	57.2	120.1
Kirby Cove	4	650	420	33.1	102.0	48.7	126.3
Lime Point	4	680	420	13.4	98.2	31.6	111.1
Cavallo Point	3	620	430	16.6	113.9	30.3	135.4
Fort Baker	3	660	450	26.4	97.9	39.2	118.3
Total sites:	8		Means:	30.3	88.0	44.4	115.1
				R_u =	7.522	R_c =	7.744
				K_u =	14.6	K_c =	27.3
				α_{95} =	15.0	α_{95} =	10.8

Marin Headlands Chert

Site location	N	T	MDT	I_u	D_u	I_c	D_c
Alex. Ave. (MH1)	5	610	425	19.5	76.3	46.2	116.2
Alex. Ave. (MH2)	10	650	480	21.2	80.5	50.8	102.8
Total samples:	15		Means:	20.6	79.0	49.4	107.7
				R_u =	12.541	R_c =	12.918
				K_u =	5.7	K_c =	6.7
				α_{95} =	17.6	α_{95} =	15.9

Marin Headlands Sandstone

Site location	N	T	MDT	I_u	D_u	I_c	D_c
Bonita Cove	11	650	390	19.0	44.1	33.6	79.0
				R_u =	10.664	R_c =	9.704
				K_u =	29.7	K_c =	7.7
				α_{95} =	8.5	α_{95} =	17.6

Marin Headlands Siltstone

Site location	N	T	MDT	I_u	D_u	I_c	D_c
Alex. Ave.	4	610	410	21.6	74.8	60.7	101.0
				R_u =	3.687	R_c =	3.973
				K_u =	9.6	K_c =	111.7
				α_{95} =	31.3	α_{95} =	8.7

Explanation: Site locations are referenced to Fig. 1. N, number of samples. T, final temperature in degrees C of thermal treatment. MDT, mean destructive temperature (°C) of main linear component of thermal demagnetization. I_u, D_u, inclination (degrees downward) and declination (degrees eastward) of main component of remanent magnetization before correction for tilt. I_c, D_c, inclination and declination after tilt correction. R, vector sum of N unit vectors. K, precision parameter (Fisher, 1953). α_{95}, semi-angle of cone of 95% confidence around mean direction (Fisher, 1953).

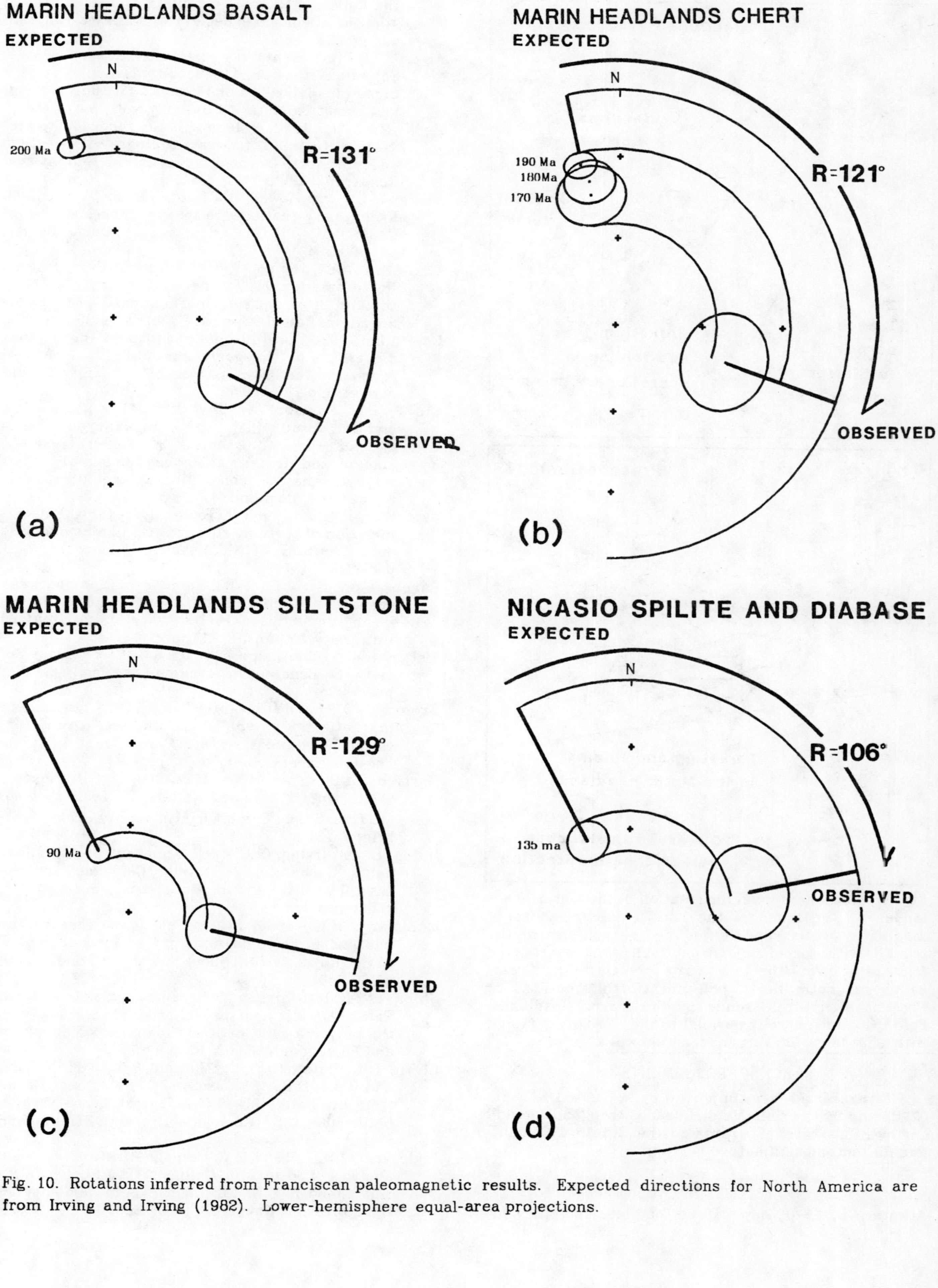

Fig. 10. Rotations inferred from Franciscan paleomagnetic results. Expected directions for North America are from Irving and Irving (1982). Lower-hemisphere equal-area projections.

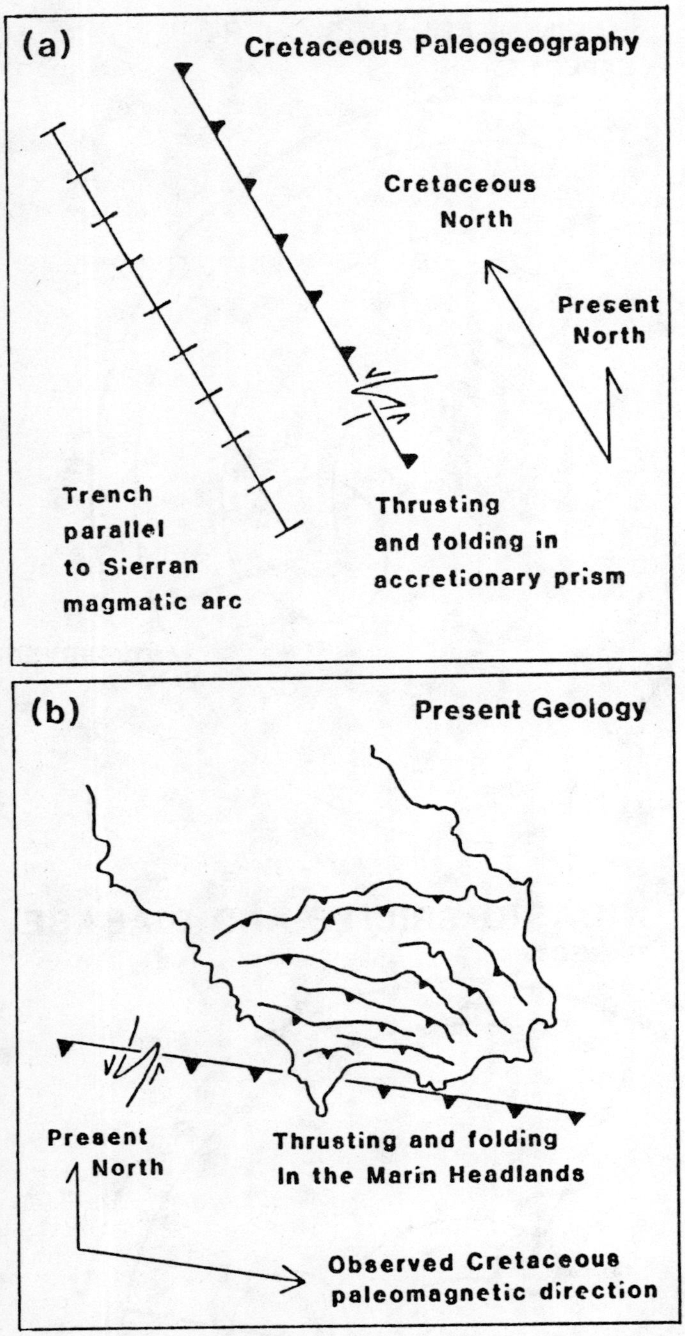

Fig. 11. Tectonic model for rotation of the Marin Headlands. a) The trend of the Late Cretaceous Sierran magmatic arc is represented by the azimuth of the trench, with associated thrust faults and overturned folds as shown. The long arrow points to the Cretaceous magnetic north pole for North America. b) Present structural trends in the Marin Headlands. Fold axes lie generally parallel to the E-W thrust faults, with the sense of vergence as shown.

ACKNOWLEDGEMENTS

This research was supported by NSF grant EAR 81-07702 and by the Gulf, Mobil, Texaco, Arco, and Aminoil corporations through grants to the Stanford Circum-Pacific Tectonics Affiliates Program.

REFERENCES CITED

Atwater, T., 1970, Implications of plate tectonics for the Cenozoic tectonic evolution of western North America: Geological Society of America Bulletin, v.81, p.3513-3536.

Alvarez, Walter, Kent, D.V., Premoli Silva, Isabella, Schweickert, R.A., 1980, Franciscan Complex limestone deposited at 17° south latitude: Geological Society of America Bulletin, v.91, p.476-484.

Beck, M.E., Jr., 1980, Paleomagnetic record of plate-margin tectonic processes along the western edge of North America: Journal of Geophysical Research, v.85, p.7115-7131.

Blake, M.C., Jr., Howell, D.G., and Jones, D.L., 1982, Preliminary tectonostratigraphic terrane map of California: U.S. Geological Survey Open-File Report 82-593.

Cox, Allan, 1980, The rotation of microplates in western North America, in, Strangway, D.W., ed., The Continental Crust and Its Mineral Deposits, Geological Society of Canada Special Paper 20, p.305-321.

Cox, Allan, 1970, Latitude dependence of the angular dispersion of the geomagnetic field: Geological Journal of the Royal Astronomical Society of London, v.20, p.253-269.

Crerar, D.A., Namson, Jay, Chyi, M.S., Williams, Loretta, and Feigenson, M.D., 1982, Manganiferous cherts of the Franciscan assemblage: I. General geology, ancient and modern analogues, and implications for hydrothermal convection at oceanic spreading centers: Economic Geology, v.77, p.519-541.

Demarest, H.H.,Jr., 1983, Error analysis for the determination of tectonic rotation from paleomagnetic data: Journal of Geophysical Research, v.88, p.4321-4328.

Engebretson, David C., 1983, Relative motions between oceanic and continental plates in the Pacific Basin: unpublished Ph.D. dissertation, Stanford University, California, 211pp.

Fisher, R.A., 1953, Dispersion on a sphere: Royal Society of London, Proceedings, series A, v.217, p.295-305.

Gromme, C.S. and Gluskoter, H.J., 1965, Remanent magnetization of spilite and diabase in the Franciscan Formation, western Marin County, California: Journal of Geology, v.73, p.74-94.

Harland, W.B., Cox, A.V., Llewellyn, P.G., Pickton, C.A.G., Smith, A.G., and Walters, R., 1982, A Geologic Time Scale: Cambridge University Press, Cambridge, 131 pp.

Irving, E. and Irving, G.A., 1982, Apparent polar wander paths, Carboniferous through Cenozoic, and the assembly of Gondwana: Geophysical Surveys, v.5, p.141-188.

McElhinny, M.W., 1964, Statistical significance of the fold test in paleomagnetism: Geophysical Journal of the Royal Astronomical Society of London, v.8, p.338-340.

Shibuya, Hidetoshi and Sasajima, Sadao, 1983, Paleomagnetism of red cherts: a case study in the Inuyama area, central Japan: Rock Magnetism and Paleogeophysics, v.10, p.87-92.

Stern, T.W., Bateman, P.C., Morgan, B.A., Newell, M.F., and Peck, D.L., 1981, Isotopic U-Pb ages of zircon from the granitoids of the Central Sierra Nevada, California: U.S. Geological Survey Professional Paper 1185.

Zijderveld, J.D.A., 1967, A.C. demagnetization of rocks: analysis of results: in Collinson, D.W., Creer, K.M., and Runcorn, S.K., eds. Methods in Paleomagnetism: Elsevier, New York, p.254-286.

GEOLOGY OF THE NICASIO RESERVOIR TERRANE, MARIN COUNTY, CALIFORNIA

Robert H. Wright
Harlan, Miller, Tait Associates
1269 Howard St., San Francisco, CA 94103

ABSTRACT

The results of petrographic and structural studies of the Franciscan Complex rocks exposed in an area extending from Mt. Tamalpais north to San Geronimo Valley support the terrane concept of Franciscan geology. The rocks include an east-northeast-dipping homoclinal sedimentary sequence (Nicasio Reservoir terrane) of Valanginian age, consisting of weakly metamorphosed, volcanic-rich feldspatholitic arenites overlying bedded chert and basalt. This sequence is diapirically intruded by serpentinized harzburgite tectonite and dunite, with minor cumulate(?) gabbro (Pine Mountain ultramafics), interpreted as a remnant of oceanic crust (ophiolite), along two parallel north-northwest-trending belts. Petrographic study of 62 samples of sandstone from the Nicasio Reservoir terrane indicates a QFL detrital framework mode of Q = 9 ± 3%, F = 36 ± 11%, L = 55 ± 12%. Sandstones are essentially K-feldspar-free, and the majority of lithic fragments (L) are plutonic (hypabyssal) and volcanic. The sandstone composition suggests an undissected volcanic magmatic arc provenance. Five new chemical analyses show the sandstones to have characteristics similar to those of sediments deposited near Andean-type plate margins.

INTRODUCTION

The area mapped for this study is located in south-central Marin County, California (fig. 1), and includes approximately 100 sq/km of portions of the Bolinas, San Geronimo, and San Rafael 7.5' quadrangles. It lies astride grassland-chaparral-oak-redwood covered northwest-trending ridges and valleys characteristic of parts of the northern Coast Ranges physiographic province.

The study is bounded on the south by the anomalous NNE-SSW trending crest of Mt. Tamalpais (elev. 784 m), and on the north by the E-W trending San Geronimo Valley (elev. 78 m), which contains the villages of Lagunitas, Forest Knolls, San Geronimo, and Woodacre. The western boundary approximately follows the NNW trending drainage of Lagunitas Creek, and the NNW-SSE trending Bolinas Ridge (elev. 504 m) lies just to the west. The eastern boundary approximately follows a line of NNW-SSE trending ridges, including White Hill (elev. 408 m), Blue Ridge (elev. 315 m), and Bald Hill (elev. 350 m), which separate the study area from the valleys to the east containing the villages Fairfax and Ross.

Portions of the study area were first mapped by Lawson and his students (1914), and subsequently by Weaver (1949). In both these earlier reports the rocks are included under "Franciscan Group undifferentiated." Lawson (1914), interpreted the structure of the area to be a syncline centered on the serpentine belt. Weaver's mapping covers only that portion located in the San Geronimo quadrangle, but his structure section shows a syncline centered in the greenstone of Bolinas Ridge to the west, and an anticline through the ridge separating Woodacre Valley from the eastern end of San Geronimo Valley.

Gluskoter (1962, 1969), established that the greenstone of Bolinas Ridge and the enclosing sedimentary units dip homoclinically to the northeast, and he shows a syncline centered on the serpentine belt. He included all the Franciscan rocks east of the greenstone in his "eastern zone," while all those west of the greenstone were included in his "western zone."

Salem Rice (unpublished field mapping, 1970-71) mapped the rocks of the crest and south flank of Mt. Tamalpais. Rice's mapping, among others, is included as part of a regional compilation of the geology of Marin and adjacent counties (Blake and others, 1974). This latter map establishes the regional framework within which the present study was initiated. The area mapped for the present study lies predominantly within the sheared unit (KJfs) of Blake and others (1974), and a syncline axis is implied by them to lie between two serpentine belts on their map. The KJfs unit is correlative with the Central melange belt of the Franciscan in the northern Coast Ranges (Blake and Jones, 1981). Some preliminary petrographic and chemical data from the study area has been presented by Blake and Wright (1976).

The concept of lithotectonic belts in the Franciscan Complex, each containing a different lithologic association, different metamorphic mineral assemblages, and fossils of different ages, is compatible with the application of plate tectonic theory to an interpretation of the evolution of the Franciscan. Hamilton (1969, 1978) interprets the Franciscan Complex as a type example of an accretionary complex composed of subparallel belts of successively younger-to-the-west, deformed oceanic materials, accreted sequentially to the continental margin by subduction over a period of time, now believed to extend into Miocene time (McLaughlin and others, 1983).

Dickinson (1971) convincingly argued that the Franciscan Complex, the Great Valley sequence, and the Sierra Nevada Batholith comprises the arc-, arc-trench gap, and trench setting typical of convergent plate boundaries. Ernst (1970) came to similar conclusions based on metamorphic mineral assemblages in the three regions.

While the plate tectonics and corollary subduction model is currently a widely accepted and convenient model to explain Franciscan-Great-Valley-Sierran relations, Blake and Jones (1981) point out that the model encounters difficulty when detailed correlations are made between different parts of the system. Current Franciscan-Great Valley discussion seems to be focusing on the provenance of age-equivalent and spatially adjacent Franciscan and Great Valley sequence rocks. Further discussion centers on whether the Franciscan is composed of distinguishable terranes, at least some of which may have had a geologic history separate from the simple accretionary complex, as conceived of by Dickinson (1971).

One line of evidence concerns the differences in grain size and modal and chemical compositions between some Franciscan graywackes and sandstones of the coeval Great Valley sequence. Blake and Jones (1981) note that differences in the Tithonian to Valanginian age Franciscan graywackes in the Yolla Bolly and sandstones of the coeval Great Valley sequence sandstones make it unlikely that they both were derived from the Sierran-Klamath source terrane. In addition, they question the material found in the Franciscan deposits came from across the fine-grained distal parts of the Great Valley sequence. They also present QFL

in Blake, M.C. Jr., ed., 1984, **Franciscan Geology of Northern California:** Pacific Section S.E.P.M., Vol. 43, p. 99-112

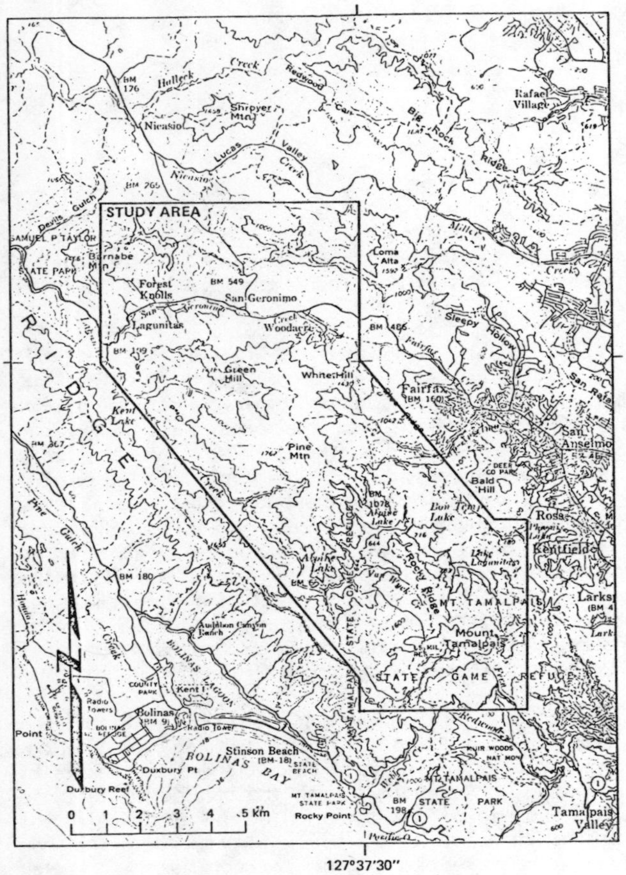

Figure 1. Location map of study area.

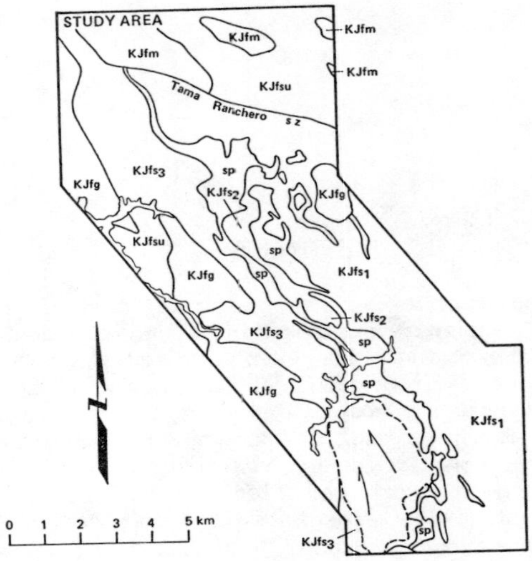

KJ_{fs1-3} Sedimentary rocks of Nicasio Resevoir terrane

K_{ffg} Greenstone of Nicasio Resevoir terrane

sp Pine Mountain ultramafics

KJ_{fsu} Undifferentiated Central Franciscan belt

KJ_{fm} Yolla Bolly terrane

Figure 2. Generalized geologic map showing distribution of geologic units in the study area.

plot of modal data for Franciscan graywackes ranging from Tithonian to Campanian, together with the petrofacies fields for the Great Valley sequence (Dickinson and Rich, 1972) and point out some major differences. The most pronounced differences are between the Franciscan graywackes of Tithonian age in the Diablo Range (Jacobson, 1978), and the nearly coeval Stony Creek petrofacies of the Great Valley sequence. They interpret this to indicate that Franciscan quartzofeldspathic sandstone was being deposited at the same time as, and necessarily with some considerable geographic separation from, the fine-grained silceous volcanogenic rocks being deposited on the Coast Range ophiolite.

A second line of evidence is the compositional and chemical differences between sandstones of the Great Valley sequence overlying ophiolite in several "western outliers," and sandstones of the coeval Great Valley sequence father east. They suggest that it is unlikely these rocks had any direct continuity and postulate that the "western outliers" may be allochthonous.

There is some paleomagnetic evidence to support large scale northward translation of Franciscan rocks (Alvarez and others, 1980), and other Late Cretaceous rocks of northern California (Champion and others, 1981). Accumulating evidence suggests that northward translations and clockwise rotation in relative plate motion between the Pacific and North American plates were important elements in the Mesozoic history of the Pacific margin of North America (Hillhouse, 1977; Beck, 1976, 1980; Packer and Stone, 1974; Stone and Packer, 1977, 1979; Jones and others, 1977, 1978b; Saleeby and others, 1978; Davis and others, 1978; Saleeby, 1981). Recent evidence that a long interval of time (40-55 m.y.) occured between deposition and subduction of some Franciscan rocks provides a time framework for such northward translations (Mattinson and Echeveria, 1980). Of major significance to interpretations of provenance of rocks in the study area is the interpretation that the greenstones of the Nicasio Reservoir terrane, which structurally and stratigraphically underlie the rocks of the study area, are allochthonous and have been rotated clockwise some 100° and translated relatively northward from about 18°N (Gromme, this volume).

Dickinson and others (1982) argue that "...continued speculation that any significant fraction of the sandy detritus in the Franciscan came from sources outside North America ...is...unwarranted." They conclude from a systematic compariston of detrital modes for Franciscan sandstone and coeval sandstones of the Great Valley sequence that compositons show a "major overlap," and that both groups of rocks were apparently derived from the same general ancestral Sierran-Klamath magnetic area source terrane. They believe that delievery of the arc-derived detritus to the Franciscan trench was accomplished by submarine canyons crossing the Great Valley forearc basin, and by delivery to the ends of the Franciscan trench with subsequent dispersal by longitudinal flow.

Dickinson and others (1982) attribute differences in compositions to derivation from different segments of the arc terrane, complex recycling processes on the trench slope, and diagenetic effects in the Franciscan. They also correctly point out that meaningful comparison of Franciscan and Great Valley compositions should be made for rocks of equivalent age and close areal proximity, even though age relationships within the Franciscan are not, as yet, as well-defined as within the Great Valley sequence.

The primary objectives of this study were to (1) determine the distribution of and structural relationships between the serpentinite and associated Franciscan sedimentary rocks in Central Marin County, and to (2) apply the petrographic techniques of Dickinson (1970; Dickinson and Rich, 1972) to the Franciscan sedimentary rocks.

To accomplish the objectives of this study, an air photo interpretation of the approximately 100 sq/km study area was made using black and white (scale 1:12,000), and color (scale 1:15,000) stereo air photographs. Field reconnaissance and detailed field mapping were conducted in the study area and vicinity during a period of 6 years (1970-76). Approximately 1,650 field data points were established. From the samples collected in the field, 204 were slabbed and stained for potassium-feldspar determination, and 608 petrographic thin sections were prepared, stained for potassium and plagioclase feldspar, and inspected under the petrographic microscope. Of the 608 thin sections, 62 were from the central portion of the study area selected for detailed petrographic modal analysis, and 5 for chemical analysis. Twenty-six chert samples were submitted for biostratigraphic age determination.

DESCRIPTION OF UNITS

Undifferentiated Sedimentary and Associated Rocks

A NNW-SSE trending, northeast dipping sequence of clastic sedimentary rocks with associated chert and greenstone occurs in the northeast portion of the study area, north of San Geronimo Valley (fig. 2). Included in this sequence are several large fault-bounded masses of higher metamorphic grade rocks (KJfm = Yolla Bolly terrane of Blake and others, this volume). All these rocks are separated from the rocks in the remainder of the study area by the WNW-ESE trending Tama Ranchero high angle shear zone (Berkland, 1969).

The clastic sedimentary rocks consists of bedded to massive, medium- to coarse-grained, predominantly feldspatholithic graywacke sandstone, interbedded with siltstone and mudstone. Where weathered, sandstones are variably fractured but often crop out in more resistant exposures within a matrix of the relatively more sheared siltstone and mudstone.

The common metamorphic mineral assemblage is quartz-albite-chlorite-phengite±prehnite±pumpellyite. Fabric ranges from textural zone 1 to textural zone 2 of Blake and others (1967). Potassium-feldspar is absent (Wright, 1974).

Included within the clastic sequence are discrete masses or "knockers" of greenstone and chert which may represent tectonically disrupted, originally interbedded chert and pillow basalts and/or tectonic inclusions derived from other terranes.

No fossils have been found in the clastic sedimentary rocks and no ages have been obtained from the cherts in this area. These rocks are herein included as undifferentiated sedimentary and associated rocks of the Central belt.

An area of undifferentiated clastic sedimentary rocks separating outcrops of the greenstone of the Nicasio Reservoir terrane (KJfg) located southwest of Big Carson Creek, is also shown on the geologic map (fig. 3) as KJfsu. The relationship of these rocks to those in the rest of the study area, particularly the Nicasio Reservoir terrane, in unknown and probably complicated by faulting. These rocks may be correlative with the sedimentary rocks of the San Bruno Mountain terrane which structurally underlies the Nicasio Reservoir terrane.

Nicasio Reservoir terrane

The greater part of the study area is underlain by a sequence of clastic sedimentary rocks that was deposited on radiolarian chert and basalt. This sequence of rocks has been called the Nicasio Reservoir terrane (Blake and others, 1982; this volume).

Greenstone and associated rocks

Portions of several large masses of altered basic intrusive and volcanic rocks, with minor intercalated chert, crop out along the western boundary of the study area (figs. 2, 3). These masses represent portions of the southern two of three elongate northwest (35°) trending masses comprising the "central spilitic zone" of Gluskoter (1962), who investigated the petrology and structure of these rocks in detail (1962, 1964, 1969; Gromme and Gluskoter, 1965). These masses extend in a linear belt from the west shoulder of Mount Tamalpais, northwest to the vicinity of the Nicasio Reservoir, a distance of approximately 25 km. Locally, these masses have an apparent thickness greater than 2,400 meters. Several smaller bodies of similar basic volcanic rocks occur in the study area north of Forest Knolls and along the eastern boundary underlying White Hill and portions of Blue Ridge and Bald Hill.

The rocks consists predominantly of massive to pillow, locally vesicular basalt flows, with lesser amounts of tuff and volcanic breccia, and intrusive, fine- to coarse-grained diabase and rare gabbro. Thin-bedded radiolarian chert in lenses up to 30 meters thick occurs locally near the base of all three of the largest masses, and locally at the top of the southernmost large mass near Oat Hill. Radiolarian chert is also associated with the apparently correlative massive to pillow, locally vesicular, basalt masses in the White Hill-Blue Ridge area.

All the rocks within these masses are altered to a greater or lesser degree and are generally termed greenstone after the greenish color of less weathered outcrops. The common mineral assemblage is relic augite±ilmenite- albite-chlorite-magnetite±apatite±epidote±laumontite± pumpellyite-±calcite±orthoclase. These minerals are stable in the sub-greenschist metamorphic facies.

Where "fresh" these rocks are dark greenish gray or olive gray to grayish olive green. In general, they are deeply weathered and light or moderate reddish brown. Deeply weathered outcrops of greenstone are sometimes difficult to distinguish from deeply weathered sandstone, particularly if pillow structure in the greenstone has been completely obliterated. Weathered greenstone characteristically break along a seemingly infinite number of very dusky red, iron oxide coated fractured surfaces, into smaller and smaller fragments. Weathered sandstone, on the other hand, generally yields "fresh" surfaces after a few hammer blows, after which its clastic character can be observed with a hand lens.

The association of greenstone with cherts, and the pillow structure of the flows, together with the absence of significant amounts of clastic sediments within the masses, suggests that most of these rocks were extruded onto the open ocean sea floor. Gluskoter (1962) presented chemical analyses on 1 albite diabase, 7 pillow basalts, 3 samples of the matrix surrounding the pillow basalts, and 1 labradorite diabase. He concluded that the range in chemical composition may, in part, represent differentiation of a parent tholeiite magma, or increasing contamination by sea water with time (see Blake and others, this volume for a discussion of origin of the volcanic rocks).

The contact relationships of the greenstone masses with the surrounding sedimentary rocks are not clear. In most areas the contacts are poorly exposed or obscured due to dense vegetative cover, deep weathering, overlying surficial and landslide deposits, and modifications by landsliding. The lower (western) contact with the structurally underlying potassium feldspar-bearing sedimentary rocks of the San Bruno Mountain terrane (Blake and others this volume) is everywhere where exposed, a sheared and disturbed zone,

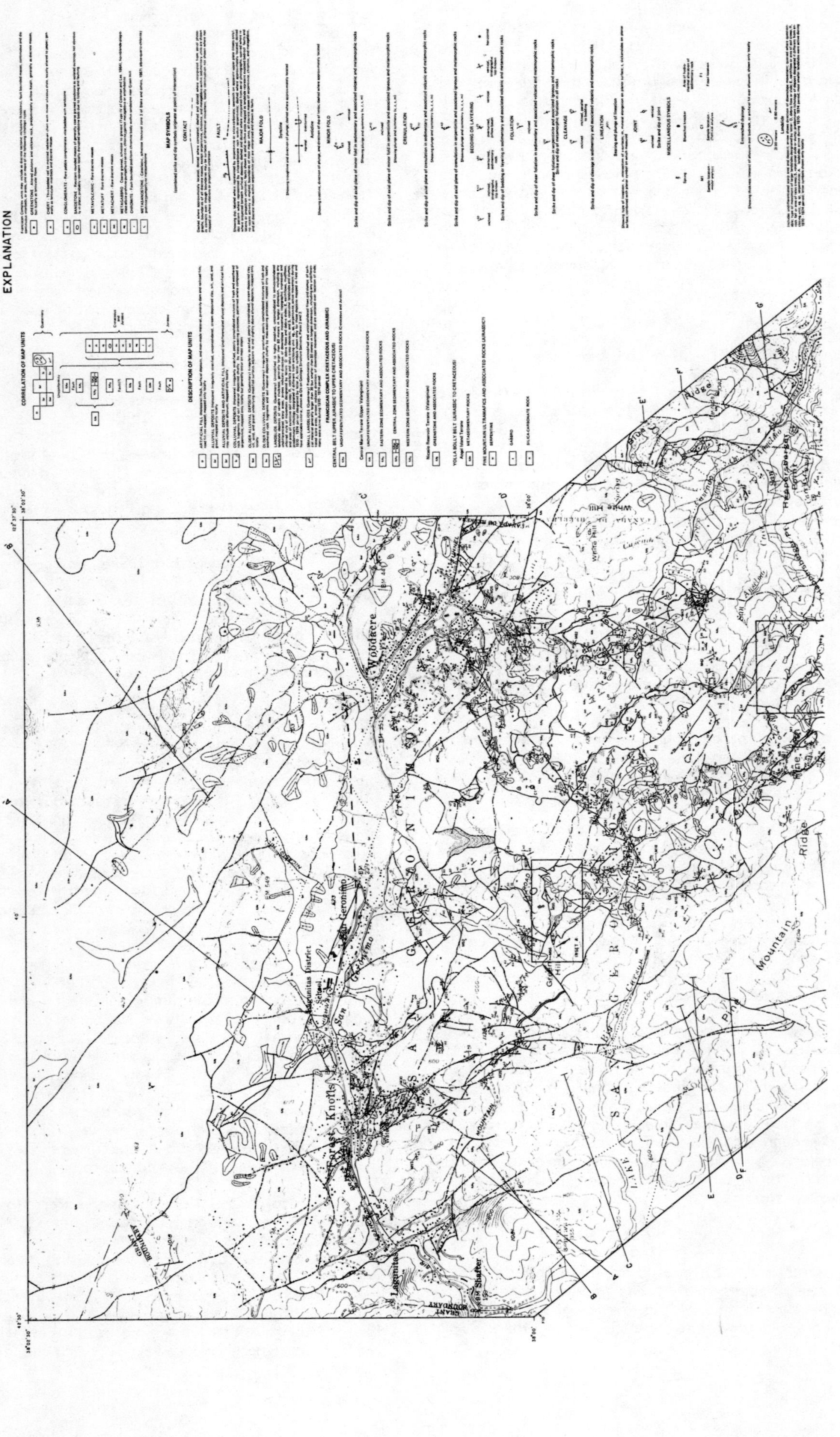
EXPLANATION
CORRELATION OF MAP UNITS
DESCRIPTION OF MAP UNITS
MAP SYMBOLS
Woodacre
Lagunitas
Knolls
Mountain
Ridge

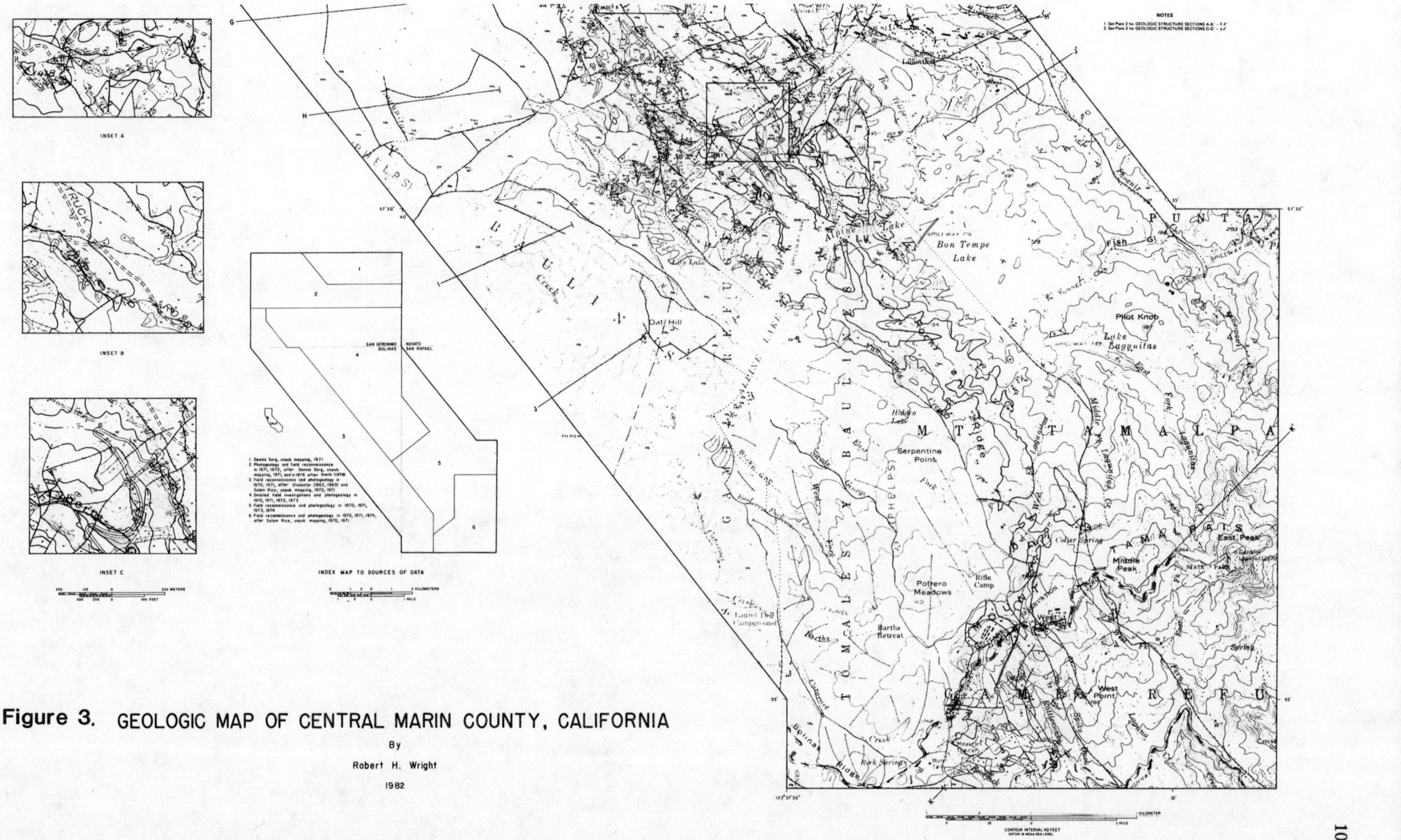

Figure 3. GEOLOGIC MAP OF CENTRAL MARIN COUNTY, CALIFORNIA

By

Robert H. Wright

1982

locally up to 30 meters wide.

Gluskoter (1962) interpreted the upper (eastern) contact of the greenstone masses to be conformable with the structurally and stratigraphically(?) overlying sedimentary rocks of his "eastern zone." In a general sense, this interpretation appears to be accurate. However, sheared serpentinite and foliated cataclastic rocks are locally present along the contact. This suggests that some small scale differential movement has taken place between the relatively more competent greenstone masses and the overlying sedimentary rocks and the contact is therefore almost everywhere interpreted as faulted.

Northeast of Oat Hill along the east side of Lily Gulch (fig. 3), thin-bedded, east- to northeast-dipping chert rests conformably on pillow basalt (greenstone), and is, in turn, conformably overlain by bedded sandstone of the Nicasio Reservoir terrane. Although the base of the sequence is separated from the main greenstone mass by some 500 meters, which includes a section of bedded sandstone ($KJfs_3$), landsliding, and probably two faults, this sequence is most reasonably interpreted as representing an undisturbed portion of the original gradational contact between the large greenstone masses and the overlying sedimentary rocks.

Sedimentary rocks

The clastic sedimentary rocks of the Nicasio Reservoir terrane consist predominantly of weakly metamorphosed, bedded to massive, fine- to coarse-grained lithic sandstone, locally tuffaceous, and interbedded siltstone and mudstone, with rare pebble conglomerate (cg). The sandstones are composed predominantly of angular to rounded grains of feldspar, quartz and lithic fragments, and can be classified as volcanic-rich feldspatholithic arenites. Potassium feldspar is absent (Wright, 1974).

Most outcrops consist of little-weathered, dark greenish-grey, medium-to coarse-grained thick bedded to massive sandstone. Although invariably moderately fractured, outcrops are generally hard and strong to locally very strong, breaking across cemented fractures into angular fragments only with several hammer blows. Where locally moderately to deeply weathered, these outcrops are greyish orange to light olive brown and friable.

Outcrops of bedded sequences of sandstone, siltstone, and mudstone are less common, but occur locally in gullies and creeks. All outcrops are invariably fractured. In some cases these bedded sequences consists of regular interbeds with individual beds ranging from a few centimeters up to a meter thick. Primary sedimentary structures are uncommon in these outcrops, with bedding often being defined by discontinuous trains of mudstone clasts and laminations, along which parting has occurred. Locally interbedded mudstone occurs in beds up to 6 centimeters thick.

More commonly, bedded sequences are rhythmically bedded with individual beds ranging from laminations to beds up to 0.3 meters thick or more, and sequences ranging up to a meter or more in thickness. Graded bedding is common in rhythmically bedded sequences, as are rip-up mudstone clasts, convolute bedding and flame structures, cross and parallel laminae, cross bedding, and concentrations of carbon. Load casts are locally present. These rhythmically bedded sequences suggest low energy turbidity current disposition.

Locally, differential movement has occurred in bedded sequences preferentially in the finer-grained siltstone and mudstones, resulting in a chaotic sequence of boudined and telescoped fragments of sandstone beds in a contorted matrix of siltstone and mudstone. Little can be interpreted from such outcrops except that they were once a bedded sequence. Similar differential movement in thick bedded to massive sandstone has resulted in resistant, rounded outcrops isolated in a subdued sheared matrix.

Most outcrops are variably fractured or jointed. In unweathered outcrops, fractures are generally cemented and tight and the rock often breaks across fractures into angular fragments. Where outcrops are weathered, fractures are often coated with iron-oxide and the rock tends to break along these fractures, although a fresh surface can usually be obtained unless the rock is completely weathered. Where fractures occur at high angles to bedding, they often combine with a bedding parallel, slip cleavage to form a rhombic-shaped pattern, which in more weathered outcrops, tend to form rounded, crescent-shaped fragments.

The sandstones of the Nicasio Reservoir terrane are weakly metamorphosed. In the field, except for the textural zone 2 cataclastic metasandstones (f_1), rock fabric is textural zone 1 of Blake and others (1967). In thin sections, most samples are isotropic, but some samples have a slight fabric with narrow anastomosing shear zones up to 2 mm wide and could be classified as textural zone 2A. The common metamorphic mineral assemblage present is quartz-albite-chlorite-phengite±pumpellyite±lawsonite±prehnite.

Five sandstone samples from the Nicasio Reservoir terrane were subjected to chemical analysis and the results are presented in table 1. The compositions have characteristics of sediments deposited near both Andean-type margins (quartz 15-65%, SiO_2 68-74%, and K_2O/Na_2O 1.0), and west Pacific-type margins (quartz 15%, SiO_2 ave. 58%, K_2O/Na_2O 1.0) of Crook (1974), and Schwab (1975), but are most similar to Andean-type margins.

Associated with the clastic sedimentary rocks are thin bedded radiolarian chert (ch) and altered pillow or flow basalt (gs) which locally occur as interbeds and discrete masses throughout the clastic sedimentary sequence. The chert is generally hard and brittle. Where iron-rich, it is generally red or mottled to dark reddish brown to greyish green. Radiolarian tests can usually be seen in red and green colored samples with a hand lens, and often can be seen with the naked eye. They are clearly visible in thin sections. Iron poor varieties and altered chert are usually greyish yellow to white, and barren of preserved radiolaria. Chert intrebedded with sandstone is usually present in thin-bedded sequences up to several meters thick. Individual beds range in thickness from a few centimeters to 150 cm or more, and are separated by thin, hard siliceous shale interbeds, locally up to a few centimeters thick. Most thin bedded chert has a characteristic "pinch and swell" structure.

The contact relationships of most lenticular chert sequences interbedded with sandstones are obscure. However, the conformable nature of chert bedding with that in the immediately overlying and underlying sandstone, together with their lateral continuity along strike for tens of meters strongly suggests that most of the lenticular chert sequences are in fact interbedded within the sandstone sequence and are not tectonic inclusions. In addition, rounded to elongate rip-up clasts of chert locally occur within the basal few meters of the overlying sedimentary rocks.

Radiolarian chert also occurs as discrete masses within the sedimentary sequence. Most of these masses are rounded to elongate and, in many, the thin bedded chert is contorted and isoclinically folded. Some of these masses include attached greenstone. It is probable that most of the masses are fragments of once locally interbedded chert that have been disrupted and dismembered by folding and faulting, and perhaps moved around by landsliding. Alternatively they may be, in part, detrital fragments and/or tectonic inclusions, possibly related to the serpentinite.

Greenstone (gs) locally occurs throughout the clastic sedimentary sequence, predominantly as discrete, rounded to elongate masses, but sometimes as lenticular, intercalated flows. Most greenstones within the discrete masses appear to be deeply weathered pillow basalts similar to those described for the Nicasio Reservoir terrane, and some have attached, thin-bedded radiolarian chert.

One body of deeply weathered greenstone, about 30 meters thick, occurs within the clastic sequence and is conformable to the bedding in the overlying and underlying sandstones over a lateral distance of about 500 meters. This body most probably originated as either a basalt flow which was extruded onto the underlying sediments or perhaps as a sill-like body intruded into semi-consolidated sediments from a fissure feeder.

It is possible that most discrete greenstone masses are fragments of once locally intercalated flows which have been disrupted and dismembered by folding and faulting. Alternatively, they may be, in part, detrital fragments and/or tectonic inclusions. The fact that the plutonic-volcanic portion of the Nicasio Reservoir terrane stratigraphically underlies the clastic sedimentary sequence, and almost certainly was a source of sand-sized detritus provided to the sequence, suggests that many of the discrete masses may be larger detrital fragments.

In addition to chert and greenstone, rare discrete masses of metavolcanic (mv), metatuff (mt), and metachert (mch), occur within the clastic sedimentary sequence. These "knockers" may be detrital and/or tectonic, possibly related to the serpentinite.

The rocks of the Nicasio Reservoir terrane within this study area are grossly separated into three belts by the NW-SE trending serpentinite of the Pine Mountain ultramafics. Where outcrops of the Nicasio Reservoir terrane are clearly separated by relatively continuous, clearly defined serpentinite belts, they have been artificially divided into three separate zones. Western zone rocks ($KJfs_3$), are those which occur southwest of the western serpentinite belt; those between the two serpentinite belts are termed central zone rocks ($KJfs_2$); while those east of the eastern serpentinite belt are termed eastern zone rocks ($KJfs_1$). This purely artificial breakdown is made only for the purpose of emphasizing the serpentinite distribution and evaluating its structural relationship to the surrounding sedimentary rocks. In a gross sense, this breakdown represents a stratigraphic and age trend in that western zone rocks are probably older and lower in the section than eastern zone rocks. In areas that are not clearly west, east or between the serpentinite belts, and particularly where relationships are complicated by faulting or where the serpentinite outcrops are discontinuous, the Nicasio Reservoir terrane rocks are undifferentiated (KJfs).

East and southeast of Pine Mountain, central zone ($KJfs_2$) rocks have been divided on the basis of lithology, outcrop patterns, and bedding attitudes into an upper ($KJfs_{2u}$) and lower ($KJfs_2$) unit. The upper unit consists of predominantly medium- to coarse-grained, thick bedded to massive sandstone in hard resistant outcrops. The lower unit consists of relatively finer grained, thinner bedded sandstone in more subdued outcrops. The contact between these two lithologies and outcrop patterns can be mapped in the field for some distance until they are faulted and cut out by serpentinite. These two subdivisions of the central zone are the only map units within the Nicasio Reservoir terrane that are implied to have a stratigraphic and structural basis.

Age

One megafossil has been reported found in the study area (Wright, 1974). A single Early Cretaceous (Valanginian) Buchia pacifica was collected by S. J. Rice near Rock Springs at the southwest corner of the study area (fig. 3). The fossil was collected from sedimentary rocks near the upper contact of the greenstone of the Nicasio Reservoir terrane. Although the contact between the greenstone and overlying sedimentary rock is probably faulted in this area, the fossil location is near the base of the sedimentary sequence and the fossil dates the base of the clastic sedimentary sequence as Valanginian.

Twenty samples of radiolarian chert from the study area and from the San Bruno Mountain terrane, west of the study area, were collected and submitted in 1973 to C. A. Pessagno, Jr., of the University of Texas at Dallas for biostratigraphic age determinations. Of the 20 samples submitted, 9 were barren and 11 provided sufficient fauna for age determinations. These initial samples from cherts in the Nicasio Reservoir terrane yielded age determinations of Tithonian (?) to Neocomian, or Tithonian (?), with one (F-3) yielding a firm Tithonian call.

Because all of the initial samples that yielded age determinations were from discrete masses which cannot clearly be said to be interbedded within, and therefore date, the sedimentary sequence, and because these initial age determinations were made during the early developmental days of the science of radiolarian biostratigraphy which has since been refined significantly, a number of additional chert samples were collected from other chert localities within the study area. These samples were submitted in 1981 to D. L. Jones of the U.S. Geological Survey for biostratigraphic age determinations. Preliminary results from these samples confirm the Valanginian age of chert deposits on the basalt of the Nicasio Reservoir terrane. However, cherts occurring as lenses or blocks in the lower part of the overlying clastic sedimentary rocks include Tithonian and older fossils (Murchey and Jones, this volume). These Tithonian and older age determinations from discrete masses of chert in the Nicasio Reservoir terrane indicate that these cannot all be considered to have been originally interbedded within the clastic sedimentary sequence and at least some must be detrital or tectonic inclusions. Sample F17 is among the oldest radiolarian chert ages yet reported from the Franciscan.

Pine Mountain Ultramafics and Associated Rocks

A discountinuous zone of serpentinized ultramafic and associated rocks occurs in central Marin County along a roughly linear trend extending from the south slope of Mount Tamalpais northwest to the Nicasio Reservoir, a distance of approximately 24 km (Blake and others, 1974). The most extensive and continuous exposures of these rocks occur between Mount Tamalpais on the south and San Geronimo Valley on the north, and they are herein informally called the Pine Mountain ultramafics after a prominent hill which they underlie.

In the study area these rocks, in a gross sense, crop out in two roughly parallel belts underlying two prominent, approximately N35°W trending ridges. However, this outcrop pattern is complicated by high-angle faulting, as well as by the apparent mobility of these rocks within the surrounding sedimentary rocks, and by colluvial and large scale mass-wasting surficial processes. These factors can account for all the occurrences of serpentinized ultramafic and associated rocks away from the main outcrops. No serpentinized ultramafic or associated rocks which can be interpreted to have a sedimentary origin were found in the study area.

The Pine Mountain ultramafics consists predominantly of partially to completely serpentinized peridotite (sp), or harzburgite tectonite and dunite, with minor associated gabbro (g), and rare podiform chromite (c). Associated with these rocks are silica-carbonate rock (sc), and metagabbro tectonic inclusions (mg). Other minor tectonic inclusions and roof pendant-like masses within the serpentinite include blocks of sandstone, greenstone (gs) and chert (ch) from the surrounding sedimentary rocks.

The exposed contact relationships of the serpentinite masses with the surrounding rocks are sharp, genrally uncomfortable, and unmetamorphosed. All boundaries of serpentinite masses are interpreted to be faults. In some areas high angle to vertical faults cut-serpentinite masses, displacing contacts. In areas where tongue-like protrusions extend from large masses into the surrounding rocks, contacts are sinuous, and steeply dipping. In general, however, contacts of larger less disrupted masses, particulary those along higher ridge tops, dip at near horizontal to low angles toward (into the slope) the center of the masses. The dips of these contacts steepen progressively with depth. These contact relationships suggest intrusion of a cold solid mass.

Two small, relatively fresh non-foliate gabbro (g) masses associated with the serpentinite occur east and north-northeast of Green Hill (fig. 3). The gabbro is a fine- to medium-grained hornblende gabbro composed of brown hornblende, clinopyroxene (augite?), plagioclase, apatite, and magnetite. Gluskoter (1962) reported the plagioclase composition as An 60-55. The chemical composition of one sample (C3) from the mass southeast of Green Hill is similar to those of cumulate gabbros from the Mesozoic Coast Range ophiolite described by Bailey and Blake (1974, table 3), and has the low alkali character of oceanic ridge equivalents.

Both gabbro masses are in partial contact with sandstone. However, there is no evidence to suggest that they were intruded along the contact between the sandstones and serpentinite. The contacts between the gabbro masses and the serpentinite are sharp, but show no evidence of tectonic movement. These two masses are interpreted to be portions of gabbro dikes cutting the serpentinite.

One small, fault-bounded podiform mass of black granular chromite was mapped just west of White Hill. The chromite body is about 20 meters wide and 100 meters long. Although enclosed within sandstone of the Nicasio Reservoir terrane, this mass is in close proximity to large serpentinite outcrops and is clearly related to the serpentinite.

Silica-carbonate rock is commonly found along the margins of serpentinite masses and along fault zones within serpentinite masses, and clearly is a post-serpentinization product. There is no evidence of hydrothermal alteration of the adjacent sedimentary rocks, suggesting that most of the contact alteration took place prior to emplacement of the silica-carbonate rock masses at their present level. Only in a road cut along Sir Francis Drake Highway near San Geronimo is silica-carbonate rock in proximity to hydrothermally altered sedimentary rocks, and at this location the relationships are complicated by high-angle faulting. No evidence for ongoing silica-carbonization was found in the study area except possibly in the vicinity of several active springs in the area south of Forest Knolls.

A number of coarsely crystalline, schistose to gneissic "high grade" metamorphic blocks are associated with the serpentinite. These blocks are ovoid to spherical and range from a few meters in diameter to up to 225 meters in longest dimension. The blocks characteristically occur within the westernmost serpentinite belt. A few occur in close proximity to and downslope of serpentinite masses, and have either been transported downslope from the serpentinite by mass wasting processes, or faulted into the sedimentary rocks.

TABLE 1 CHEMICAL ANALYSES OF SANDSTONES, NICASIO RESEVOIR TERRANE

SiO_2	TiO_2	Al_2O_3	Fe_2O_3	FeO	MnO	MgO	CaO	Na_2O	K_2O	P_2O_5	H_2O^+	H_2O^-	CO_2	Sum	PD	$\frac{K_2O}{Na_2O}$	Q%
Sample: Map No.: C5 Field No. SG72W-236 Lab. No.: W-182727																	
70.1	0.53	12.7	1.6	2.7	0.06	3.3	0.95	2.7	1.5	0.12	2.8	0.53	0.01	100	2.60	0.56	14
Sample: Map No.: C6 Field No. SG72W-7 Lab. No.: W-182728																	
69.5	0.48	13.7	3.8	1.0	0.07	2.1	1.5	3.7	0.90	0.08	2.7	0.91	0.01	100	2.51	0.24	19
Sample: Map No.: C7 Field No. SG72W-884 Lab. No.: W-182729																	
72.7	0.51	12.7	1.9	1.7	0.06	1.5	1.1	2.9	1.6	0.08	2.1	0.49	0.01	99	2.58	0.55	14
Sample: Map No.: C8 Field No. SG72W-436 Lab. No.: W-182730																	
70.4	0.64	12.5	2.8	1.9	0.05	2.7	0.66	2.3	1.3	0.13	2.8	0.80	0.01	99	2.64	0.57	16
Sample: Map No.: C9 Field No. B73W-27 Lab. No.: W-182731																	
67.7	0.68	13.3	2.5	4.0	0.08	2.9	0.59	3.0	1.0	0.12	3.2	0.46	0.02	100	2.61	0.33	14

NOTE: See Fig 3. for location of sample.
Rapid rock and powder density performed by USGS, Washington, D.C., Analyst: Hezekiah Smith.

PD = Powder Density Q% = Framework Quartz % from TABLE 2.

The "high grade" metamorphic rocks in the study area consist predominantly of metagabbros (amphibolites) equivalent to type IV-A metabasalts of Coleman and Lee (1963). These rocks are coarsely-crystalline gneissic schists containing the mineral assemblage green to brown hornblende-plagioclase (albite- prehnite)-pumpellyite-sericite-chlorite±garnet±sphene. One schistose amphibolite(?) equivalent to type IV-A metabasalt (Coleman and Lee, 1963), contains significant secondary glaucophane and lawsonite. No eclogites were found in the study area and only two small, weathered "blueschist" blocks have been found. The mineral assemblages in these rocks indicate isochemical recrystallization in a high-pressure, moderate-temperate environment (Coleman and Lee, 1963).

It is clear that these rounded blocks, commonly called "knockers," of "high grade" metamorphic rock are alien to their present environment with their metamorphic grade and fabric bearing no relation to that of the enclosing rocks. These "knockers" are generally interpreted to have been tectonically emplaced upward into their present environments.

The origin of these rocks is enigmatic, but they are thought to be fragments of old oceanic crust which was subducted in a trench environment, metamorphosed, and subsequently fragmented and tectonically emplaced into higher structural levels by faulting and/or by transport along with serpentinite (Ernst and others, 1970).

No age dates have been obtained for the Pine Mountain ultramafics. If they are considered to be a remnant of the Coast Range ophiolite they can be assigned a Jurassic age with some confidence. The age of the Coast Range ophiolite ranges from 153 to 165 m.y. (Middle Jurassic; Hopson and others, 1981) based on U/Pb isotopic ages and K/Ar ages. If, on the other hand, they are part of the dismembered basement of the Nicasio Reservoir terrane, they could be as young as early Cretaceous.

It is interesting to note that the igneous rocks of the Nicasio Reservoir terrane, which parallels to the west, and mirrors the outcrop distribution of the Pine Mountain ultramafics, corresponds to parts of the sheeted dike or sill complex member, and the upper pillowed and massive basalt with breccia member, of the reconstructed ophiolite sequence of Hopson and others (1981). The peridotite body which lies at depth below the Pine Mountain ultramafics may be the dismembered, but only slightly removed, floor of the plutonic-volcanic sequence of the Nicasio Reservoir terrane (see below). If true, fragments of this oceanic floor are now tectonically within the sedimentary pile that the floor once carried for some significant distance, if evidence to support large scale northward transport of the Nicasio Reservoir terrane is accepted.

STRUCTURE

The overall structure of the area is that of a NNW-SSE trending, east-northeast dipping homocline in the basalt of the Nicasio Reservoir terrane which passes eastward into gentle folds in the overlying sedimentary sequence.

This grossly simple structure is breached by at least two NNW-SSE trending, steeply northeast dipping to vertical faults or shear zones along which serpentinized ultramafic rock of the Pine Mountain ultramafics has been intruded. The near surface expression of these intrusions are elongate "morning glory," or wedge-like, structures which tend to cap and hold up northwest-trending ridges. Locally, fingers of the relatively mobile serpentinite intrude the surrounding country rock.

The serpentinite structures have been cut and disrupted by high angle, generally normal displacements along predominantly NE-SW, E-W, and NW-SE faults.

All such structures are in turn truncated on the northeast by the WNW-ESE trending, high-angle Tama Ranchero shear zone (Berkland, 1969). Possible right slip movement on this fault zone has apparently fragmented and dragged fragments of the Nicasio Reservoir terrane and the Pine Mountain ultramafics along its trend.

The apparent northeast curve in the gross outcrop patterns of the two serpentinite belts at the north ends near the Tama Ranchero shear zones may be due to right shear along this fault which gave rise to local drag folding about a near vertical axis. However, there is no direct evidence for such a fold on the field.

SEDIMENTARY PETROLOGY

Petrographic modes (Chayes, 1956; van der Plas and Tobi, 1965) were determined for 62 samples from the central portion of the study area, roughly between San Geronimo Valley on the north and Bon Tempe-Alpine Lake area on the south. Interpretation of detrital modes followed the method of Dickinson (1970a), without major change. Samples for point counting were collected on the basis of petrographic examination of approximately 590 thin sections of sandstone from the detailed study area. All thin sections were cut normal to bedding, where it could be recognized and stained for potassium (K) feldspar and sodium (P) feldspar using the procedure of Laniz and others (1964). Samples selected for analysis were those which were considered representative of the distribution of interbedded sandstones in the central portion of the study area, and which appeared to provide the best opportunity for approximation of original detrital modes. The latter objective was sought by avoiding samples of rocks with original compositional differences, or affected by diagenetic and metamorphic alterations. Samples from rounded, possibly detached sandstone masses were avoided. The criteria used to avoid selecting samples affected by alteration are briefly discussed below.

Samples were selected which were unweathered and lacked calcite cement. In addition, samples were selected which, in both thin section and corresponding stained slab, did not have the washed-out and smeared yellow K-feldspar stain characteristic of samples containing diagenetic potassium feldspar in the study region. Gluskoter (1962, 1964, 1969) first established that both potassium and sodium feldspar alteration occurs in some Franciscan sandstones in Central Marin County, although he also noted that K-feldspar of unequivocal detrital origin was present in his "western zone." As documented by Gluskoter (1962), the evidence of secondary orthoclase was most pronounced in the south and west of the study area. The almost complete absence of detrital K-feldspar in the study area is not thought to result from either widespread replacement of K-feldspar by albite or from complete removal of K-feldspar by intrastratal solution, although this is difficult to prove (Cowan, 1974).

TABLE 2. SUMMARY OF DETRITAL GRAIN PARAMETERS AND MODAL POPULATIONS, NICASIO RESEVOIR TERRANE

Sample		Primary Parameters and Population	Secondary Parameters			Secondary Populations		
Map No.	Field No.	QFL	Qp/Q	P/F	Lv/L	QmFLt	QpLvLs	QmPK
M1	SG72W-558	10-38-52	0.3	1.0	0.4	7-38-55	5-41-54	15-85-0
M2	SG72W-679	9-38-53	0.1	1.0	0.3	9-38-53	2-32-66	18-82-0
M3	SG72W-665	9-40-51	0.5	1.0	0.8	5-40-55	8-77-14	11-89-0
M4	SG72W-598	8-43-49	0.1	1.0	0.8	7-43-50	2-82-16	14-86-0
M5	SG72W-516	10-33-57	0.5	0.9	0.7	5-33-62	8-65-27	12-87-1
M6	SG72W-451	5-18-77	0.1	1.0	0.6	3-18-79	3-55-42	12-88-0
M7	SG72W-629	8-39-53	0.1	1.0	0.5	8-39-53	2-56-42	17-83-0
M8	SG72W-646	9-33-58	0	1.0	0.5	9-33-58	0-52-48	22-78-0
M9	SG72W-574	3-16-81	0.1	1.0	0.8	3-16-81	1-75-24	17-83-0
M10	SG72W-576	5-33-62	0.2	1.0	0.8	4-33-63	1-78-21	12-88-0
M11	SG72W-564	12-36-52	0.1	1.0	0.5	10-36-54	3-50-47	22-78-0
M12	SG72W-313	15-35-50	0.1	1.0	0.7	14-35-51	3-72-25	28-72-0
M13	SG72W-929	7-22-71	0.3	1.0	0.9	5-22-73	3-83-14	18-82-0
M14	SG72W-443	6-21-73	0.4	1.0	0.9	4-21-75	3-91-6	14-86-0
M15	SG72W-436	14-27-59	0.6	1.0	0.5	6-27-67	12-44-44	17-83-0
M16	SG72W-727	7-19-74	0.1	1.0	0.9	6-19-75	1-85-14	28-72-0
M17	SG72W-526	10-22-68	0.3	1.0	0.9	7-22-71	4-91-5	23-77-0
M18	SG72W-481	5-17-78	0.2	1.0	0.5	4-17-79	2-49-49	18-82-0
M19	SG72W-7	19-54-27	0.3	1.0	0.1	14-54-32	15-11-74	20-80-0
M20	SG72W-81	10-48-42	0.2	1.0	0.9	8-48-44	4-91-5	15-85-0
M21	SG72W-87D	9-44-47	0.2	1.0	0.9	8-44-48	3-84-13	15-85-0
M22	SG72W-93	11-47-42	0.1	1.0	0.9	10-47-43	4-85-11	17-83-0
M23	SG72W-116	9-44-47	0.1	1.0	0.8	8-44-48	1-82-17	16-84-0
M24	SG72W-131	3-27-70	0.8	1.0	0.3	1-27-72	3-81-16	2-98-0
M25	SG72W-50	9-45-46	0.1	1.0	0.7	8-45-47	1-72-27	15-85-0
M26	SG72W-725	9-26-65	0.2	1.0	0.5	7-26-67	3-61-36	25-75-0
M27	SG72W-711	9-52-39	0.1	1.0	0.9	8-52-40	1-89-10	20-80-0
M28	SG72W-793	3-28-69	0.2	1.0	0.9	2-28-70	1-89-10	7-93-0
M29	SG72W-831	7-42-51	0.2	1.0	0.6	6-42-52	3-60-37	12-88-0
M30	SG72W-884	16-34-50	0.5	0.9	0.2	8-34-58	15-17-68	19-78-3
M31	SG72W-865	9-50-41	0.1	1.0	0.8	8-50-42	3-79-18	14-86-0
M32	SG72W-851	14-48-38	0.1	1.0	0.9	13-48-39	2-84-14	21-79-0
M33	SG72W-178	8-28-64	0.3	1.0	0.6	6-28-66	3-56-41	16-84-0
M34	SG72W-342	11-44-45	0.2	1.0	0.7	9-44-47	4-70-26	16-84-0
M35	SG72W-276	9-45-46	0.1	0.9	0.8	8-45-47	1-80-19	16-82-2
M36	SG72W-901	8-36-56	0.1	1.0	0.7	7-36-57	1-66-33	18-82-0
M37	SG72W-913	4-17-79	0.8	1.0	0.9	1-17-82	4-91-5	5-95-0
M38	SG72W-236	14-58-28	0.5	0.9	0.3	7-58-35	10-27-63	20-79-1
M39	SG72W-205	9-34-57	0.2	1.0	0.9	7-34-59	3-89-8	17-83-0
M40	SG72W-198	6-44-50	0.1	1.0	0.9	5-44-51	1-92-7	11-89-0
M41	B73W-11A	9-38-53	0	1.0	0.9	9-38-53	0-95-5	20-80-0
M42	B73W-15	10-39-51	0.1	1.0	0.9	9-39-52	2-92-6	19-81-0
M43	B73W-27	14-27-59	0.3	1.0	0.1	10-27-63	14-12-74	14-86-0
M44	B73W-45	10-32-58	0.1	1.0	0.7	9-32-59	2-72-26	22-78-0
M45	B73W-55	6-29-65	0.3	1.0	0.9	4-29-67	3-84-13	13-87-0
M46	B73W-81	5-47-48	0.1	1.0	0.9	4-47-49	1-93-6	9-91-0
M47	B73W-87	4-42-54	0	1.0	0.9	4-42-54	0-87-13	9-91-0
M48	SR71-196	11-37-52	0.1	1.0	0.8	9-37-54	3-75-22	19-81-0
M49	SR71-169	9-45-46	0.1	1.0	0.9	8-45-47	3-87-10	15-85-0
M50	SR71-171	14-41-45	0.2	1.0	0.9	11-41-48	6-86-8	21-79-0
M51	SG73W-21	4-23-73	0.2	1.0	0.8	3-23-74	1-84-15	13-87-0
M52	SG72W-820	8-37-55	0.2	1.0	0.9	7-37-56	2-87-11	16-84-0
M53	SG72W-66	13-40-47	0.1	0.9	0.7	12-40-48	1-68-31	23-76-1
M54	SG72W-817	4-36-60	0	1.0	0.9	4-36-60	0-89-11	9-91-0
M55	SG72W-361	4-24-72	0.6	1.0	0.9	2-24-74	3-85-12	7-93-0
M56	SG72W-787	7-24-69	0.1	1.0	0.9	6-24-70	1-95-4	21-79-0
M57	SG72W-297	9-55-36	0	1.0	0.8	9-55-36	0-81-19	15-85-0
M58	SG72W-283	9-49-42	0	1.0	0.9	9-49-42	0-80-20	16-84-0
M59	SG73W-4	7-28-65	0.2	1.0	0.9	6-28-66	2-85-13	17-83-0
M60	SG73W-10	9-40-51	0.1	1.0	0.7	8-40-52	1-65-34	17-83-0
M61	SG73W-14	5-28-67	0.1	1.0	0.9	4-28-68	1-90-9	15-85-0
M62	SG72W-718	6-50-44	0	1.0	0.9	6-50-44	0-95-5	10-90-0
Mean and Standard Deviation:			Mean:					
		9±3-36±11-55±12	0.2	0.99	0.7	7-36-57	3-83-14	16-84-0

NOTE: See Figure 3 for location of sample.
See Tables 3 and 4 for definitions of detrital framework grain parameters and modal grain populations.

TABLE 3. CLASSIFICATION OF DETRITAL GRAIN TYPES

Sums	Symbols	Grain Types
	Q	total quartzose grains
Q = Qm + Qp	Qm	monocrystalline quartz grains
	Qp	polycrystalline quartzose lithic fragments (chert, etc.)
	F	total feldspar grains
F = P + K	P	plagioclase grains
	K	K-feldspar grains
	L	total unstable aphanitic lithic fragments
L = Lv + Ls	Lv	volcanic, hypabyssal, and meta-volcanic aphanitic lithic fragments
	Ls	sedimentary and metasedimentary aphanitic lithic fragments
Lt = L + Qp	Lt	total aphanitic lithic fragments (both unstable and quartzose)

NOTE: Classification and notation after Graham and others (1976); and Dickinson and Suczek (1979).

For each of the 62 samples selected, 500 points were counted per thin section on a 0.5 mm grid measured by a mechanical stage. With the counting grid selected to be generally as large as the largest grain encountered (van de Plas, 1959), the determined detrital modes have a maximum statistical uncertainty of ±4.5 (van der Plas and Tobi, 1965), and the calculated standard deviations can be considered an accurate measure of the theoretical counting error (Chayes, 1956), assuming a small value for operator error.

The results of the petrographic analyses of the 62 sandstone samples from the central part of the study area, are summarized in Table 2. Three primary detrital grain parameters (Q, F, and L), one primary modal population (QFL), three secondary detriatal grain parameters(Qp/Q, P/F, and Lv/Lt), and three secondary modal populations (QmFLt, QpLvLs, and QmPK) are listed in this table, and are defined in Tables 3 and 4.

As Table 2 shows, total framework quartzose grains (Q) range from 3 to 19 percent, with a mean value of 9 ± 3 percent for the 62 samples. The majority of quartzose grains are monocrystalline (Qm); the ratio of polycrystalline quartzose lithic fragments (Qp), mainly chert, to total quartzose grains (Q), ranges from 0 to 0.8 with a mean value of 0.2.

Total framework feldspar grains (F) range from 16 to 58 percent with a mean value of 36 ± 11 percent for the 62 samples. The majority of feldspar grains are plagioclase (P) (albite ± oligoclase); the ratio of plagioclase grains (P) to total feldspar grains (F) ranges from 0.9 to 1.0 with a mean value of 0.99.

As seen in the QFL plot (fig. 4), most of the range in data, the mean, and the greater part of the envelope of standard deviation fall within the field for sandstones derived from undissected magmatic arc provenance as inferred by Dickinson and Suczek (1979, fig. 1).

In the QmFLt diagram (fig. 5), the same basic primary grain population is shown in the QFL diagram. However, polycrystalline quartzose lithic fragments (Qp) are shifted to the total lithic pole (Lt), with Qm (monocrystalline quartz grains), F, and Lt are recalculated to 100 percent (Qm+F+Lt = 100). This plot tends to emphasize grain size of the source rocks, since finer-grained rocks yield more lithic fragments in the sand-size range.

TABLE 4. DETRITAL FRAMEWORK GRAIN PARAMETERS AND MODAL POPULATIONS PLOTTED ON TRIANGULAR DIAGRAMS

Diagram	Population
QFL	total framework with all quartzose grains grouped together
QmFLt	total framework with all aphanitic lithic fragments grouped together
QpLvLs	polycrystalline aphanitic lithic fragments only
QmPK	monocrystalline mineral grains only

NOTE: Framework is sum of grain types defined in Table 3.

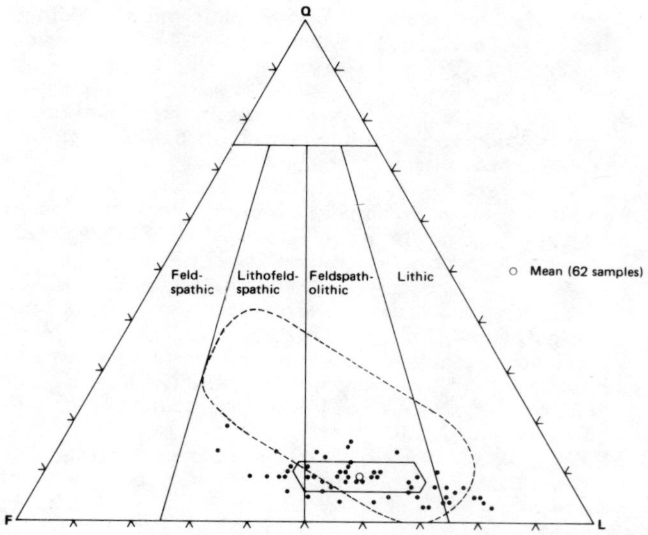

Figure 4. QFL triangular plot of detrital framework modes, Nicasio Resevoir terrane, showing mean and envolope of standard deviation. Sandstone classification after Crook (1960). Dashed line shows field for sandstones derived from magmatic arcs as inferred by Dickinson and Suczek (1979).

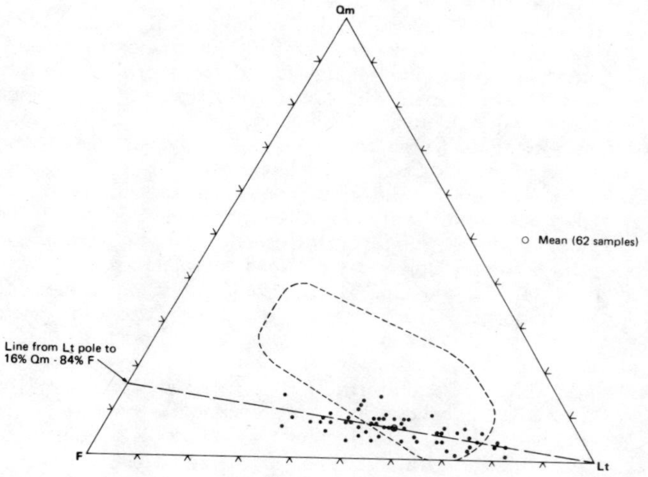

Figure 5. QmFLt triangular plot of detrital framework modes, Nicasio Resevoir terrane. Dashed line shows field for sandstones derived from magmatic arcs as inferred by Dickinson and Suczek (1979)

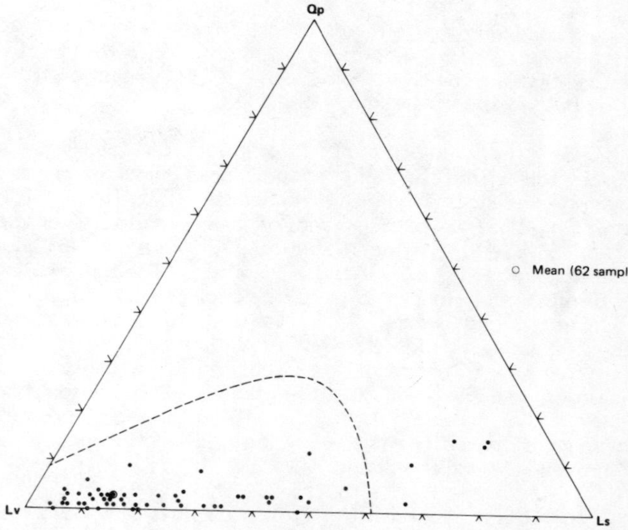

Figure 6. QpLvLs triangular plot of detrital framework modes, Nicasio Resevoir terrane. Dashed line shows field for sandstones from arc orogen sources as inferred by Dickinson and Suczek (1979)

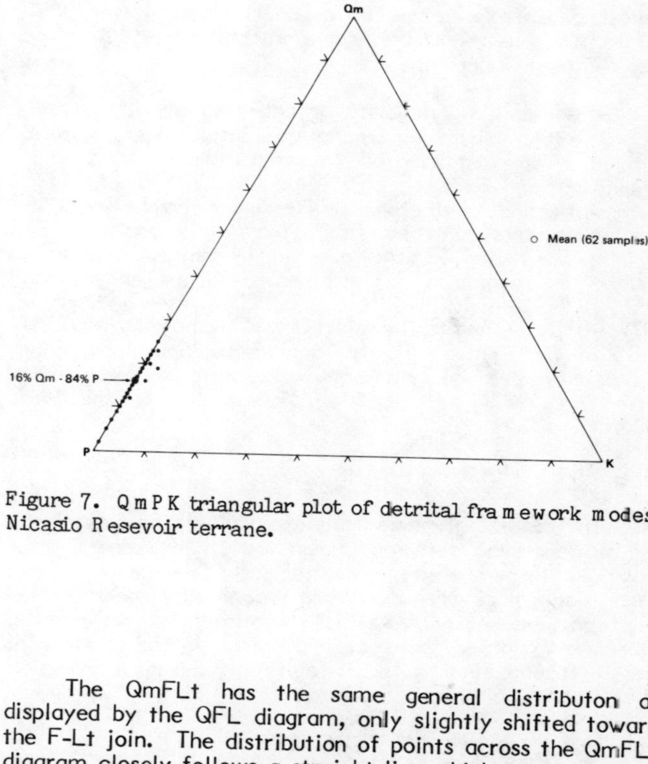

Figure 7. QmPK triangular plot of detrital framework modes, Nicasio Resevoir terrane.

Total unstable aphanitic lithic fragments (F) range from 27 to 81 percent with a mean value of 55 ± 12 percent for the 62 samples. The majority of lithic fragments are hypabyssal volcanic, and metavolcanic aphanitic lithic fragments (Lv); the ratio of these types of lithic fragments (Lv) to total unstable lithic fragments (Lt) ranges from 0.1 to 0.9 with a mean value of 0.7.

The data presented in Table 2, representing the primary (QFL) and three secondary (QmFLt, QpLvLs, QmPK) modal populations as defined in Table 4 is displayed graphically on four triangular diagrams (figs. 4-7). The four triangular diagrams chosen to display modal populations are the same as those used by Dickinson and Suczek (1979). More recently Dickinson and others (1982) used these diagrams to compare selected Franciscan and Great Valley sandstone.

The QmFLt has the same general distributon as displayed by the QFL diagram, only slightly shifted toward the F-Lt join. The distribution of points across the QmFLt diagram closely follows a straight line which passes through the QmFLt mean, and connects the Lt pole with the QmPK mean (16-84-0), shown in Table 2 and on figure 7, below. This point represents the mean proportion of quartz and a bite grains in the Nicasio Reservoir terrane population. A similar relationship has been inferred by Dickinson and others (1982, fig. 4) for the mean QmFLt framework modes of selected Franciscan suites wholly lacking K-feldspar, although their connecting line falls higher on the Qm-F join, connecting the Lt pole to 41.5 percent QM - 58.5 percent F.

The QpLvLs diagram (fig. 6) displays only a partial framework grain population, emphasizing the character of the polycrystalline components of the framework, with Qp, Lv, and Ls (Table 3) recalculated to 100 percent (Qp+Lv+Ls = 100).

As shown on figure 6, the data from the study area plot in an elongate roughly linear pattern distinctly toward the Lv pole and hugging the Lv-Ls join, with a slight inclination from the Lv pole to the Qp-Ls join.

The great majority of data points on the QpLvLs diagram fall within the field for sandstones derived from arc orogen provenances (Dickinson and Suczek, 1979, fig. 3). The mean QpLvLs framework mode of the Nicasio Reservoir terrane (3-83-14), is most similar to the mean QpLvLs framework mode (0-82-18) of the Neogene "Blue-sand" units, California Coast Ranges suite (Lerbekmo, 1961; suite no. 36 of Dickinson and Suczek, 1979, Table 1) and to the mean QpLvLs framework mode (2-89-9) of the Jurassic-Cretaceous Stony Creek petrofacies of the Great Valley sequence suite (Dickinson and Rich, 1972, suite no. 40 of Dickinson and Suczek, 1979, Table 1). Both these suites represent provenances transitional between the undissected magmatic arc provenance and the dissected magmatic arc provenance.

The QmPK diagram (fig. 7) also displays only a partial framework grain population, and emphasizes the character of the monocrystalline components of the framework, with Qm, P and K (Table 3) recalculated to 100 percent (Qm+P+K = 100).

As shown on figure 7, the data from the study area plot almost exclusively on the Qm-P join, distinctly toward the P pole. This is to be expected for sandstone lacking K-feldspar and low in quartz grains.

The mean and majority of data points on the QmPK diagram plot in the same area as the plotted means of suites derived from undissected magmatic arc provenances (Dickinson and Suczek, 1979, fig. 4). The mean QmPK framework mode of the Nicasio Reservoir terrane (16-84-0), is most similar to the mean QmPK framework mode (15-85-0) of the Jurassic Methow-Pasayten sequence, Washington undissected magmatic arc provenance suite (Tennyson and Cole, 1978; Cole, 1973; suite no. 30 of Dickinson and Suczek, 1979,, Table 1), and to the mean QmPK framework mode (11-89-0) of the Triassic North Range Beds, Southland, South Island, New Zealand undissected magmatic arc provenance suite (Boles, 1974, suite no. 32 of Dickinson and Suczek, 1981, Table 1).

SUMMARY

The results of geologic and petrologic studies of the Nicasio Reservoir terrane suggest that it represents a far travelled, dismembered fragment of early Cretaceous oceanic crust that was covered by clastic sedimentary rocks derived from an undissected volcanic magmatic arc provenance. Subsequently, these rocks were accreted to the continental margin accompanied by thrust-faulting and diapiric intrusion of the subcrustal ultramafic rocks.

REFERENCES

Alvarez, W., Kent, D. V., Silva, I. P., Schweickert, R. A., and Larson, R. A., 1980, Franciscan complex limestone deposited at 17° South paleolatitude: Geological Society of America Bulletin, Part I, v. 91, no. 8, p. 476-484.

Bailey, E. H., and Blake, M. C., Jr., 1974, Major chemical characteristics of Mesozoic Coast Range ophiolite in California: U.S. Geological Survey Journal of Research, v. 2, p. 637-656.

Beck, M. E., Jr., 1976, Discordant paleomagnetic pole positions as evidence of regional shear in the western Cordillera of North America: American Journal of Science, v. 276, p. 694-712.

-----,1980, Paleomagnetic record of plate-margin tectonic processes along the western edge of North America: Journal of Geophysical Research, v. 85, no. B12, p. 7115-7131.

Berkland, J. O., 1969, Geology o the Novato quadrangle, Marin County, California: unpublished San Jose State College M.S. thesis.

Blake, M. C., Jr., Irwin, W. P., and Coleman, R. G., 1967, Upside-down metamorphic zonation, blueschist facies, along a regional thrust in California and Oregon: U.S. Geological Survey Professional Paper 575-C, p. C1-C9.

Blake, M. C., Jr., Bartow, J. A., Frizzell, V. A., Jr., Schlocker, Julius, Sorg, Dennis, Wentworth, C. M., and Wright, R. H., 1974, Preliminary geologic map of Marin and San Francisco Counties and parts of Alameda, Contra Costa, and Sonoma Counties, California: U.S. Geological Survey Miscellaneous Field Studies Map MF-574, scale 1:62,500.

Blake, M. C., Jr., Howell, D. G., and Jones, D. L., 1982, Preliminary tectonostratigraphic terrane map of California: U.S. Geological Survey Open-File Report 82-593, 9 p., 3 maps, scale 1:750,000.

Blake, M. C., Jr., and Jones, D. L., 1981, The Franciscan assemblage and related rocks in northern California: a reinterpretation, in The Geotectonic Development of California, Ernst, W. G., ed., v. 1, p. 307-328, Prentice-Hall, Inc.

Boles, J. R. 1974, Structure, stratigraphy, and petrology of mainly Triassic rocks, Hokonui Hills, Southland, New Zealand: New Zealand Journal of Geology and Geophysics, v. 17, p. 337-374.

Champion, D. E., Howell, D. G., and Marshall, M. C., 1981, Paleomagnetic evidence for 3800 km of northwestward translation of San Miguel Island, southern California, Borderland (abs.): EOS, Transactions of American Geophysical Union, v. 62, no. 45, p. 855.

Chayes, F., 1956, Petrographic modal analysis: New York, John Wiley and Sons, Inc., 113 p.

Cole, M. R., 1973, Petrology and dispersal patterns of Jurassic and Cretaceous sedimentary rocks in the Methow River area, north Cascades, Washington: Ph.D. Dissertation, University of Washington, 110 p.

Coleman, R. G., and Lee, D. E., 1963, Glaucophane-bearing metamorphic rock types of the Cazadero area, California: Journal of Petrology, v. 4, part 2, p. 260-301.

Cowan, D. S., 1974, Deformation and metamorphism of the Franciscan subduction zone complex northwest of Pacheco Pass, California: Geological Society of America Bulletin, v. 85, no. 10, p. 1623-1634.

Crook, K. A., 1974, Lithogenesis and geotectonics: the significance of compositional variation in flysch arenites (graywackes), in Dott, R. H., Jr., and Shaver, R. G., eds., Modern and ancient geosynclinal sedimentation: Society of Economic Paleontologists and Mineralogists Special Publication, no. 19, p. 304-310.

Davis, G. A., Monger, J. W. H., and Burchfiel, B. C., 1978, Mesozoic construction of the Cordilleran "collage", central British Columbia to central California, in Howell, D. G., and McDougall, K. A., eds., Mesozoic Paleogeography of the western United States: Pacific Section, Society of Economic Paleontologists and Mineralogists, Pacific Coast Paleogeography Symposium 2, p. 1-32.

Dickinson, W. R., 1970a, Interpreting detrital modes of graywacke and arkose: Journal of Sedimentary Petrology, v. 40, p. 695-707.

-----,1971, Clastic sedimentary sequences deposited in shelf, slope, and trough settings between magmatic arcs and associated trenches: Pacific Geology, v. 3, p. 15-30.

Dickinson, W. R., and Rich, E. I., 1972, Petrologic intervals and petrofacies in the Great Valley sequence, Sacramento Valley, California: Geological Society of America Bulletin, v. 83, p. 3007-3024.

Dickinson, W. R., and Suczek, C. A., 1979, Plate tectonics and sandstone compositions: American Association of Petroleum Geologists Bulletin, v. 63, p. 2164-2182.

Dickinson, W. R., Ingersoll, R. V., Cowan, D. S., Helmold, K. P., and Suczek, C. A., 1982, Geological Society of America Bulletin, v. 93, p. 95-107.

Ernst, W. G., 1970, Tectonic contact between the Franciscan melange and the Great Valley sequence - crustal expression of a late Mesozoic Benioff Zone: Journal of Geophysical Research, v. 75, p. 886-902.

Ernst, W. G., Seki, Y., Onuki, H., and Gilbert, M. C., 1970, Comparative study of low-grade metamorphism in the California Coast Ranges and the outer metamorphic belt of Japan: Geological Society of America, Memoir 124, 276 p.

Gluskoter, H. J., 1962, Geology of a portion of western Marin County, California: Ph.D. Dissertation, University of California. Berkeley, California, 184 p.

-----,1964, Orthoclase distribution and authigenesis in the Franciscan Formation of a portion of western Marin County, California: Journal of Sedimentary Petrology, v. 34, no. 2, p. 335-343.

-----,1969, Geology of a portion of western Marin County, California: California Division of Mines and Geology, Map Sheet 11.

Gromme, C. S., and Gluskoter, H. J., 1965, Remnant magnetization of spilite and diabase in the Franciscan Formation, western Marin County, California: Journal of Geology, v. 73, no. 1, p. 74-94.

Hamilton, W., 1969, Mesozoic California and the underflow of Pacific mantle: Geological Society of America Bulletin, v. 80, p. 2409-2430.

-----,1978, Mesozoic tectonics of the western United States, in Howell, D. G., and McDougall, K. A., eds., Mesozoic Paleogeography of the western United States: Pacific Section, Society of Economic Paleontologists and Mineralogists, Pacific Coast Paleogeography Symposium 2, p. 33-70.

Hillhouse, J., 1977, Paleomagnetism of the Triassic Nikolai Greenstone, McCarthy quadrangle, Alaska: Canadian Journal of Earth Sciences, v. 14, no. 11, p. 2578-2592.

Hopson, C. A., Mattinson, J. M., and Pessagno, E. A., Jr., 1981, Coast Range ophiolite, western California, in The Geotectonic Development of California, Ernst, W. G., ed., v. 1, p. 419-510, Prentice-Hall, Inc.

Jacobson, M. I., 1978, Petrologic variations in Franciscan sandstone from the Diablo Range, California, in Howell, D. G., and McDougall, K. A., eds., Mesozoic Paleogeography of the western United States: Pacific Section, Society of Economic Paleontologists and Mineralogists, Pacific Coast Paleogeography Symposium 2, p. 401-417.

Jones, D. L., and others, 1977, Wrangellia - a displaced terrane in northwestern North America: Canadian Journal of Earth Science, v. 14, no. 11, p. 2565-2577.

Jones, D. L., Silberling, N. J., and Hillhouse, J. W., 1978b, Microplate tectonics of Alaska--significance for the Mesozoic history of the Pacific Coast of North America, in Howell, D. G., and McDougall, K. A., eds., Mesozoic Paleogeography of the western United States: Pacific Section, Society of Economic Paleontologists and Mineralogists, Pacific Coast Paleogeography Symposium 2, p. 71-74.

Laniz, R. V., Stevens, R. E., and Norman, M. B., 1964, Staining of plagioclase feldspar and other minerals with F. D. and C. red no. 1: U.S. Geological Survey Professional Paper 501-B, p. B152-B153.

Lawson, A. C., 1914, U.S. Geological Survey Geological Atlas, San Francisco Folio (no. 193).

Lerbekmo, J. F., 1961, Genetic relationship among Tertiary blue sandstones in central California: Journal of Sedimentary Petrology, v. 31, p. 594-602.

Mattinson, J. M., and Echeverria, L. M., 1980., Ortigalita Peak Gabbro, Franciscan complex: U-Pb dates of intrusion and high-pressure-low-temperature metamorphism: Geology, v. 8, no. 12, p. 589-593.

McLaughlin, R. J., Kling, S. A., Poore, R. Z., McDougall, Kristin, and Beutner, E. C., 1983, Post-Middle Miocene accretion of Franciscan rocks, northwestern California: Geological Society of America Bulletin, v. 93, p. 595-605.

Packer, D. R., and Stone, D. B., 1974, Paleomagnetism of Jurassic rocks from southern Alaska and the tectonic implications: Canadian Journal of Earth Sciences, v. 11, no. 7, p. 976-997.

Saleeby, J. B., Goodwin, S. E., Sharp, W. D., and Busby, C. J., 1978, Early Mesozoic paleotectonic-paleogeographic reconstruction of the southern Sierra Nevada region, in Howell, D. G., and McDougall, K. A., eds., Mesozoic Paleogeography of the western United States: Pacific Section, Society of Economic Paleontologists and Mineralogists, Pacific Coast Paleogeography Symposium 2, p. 311-336.

Schwab, F. L., 1975, Framework mineralogy and chemical composition of continental margin-type sandstone: Geology, v. 3, no. 9, p. 487-490.

Stone, D. B., and Packer, D. R., 1977, Tectonic implications of Alaska Peninsula paleomagnetic data: Tectonophysics, v. 37, p. 183-201.

-----,1979, Paleomagnetic data from the Alaska Peninsula: Geological Society of America Bulletin, Part I, v. 90, no. 6, p. 545-560.

Tennyson, M. E., and Cole, M. R., 1978, Tectonic significance of upper Mesozoic Methow-Pasayten sequence, northeastern Cascade Range, Washington and British Columbia, in Howell, D. G., and McDougall, K. A., eds., Mesozoic Paleogeography of the western United States: Pacific Section, Society of Economic Paleontologists and Mineralogists, Pacific Coast Paleogeography Symposium 2, p. 499-508.

van der Plas, L., 1959, Petrology of the northern Adula region, Switzerland: Leidse Geologische Mededelingen, v. 24, p. 415-598.

van der Plas, L., and Tobi, A. C., 1965, A chart for judging the reliability of point counting results: American Journal of Science, v. 263, p. 87-90.

Weaver, C. E., 1949, Geology of the Coast Ranges immediately south of the San Francisco Bay Region, California: Geological Society of America Memoir 35.

Wright, R. H., 1974, Map showing the distribution of potassium feldspar and fossils in Mesozoic rocks of Marin and San Francisco Counties and parts of Alameda, Contra Costa, and Sonoma Counties, California: U.S. Geological Survey Miscellaneous Field Studies Map MF-573.

Exposure of Central melange terrane matrix in abandoned quarry at Greenbrae, Marin County. Pampas grass is 3 meters high, for scale. Photo by E. H. Bailey, 1980.

PALEOMAGNETISM OF FRANCISCAN BASALT, MARIN COUNTY, CALIFORNIA, REVISITED

Sherman Gromme
U.S. Geological Survey
Branch of Isotope Geology
345 Middlefield Road, Mail Stop 937
Menlo Park, Calif. 94025

ABSTRACT

Paleomagnetic data from pillow lava and diabase in the Franciscan assemblage in western Marin County, California that were originally published in 1965 are here re-analyzed. The results of partial alternating-field demagnetization of specimens from some sites are included in the mean direction of magnetization; these data had not previously been incorporated. The pole position calculated here differs little from that originally published, but the precision is slightly greater and the fold test gives a somewhat more positive result. A recent age determination from radiolaria extracted from interbedded chert is Valanginian. This allows a comparison of the greatly discordant result from these rocks to be compared with the Cretaceous reference pole position for the North American craton. The comparison indicates that these rocks, termed the Nicasio terrane, have moved approximately 20° northward since their formation, and are rotated approximately 106° eastward. Such translations and rotations are typically found in pre-Tertiary rocks of the western margin of North America.

INTRODUCTION

Only two paleomagnetic studies have been published that dealt with rocks of the Franciscan assemblage of northern California and showed that the remanent magnetizations in the rocks were acquired when the rocks were formed. The more recent of these, by Alvarez and others (1980), presented paleontologic and paleomagnetic data that, taken together, indicate that limestone blocks occurring near Laytonville, California had been deposited originally at 17° south of the Cretaceous equator. These blocks of pelagic limestone are embedded in melange; their Albian-Cenomanian age, combined with an estimated time of deformation of the melange ranging from late Cretaceous to mid-Tertiary, requires a rate of northward transport of the order of 10 to 25 cm/year. In the context of present-day plate velocities, such rates are high.

The earlier of the two studies (Gromme and Gluskoter, 1965) concerned three areas of spilite and diabase in western Marin County, and was published prior to the advent of plate-tectonics concepts, although the notion of what was then termed "continental drift" was known to all paleomagnetists and was espoused by most of them. The paleomagnetic declinations in the Marin County rocks were anomalous in terms of what was then known about the history of the geomagnetic field; the pole position lay in the middle of the Atlantic Ocean at about 30° north latitude. We offered, but did not choose among, three explanatory hypotheses: (1) "Tectonic rotation.......70° around a vertical axis occurred." (2) "The non-dipole part of the geomagnetic field was temporarily predominant." (3) "There were one or more excursions of the geomagnetic pole into the Atlantic Ocean......"

The interpretation of those strange results was equivocal because the age and geologic context of the rocks were poorly known, an adequate reference apparent polar wandering path for the North American craton did not exist, and the only similar result at that time for North America was from the Eocene Siletz River Volcanics of westernmost Oregon (Cox, 1957). Subsequently it has become abundantly clear that clockwise structural rotations, manifested by eastward-deflected paleomagnetic declinations, are a characteristic of rocks on the west coasts of California, Oregon, and Washington; a recent summary has been published by Beck (1980). The purpose of this article is to update the conclusions of Gromme and Gluskoter (1965), in particular to show that hypothesis (1) was the correct one and also to show that the Marin County rocks have been displaced northward by a significant distance.

TECTONIC SETTING AND AGE

As mapped by Gluskoter (1969), the Franciscan volcanic rocks of western Marin County are shown to be enclosed in a northeast-dipping homocline of sedimentary rocks, predominantly sandstone, that strikes parallel to the San Andreas Fault, with a subsidiary anticline and syncline to the east. This block of sandstone and volcanic rock is truncated on the west by the San Andreas Fault and on the north and northeast by what Gluskoter (1969) described as a "highly disturbed and disrupted 'tectonic zone'", or what would today be called melange. A second zone of melange truncates the sandstones south of Bolinas Bay. These relationships are shown on Figure 1, which is modified from Gromme and Gluskoter (1965). It seemed at that time that the complexity of the outcrop pattern of the igneous rocks required explanation, and the two possibilities offered were that the volcanic rocks represent cross-sections through seamounts, or that an originally contiguous body of rock had been separated into three by faults for which little or no other evidence existed (Gromme and Gluskoter, 1965).

The geologic relationships between the volcanic rocks and the enclosing sedimentary rocks have been reinterpreted by Blake and others (1974). They mapped a boundary (Figure 1) structurally below the igneous rocks, beneath which the sandstones are coherent and unsheared, and above which the rock type is melange consisting of sheared sandstone and shale containing tectonic inclusions of chert, metamorphic rocks, and graywacke. Although Gluskoter (1969) did not show this boundary on his map, he noted the lithologic distinction: Gluskoter's "western zone" is the coherent sandstone, and his "transitional zone" is the part of the melange that is structurally beneath the volcanic rocks. The reinterpretation amounts to the recognition that melange is much more abundant and is generally to be expected in the Franciscan assemblage, a concept that was just beginning to be discussed when Gluskoter's mapping was done in 1960 and 1961. In the

in Blake, M.C. Jr., ed., 1984, **Franciscan Geology of Northern California:** Pacific Section S.E.P.M., Vol. 43, p. 113-119

present context, the importance of the new mapping is that all or nearly all of the margins of the igneous rocks are tectonic boundaries; in other words all of the contacts between rock types shown in Figure 1 are faults, with one exception to be discussed below. In spite of the existence of extensive melange between and around them, both the sandstone blocks and the three areas of volcanic rock are structurally coherent, have similar but not identical attitudes, and are upright with tops facing toward the northeast.

Rocks of the Franciscan assemblage have been divided into seventeen tectonostratigraphic terranes by Blake and others (1982). These terranes vary greatly in extent. One of the largest is the Central terrane, which is comprised of melange of latest Jurassic and Cretaceous age. Both the sandstone and the melange in Figure 1 are shown as belonging to the Central terrane by Blake and others (1982), but they have designated the volcanic rocks as a single small terrane, unique to itself, named the Nicasio terrane after Nicasio Lake at the south end of which the rocks are best exposed (Figure 1). The consequence of this interpretation is that paleomagnetic evidence from these rocks demonstrating large-scale tectonic movements would only apply locally, and would signify only that some tectonic inclusions in Franciscan melange have been transported over very long distances, as is also true for the limestone blocks near Laytonville (Alvarez and others, 1980).

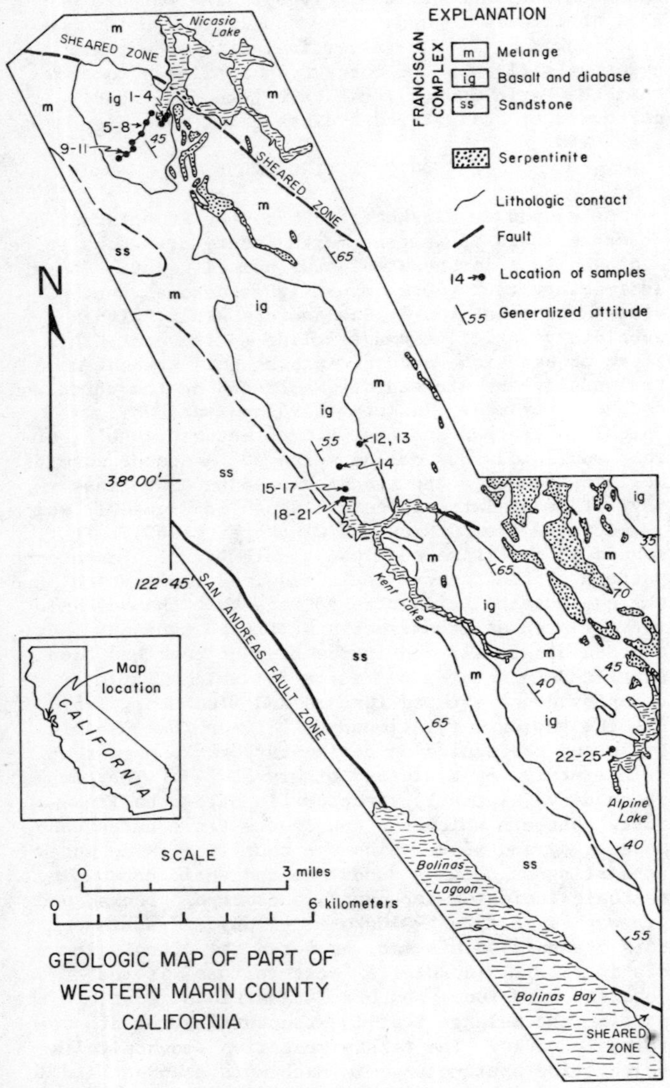

Figure 1. Geologic map of part of western Marin County, California, modified from Gromme and Gluskoter (1965), and showing the paleomagnetic sampling localities in spilitic basalt and diabase.

More recent mapping by Wright (1982) along the north shore of Alpine Lake (Figure 1) has revealed an apparently continuous stratigraphic sequence of interbedded sandstone, chert, and greenstone overlying the main body of basalt. This relationship implies that the Nicasio terrane is stratigraphically overlain by part of the Central terrane, or alternatively that the Nicasio terrane includes a significant proportion of non-igneous rocks. Northwest of Alpine Lake, however, stratigraphic relationships are entirely obscure, and at Nicasio Lake the basalt is structurally overlain by melange.

At the time the paleomagnetic investigation was done almost no fossil evidence for the age of the rocks existed, and Gromme and Gluskoter (1965) could only say that the age could be late Jurassic to late Cretaceous. Recently, radiolaria have been extracted from chert within and immediately overlying the basalt at localities 1-4 and 22-25 respectively (Figure 1); these radiolaria are Valanginian in age (B. Murchey, written communication, 1982). This age is middle lower Cretaceous, or 131-138 million years according to the time scale of Harland and others (1982).

The volcanic rocks are pillow lava, massive lava, and diabase. The lavas are spilitic, that is, the plagioclase in them is albite. Petrographic descriptions, photomicrographs, and chemical analyses have been published by Gluskoter (1962), Gromme and Gluskoter (1965), Swanson and Schiffman (1979), and Blake and others (this volume). Primary igneous features are abundant in these rocks, and although they are chemically altered (spilitized), they are not at all penetratively deformed. Excellent exposures of structurally coherent pillow lavas and diabase can be seen in the dam excavation and road cuts southwest of Nicasio Lake (Figure 1).

PALEOMAGNETIC DATA

Because of instrumental limitations and the weakness of the natural remanent magnetizations (NRM's) of most of the rocks, specimens from only seven of the twenty-five sites had been subjected to partial alternating-field (AF) demagnetization. Nonetheless the results of the demagnetization experiments largely confirmed that the NRM's were not significantly overprinted with secondary remanent magnetization components (Gromme and Gluskoter, 1965). When the work was done (in 1962), the statistics (Fisher, 1953) all had to be calculated by hand because no digital computer was available; consequently the results of the demagnetization experiments were never incorporated in the final average direction. The original specimen collection is not available, but the original laboratory data sheets have been found and the statistical calculations for each of the 25 sites were repeated and were found to be correct. Table 1 is a revised version of Table 1 of Gromme and Gluskoter (1965). Where available, the results of AF demagnetization have been substituted for NRM data in this table. Virtual geomagnetic poles (VGP's) have been calculated for each of the sites and the intensities of NRM and the rock types are also included. Omitted from Table 1 in the interest of brevity are the within-site statistical parameters, but they can be found in Tables 1 and 4 of Gromme and Gluskoter (1965).

The result of the fold test published by Gromme and Gluskoter (1965) was a reduction of the overall angular standard deviation of the site-mean NRM directions from 27.3° to 22.7°, which they asserted was a "statistically positve result". The calculation has been repeated using the data in Table 1, and the result (Table 2) is a reduction of the between-site angular standard deviation from 28.5° to 21.5°, slightly better than before. As an exercise, the significance test published by McElhinny (1964) has been made; the ratio k2/k1 is 1.79 so that the newly calculated fold test is significant at 95% confidence but not at 99% confidence. Recently McFadden and Jones (1981) have shown that McElhinny's test is incorrect and that it errs on the side of being too stringent. McFadden and Jones have given the correct form of the test, but this form requires conditions on the data that are often not met. In the first place the within-site dispersion must be the same for all sites, which is not true in the present case (see Table 1 of Gromme and Gluskoter, 1965). In the second place, for the fold test to be evaluated each site is required to have been magnetized at exactly the same (pre-folding) time in exactly the same paleofield direction. In a sequence of lava flows and intrusives such as these, however, all the sites are sure to have

Table 1. Site-mean magnetization data

Site	N	J	H	I_u	D_u	I_c	D_c	PLA	PLO	Rock type
1	3	0.12	NRM	55.4	162.9	50.7	106.0	07.8	293.2	Porphyritic diabase
2	3	1.6	100	88.0	209.5	51.8	51.1	48.4	319.0	Albite diabase
3	5	0.47	175	72.8	83.0	34.3	64.3	31.5	327.2	Albite diabase, fine grain
4	5	0.34	NRM	56.7	126.2	31.9	90.0	10.5	313.5	Pillow lava
5	6	0.033	NRM	82.9	89.9	56.3	63.6	40.5	307.8	Pillow lava
6	3	0.021	NRM	79.7	248.1	58.9	38.3	60.3	312.0	Variolitic pillow lava
7	4	0.047	NRM	70.3	349.0	39.2	22.4	65.1	0.3	Pillow lava
8	3	0.016	NRM	66.2	127.6	37.7	75.2	24.2	318.7	Pillow lava
9	4	0.019	NRM	69.9	158.7	37.3	79.8	20.5	316.3	Pillow lava
10	5	0.014	NRM	66.9	159.7	48.1	79.7	25.0	308.8	Pillow lava
11	3	0.015	NRM	57.7	145.0	38.5	93.3	10.7	308.1	Pillow lava
12	5	0.46	200	63.4	228.3	41.5	98.3	8.3	303.5	Porphyritic pillow lava
13	7	0.011	NRM	56.6	232.4	47.3	100.1	9.9	298.8	Porphyritic pillow lava
14	6	0.004	NRM	40.1	214.4	43.0	118.2	-4.4	290.6	Pillow lava
15	7	0.015	NRM	25.1	210.9	72.5	151.2	8.8	252.4	Variolitic pillow lava
16	5	0.30	75	29.4	209.3	64.9	117.9	11.4	275.4	Diabase
17	5	0.80	125	69.5	169.4	27.2	76.0	19.8	324.2	Diabase
18	5	1.7	NRM	61.5	253.0	21.5	38.2	46.6	355.3	Pillow lava
19	3	1.[illegible]	100	30.3	209.9	49.8	77.4	27.5	308.5	Diabase
20	15	1.8	150	32.4	210.2	48.1	75.4	28.2	311.0	Diabase
21	3	0.005	NRM	25.3	209.5	53.6	79.8	27.5	303.9	Pillow lava
22	3	0.005	NRM	77.9	152.0	45.2	79.6	23.8	311.1	Pillow lava
23	5	0.023	NRM	66.3	195.8	58.5	91.6	21.8	293.9	Pillow lava
24	4	0.007	NRM	63.7	104.5	27.8	80.5	16.4	321.2	Pillow lava
25	4	0.005	NRM	68.4	73.9	24.0	65.8	26.7	332.1	Pillow lava

Explanation: N, number of specimens. J, intensity of NRM in units of 10^{-3} emu/cm^3. H, peak alternating field treatment in Oersteds (NRM if no treatment). I_u, D_u, inclination (degrees downward) and declination (degrees eastward) of remanent magnetization before correction for structure. I_c, D_c, inclination and declination after structural correction. PLA, PLO, north latitude and east longitude, respectively, of virtual geomagnetic pole calculated from I_c and D_c.

slightly differing ages, so that normal geomagnetic secular variation would result in significant differences in direction from site to site even if there were no secondary magnetization at all and even if the structural corrections were perfect. In fact, for a set of paleomagnetic sites like these, there appears to be no numeric test available for correctly evaluating the significance of the fold test, so one is resigned to saying that the test gives a positive result.

The angular standard deviation of VGP's after structural correction is 24.6° (Table 2), which is rather large. McElhinny and Merrill (1975) have compiled worldwide paleomagnetic secular variation data covering the past 5 million years and have summarized them in the form of angular dispersions (standard deviations) of VGP's for 15-degree zones of site latitude. Interpolating in McElhinny and Merrill's Table 5 for a site magnetic latitude of 30° gives 15.4° for the dispersion of VGP's due to

Table 2. Overall statistics

Mean directions	N	I	D	R	k	δ	α_{95}
Uncorrected:	25	68.1	189.5	22.0292408	8.1	28.5	10.9
Corrected:	25	47.0	78.0	22.3422625	14.5	21.3	7.9

Mean paleomagnetic pole	N	PLA	PLO	R	k	δ	α_{95}
Corrected:	25	25.9	308.6	22.7810920	10.8	24.6	9.2

Explanation: N, number of sites. I, inclination in degrees downward. D, declination in degrees eastward. PLA, latitude of pole in degrees north. PLO, longitude of pole in degrees east. R, vector sum of N unit vectors. k, precision parameter (Fisher, 1953). δ, angular standard deviation. α_{95}, semiangle of 95% confidence cone around mean direction or pole (Fisher, 1953).

Table 3. Reference poles

Rock unit or locality	Age m.y.	Average Site Lat.	Average Site Long.	Pole Lat.	Pole Long.	N	α_{95}	Reference
Niobrara Formation Wyoming, Colorado, and Kansas	87-89	39.2	-105.5	65.0	186.1	46	7.7	(a)
Magnet Cove intrusives Arkansas	95-105	34.5	-92.8	65.1	186.7	19	10.9	(b)
Isachsen diabase Ellef Ringnes Island	106	78.7	-103.7	69.4	182.4	20	15.8	(c)
Monteregian Hills intrusives Quebec	100-120	45.3	-72.8	72.4	191.0	70	3.3	(d)
Mt. Ascutney intrusive Vermont	125-135	43.5	-72.5	64.7	186.1	24	2.2	(e)
Mean reference pole	87-135			67.3	186.3	5	3.4	

Explanation: Ages are radiometric except for Niobrara (see text). Lat., Long., north latitude and east longitude, respectively, in degrees. N, number of sites. α_{95}, 95% confidence cone around pole, in degrees (Fisher, 1953).
References: (a), Shive and Frerichs (1974). (b), Scharon and Hsu (1969). (c), Larochelle and Black (1963). (d), Foster and Symons (1979). (e), Opdyke and Wensink (1966).

geomagnetic secular variation alone. From the relation that the observed variance is the sum of that due to ancient secular variation and that due to all forms of experimental error, including geological mishap, the contribution of experimental error to the observed angular dispersion of VGP's is 19.2°, or appreciably more than half. The equivalent figure for directions is 17.2° (Cox, 1970). Experimental error in the present case may be divided into (1) uncertainty in structural corrections, (2) imprecision in measuring weakly magnetized rocks with a primitive magnetometer, and (3) secondary magnetizations in the specimens that were not demagnetized. The relative importance of these sources of error is not known, except that secondary magnetizations in the weakly magnetized specimens are evidently and fortunately small and randomly directed.

The recalculated averages do not differ greatly from those originally published (Gromme and Gluskoter, 1965). The between-site mean directions differ by 4°, and the uncertainty is reduced by 6%. The corresponding angular difference in pole positions is 6°. Nonetheless, inclusion of the data from partially demagnetized specimens strengthens the final result.

REFERENCE POLE

The Cretaceous Period is well known to have been a time of little apparent polar wander as viewed from North America (see for example, Irving and Irving, 1982). A reference paleomagnetic pole is here constructed from five well-established poles derived from rocks in the age range from 87 to 135 million years, all located on the North American craton. The individual poles and their mean are listed in Table 3; the mean is calculated assigning unit weight to each locality or formation, i.e. unit weight to each publication. Several of the individual poles were recalculated from the individual sources. For the Niobrara Formation (Shive and Frerichs, 1974) a composite pole was calculated as the weighted mean of poles from three localities (Pueblo, Trego, and Kemmerer). The weighting factor is R, the estimated vector sum of unit VGP's for each site, where the estimate is made using the method of Cox (1970). The same procedure was used for the pole from Mt. Ascutney in Vermont (Opdyke and Wensink, 1966), where data from two areas are summed. The poles reported for the other three areas were checked but did not have to be recalculated. Because in all of these studies statistics were reported only for directions, the circular 95% confidence cone around each pole was estimated using the method of Cox (1970). These values are shown in Table 3 for comparison only; they do not enter into the computation of the reference pole.

All of the ages cited in Table 3 are radiometric except for the Niobrara Formation. Shive and Frerichs (1974) give a paleontologically determined age range from Coniacian up into the earliest part of the Maestrichtian. It turns out that nearly all of the samples used in the present computation came from the Fort Hays Member, which is entirely Coniacian (Shive and Frerichs, 1974). According to the time scale of Harland and others (1982) the Coniacian ranges from 88.5 to 87.5 million years; these values are rounded outward in Table 3. Of all the poles in Table 3, only that from Mt. Ascutney is derived from rocks as old as the Nicasio terrane. None of the others except for the one from the Monteregian Hills differs significantly from the Mt. Ascutney pole, however, and there is no significant apparent polar wander to be seen in the 48 million years spanned by these data. Moreover, the geographic distribution of the five areas is quite broad; in Cretaceous coordinates they span 36° of latitude and 62° of longitude. The reference pole differs by 6° of arc from the one published by Irving and Irving (1982) for 130 m.y., and the uncertainty is much smaller (3.4° rather than 19°). The differences are attributable to the use by Irving and Irving of a 30 m.y. unweighted running mean to calculate their reference poles combined with the fact that the Cretaceous poles define the apex of an angle in the North American apparent polar wander path. The midpoint of the ages of Cretaceous poles in Table 3 is 111 million years. The reference pole for 110 million years given by Irving and Irving (1982) is an average of three poles, has an uncertainty of 6°, and is only 1.7° of arc from the mean pole in Table 3. Using the mean reference pole in Table 3 allows taking advantage of the Cretaceous "stillstand" in apparent polar wander to average five poles over a 48 million year window and still minimize the uncertainty in the result.

CONCLUSIONS

The relationships of the location of the Nicasio terrane to the Cretaceous reference pole and to the pole derived from the rocks of the terrane itself are shown in Figure 2. As mentioned earlier, the two poles are greatly discordant. Evaluation in tectonic terms of this discordance consists of calculating a latitude difference which is now conventionally interpreted as a manifestation of plate motion, and a rotation around a vertical axis. This rotation can only be calculated for the displaced terrane in its final (present) position, and there exists an infinity of paths by which the observed resultant rotation might have occurred.

The actual calculation is simply to solve the spherical triangle whose vertices are the site and the two poles. The included angle at the site is the rotation, and the difference between the two distances from the site to the poles is the latitude shift. The uncertainty in the latitude shift is the Pythagorean sum of the semi-angles of the confidence cones centered on the two poles. The uncertainty in the rotation is the Pythagorean sum of the semi-angles of the same two confidence cones, but subtended at the site. Hence the the uncertainty in rotation will be greater than or equal to the uncertainty in latitude shift. Following Demarest (1983), in order to convert

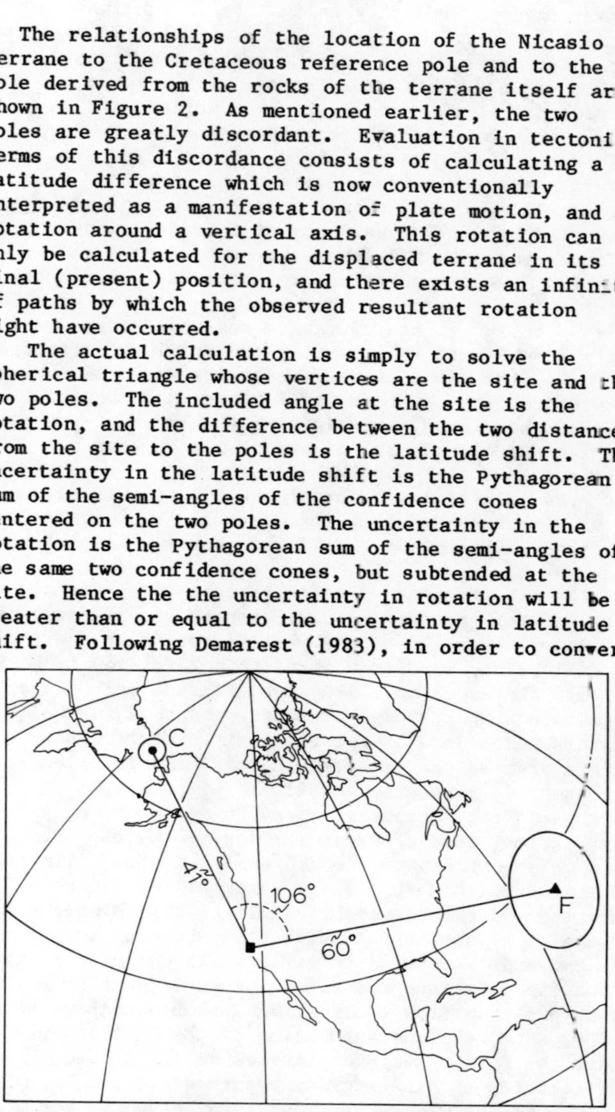

Figure 2. Equal-area projection showing the present locality of the Nicasio terrane (square), the reference paleomagnetic pole for Cretaceous time (circle marked C), and the paleomagnetic pole for the Nicasio terrane in the Franciscan Group (triangle marked F). 95% confidence circles are shown centered on the two poles. The projection is centered on the sampling locality, and the grid interval is 30°.

bivariate confidence limits to univariate confidence limits, the Pythagorean sums are multiplied by 0.78 (see also Cox and Gordon, 1984).

The values obtained for the Nicasio terrane are an eastward (clockwise) rotation of 106.4° ± 9.3°, and a northward shift of 19.7° ± 7.8°. This calculation implies that the rocks formed at 30° north latitude in the Cretaceous reference frame. Rotations and translations have also been calculated for the three igneous bodies separately; the uncertainties range from 11° to 17°, and none of the individual rotations or translations differs significantly from another. The rotation was known approximately by the original authors but was somewhat underestimated (Gromme and Gluskoter, 1965), but the 20° northward movement was not evident at that time because the reference geomagnetic field and the age of the rocks was poorly known. A similar result has recently been reported for Franciscan assemblage rocks of the Marin Headlands terrane to the south by Curry and others (1983).

All of the sites in the Nicasio terrane exhibit normal magnetic polarity in that their inclinations are all downward, and it is known with certainty from the shapes of the lava pillows that the stratigraphic section is not inverted. The age of the rocks corresponds to the middle of the M series of geomagnetic polarity chrons, approximately M9 through M12 (Harland and others, 1982). Hence given the possible indeterminacy of site of origin, the geomagnetic polarity when the rocks were formed is not known, and their downward magnetic inclinations could be interpreted to mean that they were formed in the southern Cretaceous hemisphere while the geomagnetic polarity was reversed. This would change the rotation by 180° to 74° counterclockwise and would increase the latitude shift to 79°, putting the original latitude at 30° south. Such an interpretation would be more in accordance with the results of Alvarez and others (1980), who placed the origin of the somewhat younger (Albian and Cenomanian) limestone blocks near Laytonville at 17° south latitude. In accordance with the principle of least astonishment, and because there is no direct evidence to the contrary, a northern hemisphere origin for the Nicasio terrane is preferred here.

The data from the Nicasio terrane fit perfectly into an ever-growing group of discordant paleomagnetic directions in rocks of the westernmost North American Cordillera, as was first noted in publication by Beck (1976). Later, Beck (1980) summarized a much larger signature of plate edge tectonic processes in the western Cordillera". For the western conterminous United States, the universal characteristic of this signature is clockwise rotation. Significant northward translations are seen in most but not all pre-Tertiary rocks, but in Tertiary rocks they are seen only west of the San Andreas Fault in California (Champion and others, 1984). Beck (1976) favored rotation and translation of small crustal blocks in a regime of right-lateral shear (the sense of shear on the San Andreas Fault) to explain the ubiquity of the clockwise rotations and common occurrence of anomalously shallow inclinations. Subsequently, Beck (1980) added oblique subduction of the Farallon plate under the North American plate as an attractive modification of his original hypothesis.

Gromme and Gluskoter (1965), in advancing the first of their three explanations, stated (p.92): "Tectonic rotation of all three igneous bodies as a single unit, 70° clockwise around a vertical axis, occurred. If such a movement occurred, it must have been prior to the folding of the rocks, and therefore prior to the formation of the San Andreas Fault. Thus the correspondence of this rotation to the sense of a hypothetical shear couple responsible for the observed displacement on the San Andreas Fault appears to be fortuitous. One might speculate that the rotation occurred while the Franciscan geosyncline was being filled and was in a highly mobile state." The last sentence expressed our imagination of seamounts being gradually tilted on the flank of a subsiding geosyncline, and being rotated when they slipped downward on bedding-plane faults at their bases. If one eliminates the geosyncline and substitutes oblique subduction, then this interpretation is correct in terms of what is presently known about plate interactions at the western margin of North America. The rotation is most simply considered as the result of collision of the seamounts with the overriding plate.

REFERENCES CITED

Alvarez, Walter, Kent, D.V., Premoli Silva, Isabella, Schweickert, R.A., and Larson, R.A., 1980, Franciscan Complex limestone deposited at 17° south latitude: Geological Society of America Bulletin, v.91, p.476-484.

Beck, M.E.,Jr., 1976, Discordant paleomagnetic pole positions as evidence of regional shear in the western Cordillera of North America: American Journal of Science, v.276, p.694-712.

Beck, M.E.,Jr., 1980, Paleomagnetic record of plate-margin tectonic processes along the western edge of North America: Journal of Geophysical Research, v.85, p.7115-7131.

Blake, M.C.,Jr., Bartow, J.A., Frizzell, V.A.,Jr., Schlocker, Julius, Sorg, Dennis, Wentworth, C.M., and Wright, R.H., 1974, Preliminary geologic map of Marin and San Francisco Counties and parts of Alameda, Contra Costa, and Sonoma Counties, California: U.S. Geological Survey Map MF 574.

Blake, M.C.,Jr., Howell, D.G., and Jones, D.L., 1982, Preliminary tectonostratigraphic terrane map of California: U.S. Geological Survey Open-File Report 82-593.

Champion, D.E., Howell, D.G., and Gromme, C.S., 1984, Paleomagnetic and geologic data indicating 2500 km of northward displacement for the Salinian and related terranes, California: Journal of Geophysical Research, v.89, in press.

Cox, Allan, 1957, Remanent Magnetization of lower to middle Eocene basalt flows from Oregon: Nature, v.179, p.685-686.

Cox, Allan, 1970, Latitude dependence of the angular dispersion of the geomagnetic field: Geophysical Journal of the Royal Astronomical Society of London, v.20, p.253-269.

Cox, Allan and Gordon, R.G., 1984, Paleolatitudes determined from paleomagnetic data from vertical cores: Reviews of Geophysics and Space Physics, v.22, p.47-72.

Curry, Foster, Cox, Allan, and Engebretson, D.G., 1983, Paleomagnetism of Franciscan rocks in the Marin Headlands (abstract): EOS Transactions of the American Geophysical Union, v.64, p.867

Demarest, H.H.,Jr., 1983, Error analysis for the determination of tectonic rotation from paleomagnetic data: Journal of Geophysical Research, v.88, p.4321-4328.

Fisher, R.A., 1953, Dispersion on a sphere: Royal Society of London, Proceedings, series A, v.217, p.295-305.

Foster, John and Symons, D.T.A., 1979, Defining a paleomagnetic polarity pattern in the Monteregian intrusives: Canadian Journal of Earth Sciences, v.16, p.1716-1725.

Gluskoter, H.J., 1962, Geology of a portion of western Marin County, California: unpublished Ph.D. dissertation, University of California, Berkeley,

California, 184 pp.
Gluskoter, H.J., 1969, Geology of a portion of western Marin County, California: California Division of Mines and Geology, Map Sheet 11.
Gromme, C.S. and Gluskoter, H.J., 1965, Remanent magnetization of spilite and diabase in the Franciscan Formation, western Marin County, California: Journal of Geology, v.73, p.74-94.
Irving, E. and Irving, G.A., 1982, Apparent polar wander paths, Carboniferous through Cenozoic, and the assembly of Gondwana: Geophysical Surveys, v.5, p.141-188.
Larochelle, Andre, and Black, R.F., 1963, An application of palaeomagnetism in estimating the age of rocks: Nature, v.198, p.1260-1262.
McElhinny, M.W., 1964, Statistical significance of the fold test in paleomagnetism: Geophysical Journal of the Royal Astronomical Society of London, v.8, p.338-340.
McElhinny, M.W. and Merrill, R.T., 1975, Geomagnetic secular variation over the past 5 m.y.: Reviews og Geophysics and Space Physics, v.13, p.687-708.
McFadden, P.L. and Jones, D.L., 1981, The fold test in paleomagnetism: Geophysical Journal of the Royal Astronomical Society of London, v.67, p.53-58.
Harland, W.B., Cox, A.V., Llewellyn, P.G., Pickton, C.A.G., Smith, A.G., and Walters, R., 1982, A Geologic Time Scale: Cambridge University Press, Cambridge, 131 pp.
Opdyke, N.D. and Wensink, Hans, 1966, Paleomagnetism of rocks from the White Mountain plutonic-volcanic series in New Hampshire and Vermont: Journal of Geophysical Research, v.71, p.3045-3051.
Scharon, LeRoy, and Hsu, I-Chi, 1969, Paleomagnetic investigation of some Arkansas alkalic igneous rocks: Jornal of Geophysical Research, v.74, p.2774-2779.
Shive, P.N. and Frerichs, W.E., 1974, Paleomagnetism of the Niobrara Formation in Wyoming, Colorado, and Kansas: Journal of Geophysical Research, v.79 p.3001-3007.
Swanson, S.E. and Schiffman, Peter, 1979, Textural evolution and metamorphism of pillow basalts from the Franciscan Complex, western Marin County, California: Contributions to Mineralogy and Petrology, v.69, p.291-299.
Wright, R.H., 1982, Geology of western Marin County, California: unpublished Ph.D. dissertation, University of California, Santa Cruz, California, 204pp.

SEDIMENTARY PETROLOGY OF GRAYWACKE OF THE FRANCISCAN COMPLEX IN THE NORTHERN SAN FRANCISCO BAY AREA, CALIFORNIA

A. S. Jayko and M. C. Blake Jr.
U.S. Geological Survey, 345 Middlefield Rd., MS 975
Menlo Park, California, 94020

ABSTRACT

The Franciscan Complex in the San Francisco Bay area includes several graywacke-bearing terranes that range in age from Late Jurassic to early Tertiary. Graywackes of different ages and different geographic localities have distinct quartz-feldspar-lithic (QFL) ratios which can be used for distinguishing and correlating units. Where paleontologic control on age is available, some Franciscan graywackes appear to differ in modal composition from coeval strata of the Great Valley sequence. The Yolla Bolly, Alcatraz, Novato Quarry, and Burnt Hills terranes and quartzofeldspathic facies of the Coastal terrane are all richer in quartz and feldspar than coeval strata of the Great Valley sequence. This difference suggests that the clastic detritus was not derived from the ancestral Sierran-Klamath arc which is the inferred source of the Great Valley sequence. The Franciscan graywacke may have been tectonically transported from more southerly latitudes. The Central, Marin Headlands, and Rio Nido terranes and the lithic facies of the Coastal terrane are similar to coeval strata of the Great Valley sequence. These rocks may thus have been deposited closer to the provenance of the Great Valley sequence.

INTRODUCTION

The Franciscan Complex of northern California consists of three major belts: the Coastal, Central, and Eastern (Bailey and Irwin, 1959; Irwin, 1960; Berkland and others, 1972; fig. 1). Recently, these belts have been divided into tectonostratigraphic terranes defined by differences in age, petrology, lithologic association, and metamorphic grade (Blake and others, 1982). In this report we examine the clast compositions of the various Franciscan terranes and compare them with those petrofacies of equivalent age from the Great Valley sequence.

The Franciscan Complex has long been inferred to be derived from the same source as the adjacent and coeval Great Valley sequence (Hamilton, 1969; Dickinson, 1970; Dickinson and others, 1982). Systematic variations in the lithic content of Upper Jurassic to Upper Cretaceous graywacke from the Great Valley sequence have been correlated with erosion of an inferred volcanic edifice and unroofing of the underlying Sierra Nevada batholith (Ojakangas, 1968; Dickinson and Rich, 1972; Ingersoll, 1978, Mansfield, 1979). Dickinson and others (1982) proposed that similar compositional variations should occur within equivalent-age rocks of the Franciscan Complex. Results from our study indicate both similarities and differences between the composition of graywacke of the Franciscan Complex and that of petrofacies of the coeval Great Valley sequence. This comparison suggests a more complex scenario for the origin of some Franciscan Complex graywacke than previously inferred.

This study was undertaken as a follow up to work begun in 1970 on sudivision of the Franciscan Complex in the northern San Francisco Bay area (Blake and others, 1971; Blake and Jones, 1974) that was based in part on variations in sandstone petrology recognized in reconnaissance modal analyses. In addition to our sampling, a few samples collected by E. H. Bailey, J. G. Schlocker, Clyde Wahrhaftig and R. H. Wright were included in the study. Only graywacke from the Franciscan Complex and Great Valley sequence east of the San Andreas fault are considered here.

SEDIMENTARY PETROLOGY

Samples of medium-grained graywacke were point-counted in thin section at 1-mm intervals under 50x magnification (Table 1). The parameters described by Dickinson (1970) were used except that hypabyssal intrusive fragments were included with volcanic fragments. No effort was made to distinguish between types of volcanic-rock fragments (such as felsite, dacite, or basalt) because varying degrees of alteration prevented consistent determination. Approximately 400-500 grains per thin section were counted by one operator (Jayko) on 69 samples.

Because some interpretation is required in determining grain type, duplicate counts of 10 samples were made by three other operators (Blake, H. R. Ohlin, and R. H. Wright) to determine reproducibility. As noted by earlier workers (Dickinson, 1970; Jacobson, 1976), the greatest variation occurs in the lithic-fragment portion. The maximum variations recorded between the results of four operators on the same sample were 8.1 percent in lithic clasts, 7.0 percent in quartz, and 6.2 percent in feldspar.

In most petrographic studies of the Great Valley sequence (Dickinson and Rich, 1972), the ratio quartz (Q): feldspar (F): lithic fragments (L) (QFL) is calculated with chert included in Q. Jacobson (1978), however, who compiled available modal analysis from the Franciscan included chert in L. Detrital modes from this study are plotted with chert included in Q to facilitate comparison with Great Valley sequence detrital modes. For ease of comparison, the ratios with chert included in L are included in Table B. This choice of convention is not critical in this study because chert ranges only from 0 to 12 percent. The QFL ratios proved to be the most valuable for comparisons; other modal parameters are listed in the table for reference.

Ingersoll (1978), Ingersoll and Dickinson (1981) and Dickinson and others (1982) presented the modal data from the Great Valley sequence in terms of the means and standard deviations on QFL diagrams instead of showing QFL values directly. A similar approach is used here to facilitate comparisons.

PREVIOUS WORK

Several hundred modal determinations of Franciscan Complex graywacke have been reported from scattered localities throughout northern California, including the Diablo Range, Cape Mendocino area, Clear Lake area, and Marin County. We briefly summarize

in Blake, M.C. Jr., ed., 1984, **Franciscan Geology of Northern California:** Pacific Section S.E.P.M., Vol. 43, p. 121-134

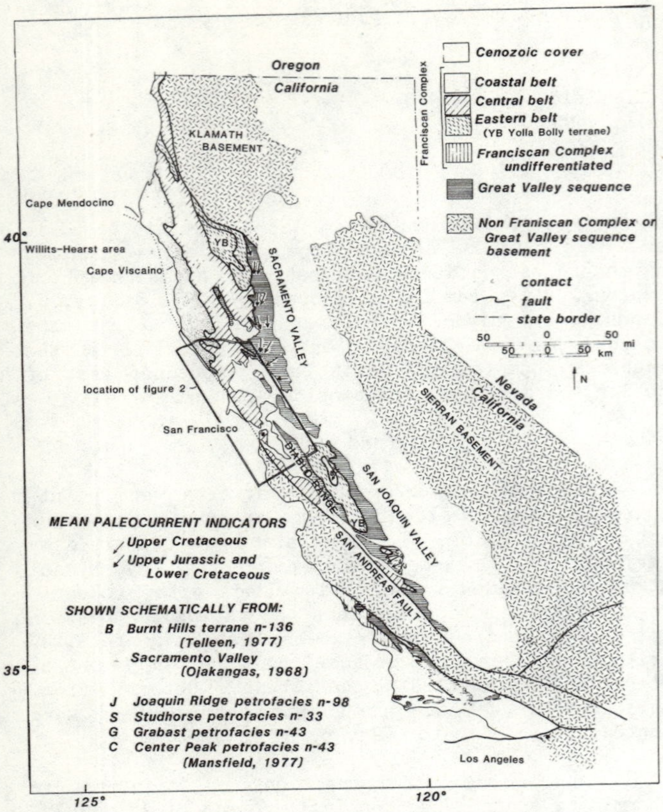

Figure 1. Geolgic sketch map of California showing locations of the Franciscan Complex and Great Valley sequence, and area of figure 2. Paleocurrent directions are shown schematically.

these data below and describe the graywacke petrofacies of the Great Valley sequence.

Franciscan Complex

The first petrologic subdivision of the Franciscan Complex was based on the potassium feldspar content of graywacke (Bailey and Irwin, 1959) which allowed the differentiation of the Franciscan Coastal belt (commonly K-feldspar rich, 5-25 percent) from the rest of the Franciscan Complex (commonly zero to trace percent). This criterion, in addition to QFL ratios, has been used in local studies in the San Francisco Bay area (Gluskoter, 1964; Berkland, 1969; Schlocker, 1974; Wright, 1982), in which several potassium feldspar-rich units were identified.

The petrologic modes of Franciscan graywacke from the Diablo Range reported by Cowan (1974) and Jacobson (1978) fall into two distinct compositional fields represented by the Burnt Hills and Yolla Bolly terranes (Blake, 1981). The Upper Cretaceous (Coniacian) Burnt Hills terrane (Blake, 1981), which includes the Burnt Hills sequence of Jacobson (1978) and the Orestimba and Mustang Peak metagraywacke unit of Cowan (1974), is quartzofeldspathic and contains less than 15 percent lithic grains. The Upper Jurassic and Lower Cretaceous (Tithonian to Valanginian) Yolla Bolly terrane (Blake, 1981), which includes the Eylar Mountain sequence and Las Aguilas Creek melange of Jacobson (1978), as well as the Zimba metagraywacke unit and the Garzas melange of Cowan (1974), are more lithic and less quartzose than the younger Burnt Hills sequence.

Other published QFL data on Franciscan graywacke of the northern Coast Ranges are scarce. Suppe (1973) reported modal data for eight samples of his Hellhole Graywacke, a Tithonian to Valanginian unit correlated with the Yolla Bolly terrane in the northern Coast Range (Blake and others, 1982). Jacobson (1978) summarized the unpublished modal data of Kleist (1974) from the Franciscan Coastal belt near Cape Viscaino. Although Dickinson and others (1982) reported means of modal data from four Franciscan units, their sample locations and raw modal data were not published. Early work by McNitt (1968) in the Clear Lake area is reviewed by McLaughlin (this volume), and Soliman's (1965) modal data from the Diablo Range were reinvestigated by Jacobson (1976, 1978).

Great Valley Sequence

Petrologic studies of the Great Valley sequence are numerous, and several distinct petrofacies corresponding to particular time periods and geographic areas have been recognized (Ojakangas, 1968; Dickinson and Rich, 1972; Ingersoll, 1978; Mansfield, 1979; Ingersoll and Dickinson, 1981). Detrital modes of Great Valley sequence rocks in the Sacramento Valley generally differ from those in the San Joaquin Valley to the south. This difference is thought to be related to differences in sediment-transport direction in different parts of the Mesozoic sedimentary basin (Ojakangas, 1968; Dickinson and others, 1982).

The Upper Jurassic and Lower Cretaceous (Neocomian) part of the Great Valley sequence consists of three geographically distinct petrofacies, from north to south: (1) the Platina (Ingersoll and Dickinson, 1981) of the northernmost Sacramento Valley, which unconformably overlies Klamath basement; (2) the Stoney Creek (Ojakangas, 1968; Dickinson and Rich, 1972) of the Sacramento Valley, which unconformably overlies the Coast Range ophiolite; and (3) the Gravelly Flat (Mansfield, 1979) of the San Joaquin Valley, which is inferred to unconformably overlie the Coast Range ophiolite. The Aptian and Albian part of the section consists of the Lodoga petrofacies in the Sacramento Valley (Dickinson and Rich, 1972; Ingersoll and Dickinson, 1981) and the Center Peak and Grabast petrofacies in the San Joaquin Valley (Mansfield, 1979). See Ingersoll (1978), for a map showing the general locations of these Great Valley sequence petrofacies.

Petrofacies of Upper Cretaceous strata of the Great Valley sequence generally are more consistent in composition from the Sacramento and San Joaquin Valleys than are the older petrofacies. These younger petrofacies include, from oldest to youngest, the Boxer, Cortina, and Rumsey petrofacies of the Sacramento Valley (Ingersoll and Dickinson, 1981), which correspond to the Grabast and Studhorse, Upper and Lower Los Gatos, and Joaquin Ridge petrofacies of the San Joaquin Valley (Mansfield, 1979).

The Boxer, Cortina, and Rumsey petrofacies all contain high- and low-lithic variations. These variants are not systematically related to either stratigraphic or geographic position (Dickinson and Rich, 1972; Ingersoll, 1978; Ingersoll and Dickinson, 1981). The means and standard deviations of modal data for the high- and low-lithic variants (Ingersoll, 1978; Ingersoll and Dickinson, 1981; Dickinson and others, 1982) and the average of the variants (Ingersoll and Dickinson, 1981) define three commonly distinct QFL fields for each Upper Cretaceous

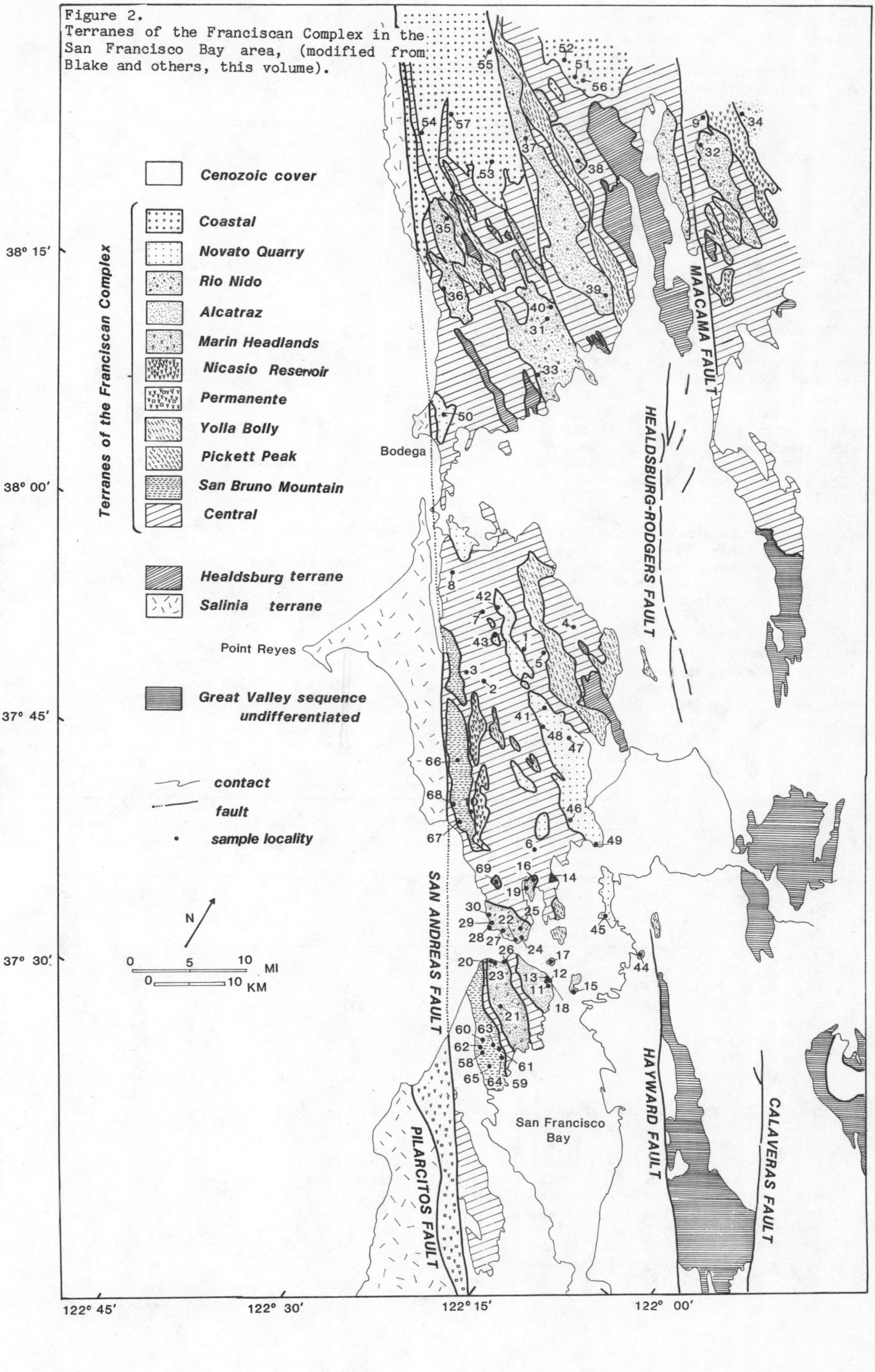

Figure 2.
Terranes of the Franciscan Complex in the San Francisco Bay area, (modified from Blake and others, this volume).

petrofacies. Because both the high- and low-lithic variants and the average of the two are recently used, we compare the Franciscan rocks to both variants here. Tectonic relations inferred from modal data from the Great Valley sequence and basin analyses are discussed extensively elsewhere and are not repeated here (Ojakangas, 1968; Dickinson and Rich, 1972; Ingersoll, 1978; Mansfield, 1979; Ingersoll and Dickinson, 1981; Dickinson and others, 1982).

RESULTS

A total of 12 Franciscan terranes of Upper Jurassic to Tertiary age have been mapped in the San Francisco Bay area (Blake and others, this volume, fig. 2). Samples from seven of these terranes, the Central, Alcatraz, Marin Headlands, Rio Nido, Novato Quarry, Coastal, and San Bruno Mountain were point-counted and compared with petrofacies of equivalent age from the Great Valley sequence, as well as with samples of Franciscan graywacke for which modal data have been published.

Graywacke clasts from the Pickett Peak terrane are too modified by recrystallization to count. The composition of the Nicasio Reservoir terrane is described by Wright (this volume), and of the Yolla Bolly terrane (Eylar Mountain sequence) by Jacobson (1978). Graywacke of the Permanente terrane has not yet been adequately mapped (Blake and others, this volume) to permit confident sampling, and so it is not included in this study. Table 1 and figure 3 summarize the modal data for each terrane; tables 2 and 3 list the complete modal data. Blake and others (this volume) describe the age, lithology, and field relation of each terrane; we describe the petrology here. Table 4 shows a general time-stratigraphic chart for the units described below.

The Central belt melange matrix of the Franciscan Complex consists of thin beds of graywacke interbedded in voluminous argillite. Fossils found in the Central belt melange consist of Tithonian to Valanginian megafossils, radiolarians and dinoflagellates found in sandstone and shale of Marin and Sonoma Counties (Bailey and others, 1964; Dudley, 1972; Wright, 1974;

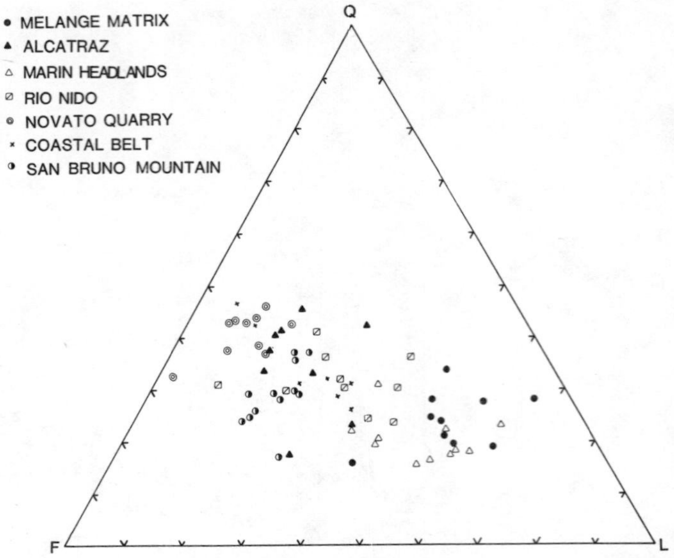

Figure 3. QFL ratios of 69 samples of graywacke of the Franciscan Complex from various terranes in the San Francisco Bay area.

Blake and others, this volume; Murchey and Jones, this volume). The melange matrix that surrounds most of the terranes throughout the San Francisco Bay area has a limited compositional range. The mean QFL ratio (24-22-54) of this matrix is the most lithic of the Franciscan units of this study, and is equivalent to the Eastern melange zone of Gluskoter (1964) from western Marin County. The graywacke of the Central Franciscan belt matrix lacks detrital potassium feldspar (Wright, 1974) and averages about 6 percent chert detritus. A comparison of modal data from graywacke of the Central belt with petrofacies of equivalent age from the Great Valley sequence (fig. 4a) shows that the Central belt graywacke is compositionlly similar to the Stoney Creek petrofacies (Dickinson and Rich, 1972; Ingersoll, 1978).

TABLE 1

Average modal data from Franciscan terrane of the San Francisco Bay Region

Terrane	(n)	QFL	Q-F-L^{SD}	Qp/Q	Lv/L	M/Qm	Q/Q+L	cht	M
Melange	(10)	24-22-54	5-9-7	.09	.86	28.6	46.8	6.4	2
Alcatraz	(9)	35-41-24	9-7-7	.11	.83	21.3	43.0	2.8	4
Marin Headlands	(11)	20-28-52	5-5-8	.16	.81	24.6	34.8	3.1	4
Rio Nido	(10)	31-39-30	5-8-10	.05	.82	19.9	42.0	4.8	2
Novato Quarry	(10)	41-48-11	4-7-4	.08	.78	15.7	44.6	1.4	3
Coastal Belt (qf)	(2)	45-46-9	2-1-3	.06	.78	15.2	48.8	.3	4
Coastal Belt (l)	(5)	30-39-31	2-3-4	.03	.83	23.1	40.3	3.5	5
San Bruno Mtn.	(12)	29-49-22	6-6-3	.04	.89	29.9	34.7	1.2	5

QFL the average ratio of quartz:feldspar:lithic clasts, in percent
(n) number of samples counted
QFL_{SD} standard deviation of QFL
Qp/Q ratio of polycrystalline to monocrystalline quartz
Lv/L proportion of lithic volcanic to total lithic clasts
M/Qm proportion of mica to monocrystalline quartz
Q/Q+L percent quartz to toal quartz plus lithics
chert total chert in percent
%M percent total mica
(qf) quartzofeldspathic facies of the Coastal belt
(l) lithic facies of the Coastal belt

Table 2
Modal data from graywackes of the San Francisco Bay Area

Sample #	Total Feldspar	Quartz mono/poly	Chert	Volcanic and hypabyssal rock fragments	Sedimentary rock fragment	Metamorphic rock fragment	Mica	Chlorite	Opaque	Single mineral	Matrix	Calcite	Other	Total
Central belt														
1.	75	45/7	39	196	0	0	18	2	6	0	33	0	0	421
2.	23	66/0	44	222	16	15	1	4	4	0	25	0	0	420
3.	112	92/1	4	235	2	2	11	3	5	5e	17	0	0	489
4.	75	61/8	12	248	11	10	5	6	3	3e	18	0	0	460
5.	102	56/9	27	202	6	3	29	10	11	0	22	0	1	478
6.	106	91/5	9	208	1	0	2	5	2	4e	26	0	0	459
7.	57	73/14	19	150	0	8	15	2	4	5e	37	0	0	384
8.	106	60/4	23	240	4	1	3	7	4	0	20	0	0	472
9.	196	48/2	22	173	13	0	15	16	3	1g	48	0	4	545
10.	62	57/8	50	230	17	2	3	8	0	0	22	0	4	463
Alcatraz														
11.	185	111/7	11	57	1	6	26	3	2	2e	26	0	0	437
12.	150	115/14	10	39	0	7	29	11	2	3e,2px	55	0	0	437
13.	184	151/1	1	63	3	0	14	6	1	0	32	7	0	463
14.	221	164/4	11	127	0	2	13	1	7	0e	23	4	0	586
15.	122	146/19	26	125	8	2	18	6	4	2	26	0	0	504
16.	227	60/9	7	123	0	3	16	3	6	3	31	3	5	496
17.	119	95/34	19	40	2	7	9	12	3	7	29	0	0	376
18.	145	132/6	5	54	1	0	14	5	2	0	16	0	0	380
19.	169	83/6	9	160	1	0	19	2	4	7e	44	3	29	538
Marin Headlands														
20.	68	51/21	42	306	0	4	12	1	3	3e	13	0	1	525
21.	81	38/3	32	168	10	0	4	1	6	2e	49	5	0	399
22.	145	42/10	29	255	9	1	20	5	5	1e	30	7	3	562
23.	108	43/19	27	281	12	3	15	6	8	2e	39	17	5	585
24.	127	45/7	10	201	5	0	6	3	7	7	49	0	0	467
25.	179	92/3	6	136	31	2	19	1	8	4e,	26	0	0	507
26.	139	71/0	9	162	2	3	10	3	14	0	14	0	0	418
27.	115	57/0	27	210	53	6	4	8	9	8e,1cz	22	17	1	538
28.	182	53/12	44	250	17	3	20	17	11	0	36	3	2	650
29.	163	145/9	8	134	63	7	26	7	10	6e,1cz	33	2	2	616
30.	143	73/6	18	251	68	4	5	2	8	0	23	3	2	606
Rio Nido														
31.	155	98/4	11	201	7	2	17	9	8	5e	36	3	1	557
32.	154	84/7	10	151	10	1	11	7	10	2e	26	2	1	477
33.	162	171/0	14	93	8	1	11	2	1	2e	52	0	1	518
34.	158	93/4	26	135	6	1	15	3	2	8e,1sph	45	2	6	503
35.	101	95/3	9	137	11	0	3	14	9	4e	38	0	3	427
36.	162	108/8	29	108	3	1	9	15	8	5e	39	8	1	504
37.	150	126/11	8	102	3	2	11	8	3	2e	40	0	0	466
38.	146	115/6	32	145	24	9	14	17	12	4e	23	0	4	551
39.	202	113/0	12	89	7	2	21	14	1	12e,	40	0	2	516
40.	286	143/6	1	44	9	0	9	5	1	2e,2s	31	0	0	539
Novato Quarry														
41.	174	135/17	3	21	2	3	18	21	2	0	42	0	0	438
42.	190	129/24	2	47	0	2	17	2	6	2	50	0	0	471
43.	156	134/2	6	30	1	0	6	2	1	1e	36	0	0	373
44.	156	133/11	8	33	1	0	9	2	1	4	44	0	0	400
45.	180	136/11	7	13	0	0	24	6	4	8	40	0	0	429
46.	126	120/9	8	23	1	3	20	5	4	4	54	0	0	373
47.	233	144/18	3	38	2	2	20	6	4	1e	19	5	11	506
48.	151	149/9	5	55	3	2	26	12	8	0	31	8	0	459
49.	200	145/10	3	65	0	1	5	1	3	1e	25	0	0	459
50.	331	158/4	4	13	5	0	11	3	0	0	23	0	0	548
Coastal belt														
51.	228	206/5	3	56	3	1	43	0	4	3e,1sph	18	0	0	571
52.	209	200/6	0	25	2	1	16	3	3	1h,4e	14	17	0	502
53.	182	103/11	12	96	3	0	6	1	2	2e	31	10	0	459
54.	154	106/2	21	106	7	1	23	7	6	3e,1hb	39	0	2	478
55.	198	110/12	14	143	33	12	37	6	8	10e,3hb	16	0	0	602
56.	162	122/5	16	146	12	11	18	14	4	3e,	39	0	0	556
57.	169	109/3	13	104	36	1	16	9	10	5e,	37	0	2	514
San Bruno Mountain														
58.	199	158/23	8	96	2	10	19	2	1	0	11	18	0	547
59.	191	150/4	5	76	3	6	26	3	12	6e/1sph	35	0	0	518
60.	177	85/6	10	80	0	2	24	4	3	2e/1hb	6	0	0	399
61.	134	81/6	0	65	1	3	27	11	15	3e	47	0	2	394
62.	254	69/1	10	125	0	0	8	11	1	4e	24	0	0	507
63.	277	109/3	3	83	3	3	31	6	6	0	28	0	0	552
64.	291	121/3	5	96	2	3	45	5	6	0	34	9	0	620
65.	238	108/1	1	82	3	0	21	5	6	2e,	33	17	0	517
66.	207	97/6	9	57	6	2	32	3	0	1e	43	0	1	464
67.	173	106/5	1	97	0	0	9	9	8	7e	30	5	3	453
68.	183	144/7	8	83	3	3	19	3	4	4e,	26	6	1	494
69.	193	113/0	2	74	6	1	13	3	2	0	49	0	0	456

e = epidote, px = pyroxene, sph = sphene, hb - hornblende, g = green amphibole cx - clinozoisite,

Table 3
QFL and other ratios from Table 2

Sample #	A			B			Q+F+L+cht						
	Q	F	L+cht	Q+cht'	F'	L'	total	%Matrix	M/Qm	Lv/L	Q/F	Qp/Q	Q/Q+F
							Central belt						
1.	16	23	61	28	23	48	87	8	44	.83	69	.15	41
2.	17	6	77	28	6	66	93	6	8	.74	287	.00	74
3.	20	25	55	21	25	54	93	4	15	.96	83	.01	45
4.	16	18	66	19	18	64	85	4	18	.88	92	.13	48
5.	18	24	58	24	24	52	89	5	70	.84	64	.16	42
6.	23	25	52	25	25	50	91	6	8	.95	91	.05	47
7.	28	18	54	34	18	48	84	10	23	.84	152	.19	60
8.	14	24	61	20	24	56	93	4	17	.89	60	.06	38
9.	11	43	46	16	43	41	83	9	65	.83	25	.04	20
10.	16	15	69	28	15	57	90	5	19	.76	105	.14	51
							Alcatraz						
11.	31	49	20	34	49	17	87	6	26	.76	64	.06	39
12.	37	44	19	40	44	16	79	13	35	.69	86	.12	44
13.	38	45	17	38	45	17	87	7	13	.94	83	.01	45
14.	31	41	28	33	41	26	92	4	8	.90	76	.02	43
15.	36	27	37	42	27	31	89	5	16	.82	135	.13	57
16.	16	53	31	18	53	29	86	6	32	.25	304	.15	23
17.	40	36	24	45	36	19	87	8	22	.58	108	.35	52
18.	40	42	18	41	43	16	91	4	14	.91	95	.04	49
19.	21	39	40	23	39	38	80	8	25	.94	53	.07	34
							Marin Headlands						
20.	15	14	71	23	14	63	94	2	25	.86	106	.41	51
21.	13	24	63	22	24	54	87	12	13	.80	47	.07	31
22.	11	29	60	16	29	54	87	5	59	.86	29	.23	26
23.	13	22	65	18	22	60	84	7	49	.87	57	.44	36
24.	13	32	55	16	32	52	85	10	20	.93	41	.15	29
25.	21	40	39	22	40	37	89	5	22	.77	53	.03	34
26.	18	36	46	21	36	43	92	3	18	.92	51	.0	34
27.	12	25	63	18	26	57	87	4	21	.70	50	.0	33
28.	12	32	56	19	32	48	86	5	70	.79	36	.22	26
29.	29	31	40	31	31	39	86	5	23	.63	94	.06	48
30.	14	25	61	17	25	57	93	4	10	.73	55	.08	36
							Rio Nido						
31.	21	32	46	24	32	44	86	6	26	.91	66	.04	40
32.	22	37	41	24	37	41	87	5	21	.87	59	.08	37
33.	38	36	26	41	36	23	87	10	8	.80	106	.0	51
34.	23	37	40	26	38	34	84	9	19	.4	61	.04	38
35.	27	29	44	30	28	42	83	9	18	.87	97	.03	49
36.	29	47	34	36	40	23	79	8	22	.76	72	.07	42
37.	34	37	37	36	37	27	86	9	15	.88	91	.09	48
38.	26	31	43	32	31	37	86	4	27	.69	83	.05	45
39.	27	47	26	29	47	23	82	8	32	.80	56	.00	36
40.	30	59	11	31	58	11	91	6	10	.81	52	.04	34
							Novato Quarry						
41.	43	49	9	43	49	8	81	10	29	.72	87	.12	43
42.	38	47	15	38	47	14	85	11	15	.92	80	.18	45
43.	41	47	12	43	47	10	89	10	6	.81	87	.01	46
44.	41	45	13	44	45	11	85	11	8	.78	92	.08	48
45.	42	52	6	44	52	4	81	9	22	.65	82	.08	44
46.	43	42	15	46	42	12	80	14	21	.65	102	.07	50
47.	37	53	10	38	53	9	87	4	18	.84	69	.12	41
48.	41	39	19	43	39	17	88	7	25	.84	105	.06	51
49.	36	47	17	37	47	16	92	5	4	.94	77	.07	44
50.	32	65	3	33	65	2	93	4	9	.59	49	.02	33
							Coastal belt						
51.	42	46	12	43	46	11	87	3	21	.88	92	.02	48
52.	47	47	6	47	47	6	88	3	9	.93	98	.03	49
53.	28	45	27	31	45	24	89	7	7	.86	63	.11	38
54.	27	39	34	32	39	29	83	8	28	.78	70	.02	41
55.	23	38	39	26	39	36	87	3	38	.70	62	.11	38
56.	28	35	37	31	35	33	82	7	26	.78	78	.04	44
57.	26	39	35	29	39	32	85	7	23	.75	66	.03	40
							San Bruno Mountain						
58.	36	39	25	37	39	24	93	2	13	.87	91	.14	48
59.	35	43	23	36	43	22	86	7	19	.84	81	.03	45
60.	25	49	26	28	49	23	90	1	33	.87	51	.07	34
61.	30	46	24	30	47	24	74	12	47	.94	65	.07	39
62.	15	55	30	17	55	28	91	5	27	.92	272	.01	22
63.	23	58	19	24	58	18	87	5	34	.87	40	.03	29
64.	24	56	20	25	56	19	84	5	41	.60	43	.02	30
65.	25	55	20	25	55	20	84	6	24	.95	46	.0	31
66.	27	54	19	29	54	17	83	9	36	.70	55	.06	33
67.	29	45	26	29	45	25	84	7	17	.90	61	.05	39
68.	35	42	23	37	42	21	87	5	15	.56	82	.05	45
69.	29	49	22	29	40	21	85	11	14	.92	58	.00	37

*See table 1 for heading explanation.
A QFL ratio for sample with chert in the quartz apex (method of Jacobson, 1978)
B QFL ratio for sample with chert in the lithic apex (method of Ingersoll, 1978, 1981; Dickinson and others, 1982)

TABLE 4

TIME-STRATIGRAPHY CORRELATION OF UNITS DESCRIBED PETROGRAPHICALLY

		GREAT VALLEY SEQUENCE		FRANCISCAN COMPLEX		
		SACRAMENTO VALLEY	SAN JOAQUIN VALLEY	SAN FRANCISCO BAY AREA	DIABLO RANGE	NORTHERN COAST RANGE
UPPER CRETACEOUS	EUROPEAN STAGES	Dickinson and Rich, 1972	Mansfield, 1979	Blake and Howell, this volume	Jacobson, 1978	Suppe, 1973 Dickinson and others, 1982
	MAASTRICHTIAN			Coastal belt ?		Coastal belt ?
	CAMPANIAN	Rumsey	Joaquin Ridge	Novato Quarry		
	SANTONIAN		Upper Los Gatos Creek			
	CONIACIAN	Cortina			Burnt Hills	
	TURONIAN		Lower Los Gatos Creek	Rio Nido		
	CENOMANIAN	Boxer	Studhorse	Marin Headlands		Willits Hearst ???
LOWER CRETACEOUS	ALBIAN	Lodoga	Grabast Center Peak			
	APTIAN		Gravelly Flat			
	BARREMIAN (NEOCOMIAN)	Platina				
	HAUTERIVIAN (NEOCOMIAN)					
	VALANGINIAN (NEOCOMIAN)	Stoney Creek		Alcatraz	Yolla Bolly	Yolla Bolly (Hellhole Graywacke)
	BERRIASIAN (NEOCOMIAN)			Central		
UPPER JURASSIC	TITHONIAN					

San Bruno Mountain (age unknown)

The Lower Cretaceous (Valanginian) Alcatraz terrane consists of quartzofeldspathic graywacke that averages a trace to 3 percent potassium feldspar, and as much as 15 percent locally (Wright, 1974; Schlocker, 1974). Graywacke and shale of the Alcatraz terrane have yielded several fossils, including a Cretaceous Inoceramus ellioti (Gabb, 1864), an Early Cretaceous Lucina alcatrazis (Anderson, 1938), and more recently a Pleuromya, and Buchia pacifica (Armstrong and Gallagher, 1977). The mean QFL ratio (35-42-23), is more quartzofeldpathic than that of any age-equivalent petrofacies of the Great Valley sequence (fig. 4a).

Graywacke of the Yolla Bolly terrane (Blake, 1981) in the Diablo Range, as studied by Jacobson (1978), overlaps that of the Alcatraz terrane. The Yolla Bolly terrane includes the Tithonian to Valanginian Eylar Mountain sequence (Jacobson, 1978; Murchey and Jones, this volume) and the Zimba metagraywacke unit (Cowan, 1978). Metamorphic K-Ar ages for the Yolla Bolly terrane range from 76-122 m.y. (Suppe and Armstrong, 1972), and one U-Pb determination on metamorphic sphene yielded an age of 92 m.y. (Mattinson and Echeverria, 1980). These metamorphic ages indicate that the sedimentary rocks are at least older than Cenomanian. In general, graywacke of the Yolla Bolly terrane in the Diablo range is similar in composition to that of the Alcatraz terrane and is more quartzofeldspathic than coeval the Great Valley sequence. The Tithonian to Valanginian Hellhole Graywacke of the Franciscan Complex (Suppe, 1973) also overlaps in composition with coeval petrofacies of the Great Valley sequence; this graywacke was correlated with the Yolla Bolly terrane of the northern Coast Ranges by Blake and others (1982).

The Marin Headlands terrane consists of middle Cretaceous lithic graywacke that unconformably overlies Lower Jurassic to middle Cretaceous pillow lava and chert of oceanic origin (Murchey and others, 1983, this volume; Karl, this volume). Two fossils of upper Albian and Cenomanian age have been found in graywacke of the Marin Headlands terrane on the northern San Francisco Penninsula and the Marin Headlands (Bailey, Irwin and Jones, 1964; p. 117, nos. 10 and 11). Figure 4b compares the Marin Headlands terrane with the coeval Albian Center Peak petrofacies, and Albian to Cenomanian Grabast petrofacies west of the San Joaquin Valley (Mansfield, 1979), and the Aptian and Albian Lodoga petrofacies of the Sacramento Valley (Ingersoll and Dickinson, 1981). The composition of graywacke of the Marin Headlands terrane, QFL ratio (21-28-52) overlaps that of the Center Peak petrofacies (Mansfield, 1979) of the San Joaquin Valley. Figure 4c compares the Marin Headlands terrane with the Cenomanian Boxer petrofacies west of the Sacramento Valley (Dickinson

Figure 4. Comparison of units of the Great Valley sequence (solid lines) and Franciscan Complex (dashed lines) for select time periods. Dots indicate units that have published mean QFL ratios but no standard deviations; n, number of samples counted in each field. Compositional field for each unit was constructed by plotting standard deviation of QFL ratio about the mean (Ingersoll and Dickinson, (1981). References: 1, Ingersoll, (1978); 2, Jacobson, (1978); 3, Mansifeld, (1979); 4, Ingersoll and Dickinson, (1981); 5, Dickinson and others, (1982); 6) This study; 7) Cowan, (1974); 8) Kleist, (1974); 9) Suppe, (1973).

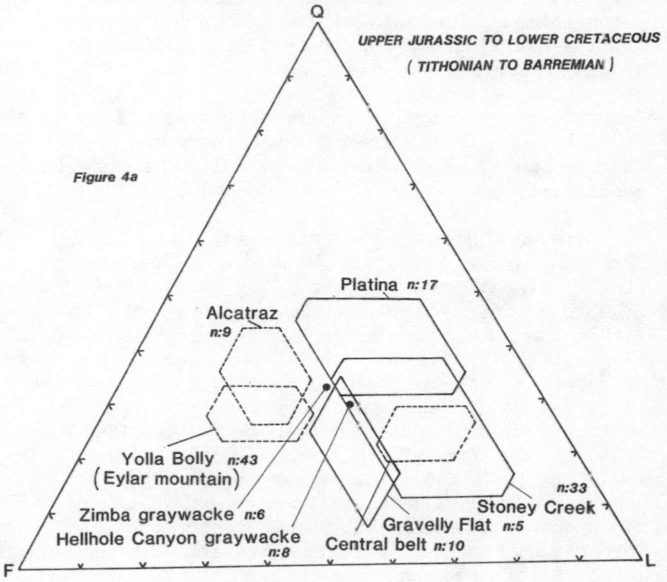

Figure 4a. Franciscan Complex: Central belt, ref. 6; Alcatraz terrane, ref. 6; Yolla Bolly terrane, ref. 2; Mustang Peak metagraywacke unit, mean of six samples, ref. 7; Zimba metagraywacke unit, mean of six samples, ref. 7; Hellhole Graywacke mean of eight samples, ref. 9; Great Valley sequence: Platina petrofacies, ref. 4; Stoney Creek petrofacies, ref. 4: Gravelly Flat, ref. 3.

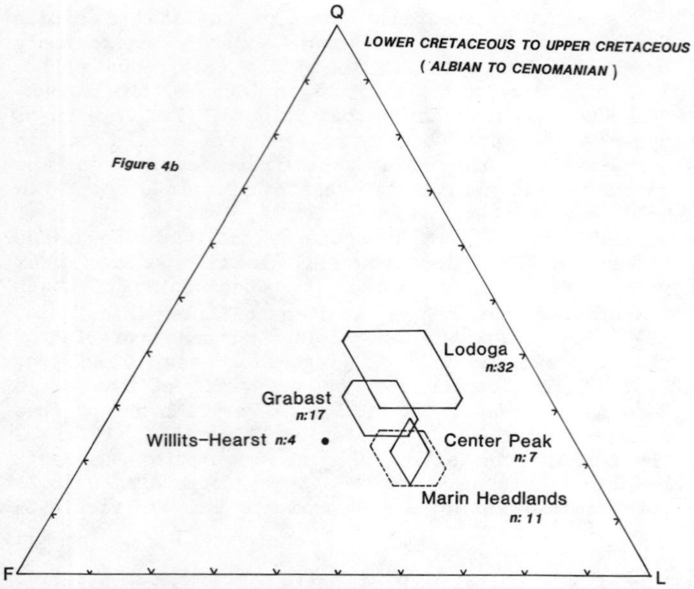

Figure 4b. Franciscan Complex: Marin Headlands terrane, ref. 6; Great Valley sequence: Center Peak petrofacies, ref. 3; Grabast petrofacies, ref. 3; Lodoga, ref. 4. Note: Marin Headlands terrane (Aptian-Cenomanian is compared with the Albian Great Valley sequence in fig. 4b and with the Cenomanian part of the Great Valley sequence in fig. 4c).

and others, 1981; Ingersoll, 1978; Ingersoll and Dickinson, 1981) and the Studhorse petrofacies west of the San Joaquin Valley (Mansfield, 1979). The graywacke of the Marin Headlands terrane overlaps in detrital mode the high-lithic Boxer petrofacies and slightly overlaps in standard deviation the Studhorse petrofacies. The mean of four point counts from the Willits-Hearst area in the northern California Coast Ranges (reported in Dickinson and others, 1982) is shown for comparison in figure 4b. An Aptian-Cenomanian age range is inferred from sandstone in the vicinity (J. O. Berkland, oral. commun., 1982) and just to the north (Gucwa, 1976), but is very poorly constrained.

The Rio Nido terrane is tentatively assigned a Turonian age on the basis of a single fossil locality (loc. 7 of Bailey and others, 1964, p. 117). It has a QFL ratio of 31-39-30, which overlaps both the low-lithic Cortina petrofacies of the Sacramento Valley (Ingersoll, 1978) and the Lower Los Gatos Creek petrofacies of the San Joaquin Valley (Mansfield, 1979), as well as being identical to that of the lithic variant of the Coastal belt described below. Although the Coniacian, Burnt Hills terrane (Blake, 1981) of the Diablo Range overlaps in age with the Upper Los Gatos Creek and Cortina petrofacies, it is more quartzofeldspathic than either the Rio Nido terrane of the San Francisco Bay area or the coeval Great Valley sequence (fig. 4d).

The Campanian Novato Quarry terrane consists of quartzofeldspathic graywacke with a QFL ratio of 41-48-11 (fig. 4e). This composition is consistent from several different slabs and from beds containing three fossil localities (loc. 17 of Bailey and others, 1964, p. 116; Blake and others, this volume). This graywackes is similar to graywacke of the Coniacian Burnt Hills terrane (QFL ratio 39-52-9, fig. 4d, Jacobson 1978), and to the quartzofeldspathic facies

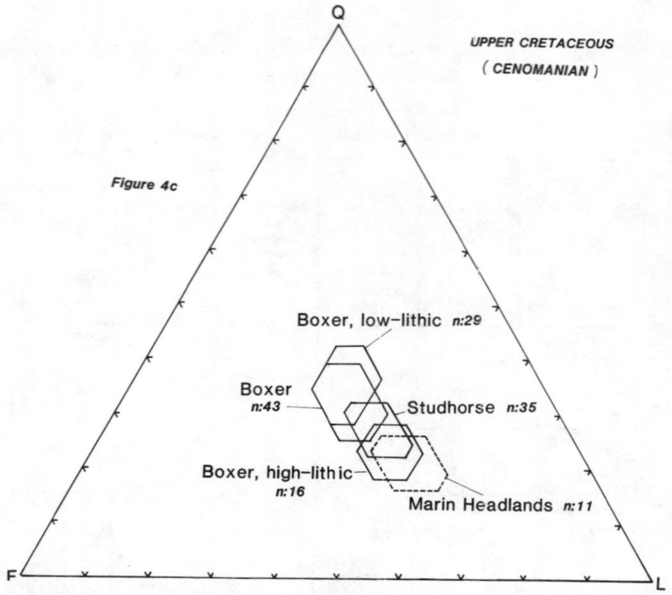

Figure 4c. Franciscan Complex: Marin Headlands terrane, ref. 6; Willits-Hearst locality, ref. 5; Great Valley sequence: Boxer petrofacies, ref. 4; Boxer high- and low-lithic petrofacies, refs. 1 and 4; Studhorse petrofacies, ref. 3.

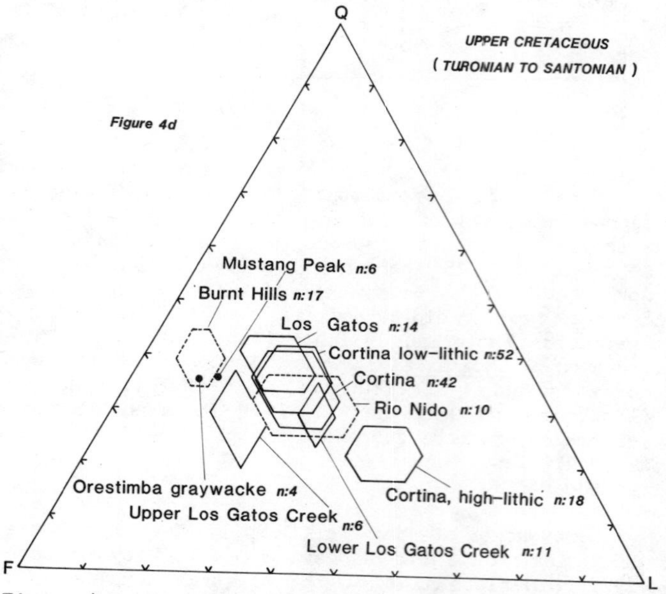

Figure 4d. Franciscan Complex: Burnt Hills terrane, ref. 2; Orestimba metagraywacke unit, ref. 7; Rio Nido terrane, ref. 6; Great Valley sequence: Upper Los Gatos Creek petrofacies, ref. 3; Lower Los Gatos Creek petrofacies, ref. 3; Los Gatos petrofacies, ref. 4; Cortina petrofacies, ref. 4; Cortina, low- and high-lithic petrofacies, refs. 1 and 5.

of the Coastal belt. The Novato Quarry terrane locally contains more than 10 percent potassium feldspar, commonly 3-5 percent. Although Upper Cretaceous graywacke of the Sacramento Valley (Rumsey, Cortina, and Boxer petrofacies) commonly contains 10-30 percent potassium feldspar (Bailey and Irwin, 1959), the Novato Quarry terrane is significantly less lithic (fig. 4e).

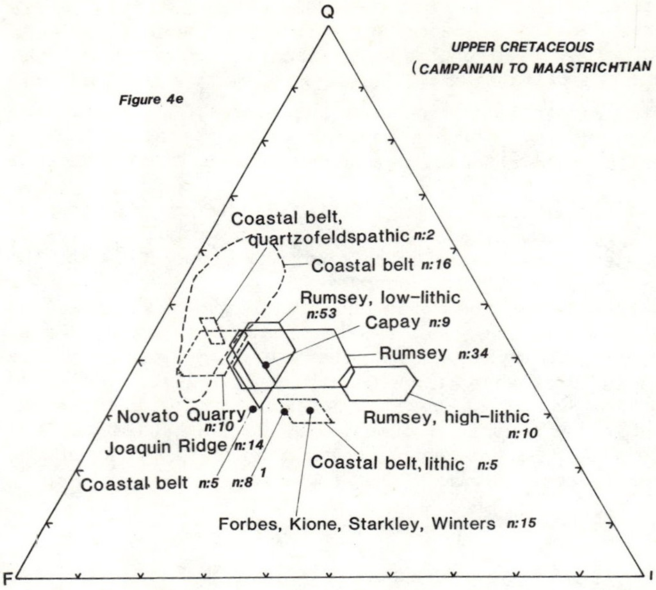

Figure 4e. Franciscan Complex: Novato Quarry terrane, ref. 6; Coastal belt, quartzofeldspathic, ref. 6; Coastal belt, lithic, ref. 6; Coastal belt[1], ref. 5; Coastal belt, ref. 7 (range of sample variation circled rather than standard deviation); Great Valley sequence: Rumsel petrofacies, ref. 4; Rumsey low- and high-lithic petrofacies, ref 1 and 5; Joaquin Ridge petrofacies, ref. 3; Forbes, Kione, Starkley, Winters Formations, ref. 5; Capay Formation, ref. 5.

Two modal variants occur in the Coastal belt of the San Francisco Bay area. The first variant is a quartzofeldspathic facies QFL ratio (45-46-9), similar to the Coastal belt graywacke reported by Kleist (1974) from Mendocino County; the second variant is more lithic (QFL ratio of 30-39-31). Both facies are inferred to be uppermost Cretaceous to early Tertiary by correlation with dated graywacke farther to the north (Blake and Jones, 1974; Evitt and Pierce, 1975), although no fossil control is available for the samples studied. The two petrofacies are not yet defined geographically or stratigraphically and may intertongue similarly to the high- and low-lithic variants of the Rumsey, Cortina, and Boxer petrofacies of the Great Valley sequence in the Sacramento Valley (Dickinson and Rich, 1972; Ingersoll, 1978). Figure 4e compares Coastal belt modal data from this study, Dickinson and others (1982), and Kleist (1974) with those on uppermost Cretaceous and Eocene graywacke of the Great Valley sequence. The lithic variant of the Coastal belt is nearly identical to the Campanian and Maestrichtian sandstone of the Forbes, Kione, Winter, and Starkley Formations (Dickinson and others, 1982), whereas the quartzofeldspathic facies does not appear to have a correlative. Sandstone of the Eocene Capay Formation of the Sacramento Valley is intermediate in composition between the two Coastal belt facies. The Upper Cretaceous Burnt Hills terrane, the Novato Quarry terrane, and the quartzofeldspathic facies of the Coastal belt are all nearly identical in composition, and less lithic than the coeval Great Valley sequence.

The San Bruno Mountain terrane consists of three blocks of graywacke of unknown age, with a QFL ratio of 29-49-22. The graywacke contains abundant detrital potassium feldspar and is locally cut by veins of adularia and epidote. One of the slabs of graywacke included in the San Bruno Mountain terrane (Bolinas Ridge locality) was previously studied by Gluskoter (1964), who correlated it with the Coastal belt because it contains abundant detrital potassium feldspar. The San Bruno Mountain terrane is similar in composition only to the Santonian Upper Los Gatos petrofacies of the San Joaquin Valley (Mansfield, 1979) but because no age control exists, inferences cannot be drawn from this correlation (fig. 4f). The San Bruno Mountain terrane considerably overlaps the composition of the Yolla Bolly terrane in the Diablo Range, and slightly overlaps the Alcatraz and Rio Nido terranes (figures 4a and 4d).

DISCUSSION

The results of our study confirm the conclusion of Jacobson, (1978) that distinct petrologic modes characterize many Franciscan Complex units, and that these modes can be used to help differentiate terranes of the Franciscan Complex. The data also indicate that some Franciscan terranes of the San Francisco Bay area were not derived from the same source as graywacke of equivalent age in the Great Valley sequence.

Upper Jurassic and Lower Cretaceous Strata

Within the Diablo Range and the San Francisco Bay area, Upper Jurassic and Lower Cretaceous graywackes of the Franciscan Complex define two distinct modal fields that do not overlap in composition. Graywacke from the Central belt melange matrix (Tithonian to Valanginian) and the Marin Headlands terrane (Albian to Cenomanian) are both relatively lithic and constrast strongly with those of the more quartzofeldspathic Alcatraz (Valanginian) and Yolla Bolly (Tithonian to Valanginian) terranes. The lithic graywacke of the Central belt melange overlaps entirely in composition the coeval Great Valley sequence petrofacies. This similarity in age and composition is consistent with the idea proposed by Dickinson and others (1982), among others, that both units were derived from the same source. They may even, as Blake and Jones (1974) suggested, represent tectonized basal rocks of the Great Valley sequence.

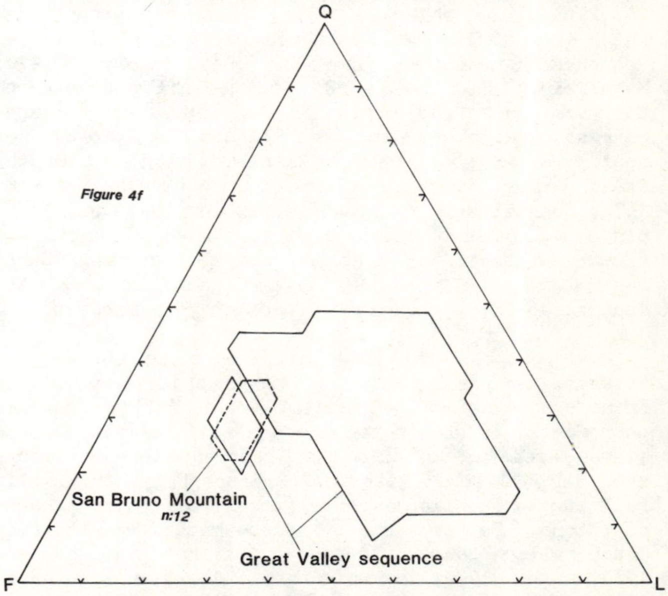

Figure 4f. The Franciscan Complex: San Bruno Mountain terrane of unknown age is compared with aggregate of all Great Valley sequence petrofacies generalized from figures 4a-e.

The Alcatraz and Yolla Bolly terranes of the Diablo Range, however, exhibit detrital modes that are distinct from the coeval Great Valley sequence. The abundant fossil remains in the Alcatraz terrane, as well as the Cenomanian or older radiometric ages for the Yolla Bolly terrane, suggest that the Franciscan Complex terranes are assigned appropriate paleontologic ages.

Jacobson (1978) argued that the Eylar Mountain sequence (Yolla Bolly terrane) was derived from the Sierran-Klamath arc, even though the modal composition differs from that of the coeval Great Valley sequence. He attributed this difference to selective sorting during sediment transport and to metamorphic destruction of rock fragments into matrix. Destruction of metamorphic-rock fragments does not appear to be a cause for the low lithic content of the Alcatraz terrane, because this terrane is unmetamorphosed and contains potassium feldspar (Wright, 1974). The Alcatraz terrane also averages only 7 percent matrix, not enough to significantly shift the modal field. The matrix content of the graywacke of the Eylar Mountain sequence (Yolla Bolly terrane) is greater than that of the Alcatraz terrane and averages 16 percent (Jacobson, 1976). If all the matrix of the graywacke of the Eylar Mountain sequence were inferred to have been formed from rock fragments, the mean QFL mode would fall at the quartzofeldspathic edge of the Gravelly Flat compositional field, where it still would show a composition generally less lithic than any coeval Great Valley sequence petrofacies.

Selective sorting during transport (Jacobson, 1978) does not appear to be the mechanism responsible for the compositional variations observed between these Franciscan Complex and Great Valley sequence graywackes, because both lithic and quartzofeldspathic petrofacies occur within the same time period.

Longitudinal current transport from different parts of the ancestral Sierran-Klamath arc has also been suggested as a means of producing the compositional differences between the coeval Franciscan Complex and Great Valley sequence (Dickinson and others, 1982). Transport directions, which are fairly well studied in the Great Valley sequence (Ojakangas, 1968; Swe and Dickinson, 1970; Mansfield, 1979; figure 1), are generally from north to south and from northeast to southwest. Scarce current indicators in the Franciscan Complex are southward to southwestward similar to those in the Great Valley sequence (Burnt Hills terrane,Telleen, 1977; Coastal belt near Cape Mendocino, Bachman, 1978; Blake and others, this volume). However, until the extent of tectonic rotations can be assessed (Curry and others, this volume; Gromme, this volume), the Franciscan Complex current directions are suspect.

We propose that the difference in composition between the Yolla Bolly and Alcatraz terranes of the Franciscan Complex and coeval Great Valley sequence petrofacies is due to two factors: (1) derivation from source areas other than the inferred Sierran-Klamath arc, and (2) to longitudinal transport by strike-slip faulting or oblique convergence to their present positions. On the basis of paleomagnetic results from other allochthonous terranes (Alvarez and others, 1980; Beck, 1980; Gromme, this volume) and inferred plate-motion directions (Engebretson, 1983), some Franciscan graywacke was likely deposited along the western Pacific margin, probably at more southerly latitudes. The similarity in composition between the matrix graywacke of the Upper Jurassic and Lower Cretaceous Franciscan Central belt and the coeval Great Valley sequence suggests that these rocks were deposited near one another (and by inference to the Sierran-Klamath arc, Ojakangas, 1968; Ingersoll and Dickinson, 1981; Dickinson and others, 1982) and were part of the buttress against which the allochthonous Yolla Bolly and Alcatraz terranes were later accreted.

In general, the Marin Headlands terrane is compositionally similar to the San Joaquin Valley's, Center Peak petrofacies and may overlap with the lithic-rich Boxer petrofacies. This similarity suggests that the Marin Headlands terrane was near the same provenance that provided detritus to the coeval Great Valley sequence. The age of graywacke of the Marin Headlands terrane, which overlies chert and pillow basalt of oceanic origin (Murchey and others, 1983; Karl, this volume), can be used to infer the general time at which oceanic crust approached the continental margin of North America.

Upper Cretaceous Strata

The Upper Cretaceous graywacke of the San Francisco Bay area and the Diablo Range fall into two nonoverlapping QFL compositional fields. The more lithic of the two represented by the Rio Nido terrane and the lithic facies of the Coastal belt, are similar in composition to the Alcatraz and Yolla Bolly terranes of the Diablo Range. They also overlap in composition with the coeval Great Valley sequence petrofacies. The more quartzofeldspathic graywackes, represented by the Burnt Hills and Novato Quarry terranes and the quartzofeldspathic facies of the Coastal belt, are distinct from the coeval Great Valley sequence petrofacies.

One of the key factors in interpreting the tectonic implications of graywacke of Late Cretaceous age is that both lithic and quartzofeldspathic petrofacies are present within the Coastal belt. The occurrence of these two petrofacies within a single unit suggests that they were deposited in an area that was deriving sediment from two different sources. The source for the lithic facies may have been the Sierran-Klamath arc, by inference and correlation with the Great Valley sequence, whereas the quartzofeldspathic facies was likely derived from a more cratonic provenance. The Burnt Hills and Novato Quarry terranes would also represent sediment that was derived from a more cratonic source than the Sierran-Klamath arc. The Rio Nido terrane (and the San Bruno Mountain (?) terrane), because of its compositional similarity to the Great Valley sequence, may represent slope-basin sediment deposited on the Franciscan Central belt and, like the coeval Great Valley sequence, have been derived from the Sierran-Klamath arc.

In conclusion, the modal data appears to be consistent with the inference of multiple provenances for the sediment that formed the graywacke of the Franciscan Complex. Some of the terranes, such as the Central, Marin Headlands, and Rio Nido, may have been derived from local Sierran and Klamath sources as inferred for graywacke of the coeval Great Valley sequence graywackes. Other terranes, such as the Alcatraz, Yolla Bolly, Burnt Hills, and Novato Quarry, may have been tectonically transported from more distant areas and subsequently accreted to the continental margin.

ACKNOWLEDGMENTS

This manuscript was greatly improved by critical reviews by Victor Seiders, Virgil Frizzell, and Carl Wentworth of the U.S. Geological Survey.

REFERENCES

Alvarez, Walter, Kent, D. P., Premoli-Silva, Isabella, Schweikert, R. A., Larson, R. A., 1980, Franciscan complex limestone deposited at 17° south paleolatitude: Geological Society of America Bulletin, v. 91, p. 476-484.

Anderson, F. M., 1938, Lower Cretaceous deposits in California and Oregon: Geological Society of America Special Paper 16, 339 p.

Armstrong, C. F., and Gallagher, Kathy, 1977, Fossils from the Franciscan assemblage, Alcatraz Island: California Geology, v. 30, p. 134.

Bachman, S. B., 1978, A Cretaceous and early Tertiary subduction complex, Mendocino coast, northern California, in Howell, D. G., and McDougall, K. A., eds., Mesozoic Paleogeography of the western United States: Pacific Section , Society of Economic Paleontologists and Mineralogists, v. 2, p. 419-430.

Bailey, E. H., and Irwin, W. P, 1959, K-feldspar content of Jurassic and Cretaceous graywackes of the Northern Coast Ranges and Sacramento Valley, California: American Association of Petroleum Geologists Bulletin, v. 43, p. 2797-2809.

Bailey, E. H., Irwin, W. P., and Jones, D. L., 1964, Franciscan and related rocks and their significance in the geology of western California: California Division of Mines and Geology Bulletin 183, 177 p.

Beck, M. E., Jr., 1980, Paleomagnetic record of plate-margin tectonic processes along the western edge of North America: Journal Geophysical Research, v. 85, p. 7115-7131.

Berkland, J. O., 1969, Geology of the Novato quadrangle, Marin County, California: unpublished M.S. thesis, San Jose State College, scale 1:24,000.

Berkland, J. O., Raymond, L.A., Kramer, J. C., Moores, E., M., and O'Day, M., 1972, What is Franciscan?: American Association of Petroleum Geologists, v. 56, p. 2295-2302.

Blake, M. C., Jr., 1981, Geologic transect of the northern Diablo Range, California, in Frizzel, V. A. Jr., ed., Upper Mesozoic Franciscan rocks and Great Valley sequence, Central Coast Range, California, Pacific Section, Society of Economic Paleontologists and Mineralogists, Annual metting field trip guide, p. 35-43.

Blake, M. C., Jr., Bartow, J. A., and Frizzell, V. A., Jr., Schlocker, Julius, Sorg, Dennis, Wentworth, C. M., and Wright, R. H., 1971, Preliminary geologic map of Marin and San Francisco Counties, and parts of Alameda, Contra Costa, and Sonoma Counties: U.S. Geological Survey Miscellaneous Field Studies Map, MF-574.

Blake, M. C., Jr., and Jones, D. L., 1974, Origin of Franciscan melanges in northern California, in Dott, R. H., Jr. and Shaver, R. H., eds., Modern and ancient geosynclinal sedimentation: Society of Economic Paleontologists and Mineralogists Special Publication No. 19, p. 345-357.

Blake, M. C., Jr., Howell, D. G., and Jones, D. L., 1982, Preliminary Tectonostratigraphic terrane map of California: U. S. Geological Survey Open File Report OF 82-593, scale 1:750,000.

Cowan, D. S., 1974, Deformation and metamorphism of the Franciscan subduction zone complex northwest of Pacheco Pass, California: Geological Society of America Bulletin, v. 85, p. 1623-1634.

Dickinson, W. R., 1970, Interpreting detrital modes of graywacke and arkose: Journal of Sedimentary Petrology, v. 40, p. 695-707.

Dickinson, W. R., and Rich, E. I., 1972, Petrologic intervals and petrofacies in the Great Valley sequence, Sacramento Valley, California: Geological Society of America Bulletin, v. 83, p. 3007-3024.

Dickinson, W. R., Ingersoll, R. V., Cowan, D. S., Helmold, K. P., and Suczek, C. A., 1982, Provenance of Franciscan graywackes in coastal California: Geological Society of America Bulletin, v. 93, p. 95-107.

Dudley, P. P., 1972, Comments on the distribution and age of high-grade blueschists, associated eclogites, and amphibolites from the Tiburon Peninsula, California: Geological Society of America Bulletin, v. 83, p. 3497-3500.

Engebretson, D. C., 1982, Relative motions between oceanic and continental plates in the Pacific basin: Ph.D. thesis, Stanford University, 150 p.

Evitt, W. R., and Pierce, S. T., 1975, Early Tertiary ages from the Coastal Belt of the Franciscan Complex, northern California: Geology, v. 3, p. 433-436.

Gabb, W. M., 1864, Description of the Cretaceous fossils: California Geological Survey, Paleontology, v. 1, p. 57-217.

Gluskoter, H. J., 1964, Orthoclase distribution and authegenesis in the Franciscan Formation of a portion of western Marin County, California: Journal of Sedimentary Petrology, v. 34, p. 335-343.

Gucwa, P. R., 1975, Middle to Late Cretaceous sedimentary melange, Franciscan Complex, northern California: Geology, v. 3, p. 105-108.

Hamilton, Warren, 1969, Mesozoic California and the underflow of Pacific mantle: Geological Society of America Bulletin, v. 80, p. 2409-2430.

Ingersoll, R. V., 1978, Petrofacies and petrologic evolution of the Late Cretaceous fore-arc basin, northern and central California: Journal of Geology, v.86, p. 335-353.

Ingersoll, R. V., and Dickinson, W. R., 1981, Great Valley (Group) Sequence, Sacramento Valley, California, in Frizzell, Virgil, ed., Upper Mesozoic Franciscan and Great Valley sequence, Central Coast Range, California: Pacific Section, Society of Economic Paleontologists and Mineralogists, Annual meeting field trip guide, p. 1-33.

Irwin, W. P., 1960, Geologic reconnaissance of the northern Coast and Klamath Mountains, California, with a summary of the mineral resources: California Division of Mines and Geology Bulletin 179, 80 p.

Jacobson, M. I., 1976, Petrologic variations in Franciscan sandstone from the Diablo Range, California: University of California, Berkeley, M.S. thesis, 173 p.

Jacobson, M. I., 1978, Petrologic variations in Franciscan sandstone from the Diablo Range, California, in Howell, D. G., and McDougall, K. A., eds., Mesozoic paleogeography of the western United States: Society of Economic Paleontologists and Mineralogists, Pacific Section, Pacific Coast Paleogeography Symposium 2, p. 401-417.

Kleist, J. R., 1974, Geology of the Coastal Belt Franciscan Complex, near Ft. Bragg, California: Austin University of Texas, Ph.D. thesis, 133 p.

Mansfield, C. F., 1979, Upper Mesozoic subsea fan deposits in the southern Diablo Range, California: Record of the Sierra Nevada magmatic arc: Geological Society of America Bulletin, v. 90, p. 1025-1046.

Mattinson, J. M., and Echeverria, L. M., 1980, Ortigalita Peak gabbro, Franciscan Complex: U-Pb date of intrusion and high-pressure-low temperature metamorphism: Geology, v. 8, p. 589-593.

McNitt, J. R., 1968, Geology of the Kelseyville quadrangle, Sonoma, Lake, and Mendocino Counties, California: California Division of Mines and Geology Map Sheet 5.

Murchey, B., Jones, D. L. and, Cousin, R. I., 1983, Age and significance of chert in the Franciscan Complex in the San Francisco Bay area: Eos, Transactions of the American Geophysical Union, v. 64, p. 867.

Ojakangas, R. W., 1968, Cretaceous sedimentation, Sacramento Valley, California: Geological Society of America Bulletin, v. 79, p. 973-1008.

Schlocker, Julius, 1974, Geology of the San Francisco North Quadrangle, California: U.S. Geological Survey Professional Paper 782, 109 p.

Silberling, N.J., and Jones, D.L., 1984, Lithotectionic terrane maps of the North American Cordillera: U.S. Geological Survey Open File Report OF 84-523, scale 1:2,500,000.

Soliman, S. M., 1965, Geology of the east half of the Mount Hamilton quadrangle, California: California Division of Mines and Geology Bulletin 185, 32 pp.

Suppe, John, 1973, Geology of the Leech Lake-Ball Mountian region, California: University of California Publication in Geological Sciences, v. 107,
82 p.

Suppe, John, and Armstrong, R. L., 1972, Potassium-argon dating of Franciscan metamorphic rocks, American Journal of Science, v. 272, p. 217-233.

Swe, Win, and Dickinson, W. R., 1970, Sedimentation and thrusting of late Mesozoic rocks in the Coast Range near Clear Lake, California: Geological Society America Bulletin, v. 81, p. 165-188.

Telleen, K. E., 1977, Paleocurrents in part of the Franciscan Complex, California: Geology, v. 5, p. 49-51.

Wright, Robert, 1974, Map showing the distribution of potassium feldspar and fossils in Mesozoic rocks of Marin and San Francisco Counties and parts of Alameda, Contra Costa, and Sonoma Counties, California: U. S. Geological Survey Miscellaneous Field Studies MF-573, scale 1:250,000

Wright, Robert, 1982, Geology of western Marin County: University of California, Santa Cruz, Ph.D. thesis, 204 p.

Sandstone dike in thin-bedded turbidites of the Great Valley sequence, Dry Creek, Tehama county. Photo by E. H. Bailey, 1962.

CLAST COMPOSITIONS OF UPPER MESOZOIC CONGLOMERATES OF THE CALIFORNIA COAST RANGES AND THEIR TECTONIC SIGNIFICANCE

Victor M. Seiders and Charles D. Blome
U.S. Geological Survey
345 Middlefield Road
Menlo Park, CA 94025

ABSTRACT

In three areas of the California Coast Ranges, conglomerates of the Franciscan assemblage, including rocks in metamorphosed, coherent sequences, are similar to conglomerates of the Great Valley sequence of the same region. In the northern Coast Ranges, conglomerates from the Great Valley sequence are of three types, each with a fixed age range: 1) Tithonian-Valanginian conglomerates are rich in chert, 2) some Valanginian conglomerates are rich in felsic volcanic rock pebbles, and 3) Valanginian and younger conglomerates are rich in both chert and felsic volcanic rock clasts. Within the Tithonian-Valanginian interval, chert-rich conglomerate was deposited mainly in the eastern part of the area, volcanic-rich conglomerate was deposited mainly farther west, and mixed-clast conglomerate was deposited still farther west. The same three types of conglomerate in the Franciscan assemblage have a similar geographic distribution but there is no close age control. Conglomerates of the Great Valley sequence and the Franciscan assemblage of the Diablo Range are largely a mixture of chert and felsic volcanic rock clasts with subordinate quartzite. In the central California coastal region, conglomerates from both units are composed mainly of chert and quartzite. Quartzite content in both rock units increases from the northern Coast Ranges, to the Diablo Range, to the central coast area.

Radiolarians identified from chert pebbles in the Franciscan assemblage and the Great Valley sequence in northern California are largely of Triassic and Jurassic age and are similar in age and faunal content to radiolarians from chert in the bedrock of the Klamath Mountains.

These data suggest that within each of the three areas the conglomerates of the Franciscan assemblage and the Great Valley sequence had essentially the same source areas. The southward increase in quartzite clasts probably reflects the fact that the northwest trend of Mesozoic structures truncates the old northeast trend of the Paleozoic continental margin. The northern Mesozoic deposits tapped a source area that was more oceanic in character and poor in quartzite , whereas the southern deposits were drawn from regions deeper within the continent that were richer in quartzite. The distinctiveness of the conglomerates from the three regions suggests that there have been no very large tectonic dislocations of the conglomerates within each region. The western outliers of the Great Valley sequence, however, may have been displaced about 60 km northward along the Hayward fault. A 180 km long sliver between the Golden Gate and Gilroy contains Franciscan assemblage conglomerates that contrast in composition with adjacent regions and may have been moved about 300 km northward on an early trace of the San Andreas fault.

INTRODUCTION

The source of the terrigenous detritus in sedimentary rocks of the Franciscan assemblage has long been a subject of debate (see Dickinson and others (1982) for an excellent historical review). Most workers have either compared or contrasted the sedimentary rocks of the sparsely fossiliferous, tectonized, and metamorphosed Franciscan assemblage with coeval rocks in the stratigraphically ordered and well-dated Great Valley sequence that lies mainly to the east. On the one hand, Dickinson and others (1982), and many workers before them, have pointed to similarities in petrology between sandstones of the Franciscan assemblage and those of the coeval Great Valley sequence, suggesting that the provenance of both was the ancestral Sierran-Klamath region to the east and north. The conceptual model of Dickinson and others (1982) and earlier workers suggests that the sedimentary rocks of the Franciscan assemblage were deposited west of the Great Valley forearc trough, where they were later tectonized as the Pacific plate was subducted beneath North America. The inferred sediment dispersal paths of Dickinson and others (1982) imply that the detritus in the Franciscan was derived from areas just north and south of the source terrain of the Great Valley sequence, thereby explaining minor differences in petrology between sandstones of the two units. Blake and Jones (1974, 1978, 1981), however, have called attention to differences in grain size, petrology, and bulk chemistry of the sandstones and have suggested instead that some Franciscan sandstones had source areas quite far removed from those of the Great Valley sequence. Blake and Jones (1974) proposed a duel origin for Franciscan detrital rocks, suggesting that the matrix of Franciscan melanges is correlative with the mudstone-rich lower part of the Great Valley sequence, whereas certain stratigraphically intact Franciscan sequences are petrographically distinct and must have had a remote source.

Our attention has focused on the conglomerates, a relatively rare but widespread component of the sedimentary rocks of the two units. We have studied the lithologic compositions and radiolarian biostratigraphy of pebbles from conglomerates widely distributed in the California Coast Ranges. Our results show that the clast compositions of the conglomerates are more distinctive than sandstone petrology. They provide not only a means for comparison between rocks of the Franciscan assemblage and the Great Valley sequence, but also show distinctive differences between several regions of California.

The wide sampling of conglomerates necessary to document regional patterns has been made possible by two factors. First, we discovered that etching whole pebbles in hydrofluoric acid greatly facilitated clast identification. With this method, many conglomerates provided acceptable material even though identification of pebbles on the outcrop was impossible. Second, we were directed to conglomerate localities by many collegues with broad knowledge of the California Coast Ranges.

METHODS AND DEFINITIONS

Pebbles mainly in the size range of 0.5 to 6 cm, largely 0.5 to 3 cm, were collected for identification. Coarse-grained granitic rocks are largely excluded through this choice of clast size because such rocks usually do not form small pebbles. Moreover, clast compositions for this size range are likely to differ from most casually estimated compositions because the latter tend to emphasize the abundance of rock

in Blake, M.C. Jr., ed., 1984, **Franciscan Geology of Northern California:** Pacific Section S.E.P.M., Vol. 43, p. 135-148

types in the larger clasts. In a previous study, Seiders (1983) measured each identified pebble, but no size distinction between rock types was detected and measurement was discontinued. Rare samples in the present study show a size differentiation that was clearly evident without measurement. Care was taken to avoid collecting samples with differentially weathered pebbles, although most of the pebbles in this study are of types that are quite resistant to weathering. Metamorphosed conglomerates that do not disaggregate were sawed and the slabs examined.

Pebbles or small slabs were etched overnight in the standard 5 percent solution of hydrofluoric acid used for extracting radiolarians from chert. The solution was usually upgraded slightly by the addition of a "dash" of concentrated acid. The pebble surface is thus freed of any weathering film and imparted with a micro-relief that aids rock identification. An opaque white crust, probably calcium fluoride, forms on some rocks, especially limestone and some volcanic pebbles. On most rocks, the crust is at least partly removed by ultrasonic cleaning. It is best to separate limestone clasts before etching. After ultrasonic cleaning and drying the pebbles were examined under a low power microscope.

The distinction between chert and dense felsic volcanic rocks is often made possible by the presence of radiolarians in the former or phenocrysts in the latter. With experience, the differentiation can be made on the basis of subtle differences in groundmass texture. Some pebbles were freshly broken to aid identification, but the etched surface is usually best. Most difficult is the distinction between volcanic rock types, a problem not unique to pebble identifications.

In Tables 1-6, "mudstone" as used herein is largely a dark-gray argillite that is siliceous and gradational into chert; in some samples it includes tan silty mudstone that may be of intraformational origin. "Sandstone" refers to rocks containing detrital grains other than quartz, chiefly feldspar but also chert and rock fragments. "Quartzite" refers to rocks composed chiefly of quartz grains. "Sandstone" and "quartzite" each include rocks with the grain size of silt to small pebbles. "Quartz" refers to vein quartz. Some pebbles composed mainly of quartz have a peculiar granular texture that differs from typical vein quartz and were counted as "other", a catagory that includes both unidentified and miscellaneous rocks.

Frequent mention is made of Late Jurassic and Cretaceous stage names and they are listed in Figure 4.

CLAST COMPOSITIONS

Conglomerates from the Franciscan assemblage and the Great Valley sequence show contrasting clast compositions between the four regions: 1) northern Coast Ranges, 2) Diablo Range, 3) central California coast, and 4) Golden Gate-Gilroy block (Fig. 1). The locations of the sampled conglomerates are shown in Figures 1 and 2 and grid coordinates of the localities are given in Appendix 1.

Northern Coast Ranges

From both the Great Valley sequence and the Franciscan assemblage our data include conglomerates rich in chert, volcanic rock clasts, and both chert and volcanic rock clasts.

Chert-rich conglomerates

Conglomerates of the Great Valley sequence that are rich in chert pebbles crop out along the west side of the Sacramento Valley (Fig. 2, localities 3-6) where they are of Tithonian and Berriasian age (Seiders, 1983; Bertucci, 1983). Chert-rich conglomerate of the Great Valley sequence also occurs in an outlier of Valanginian age resting on the basement of the Klamath Mountains (Fig. 2, locality 1; Jones and Irwin, 1971). Twenty-five km to the west, at Black Lassic (Fig. 2, locality 2), a chert-rich conglomerate of the Great Valley sequence occurs in a klippe lying on the Franciscan assemblage (Irwin and Dennis, 1979; Kaplan, 1983).

Chert-rich conglomerate was found in the Franciscan assemblage only in the area east and southeast of Covelo (Fig. 2, localities 7-14). Some of these localities occur as blocks in melange. Two localities (7 and 9), however, are within the Hellhole Graywacke of Suppe (1973), which is a metamorphosed, relatively coherent sequence containing lawsonite and aragonite. One of the localities (9) has yielded a Valanginian fossil (Ghent, 1963), and the unit ranges in age from Tithonian to Valanginian (Suppe, 1973).

The chert-rich conglomerate of the Great Valley sequence and the Franciscan assemblage are nearly identical in composition (Tables 1 and 6), with mean chert contents of 88.5 and 88.2 percent, respectively, and only minor amounts of other constituents, chiefly mudstone (5.7 and 6.7 percent) and felsic volcanic rocks (3.1 percent each). The similarity in clast composition is also shown in a triangular diagram plotting chert, volcanic rocks, and quartzite (Fig. 3). The chert-rich conglomerates of the Great Valley sequence and the Franciscan assemblage plot close together within a distinct field near the chert corner.

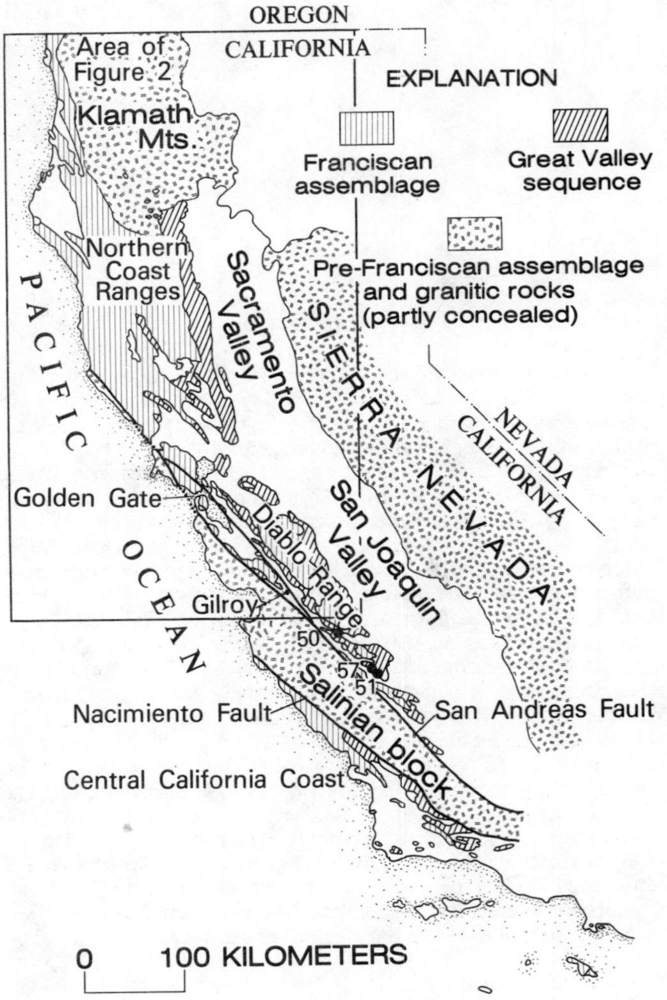

Figure 1. Geologic sketch map of northern and central California showing the area of Figure 2 and the location of conglomerates sampled outside the area of Figure 2.

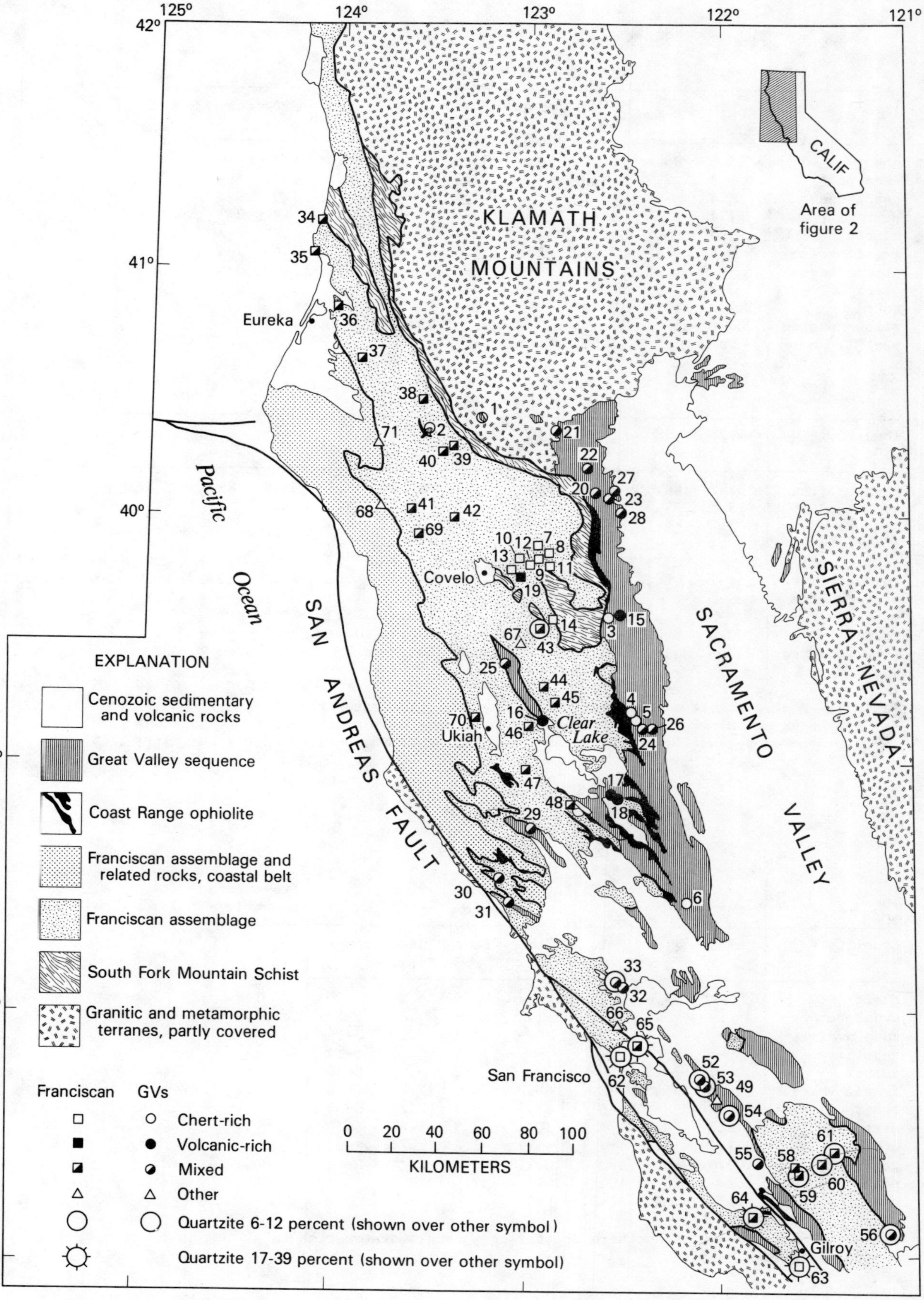

Figure 2. Geologic sketch map of northwestern California showing the location of conglomerate samples.

	Great Valley sequence								
locality #	1	2	3	4	5	6	mean	std. dev.	range
chert	85.5	85.6	84.1	89.6	90.5	95.6	88.5	4.3	84-96
mudstone	5.3	7.9	9.3	6.8	2.2	2.8	5.7	2.8	2-9
sandstone	.4	2.3	1.8	-	.4	-	.8	1.0	0-2
quartzite	1.4	1.4	-	-	-	-	.5	.7	0-1
quartz	1.8	.5	.3	-	.4	-	.5	.7	0-2
felsic volcs.	2.5	1.4	3.9	3.6	5.6	1.6	3.1	1.6	1-6
microgranite	-	.5	.3	-	-	-	.1	.2	0-5
other	3.2	.5	.3	-	.9	-	.8	1.2	0-3
total	100.1	100.1	100.0	100.0	100.0	100.0	100.0		
pebs. counted	283	215	338	222	232	253	257		
age	Val.	-	Tith.	Tith.	Berr.	-			

	Franciscan assemblage										
locality #	7	8	9	10	11	12	13	14	mean	std. dev.	range
chert	82.7	84.6	88.5	88.8	91.9	87.4	87.3	94.5	88.2	3.7	83-92
mudstone	9.1	7.4	4.9	6.2	5.6	7.9	8.2	4.1	6.7	1.7	5-9
sandstone	3.7	-	(-)	-	.7	-	.8	-	.6	1.3	0-4
quartzite	.4	-	-	-	1.1	-	.8	.5	.3	.4	0-1
quartz	-	1.1	.5	3.7	-	.3	.3	-	.7	1.2	0-4
felsic volcs.	4.1	6.2	5.6	1.3	.7	4.5	2.1	.5	3.1	2.2	1-6
microgranite	-	-	.3	-	-	-	.3	-	.1	.1	0-.3
other	-	.6	.3	-	-	-	.3	.5	.2	.2	0-.6
total	100.0	99.9	100.1	100.0	100.0	100.1	100.1	100.1	99.9		
pebs. counted	243	351	390	160	270	381	373	220	298		
age	-	-	Val.	-	-	-	-	-			

Table 1. Pebble counts of northern Coast Ranges conglomerates rich in chert pebbles. Sample 9, which was counted in sawed slabs, contains a small amount of sandstone that was ignored because of the difficulty in distinguishing between clasts and matrix.

	Great Valley sequence							Franciscan assemblage
Locality #	15	16	17	18	mean	std. dev.	range	19
chert	-	1.6	-	-	.4	.8	0-2	1.4
mudstone	1.9	.6	-	1.2	.9	.8	0-2	.5
conglomerate	-	.3	-	-	.1	.1	0-.3	-
quartzite	-	.3	-	-	.1	.1	0-.3	-
quartz	.4	.3	-	-	.2	.2	0-.4	-
felsic volcs.	85.1	85.5	86	60.2	79.2	12.7	60-86	90.1
other volcs.	5.8	-	6	27.7	9.9	12.2	0-30	.5
microgranite	5.7	8.8	7	10.1	7.9	1.9	6-10	7.1
other	1.1	2.5	1	.9	1.4	.7	1-2	.5
total	100.0	99.9	100	100.1	100.1			100.1
pebs. counted	262	318	100	347	257			212
age	Val.	Val.	Val.	Val.				-

Table 2. Pebble counts of northern Coast Ranges conglomerates rich in volcanic rock pebbles.

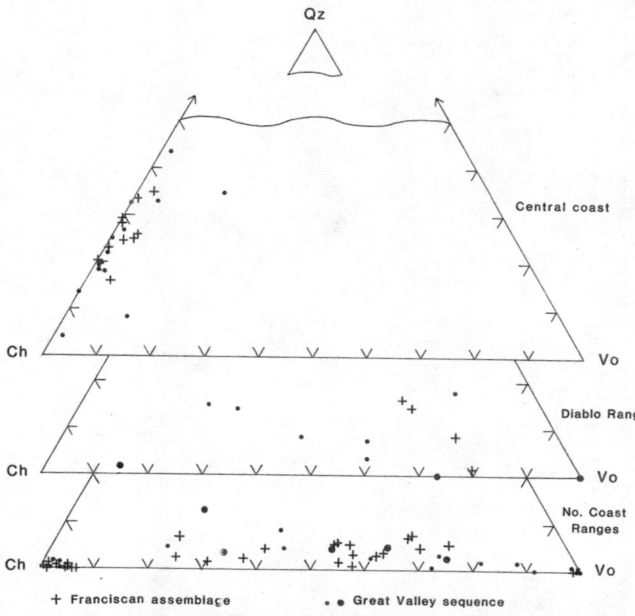

Figure 3. Triangular diagrams plotting the contents of chert (Ch), volcanic rocks (Vo), and quartzite (Qz) in conglomerates from three regions of California. In the northern Coast Ranges plot, large dots are conglomerates from the western outliers and small dots are other conglomerates of the Great Valley sequence. In the Diablo Range plot, large dots are conglomerates in mudstone in the lower part of the Great Valley sequence and small dots are other conglomerates of the Great Valley sequence.

Conglomerates rich in felsic volcanic rock clasts

Great Valley sequence conglomerate rich in felsic volcanic rock clasts was first described quantitatively in northern California by Bertucci (1983) from Valanginian rocks on the west side of the Sacramento Valley near Elk Creek (Fig. 2, locality 15). Similar conglomerate in the Great Valley sequence occurs with Valanginian fossils in the Middle Mountain area northwest of Clear Lake (Fig. 2, locality 16; J. O. Berkland, oral commun., 1983) and southeast of Clear Lake (Fig. 2, localities 17 and 18; Swe and Dickinson, 1970).

We found only one example of this type of conglomerate in the Franciscan assemblage. This is locality 19, east of Covelo (Fig. 2), where it occurs in the Poison Rock melange of Lehman (1974) at the southwest edge of the area containing chert-rich conglomerates. According to Jordan (1978), conglomerate rich in felsic volcanic rock clasts in this area is largely restricted to the Poison Rock melange of Lehman (1974). The conglomerate at locality 19 contains a few clasts of limestone with scraps of Buchia of possible Tithonian age (Jordan, 1978), which implies that the conglomerate is Tithonian or younger.

The volcanic-rich conglomerate of the Franciscan assemblage is similar to the mean of four Great Valley sequence conglomerates (Tables 2 and 6). All are composed mainly of volcanic rock clasts (90.6 and 89.1 percent, respectively) with subordinate microgranitic rocks (7.1 and 7.9 percent) and very little else. These conglomerates plot very near the volcanic rock corner of the triangular diagram for chert, volcanic rocks, and quartzite (Fig. 3).

Conglomerates rich in both chert and volcanic rock clasts

Conglomerates with a mixture of chert and volcanic rock clasts were sampled in the Great Valley sequence along the west side of the Sacramento Valley from beds ranging in age from late Valanginian to Turonian (Fig. 2, localities 20-24 and 26-28). A mixed clast conglomerate occurs in the Middle Mountain area in rocks of probable Albian age (Fig. 2, locality 25; J. O. Berkland, oral commun., 1983).

The western outliers (Blake and Jones, 1974) of the Great Valley sequence occur in apparent klippen between Ukiah and San Francisco (Fig. 2, localities 29-32) and contain mixed clast conglomerates, in part with Valanginian fossils (Berkland, 1969; Blake and Jones, 1978 and 1981). Blake and Jones (1978 and 1981) noted that conglomerates of the western outliers are rich in felsic volcanic rock clasts and the associated sandstones are rich in potassium feldspar. They stated that coeval beds of the Great Valley sequence farther east contained chert-pebble conglomerates and sandstones poor in potassium feldspar. For these reasons they proposed that the western outliers had been transported northward an unknown distance by strike-slip faulting. As noted earlier, however, the chert-rich conglomerates on the west side of the Sacramento Valley are of Tithonian and Berriasian age, not Valanginian. Valanginian conglomerates in the Great Valley sequence east of the western outliers are, in fact, rich in felsic volcanic clasts (Fig. 2, localities 15-18) and a small upper Valanginian conglomerate near Colyear Springs (Fig. 2, locality 20; Jones and Bailey, 1973; Bailey and Jones, 1973) has a mixed clast composition. There is a close similarity in clast composition between the conglomerates of the western outliers and other northern California conglomerates (see below). Thus, although we present evidence for some northward translation below, we believe that the western outliers are largely autochthonous and the conglomerates should be included with those of the northern Coast Ranges.

In the Franciscan assemblage, 17 conglomerates of mixed clast composition were collected over a wide area extending from north of Eureka to south of Clear Lake (Fig. 2, localities 34-48, 69-70). Most of these localities are either in displaced blocks in melange or are from uncertain structural settings. Locality 40, however, is from a Franciscan map unit reported to have relatively coherent strata (Irwin and others, 1974) and localities 43 and 70 are from units with generally well-bedded sandstone but including displaced blocks (Lehman, 1974; Orchard, 1979). The map unit containing locality 43 has yielded a Cenomanian(?) fossil (Blake and Jones, 1974).

Pebble counts of the nine mixed clast conglomerates from the west side of the Sacramento Valley (including one from the Middle Mountain area) and the five conglomerates from the western outliers are given in Table 3. The Sacramento Valley conglomerates, compared with the western outliers, are somewhat poorer in chert (mean 24.8 percent compared to 36.7 percent) and richer in felsic volcanic rocks (mean 48.2 percent compared to 34.6 percent). Overall, however, the clast compositions of the two groups are rather similar. A very close similarity should not be expected, since the western outliers may all be Valanginian and the Sacramento Valley conglomerates range from Valanginian to Turonian in age.

Clast compositions of the 17 mixed clast conglomerates of the Franciscan assemblage are given in Table 3; the mean composition is compared with means for the western outliers and Sacramento Valley conglomerates in Table 6 (columns E, F, and G). There is a close similarity in practically all constituents between the mean of the Franciscan conglomerates and that of the conglomerates of the western outliers. Chance in sampling may have contributed something to this similarity, since a considerable range in composition does occur, as can be seen graphically in Figure 3. Nevertheless, all three groups occupy essentially the same field in Figure 3, except that some Great Valley sequence conglomerates are richer in volcanic rock clasts than any Franciscan samples.

	Great Valley sequence																			
	west side of Sacramento Valley and Middle Mountain area												western outliers							
Locality #	20	21	22	23	24	25	26	27	28	mean	std. dev.	range	29	30	31	32	33	mean	std. dev.	range
chert	57.3	28.3	11.8	36.3	31.3	9.7	19.6	6.6	22.5	24.8	15.8	7-57	36	27.5	19.6	52.6	47.9	36.7	13.7	20-53
mudstone	1.8	6.9	20.2	6.5	10.7	2.8	8.1	7.1	7.4	7.9	5.3	2-20	6	10.2	5.5	4.6	6.1	6.5	2.1	5-10
sandstone	2.8	6.9	6.3	14.6	6.7	1.2	7.8	3.1	5.2	6.1	3.9	1-15	-	4.2	1.8	5.4	6.3	3.5	2.6	0-6
quartzite	3.7	1.7	1.3	3.1	5.0	1.6	2.4	-	.3	2.1	1.6	0-5	4	3.9	1.8	3.0	9.6	4.5	3.0	2-10
quartz	4.6	1.2	.8	4.3	3.5	1.2	2.4	-	1.8	2.2	1.6	0-5	-	1.0	3.6	3.5	8.0	3.2	3.1	0-8
felsic volcs.	16.5	45.1	54.6	29.2	22.9	78.2	58.1	69.9	59.4	48.2	21.4	16-78	32	43.6	54.5	25.3	17.6	34.6	14.7	18-55
other volcs.	-	.6	-	.3	.5	-	-	.9	-	.3	.3	0-1	10	7.1	6.2	1.1	.6	5.0	4.0	1-10
micrograniite	3.7	1.7	3.8	2.2	1.7	4.0	1.7	10.2	.9	3.3	2.8	1-10	5	1.4	4.7	2.5	1.7	3.1	1.7	1-5
limestone	-	-	-	-	17.4	-	-	-	-	1.9	5.8	0-17	-	-	-	-	-	-	-	-
other	9.6	7.5	1.3	3.4	.2	1.2	-	2.2	2.5	3.1	3.3	0-10	7	1.2	2.2	1.9	2.2	2.9	2.3	1-7
total	100.0	99.9	100.1	99.9	99.9	99.9	100.1	100.0	100.0	99.9			100	100.1	99.9	99.9	100.0	100.0		
pebs. counted	219	173	238	322	402	321	296	226	325	280			100	590	275	367	363	339		
age	Late Val.	Haut.	Haut.- Barr.	Haut.- Barr.	Apt.- Alb.	Alb.	Cen.	Cen.	Tur.				Val.	Val.	Val.	Val.	Val.			

	Franciscan assemblage																			
locality #	34	35	36	37	38	39	40	41	42	43	44	45	46	47	48	69	70	mean	std. dev.	range
chert	21.6	29.0	44.0	32.9	52.6	41.1	52.6	33.2	36.5	25.2	32.1	29.8	36	33.2	61.1	23.3	18.3	35.4	11.7	18-61
mudstone	11.3	8.5	7.7	10.0	3.9	14.8	6.6	6.7	7.6	4.3	9.9	4.6	4	6.3	6.1	5.4	5.8	7.3	2.9	4-11
sandstone	1.3	1.6	3.7	5.9	5.6	6.9	6.6	5.2	2.0	-	4.2	2.8	3	7.2	3.3	2.4	3.0	3.8	2.1	0-7
quartzite	5.6	2.4	3.4	4.6	1.0	1.4	1.9	.7	1.3	6.2	2.7	3.2	5	4.5	5.8	4.1	4.3	3.4	1.8	1-6
quartz	3.0	.4	2.2	2.7	8.9	5.7	9.1	6.0	4.8	1.9	3.1	2.8	1	.9	1.9	1.1	1.3	3.4	2.7	0-9
felsic volcs.	48.9	42.7	29.8	38.5	23.5	22.9	16.8	44.0	44.6	57.6	44.7	51.4	39	41.7	19.4	52.6	53.3	39.5	12.6	17-57
other volcs.	1.7	6.0	1.2	2.2	-	1.4	-	1.1	.3	-	-	.9	3	3.6	-	6.0	7.8	2.1	2.4	0-8
micrograniite	5.6	6.5	3.1	2.2	2.6	2.2	.3	2.2	.8	1.9	.4	1.4	9	.9	1.9	1.1	2.0	2.6	2.3	0-9
limestone	-	-	-	-	-	1.6	5.8	-	-	-	-	-	-	-	-	-	-	.4	1.4	0-6
other	.9	2.8	4.9	1.0	2.0	2.0	.3	.7	2.3	2.9	3.1	3.2	1	1.8	.3	4.1	4.1	2.2	1.4	0-5
total	99.9	99.9	100.0	100.0	100.1	100.0	100.0	99.8	100.2	100.0	100.2	100.1	101	100.1	99.8	100.1	99.9	100.1		
pebs. counted	231	248	325	410	306	494	363	268	395	210	262	218	107	223	360	369	394	305		

Table 3. Pebble counts of northern Coast Ranges conglomerates with both chert and volcanic rock clasts.

All types of northern California conglomerates are characteristically low in quartzite clasts (Fig. 3). The mean quartzite content of all mixed clast conglomerates is 3.1 percent. Only one Franciscan assemblage conglomerate exceeds 6 percent quartzite (6.2 percent; Table 3, locality 43). Likewise, only one Great Valley sequence conglomerate (locality 33 of the western outliers; Table 3, Fig. 2) has more than 6 percent quartzite. This quartzite content of 9.6 percent is the highest in the northern Coast Ranges and it is one of the two most southern localities within that area.

Distribution of conglomerate types

The geographic distribution and age of the three types of conglomerate in the Great Valley sequence of the northern Coast Ranges has been described above and but needs to be reviewed here. The distribution is shown diagramatically in Figure 4. Chert-rich conglomerate occurs in Tithonian and Berriasian strata along the west side of the Sacramento Valley and in Valanginian rocks in small outliers on the basement of the Klamath Mountains. Valanginian conglomerate rich in volcanic rock clasts occurs locally on the west side of the Sacramento Valley and more extensively in the Clear Lake-Middle Mountain area to the west. Conglomerate rich in both chert and volcanic rock clasts occurs in Valanginian beds of the western outliers and in a small lens in Valanginian rocks on the west side of the northern part of the Sacramento Valley. Conglomerate with both chert and volcanic rock clasts also occurs in younger Cretaceous beds on the west side of the Sacramento Valley and in the Middle Mountain area. In the Valanginian-Barremian interval, however, mixed clast conglomerates on the west side of the Sacramento Valley are largely restricted to the northern end of the valley. Farther south rocks of this age are largely mudstone and thin-bedded turbidites in the upper part of the Stony Creek Formation of Ingersoll and others (1977). We interpret these observations to indicate that during the Tithonian-Valanginian interval, the site of conglomerate deposition shifted westward with time and there was a concomitant change in clast composition from chert-rich, to volcanic-rich, to mixed chert and volcanic clasts.

Western outliers	Clear Lake-Middle Mtn.	Klamath Mts. outliers	West side Sacramento Valley	Stage	Series
			◑◑◑◑	Turonian	Upper Cretaceous
			◑◑◑◑	Cenomanian	
	◑◑◑◑		◑◑◑◑	Albian	Lower Cretaceous
			◑◑◑◑	Aptian	
			◑◑	Barremian	
			◑◑	Hauterivian	
◑◑◑◑	●●●●	○	◑●●	Valanginian	
			○○○○	Berriasian	
			○○○○	Tithonian	Upper Jurassic

Figure 4. Diagram showing the distribution in time and space of the three types of conglomerate in the Great Valley sequence of the northern California Coast Ranges. Open circles: chert-rich conglomerates. Solid circles: felsic volcanic-rich conglomerates. Half-filled circles: conglomerates rich in both chert and felsic volcanic clasts. Conglomerates younger than those shown occur on the west side of the Sacramento Valley and in the western outliers but were not studied.

Conglomerates of the Franciscan assemblage show a similar geographic distribution of clast compositions but lack close age control. Chert-rich conglomerates are the easternmost and occur in the area east and southeast of Covelo. The only felsic volcanic-rich conglomerate that we have found in the Franciscan occurs at the west edge of the area containing chert-rich conglomerate. All mixed clast conglomerates of the Franciscan lie to the west-southwest.

Other conglomerates

We found four northern Coast Ranges conglomerates with clast compositions that differ in some respects from any of the types described above. One Franciscan assemblage conglomerate (locality 66, Table 5) is quite low in chert (6.5 percent), contains 8.0 percent limestone, and 23.0 percent miscellaneous and unidentified clasts. Many of the latter may be felsic volcanics but textures are blurred and some clasts seem to be metamorphosed. One pebble of black chert yielded Valanginian radiolarians (Table 7). The age of this pebble contrasts sharply with the ages of pebbles in other conglomerates (mainly Triassic and Jurassic). Another Franciscan conglomerate (locality 67, Table 5) has a clast composition like the mixed clast type except that 32.7 percent of the clasts are blueschists. The conglomerate has a metamorphic texture and we presume that the pebbles of blueschist formed in place from other rock types rather than being incorporated as preexisting blueschist clasts. Locality 68 (Table 5) is a conglomerate in the lower Tertiary Yager Formation (Evitt and Pierce, 1975), yet the clast composition is similar to typical Mesozoic conglomerates of the northern Coast Ranges. Some of the chert pebbles are reddish brown, which is rare in Mesozoic conglomerates. Another Yager conglomerate (Table 5, locality 71) is unusually rich in limestone and quartz clasts.

Conglomerate containing recycled Franciscan detritus (Bailey and others, 1964; Cowan and Page, 1975) was not collected for this study because no counterpart occurs in the Great Valley sequence for comparison.

Diablo Range

In the Diablo Range the Great Valley sequence typically consists of a lower part that is rich in mudstone, rarely with lenses of conglomerate, and an upper part in which extensive units of sandstone and conglomerate are interbedded with mudstone.

We sampled pebbles from the lower mudstone-rich unit of the Great Valley sequence in three places (Figs. 1 and 2, localities 49-51). The pebble counts (Table 4) show a wide range of values of chert and felsic volcanic rocks and each is rich in clasts of intraformational origin. Even when the latter clasts are excluded, it is difficult to generalize the clast composition of conglomerates from this part of the section. There is little or no quartzite, however.

In many parts of the Diablo Range the oldest conglomerate in the Great Valley sequence above the basal mudstone unit is in the Panoche Formation of Late Cretaceous age (Dibblee, 1973, 1981). East and southeast of San Francisco Bay (Fig. 2, localities 52-55), however, the Oakland Conglomerate of Crittenden (1951) and Robinson (1956) has yielded Early Cretaceous Valanginian Buchias in a few places (Crittenden, 1951; Hall, 1958). (These rocks are not correlative with the type Oakland Conglomerate, which is Upper Cretaceous (Case, 1968)). Four pebble counts from the Oakland Conglomerate of Crittenden (1951) and Robinson (1956), along with counts from two other conglomerates of Cenomanian(?) (Schilling, 1962) and Hauterivian-Albian age (Rose and Colburn, 1963; Mansfield, 1979), are given in Table 4 (localities 52-57). The mean of these six counts (Table 6, column I) is similar to clast compositions of mixed clast conglomerates of the northern Coast Ranges (columns E, F,

	lower part of Great Valley sequence											
	lenses in mudstone			extensive conglomerates								
				Oakland Conglomerate of Crittenden (1951) and Robinson (1956)							Panoche Fm.	Center Peak Cgl.
Locality #	49	50	51	52	53	54	55	mean	std. dev.	range	56	57
chert	–	(15)	74.9	43.8	49	35.4	33.2	40.3	7.4	33–49	31.4	9.9
mudstone	9.3	22	7.4	7.1	9	17.3	5.7	9.8	5.2	6–17	4.7	5.7
sandstone	–	26	1.4	4.9	5	5.9	2.9	4.7	1.3	3–6	2.8	17.7
quartzite	–	–	1.8	11.1	12	6.3	3.3	8.2	4.1	3–12	6.7	12.0
quartz	–	1	1.4	4.0	1	.4	.4	1.5	1.7	0–4	.6	1.9
felsic volcs.	74.4	36	12.4	22.7	17	28.7	47.1	28.9	13.1	17–47	46.7	45.0
other volcs.	2.2	–	–	.2	2	4.3	4.1	2.6	1.9	0–4	3.3	.7
microgranite	2.2	–	.2	2.0	–	–	–	.5	1.0	0–2	1.4	4.0
limestone	10.6	–	(–)	–	–	–	–	–			–	–
other	1.3	1	.5	4.2	5	1.6	3.3	3.5	1.5	2–5	2.5	3.1
total	100.0	100.0	100.0	100.0	100.0	99.9	100.0	100.0			100.1	100.0
pebs. counted	227	113	443	450	131	254	244	270			360	424
age	Tith.	Tith.–Val.	Val.–Haut.	–	–	Val.	Val.				Cen.	Haut.–Alb.

	Franciscan assemblage						
Locality #	58	59	60	61	mean	std. dev.	range
chert	14.1	17.3	17.1	20.3	17.2	2.5	14–20
mudstone	16.7	4.6	3.6	5.8	7.7	6.1	4–17
sandstone	5.2	22.9	3.3	1.5	8.2	9.9	1–23
quartzite	1.1	11.1	7.2	11.9	7.8	4.9	1–12
quartz	–	1.9	.7	.6	.8	.8	0–2
felsic volcs.	55.9	40.6	63.8	43.6	51.0	10.8	41–64
other volcs.	1.1	–	.3	7.8	2.3	3.7	0–8
microgranite	1.9	–	1.3	.3	.9	.9	0–2
other	4.1	1.5	2.6	8.1	4.1	2.9	1–8
total	100.1	99.9	99.9	99.9	100.0		
pebs. counted	270	323	304	344	310		

Table 4. Pebble counts of conglomerates from the Diablo Range. In sample 50 the identification of many chert clasts is uncertain. Locality 51 contains abundant intraformational limestone clasts; the data are recalculated limestone-free.

	Franciscan assemblage Golden Gate–Gilroy block				other conglomerates			
locality #	62	63	64	65	66	67	68	71
chert	72.6	51.4	37.6	26.3	6.5	16.6	48.7	20.8
mudstone	4.4	3.3	1.2	9.0	7.3	6.2	2.2	14.9
sandstone	1.9	1.2	.8	5.9	7.3	4.3	2.0	4.0
quartzite	17.2	38.7	22.7	21.1	3.8	3.3	2.4	2.6
quartz	–	2.8	1.2	.3	.4	.9	6.1	13.5
felsic volcs.	3.1	1.9	30.6	28.7	38.7	31.8	33.9	6.9
other volcs.	–	–	5.1	1.7	–	.5	–	–
microgranite	–	–	.8	1.0	5.0	.5	4.5	–
limestone	–	–	–	–	8.0	.5	–	32.7
blueschist	–	–	–	5.2	–	32.7	–	–
other	.7	.7	–	.7	23.0	2.8	.2	4.6
total	99.9	100.0	100.0	99.9	100.0	100.1	100.0	100.0
pebs. counted	413	426	255	289	261	211	507	303

Table 5. Pebble counts of conglomerates of the Franciscan assemblage from the Golden Gate–Gilroy block (62–65), other conglomerates of the Franciscan assemblage (66–67), and the Yager Formation (68, 71).

cgl. type	chert-rich		volc.-rich		chert and volc. clasts					chert and quartzite		
rock unit	F	GVs	F	GVs	F	GVs	GVs	F	GVs	F	GVs	GVs
	A	B	C	D	E	F	G	H	I	J	K	L
chert	88.8	88.5	1.4	.4	35.4	36.7	24.8	17.2	33.8	60.4	66.1	45.6
mudstone	6.7	5.7	.5	.9	7.3	6.5	7.9	7.7	8.3	5.5	7.8	3.6
sandstone	.6	.8	-	.1	3.8	3.5	6.1	8.2	6.5	.3	3.5	1.2
quartzite	.3	.5	-	.1	3.4	4.5	2.1	7.8	8.6	21.1	20.0	36.1
quartz	.7	.5	-	.2	3.4	3.2	2.2	.8	1.4	9.4	1.1	9.9
felsic volcs.	3.1	3.1	90.1	79.2	39.5	34.6	48.2	51.0	34.5	1.8	1.0	1.6
other volcs.	-	-	.5	9.9	2.1	5.0	.3	2.3	2.4	-	-	-
microgranite	.1	.1	7.1	7.9	2.6	3.1	3.3	.9	1.2	-	.1	-
limestone	-	-	-	-	.4	-	1.9	-	-	-	-	-
other	.2	.8	.5	1.4	2.2	2.9	3.1	4.1	3.3	1.4	.4	2.0
total	99.9	100.0	100.1	100.1	100.1	100.0	99.9	100.0	100.0	99.9	100.0	100.0
# of localities	8	6	1	4	17	5	9	4	6	9	10	1

Table 6. Mean clast compositions of conglomerates of the Franciscan assemblage (F) and the Great Valley sequence (GVs), California. A - Northern Coast Ranges Franciscan assemblage. B - Northern Coast Ranges Great Valley sequence. C - Northern Coast Ranges Franciscan assemblage. D - Valanginian conglomerates, northern Coast Ranges Great Valley sequence. E - Northern Coast Ranges Franciscan assemblage. F - Western outliers, northern Coast Ranges Great Valley sequence. G - Upper Valanginian - Turonian conglomerates from the west side of the Sacramento Valley and the Middle Mountain area, northern Coast Ranges Great Valley sequence. H - Franciscan assemblage of the Diablo Range. I - More extensive conglomerates of the Great Valley sequence of the Diablo Range. J - Franciscan assemblage, central California coast. K - Centrally located Valanginian conglomerates of the central California coast Great Valley sequence. L - Tithonian conglomerate of the central California coast Great Valley sequence.

and G) except that the quartzite content is higher (8.6 percent compared to 3.1 percent). One (Table 4, locality 55) has a quartzite content of 3.3 percent, which is similar to the values from northern California.

Four conglomerates were sampled from the Franciscan assemblage of the Diablo Range. Two (Fig. 2, localities 58 and 59) are from bodies of conglomerate east of Mt. Hamilton mapped by Soliman (1965) and included by Crawford (1975) in his Eylar Mountain sequence. This unit has produced a few Tithonian fossils (Crawford, 1975; Blake and Jones, 1981), although none are near the conglomerate localities. Locality 60 was mapped by Maddock (1964) in an area shown as melange by Teleen (1977). Stratigraphic assignment of locality 61 is not clear but it seems to be in a metamorphosed, relatively intact sequence, based upon the foliation in the conglomerate and the fact that it was traced by Maddock (1964) along strike for a mile. Clast compositions of the Franciscan assemblage conglomerates (Table 4, localities 58-61) are similar to those in the section above the basal mudstone unit of the Great Valley sequence of the Diablo Range (Table 4, localities 52-57), but are generally poorer in chert and richer in felsic volcanic rock clasts. As in the Great Valley sequence, the conglomerates are mostly rich in quartzite clasts (mean 7.8 percent) although one is low (locality 58, with 1.1 percent). The range of clast compositions from the Diablo Range is shown graphically in Figure 3.

Central California coast

Clast compositions from the central California coastal area were given by Seiders (1983). Mean compositions, supplemented by additional observations, are given in Table 6, columns J, K, and L. Most Great Valley sequence conglomerates sampled seem to be of Valanginian age and have clast compositions rich in chert (mean 66.1 percent) and quartzite (mean 20.0 percent). A single Tithonian conglomerate (Table 6, column L) is less rich in chert (45.6 percent) and has more quartzite (36.1 percent) and quartz (9.9 percent). Conglomerates of the Franciscan assemblage (Table 6, column J) are unfossiliferous but are similar to the Valanginian conglomerates of the Great Valley sequence except that they are richer in quartz pebbles (9.4 percent compared to 1.1 percent). Franciscan and Great Valley sequence conglomerates with contrasting quartz contents are present along strike for 100 km (Fig. 5). The most southerly Great Valley sequence conglomerate, however, is closer to the Franciscan assemblage conglomerates in quartz content (6.0 percent), and the two most northern Franciscan conglomerates are relatively poor in quartz (5.6 and 3.6 percent), more like conglomerates of the Great Valley sequence. A line drawn on a quartz pebble content of 6 percent in the Franciscan assemblage is located nearly 200 km north of a corresponding line in the Great Valley sequence (Fig. 5). This displacement does not result from strike-slip faulting, however, because the Great Valley sequence conglomerates occur in thrust sheets resting on the Franciscan assemblage. The juxtaposition of the contrasting compositions may have resulted from northward tectonic transport of the Franciscan assemblage through oblique sub-

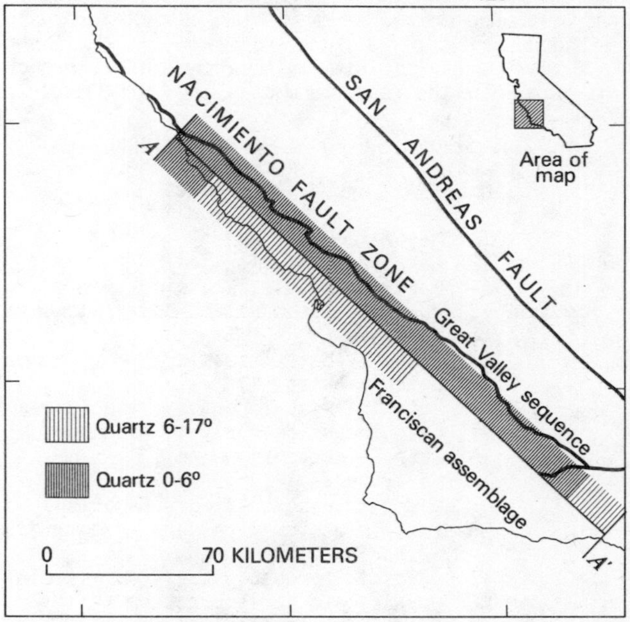

Figure 5. Sketch of the central California coastal region showing the distribution along strike of quartz pebble contents in conglomerates of the Great Valley sequence and the Franciscan assemblage. The Great Valley sequence overlies the Franciscan assemblage in thrust sheets. Quartz pebble contents along strike are projected onto line A-A' from both stratigraphic units.

duction or could be the result of a sediment distribution path that transported the Franciscan detritus northward and outboard of the Great Valley sequence (Seiders, 1983).

Most of the sampled Upper Jurassic and Lower Cretaceous conglomerates of the central coast area contain very few volcanic rock clasts. In Figure 3, volcanic rock clasts make up more than 10 percent of the total chert-volcanic-quartzite clasts in only two of the samples. A Cenomanian conglomerate rich in volcanic rock clasts was reported by McLean and others (1977) from the southern part of the area, but its distribution elsewhere is not known.

Golden Gate-Gilroy block

Four Franciscan assemblage conglomerates were obtained from the Golden Gate-Gilroy area. Locality 62, near the Golden Gate (Wahrhaftig, 1984; Fig. 2), appears to be from a tectonically displaced block. Locality 65 is in an intact metamorphosed sequence on Angel Island (Bloxam, 1960; Schlocker, 1974; Wahrhaftig, 1984). Localities 64 and 63 are in uncertain structural settings in the Franciscan at New Almaden (Bailey and Everhart, 1964) and near Gilroy (Allen, 1946).

The clast compositions from the Golden Gate-Gilroy block (Table 5), like those of the central coast, are characterized by abundant quartzite clasts (range 17-39 percent). Locality 62 shows no vein quartz and is similar to Valanginian conglomerates of the central coast Great Valley sequence. Locality 63 has a clast composition richer in quartzite (38.7 percent) and is more like the single Tithonian conglomerate of the central coast Great Valley sequence (quartzite 36.1 percent, Table 6, column L), except that quartz is lower (2.8 percent compared to 9.9 percent). The other two conglomerates (localities 64 and 65) are richer in volcanic rock clasts (35.7 and 30.4 percent) than any sampled central coast conglomerate.

Radiolarians from chert pebbles

Chert pebbles showing potentially identifiable radiolarians were set aside and the radiolarians extracted utilizing the method described by Pessagno and Newport (1972). Radiolarians were identified from 15 localities in the Franciscan assemblage and the Yager Formation, and from 6 localities in the Great Valley sequence of the northern Coast Ranges (Table 7). Radiolarian faunas are largely the same in both groups of rocks, and are chiefly Late Triassic (Karnian and Norian) with some Early and Middle Jurassic faunas. Although space does not permit a detailed faunal listing of all the radiolarian taxa from these chert clasts, all the Late Triassic radiolarians belong to the following genera: Betraccium Pessagno and Blome (1980), Capnodoce DeWever (1979; amended Blome, 1983), Capnuchosphaera DeWever (1979; amended Blome, 1983), Castrum Blome (1984), Corum Blome (1984), Ferresium Blome (1984), Pseudostylosphaera Nakaseko and Nishimura (1979), Sarla Pessagno (1979), Triassocampe Dumitrica, Kozur, and Mostler (1980), and Yeharaia Nakaseko and Nishimura (1979). Characteristic Early and Middle Jurassic radiolarian taxa from the chert clasts belong to the following genera: Archaeodictyomitra Pessagno (1977), Canutus Pessagno and Whalen (1982), Canoptum Pessagno and Poisson (1981), Droltus Pessagno and Whalen (1979), Hsuum Pessagno (1977), Lupherium Pessagno and Whalen (1982), Perispyridium Pessagno and Blome (1982), Trillus Pessagno and Blome (1980), Turanta Pessagno and Blome (1982), and Zartus Pessagno and Blome (1980).

Three other localities from the Great Valley sequence reported on by Seiders and others (1979) contained similar radiolarian faunas except that some Middle Triassic radiolarians were also present. In the central California coastal region, Late Triassic to Early and Middle Jurassic radiolarians occur in pebbles of both the Franciscan assemblage and the Great Valley sequence. A few Franciscan localities also contained pebbles with Paleozoic fossils (Seiders and others, 1979; Seiders, 1983).

Both the Triassic and Jurassic radiolarian faunas found in the pebbles and their consequent age determinations are very similar to those in bedded chert of the Klamath Mountains (Irwin and others, 1982; Blome and Irwin, 1983). We found no Paleozoic (Permian) radiolarians in the chert pebbles of the northern Coast Ranges, although chert of this age occurs sparsely through the Klamaths. Virtually all the other Klamath Mountains radiolarian chert is either Middle Triassic (Ladinian), Late Triassic (Karnian and Norian), or Early Jurassic in age. Triassic chert has been found in the Sierra Nevada in a few places (Irwin and others, 1978; R.A. Schweickert, 1982, written commun.), but the metamorphism accompaning extensive Cretaceous plutonism in the Sierras

Franciscan assemblage

Loc. no.	Age(s) of pebbles
11	Mesozoic
12	Late Triassic (Karnian), Early to Middle Jurassic
13	Jurassic (probably Bajocian/Bathonian)
25	Mesozoic
34	Probably Late Triassic
35	Late Triassic
36	Late Triassic (probably late Karnian)
37	Late Triassic (early to mid Karnian)
38	Late Triassic (Karnian), Early Jurassic, Middle Jurassic (middle Bajocian)
39	Late Triassic (late Karnian to middle Norian)
43	Jurassic (probably Early Jurassic)
46	Jurassic to Early Cretaceous
48	Late Triassic (early Karnian)
66	Early Cretaceous (Valanginian)
68	Early Jurassic (Pliensbachian), ?Middle Jurassic

Great Valley sequence

Loc. no.	Age of pebbles
2	Late Triassic (middle to late Karnian)
21	Late Triassic (Norian, probably middle Norian)
30	Middle Jurassic (middle Bajocian)
32	Early to Middle Jurassic
33	Late Triassic (Karnian to middle Norian)
52	Late Triassic (late Karnian to early Norian)

Table 7. Ages of radiolarians from chert pebbles in conglomerates, northern Coast Ranges. In most samples, pebbles were lumped for radiolarian extraction; in some samples more than one age of radiolarians was found.

and to the south may have destroyed the radiolarians that would date other bodies of chert.

TECTONIC SIGNIFICANCE

In the northern Coast Ranges, the Diablo Range, and the central California coastal area, conglomerates of the Franciscan assemblage and the Great Valley sequence have clast compositions that are characteristic of each region. Within each region the clast compositions of Franciscan and Great Valley sequence conglomerates are similar and contrast with other regions. These relations suggest that in each area the Franciscan and Great Valley sequence conglomerates were derived from similar source areas.

Our data do not support a dual origin of Franciscan detrital rocks in which the matrix of Franciscan melanges is correlated with the Great Valley sequence whereas intact Franciscan sequences have an exotic source (Blake and Jones, 1974). Pebble counts from Franciscan conglomerates in both coherent sequences and displaced blocks within melange units are similar to counts of conglomerates nearby in the Great Valley sequence. Our sample localities are widely spaced and thus we cannot rule out the existance of any exotic detrital sedimentary rocks. If such rocks are present they must be of minor volume or differ only slightly from the indigenous rock types. No starkly contrasting rock types have been found and all Franciscan fossils are within the range of nearby indigenous rocks. Much of the evidence so far presented in support of an exotic origin for detrital rocks of the Franciscan assemblage concentrates on relatively minor differences that can be explained within the framework of facies controlled depositional environments and dispersal paths (Dickinson and others, 1982). Other apparent contrasts might be resolved through the reinterpretation of local geologic relations. For example, the Eylar Mountain sequence in the Franciscan assemblage has yielded some Tithonian fossils, yet both the sandstone petrology (Jacobson, 1978) and our pebble counts seem more characteristic of younger sedimentary rocks of the Great Valley sequence of the same region. One alternative to interpreting the Eylar Mountain sequence as exotic is to postulate that the sparce and poorly preserved Tithonian fossils do not come from the rocks studied petrographically. This seems quite possible in view of the structural complexity of the Franciscan assemblage.

The distinctiveness of the compositions of the conglomerates in different parts of the Coast Ranges suggests that there have been no very large tectonic dislocations within each region that affected the conglomerates. In the northern Coast Ranges, Blake and Jones (1978 and 1981) have suggested that the western outliers have been translated northward by strike-slip faulting. Most of our pebble counts of the western outliers are low in quartzite clasts and are like other conglomerates of mixed clast composition in the northern Coast Ranges. One of the most southern of the western outliers, however, has a quartzite content of 9.6 percent and a clast composition (Table 3 and Fig, 2, locality 33) similar to conglomerates east of San Francisco (Table 3 and Fig. 3, localities 52-54). These localities lie east of the Hayward fault and about 60 km to the south. These data are consistent with as much as 60 km of right-slip on the Hayward fault but much larger values are unlikely. Prowell (1974), based on data of G.L. Curtis, estimated 45 km of right-slip on the Hayward fault in the past 8 m.y.

The clast compositions of upper Mesozoic conglomerates from the several regions of California are presented diagramatically in Figures 6 and 7. Quartzite content is key in characterizing the compositions. Quartzite generally increases from the northern Coast Ranges, to the Diablo Range, and to the central coast area and the Golden Gate-Gilroy block. Mixed clast conglomerates of the northern Coast Ranges have a mean quartzite content of 3.1 percent. The same type conglomerate in the Diablo Range, excluding the lenses in mudstone in the lower part of the Great Valley sequence (which are lower), contains 8.2 percent quartzite. All central coast conglomerates average 21.3 percent quartzite and the four conglomerates from the Golden Gate-Gilroy block average 24.9 percent quartzite. The southward increase in quartzite shown in the first three regions can be explained as an expression of the fact that northwest-trending Mesozoic structures truncate the old northeast trend of the Paleozoic continental margin (Burchfiel and Davis, 1972). Thus, the northern Mesozoic deposits tapped a source

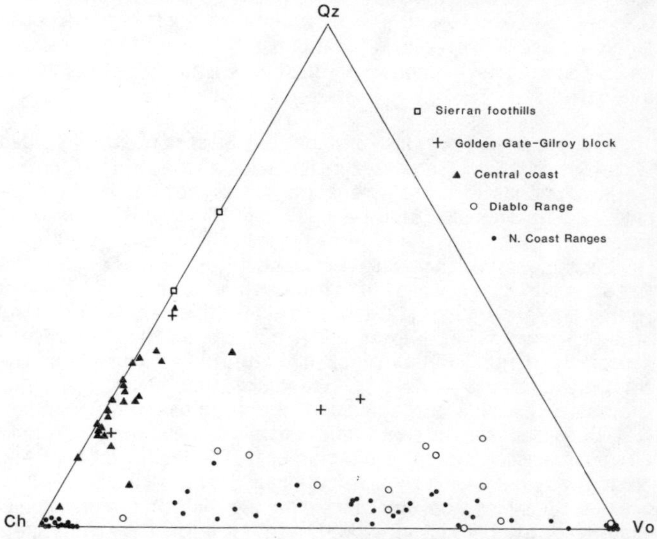

Figure 6. Triangular diagram plotting the contents of chert (Ch), volcanic rocks (Vo), and quartzite (Qz) in conglomerates from several regions of California.

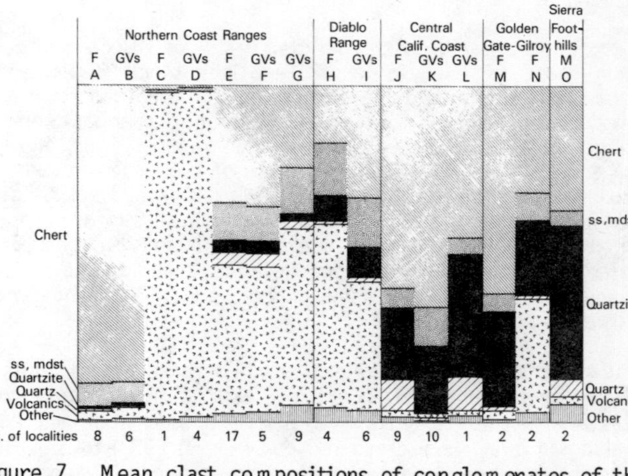

Figure 7. Mean clast compositions of conglomerates of the Franciscan assemblage (F), the Great Valley sequence (GVs), and the Mariposa Formation (M) from five regions of California. "Volcanics" are mainly felsic volcanic rocks but include other volcanics and micro-granitic rocks. A and B - Chert-rich conglomerates. C and D - Volcanic-rich conglomerates. E - Mixed-clast Franciscan assemblage. F - Mixed-clast Great Valley sequence, western outliers. G - Mixed-clast Great Valley sequence, west side of Sacramento Valley and Middle Mountain area. H - Mixed-clast Franciscan assemblage. I - Mixed-clast Great Valley sequence, more extensive conglomerates. J - Franciscan assemblage. K - Centrally located Valanginian conglomerates, Great Valley sequence. L - Tithonian conglomerate, Great Valley sequence. M - Conglomerates rich in chert and quartzite. N - Conglomerates rich in chert, quartzite, and volcanic clasts. O - Conglomerates of the Mariposa Formation.

area that was more oceanic in character and poor in quartzite, whereas the southern deposits were drawn from regions deeper within the continent and richer in quartzite.

Clast compositions in the Golden Gate-Gilroy block appear to be out of place with respect to adjacent regions. The block has most likely been translated northward along an early trace of the San Andreas fault or its proto-San Andreas predecessor. Although a precise reconstruction is not yet possible, at least 300 km of right slip is required to move this block from a position near the southern end of the Diablo Range. Cretaceous and Tertiary sedimentary rocks lie on the Golden Gate-Gilroy block (Jennings, 1977) and their bearing on the timing and amount of displacement now needs to be evaluated.

Clast compositions of conglomerates in the Golden Gate-Gilroy block show affinities with those of the central California coastal area, a region now situated on the opposite side of both the San Andreas fault and the granitic and metamorphic rocks of the Salinian block (Fig. 1). Dickinson (1983) suggested that the Franciscan assemblage and the Great Valley sequence of the central California coastal region may have been juxtaposed against the Salinian block by mid-Cretaceous left-slip of about 560 km on the Nacimiento fault. Both the Salinian block and the rocks outboard of the Nacimiento fault were later translated northward by about the same distance through San Andreas movement. Seiders (1983), based on inferred sediment distribution paths and compositions of selected conglomerates, speculated that the central coast rocks may have originated 300 km north of their present position before participating in the translations suggested by Dickinson (1983). In view of the many new observations from the northern Coast Ranges and the Diablo Range, this now seems unlikely. It is now more probable that these rocks originated near their present position, as proposed by Dickinson (1983), or somewhat farther south, but the position is not tightly constrained by our data.

The similarity of the conglomerates of the central coast with those of the Golden Gate-Gilroy block is important because it provides a tie across the San Andreas fault that reinforces the progressive compositional changes seen in conglomerates to the east. This adds another point of comparison between the Mesozoic rocks of the central coast and those east of the San Andreas fault (Dickinson, 1983; Seiders, 1983). The many similarities between these two areas speak strongly against the very large translations suggested by paleomagnetic results for the Salinian block and more outboard rocks (Champion and others, 1981; McWilliams and Howell, 1982).

Two conglomerates were studied from the foothills of the central Sierra Nevada (Figs. 6 and 7). These conglomerates, from the Mariposa Formation of Oxfordian and Kimmeridgian age, were presumed by Seiders (1983) to represent detritus supplied from the central Sierra Nevada during that time. These conglomerates have the greatest quartzite content of any we have examined thus far and contrast with younger conglomerates of the adjacent Diablo Range. Could they, too, have been translated northward from a more southern place of origin?

Our observations are widely spaced, reconnaissance in nature, and point out some interesting geological problems that require further study. A new and unexpected conclusion is that in much of the northern Coast Ranges, Tithonian-Valanginian conglomerates of the Great Valley sequence were deposited progressively farther west and change in composition with decreasing age, and that a similar geographic distribution occurs in conglomerates of the same type in the Franciscan assemblage. Many more observations are needed to define the geographic limits of conglomerate types within the Franciscan. Additional fossils must be found to date the Franciscan conglomerates and sedimentological studies are required to formulate a comprehensive depositional model which could then be integrated with the tectonic history. Our data from the Diablo Range are very sparce and more observations are needed from both the Great Valley sequence and the Franciscan assemblage. We urge future workers to count pebbles and solicite from our readers the locations of conglomerates known to them.

ACKNOWLEDGEMENTS

We are grateful to the following people for guidance to conglomerate localities: W.P. Irwin, M.C. Blake, Jr., J.C. Maxwell and his former students S.D. Etter, J.R. Kleist, and D.H. Lehman, J.O. Berkland, A.S. Jayko, Clyde Wahrhaftig, J.C. Yount, J.A. Bartow, E.E. Brabb, S.A. Graham, D.L. Jones, T.A. Kaplan, M.E. Maddock, C.F. Mansfield, III, R.J. McLaughlin, B.L. Murchey, L.A. Raymond, E.I. Rich, Julius Schlocker, and John Suppe. Three conglomerate localities were gleaned from an unpublished thesis by Fuller (1977). The critical comments of A.S. Jayko and C.M. Wentworth improved the manuscript.

REFERENCES CITED

Allen, J.E., 1946, Geology of the San Juan Bautista quadrangle, California: California Division of Mines and Geology Bulletin 133, 75 p.

Bailey, E.H., Irwin, W.P., and Jones, D.L., 1964, Franciscan and related rocks, and their significance in the geology of western California: California Division of Mines and Geology Bulletin 183, 177p.

Bailey, E.H., and Everhart, D.L., 1964, Geology and quicksilver deposits of the New Almaden district, Santa Clara County, California: U.S. Geological Survey Professional Paper 360, 206 p.

Bailey, E.H., and Jones, D.L., 1973, Preliminary lithologic map, Colyear Springs quadrangle, California: U.S. Geological Survey Miscellaneous Field Studies Map MF-516, scale 1:48,000.

Berkland, J.O., 1969, Geology of the Novato Quadrangle, Marin County, California: San Jose, California, San Jose State University, M.S. thesis, 146 p.

Bertucci, P.F., 1983, Petrology and provenance of the Stony Creek Formation, northwestern Sacramento Valley, California, in Bertucci, P.F., and Ingersoll, R.V., eds., Guidebook to the Stony Creek Formation, Great Valley Group, Sacramento Valley, California: Los Angeles, California, Society of Economic Paleontologists and Mineralogists, Pacific Section, p. 1-16.

Blake, M.C., Jr., and Jones, D.L., 1974, Origin of Franciscan melanges in northern California, in Dott, R.H., Jr., and Shaver, R.H., eds., Modern and ancient geosynclinal sedimentation: Society of Economic Paleontologists and Mineralogists Special Publication No. 19, p. 345-357.

_____ 1978, Allochthonous terranes in northern California?--a reinterpretation, in Howell, D.G., and McDougall, K.A., eds., Mesozoic paleogeography of the western United States: Society of Economic Paleontologists and Mineralogists, Pacific Section, Pacific Coast Paleogeography Symposium 2, p. 397-400.

_____ 1981, The Franciscan assemblage and related rocks in northern California: A reinterpretation, in Ernst, W.G., ed., The geotectonic development of California; Ruby Volume I: Englewood Cliffs, New Jersey, Prentice-Hall, p. 307-328.

Blome, C.D., 1983, Upper Triassic Capnuchosphaeridae and Capnodocinae (Radiolaria) from east-central Oregon: Micropaleontology, v. 29, p. 11-49.

_____ 1984, Upper Triassic Radiolaria and radiolarian zonation from western North America: Bulletins of American Paleontology, v. 85, no. 318, p. 1-88.

Blome, C.D., and Irwin, W.P., 1983, Tectonic significance of Late Paleozoic and Mesozoic radiolarians from the North Fork terrane, Klamath Mountains, California, in

Stevens, C.H., ed., Pre-Jurassic rocks in western North American suspect terranes: Los Angeles, California, Society of Economic Paleontologists and Mineralogists, Pacific Section, p. 77-89.

Bloxam, T.W., 1960, Jadeite rocks and glaucophane schists from Angel Island, San Francisco Bay, California: American Journal of Science, v. 258, p. 555-573.

Burchfiel, B.C., and Davis, G.A., 1972, Structural framework and the evolution of the southern part of the Cordilleran orogen, western United States: American Journal of Science, v. 272, p. 97-118.

Case, J.E., 1968, Upper Cretaceous and Lower Tertiary rocks, Berkeley and San Leandro Hills, California: U.S. Geological Survey Bulletin 1251-J, 29 p.

Champion, D.E., Gromme, C.S., and Howell, D.G., 1981, Paleomagnetic study of the Upper Cretaceous Pigeon Point Formation, in Frizzell, Virgil, ed., Upper Cretaceous and Paleocene turbidites, central California coast: Society of Economic Paleontologists and Mineralogists, Pacific Section, Field trip guide, v. 4, p. 53-55.

Cowan, D.S., and Page, B.M., 1975, Recycled Franciscan material in Franciscan melange west of Paso Robles, California: Geological Society of America Bulletin, v. 86, p. 1089-1095.

Crittenden, M.D., Jr., 1951, Geology of the San Jose-Mount Hamilton area, California: California Division of Mines and Geology Bulletin 157, 74 p.

DeWever, P., 1979, Systematic Section, in DeWever, P., Sanfillipo, A., Riedel, W.R., and Gruber, B., Triassic radiolarians from Greece, Sicily, and Turkey: Micropaleontology, v. 25, p. 75-109.

Dibblee, T.W., Jr., 1973, Stratigraphy of the southern Coast Ranges near the San Andreas fault from Cholame to Maricopa, California: U.S. Geological Survey Professional Paper 764, 45 p.

______ 1981, Upper Mesozoic rock units in the central Diablo Range between Hollister and New Idria and their depositional environments, in Frizzell, Virgil, ed., Geology of the central and northern Diablo Range, California: Society of Economic Paleontologists and Mineralogists, Pacific Section, Field trip guide, p. 13-20.

Dickinson, W.R., 1983, Cretaceous sinistral strike slip along Nacimiento fault in coastal California: American Association of Petroleum Geologists Bulletin, v. 67, p. 624-645.

Dickinson, W.R., Ingersoll, R.V., Cowan, D.S., Helmold, K.P., and Suczec, C.A., 1982, Provenance of Franciscan graywackes in coastal California: Geological Society of America Bulletin, v. 93, p. 95-107.

Dumitrica, P., Kozur, H., and Mostler, H., 1980, Contributions to the radiolarian fauna of the Middle Triassic of the southern Alps: Geologisch-Palaeontologische Mitteilungen Innsbruck, v. 10, p. 1-46.

Evitt, W.R., and Pierce, S.T., 1975, Early Tertiary ages from the coastal belt of the Franciscan complex, northern California: Geology, v. 3, p. 433-436.

Fuller, S.R., 1977, The origin of the major constituents of the Franciscan melange conglomerates in Humbolt County, California: Arcata, California, Humbolt State University, senior thesis, 38 p.

Ghent, E.D., 1963, Fossil evidence for maximum age of metamorphism in part of the Franciscan Formation, northern Coast Ranges: California Division of Mines and Geology Special Report 82, 41 p.

Hall, C.A., 1958, Geology and paleontology of the Pleasanton area, Alameda and Contra Costa Counties, California: University of California, Department of Geological Sciences Bulletin, v. 34, p. 1-90.

Ingersoll, R.V., Rich, E.I., and Dickinson, W.R., 1977, Great Valley sequence, Sacramento Valley: Geological Society of America, Cordilleran Section, Field Guide, 72 p.

Irwin, W.P., and Dennis, M.D., 1979, Geologic structure section across southern Klamath Mountains, Coast Ranges, and seaward of Point Delgada, California: Geological Society of America Map and Chart Series MC-28D, scale 1:250,000.

Irwin, W.P., Jones, D.L., and Blome, C.D., 1982, Map showing sampled radiolarian localities in the western Paleozoic and Triassic belt, Klamath Mountains, California: U.S. Geological Survey Miscellaneous Field Studies Map MF-1399, scale 1:250,000.

Irwin, W.P., Jones, D.L., and Kaplan, T.A., 1978, Radiolarians from pre-Nevadan rocks of the Klamath Mountains, California and Oregon, in Howell, D.G., and McDougall, K.A., eds., Mesozoic paleogeography of the western United States: Society of Economic Paleontologists and Mineralogists, Pacific Section, Pacific Coast Paleogeography Symposium 2, p. 303-310.

Irwin, W.P., Wolfe, E.W., Blake, M.C., Jr., and Cunningham, C.G., Jr., 1974, Geological map of the Pickett Peak quadrangle, Trinity County, California: U.S. Geological Survey Geologic Quadrangle Map GQ-1111, scale 1:62,500.

Jacobson, M.I., 1978, Petrologic variations in Franciscan sandstone from the Diablo Range, California, in Howell, D.G., and McDougall, K.A., eds., Mesozoic paleogeography of the western United States: Society of Economic Paleontologists and Mineralogists, Pacific Section, Pacific Coast Paleogeography Symposium 2, p. 401-417.

Jennings, C.W., 1977, Geologic map of California: California Division of Mines and Geology, scale 1:750,000.

Jones, D.L., and Bailey, E.H., 1973, Preliminary biostratigraphical map, Colyear Springs quadrangle, California: U.S. Geological Survey Miscellaneous Field Studies Map MF-517, scale 1:48,000.

Jones, D.L., and Irwin, W.P., 1971, Structural implications of an offset Early Cretaceous shoreline in northern California: Geological Society of America Bulletin, v. 82, p. 815-822.

Jordan, M.A., 1978, Geology of the Round Valley-Sanhedrin Mountain area, northern California Coast Ranges: Austin, Texas, University of Texas, Ph.D. thesis, 174 p.

Kaplan, T.A., 1983, The Lassic outlier—an outlier of Coast Range ophiolite: San Jose, California, San Jose State University, M.S. thesis, 96 p.

Lehman, D.H., 1974, Structure and petrology of the Hull Mountain area, northern California Coast Ranges: Austin, Texas, University of Texas, Pd.D. thesis, 126 p.

Maddock, M.E., 1964, Geology of the Mt. Boardman quadrangle, Santa Clara and Stanislaus Counties, California: California Division of Mines and Geology Map Sheet 3, scale 1:62,500.

Mansfield, C.A., III, 1979, Upper Mesozoic subsea fan deposits in the southern Diablo Range, California: Record of the Sierra Nevada magmatic arc: Geological Society of America Bulletin, Part I, v. 90, p. 1025-1046.

McLean, Hugh, Howell, D.C., and Vedder, J.G., 1977, An unusual Upper Cretaceous conglomerate in the central San Rafael Mountains, Santa Barbara County, California, in Howell, D.G., Vedder, J.G., and McDougall, K., eds., Cretaceous geology of the California Coast Ranges, west of the San Andreas fault: Society of Economic Paleontologists and Mineralogists, Pacific Section, Pacific Coast Paleogeography Field Guide 2, p. 79-83.

McWilliams, M.O., and Howell, D.G., 1982, Exotic terranes in western California: Nature, v. 297. p. 79-83.

Nakaseko, K., and Nishimura, A., 1979, Upper Triassic Radiolaria from southwest Japan: Osaka University, College of General Education, Science Reports, v. 29, no. 2, p.61-109.

Orchard, D.M., 1979, Geology of the Robinson Creek-Ukiah area, northern Coast Ranges, California: Austin, Texas, University of Texas, M.A. thesis, 88 p.

Pessagno, E.A., Jr., 1977, Upper Jurassic Radiolaria and radiolarian biostratigraphy of the California Coast Ranges: Micropaleontology, v. 23, p. 56-113.

______ 1979, Systematic Paleontology, in Pessagno, E.A., Jr., Finch, J.W., and Abbott, P.L., Upper Triassic Radiolaria from the San Hipolito Formation, Baja California: Micropaleontology, v.25, p. 160-197.

Pessagno, E.A., Jr., and Blome, C.D., 1980, Upper Triassic and Jurassic Pantanellinae from California, Oregon, and British Columbia: Micropaleontology, v. 26, p. 225-273.

_____1982, Bizarre Nassellariina (Radiolaria) from the Middle and Upper Jurassic of North America: Micropaleontology, v. 28, no. 3, p. 289-318.

Pessagno, E.A., Jr., and Newport, R.L., 1972, A technique for extracting Radiolaria from radiolarian cherts: Micropaleontology, v. 18, p. 231-234.

Pessagno, E.A., Jr., and Poisson, A., 1981, Lower Jurassic Radiolaria from the Gumuslu allochthon of southwestern Turkey (Taurides Occidentales): Ankra, Turkey, Maden Tetkik ve Arama Enstitusu Yayinlarindan, Bulletin no. 92, 1979, p. 47-69.

Pessagno, E.A. Jr., and Whalen, P.A., 1982, Lower and Middle Jurassic Radiolaria (multicyrtid Nassellariina) from California, east-central Oregon, and the Queen Charlotte Islands, British Colombia: Micropaleontology, v. 28, p. 111-169.

Prowell, D.C., 1974, Geology of selected Tertiary volcanics in the central Coast Ranges of California and their bearing on the Calavaras and Hayward fault problems: Santa Cruz, California, University of California, Santa Cruz, Ph.D. thesis, 182 p.

Robinson, G.D., 1956, Geology of the Hayward quadrangle, California: U.S. Geological Survey Geologic Quadrangle Map GQ-88, scale 1:24,000.

Rose, R.L., and Colburn, I.P., 1963, Geology of the east-central part of the Priest Valley quadrangle, California, in Guidebook to the geology of Salinas Valley and the San Andreas fault: American Association of Petroleum Geologists and Society of Economic Paleontologists and Mineralogists, Pacific Sections, p. 38-45.

Schilling, F.A., Jr., 1962, The Upper Cretaceous stratigraphy of the Pacheco Pass quadrangle, California: Stanford, California, Stanford University, Ph.D thesis, 153 p.

Schlocker, Julius., 1974, Geology of the San Francisco North quadrangle, California: U.S. Geological Survey Professional Paper 782, 109 p.

Seiders, V.M., 1983, Correlation and provenance of upper Mesozoic chert-rich conglomerate, California: Geological Society of America Bulletin, v. 94, p. 875-888.

Seiders, V.M., Pessagno, E.A., and Harris, A.G., 1979, Radiolarians and conodonts from pebbles in the Franciscan assemblage and the Great Valley sequence of the California Coast Ranges: Geology, v. 7, p. 37-40.

Soliman, S.M., 1965, Geology of the east half of the Mount Hamilton quadrangle, California: California Division of Mines and Geology Bulletin 185, 32 p.

Swe, Win, and Dickinson, W.R., 1970, Sedimentation and thrusting of late Mesozoic rocks in the Coast Ranges near Clear Lake, California: Geological Society of America Bulletin, v. 81, p. 165-188.

Suppe, John, 1973, Geology of the Leech Lake Mountain-Ball Mountain region, California: University of California Publications in Geological Sciences, v. 107, 82 p.

Teleen, K.E., 1977, Paleocurrents in part of the Franciscan Complex, California: Geology, v. 5, p. 49-51.

Wahrhaftig, Clyde, 1984, A streetcar to subduction (revised edition): Washington, D.C., American Geophysical Union, 76 p.

APPENDIX 1

Locations of sampled conglomerates

Except as noted, locations are given by 7-1/2 minute quadrangle and 1000-meter Universal Transverse Mercator grid.

1. Naufus Creek, 4,472,560N, 478,480E. 2. Blocksburg 15', 4,465,200N, 452,000E. 3. Elk Creek, 4,380,960N, 533,930E. 4. Wilbur Springs 15', 4,340,800N, 546,910E. 5. Wilbur Springs 15', 4,335,710N, 548,400E. 6. Capell Valley, 4,254,720N, 571,280E. 7. Buck Rock, 4,414,780, 504,310E. 8. Mendocino Pass, 4,411,000N, 507,740E. 9. Mendocino Pass, 4,409,380N, 504,310E. 10. Newhouse Ridge, 4,406,610N, 495,330E. 11. Mendocino Pass, 4,406,390N, 506,300E. 12. Newhouse Ridge, 4,404,290N, 499,190E. 13. Newhouse Ridge, 4,403,230, 491,850E. 14. Kneecap Ridge, 4,383,150N, 510,940E. 15. Elk Creek, 4,384,560N, 540,530E. 16. Lakeport 15', 4,335,500N, 507,190E. 17. Lower Lake, 4,303,220N, 537,190E. 18. Middletown, 4,300,500N, 540,990E. 19. Newhouse Ridge, 4,400,630N, 494,710E. 20. Lowrey. Composite sample: 4,438,010N, 532,250E and 4,437,750N, 532,360E. 21. Chanchelulla Peak 15', 4,463,920N, 513,540E. 22. Colyear Springs 15', 4,450,630N, 526,350E. 23. Lowrey, 4,436,750N, 535,500E. 24. Wilbur Springs 15', 4.334,280N, 550,480E. 25. Potter Valley 15', 4,360,670N, 489,720N. 26. Wilbur Springs 15', 4,333,840N, 555,150E. 27. Lowrey, 4,438,850N, 538,150E. 28. Lowrey, 4,430,340N, 541,750E. 29. Geyserville, 4,287,800N, 501,330E. 30. Fort Ross, 4,265,140N, 488,110E. 31. Duncans Mills, 4,254,350N, 491,750E. 32. Novato, 4,217,460N, 542,850E. 33. Novato, 4,217,400N, 542,250E. 34. Rodgers Peak, 4,561,270N, 407,040E. 35. Trinidad, 4,546,400N, 403,350E. 36. Arcata South, 4,520,900N, 413,910E. 37. Iaqua Buttes 15', 4,498,500N, 425,910E. 38. Blocksburg 15', 4,480,050N, 452,800E. 39. Pickett Peak 15', 4,459,420N, 466,670E. 40. Pickett Peak 15', 4,457,280N, 461,650E. 41. Jewett Rock, 4,431,600N, 448,500E. 42. Updegraff Ridge, 4,427,550N, 467,760E. 43. Hull Mountain, 4,378,320N, 504,160E. 44. Lake Pillsbury 15', 4,351,600N, 507,520E. 45. Lakeport 15', 4,344,270N, 512,740E. 46. Lakeport 15', 4,333,410N, 500,150E. 47. Hopland, 4,314,580N, 499,720E. 48. The Geysers, 4,298,800N, 519,280E. 49. Hayward, 4,166,750N, 583,760E. 50. Llanada, 4,051,490N, 681,170E. 51. Sherman Peak, 4,006,940N, 715,590E. 52. Hayward, 4,175,990N, 577,640E. 53. Hayward, 4,173,760N, 579,270E. 54. Niles, 4,159,430N, 589,910E. 55. Calaveras Resevoir, 4,139,200N, 604,140E. 56. Pacheco Pass, 4,107,930N, 666,390E. 57. Sherman Peak, 4,008,100N, 714,640E. 58. Isabel Valley, 4,135,180N, 622,600E. 59. Isabel Valley, 4,134,790N, 622,720E. 60. Mt. Stakes, 4,134,490N, 634,340E. 61. Mt. Boardman, 4,143,250N, 639,900E. 62. Point Bonita, 4,186,310N, 542,580E. 63. Chittenden, 4,090,710N, 624,930E. 64. Santa Teresa Hills, 4,114,270N, 603,090E. 65. San Francisco North. Composite sample: 4,190,540N, 549,430E and 4,190,580N, 550,920E. 66. San Rafael, 4,299,580N, 542,520E. 67. Potter Valley 15', 4,370,220N, 495,100E. 68. Garberville 15', 4,432,570N, 433,250E. 69. Leggett 15', 4,420,930N, 450,850E. 70. Ukiah 15', 4,336,840N, 478,790E. 71. Weott 15', 4,463,180N, 434,200E.

FORAMINIFERS FROM CRETACEOUS LIMESTONE OF THE FRANCISCAN COMPLEX, NORTHERN CALIFORNIA

W. V. Sliter
U.S. Geological Survey
Menlo Park, California 94025

ABSTRACT

Tectonically incorporated blocks of Cretaceous pelagic limestone within the Franciscan Complex of northern California contain low-latitude planktic foraminifers that range in age from early late Albian to middle Maestrichtian. Blocks from four localities in the Coastal terrane range in age collectively from early Campanian to middle Maestrichtian (83 to 69 Ma). Three localities in the Central terrane are older and range collectively from late Albian (Vraconian) to early Coniacian (~101.4 to ~88 Ma). Nine localities from the Permanente terrane are similar in age and range collectively from early late Albian to possibly early Coniacian (105 to 88 Ma). The age ranges support the concept of accretion of successively younger oceanic material within the Franciscan Complex.

Franciscan limestone apparently formed as sediment capping volcanic topographic highs on the eastern flank of the Pacific-Farallon Ridge or other active spreading ridges. The thicker limestone sequences represent oceanic seamount or plateau deposits, whereas other smaller outcrops perhaps were associated with active spreading ridges and offridge seamounts. Deposition of Franciscan limestone took place at bathyal water depths above the carbonate compensation depth (CCD). The majority of Franciscan pelagic limestones were deposited at depths of 1000 to 1500 m as suggested by rare benthic foraminifers. Within the Permanente terrane, seamount subsidence is recorded in the pre-late Albian through Cenomanian sequence by a change from oolitic and bioclastic limestone to foraminifer-nannofossil limestone.

The thicker Franciscan seamount limestone presumably accumulated within or adjacent to the equatorial high-productivity zone. Latitudinal limits of approximately 20° north or south latitude are suggested for middle Cretaceous deposition. Seamounts capped by thick limestone deposits probably originated south of the paleoequator and migrated northward beneath the productive zone as part of the plate. Pre-Albian(?) bituminous limestone in the Permanente terrane may have been deposited during the late Barremian through Albian anoxic event and within an intensified oxygen-minimum zone.

INTRODUCTION

Cretaceous limestone is a persistent though rare lithologic constituent of the Franciscan Complex of northern California. The dominant Franciscan lithology is graywacke with lesser amounts of argillite, mafic volcanic rocks, chert, serpentinite and metamorphic rocks (Bailey and others, 1964). The limestone occurs as isolated blocks or in bedded sequences with mafic volcanic rocks and chert and typically is tectonically associated with Franciscan melange. Regionally, the limestone outcrops often are arranged in linear zones that generally parallel the major north-south terrane boundaries as defined by Blake and others (1982). Limestone blocks within the Franciscan Complex contain low-latitude Cretaceous foraminifers that range in age from at least late Albian to middle Maestrichtian. The faunas are diverse and moderately well preserved in nonrecrystallized sections and are well suited for correlation with the international biostratigraphic zonation.

The westward growth of California has involved the accretion of geologically unrelated allochthonous terranes (Irwin, 1972; Blake and Jones, 1978; Blake and others, 1982) that were juxtaposed and imbricated with contemporaneous parautochthonous continental margin deposits through plate subduction (Dickinson, 1970, 1981; Ernst, 1970, 1984; Ingersoll, 1978, 1979). The Franciscan Complex thus records periods of tectonic accumulation and disruption of a subduction accretion prism that are expressed as a complex pattern of tectonostratigraphic terranes.

Numerous fault-bounded terranes are now recognized in California (Coney and others, 1980; Ben-Avraham and others, 1981; Blake and others, 1982). The Franciscan Complex of the California Coast Ranges is differentiated into three broad lithologic units, the Eastern, Central and Coastal belts (Irwin, 1960). The Eastern belt and its included terranes range in age from Late Jurassic to middle Cretaceous (Blake and Jayko, 1983) and are not a part of this study. The Central belt corresponds to the Central terrane of Blake and others (1982). The Coastal belt consists of several terranes, the largest of which is the Coastal terrane (McLaughlin and others, 1982). Three terranes pertain to this study, the larger Coastal and Central terranes, and the smaller Permanente terrane that may represent a subterrane of the Central melange (Fig. 1).

Driving the accretion and subduction processes is the northward drift of the Pacific basin (Heirtzler and others, 1968; Larson and Chase, 1972; Morgan, 1972; Winterer, 1973; Lancelot, 1978; Engebretson and others, 1981) superimposed on the east-west closure of the Pacific basin by the encroaching North American plate (Atwater, 1970; Pitman and others, 1974; Coney, 1978; Engebretson and Cox, 1982; Henderson and others, 1984). In addition, the northward translation of terranes within the California Coast Ranges along longitudinal faults primarily during the Cenozoic represents an added complexity to interpreting the regional geology (Bailey and others, 1964; Champion and others, 1980; Hamilton, 1978; Howell and Vedder, 1978; Howell and others, 1980; McLaughlin and others, 1982).

Despite the rarity of the limestone in the Franciscan Complex, the inherent paleontologic and lithologic characteristics make it an important element in unraveling the age, depositional and tectonic history of the allochthonous units and the

in Blake, M.C. Jr., ed., 1984, **Franciscan Geology of Northern California:** Pacific Section S.E.P.M., Vol. 43, p. 149-162

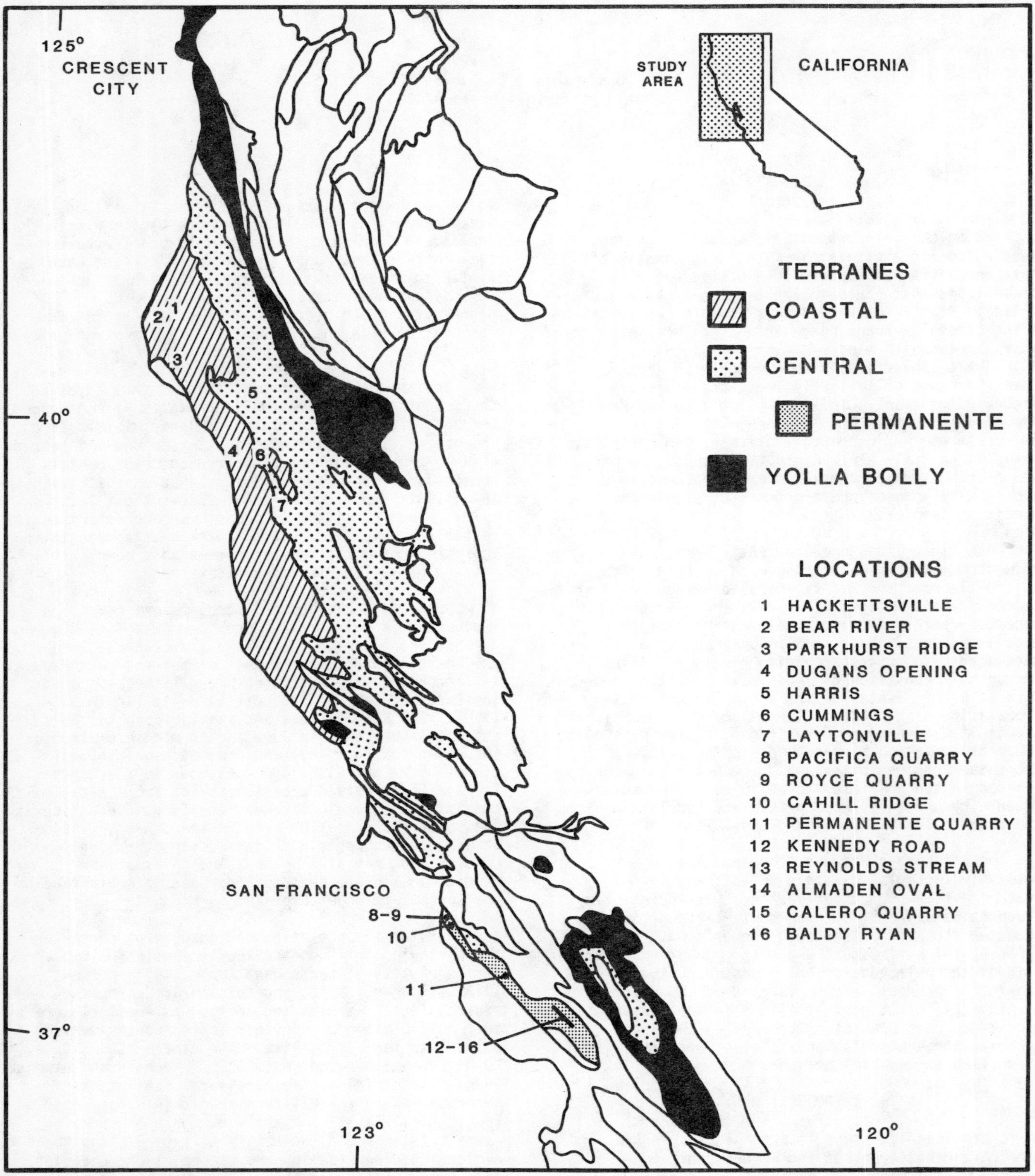

Figure 1. Generalized tectonic map of northern California showing distribution of the major tectonostratigraphic terranes (after Blake and others, 1982). Also shown are localities of foraminiferal limestone listed in Table 1.

contrasting continental margin deposits.

Reported here is a reconnaissance study of foraminifers from 16 limestone localities from the Central, Coastal and Permanente terranes of the Franciscan Complex (Fig. 1).

PREVIOUS FORAMINIFERAL STUDIES

The first studies of foraminifers from Franciscan limestone were made by Thalmann for samples collected from the Permanente Quarry in Santa Clara County (1942) and from Laytonville in Mendocino County (1943). A Turonian to possibly Santonian age was reported for foraminifers identified in thin section from these and other outcrops in California and Oregon. Cushman and Todd (1948) ascribed an Early Cretaceous age to poorly preserved, free specimens collected by Edgar H. Bailey and Donald L. Everhart of the U.S. Geological Survey from limestone-tuff interbeds of the New Almaden district, Santa Clara County. Subsequently, Glaessner (1949) revised the age of this deposit to "late Lower Cretaceous (Albian)".

Church (1952) reported a probable "Middle to Upper Cenomanian" age for freed specimens identified by H. Bolli and collected from the type area of the Calera Limestone at Rockaway Beach, San Mateo County. Kupper (1955) recollected and identified foraminifers from the New Almaden locality of Cushman and Todd (1948) and suggested an "early or medial Cenomanian" age for the deposit.

Thalmann (in Irwin, 1957) revised his earlier age interpretations and considered the Calera Limestone at Permanente Quarry to be Cenomanian in age and dated "other bodies of limestones nearby" and the limestone near Laytonville as "Upper Albian to basal Cenomanian" in age.

New Almaden again was reported on by Loeblich and Tappan (1961) and was regarded as middle to late Cenomanian in age. Bailey and others (1964) listed thin-section identifications by Andrew Marianos for various Franciscan limestone localities that included a Turonian age for the New Almaden locality. Laytonville was reexamined by Alvarez and others (1980) and thin-section identifications made by I. Premoli Silva were of late Albian to late Cenomanian age. More recently, Berry (1982) reported a Cenomanian fauna in thin section from the Central terrane near Covelo and Sliter (1984) gave age ranges for foraminifers from the Coastal, Central and Permanente terranes.

FORAMINIFERAL ZONATION

The zonal scheme used in this paper and the corresponding biostratigraphic events are shown in Figure 2. The zonation is a composite of those of Longoria (1974), Sigal (1977), Robaszynski and Caron (1979) and Premoli Silva and Sliter (1981). Biostratigraphic events correspond to the above references plus Robaszynski and others (1984). The geologic time scale is that of Harland and others (1982). Identifications of foraminifers in thin section has profited from the examination of thin sections from Gubbio, Italy collected at 1-m intervals by I. Premoli Silva and her colleagues through the remarkably continuous section of pelagic limestone. Results of the detailed lithostratigraphy, paleontology and magnetic stratigraphy from Gubbio were reported earlier (Arthur and Fischer, 1977; Premoli Silva, 1977; Alvarez and others, 1977; Lowrie and others, 1980).

LIMESTONE LOCALITIES AND THEIR AGES

Reported here are preliminary age determinations based on 144 samples from both measured sections and isolated sites from localities listed in Table 1 (see Fig. 1). Of the three terranes sampled, four localities are from the Coastal terrane, three are from the Central terrane, and nine are from the Permanente terrane. For this study the age ranges should be considered minimal as the ongoing study will undoubtedly extend the ranges at some of the localities into both older and younger age zones with more complete sampling. Selected diagnostic species from the assemblages are illustrated.

Coastal Terrane

The western Coastal terrane differs from eastern Franciscan terranes in containing more arkosic sandstone with lesser amounts of argillite, volcanic rocks, chert, and red and gray limestone (Bailey and Irwin, 1959; Bailey and others, 1964). Although rocks of the Coastal terrane are locally sheared and complexly folded, the overall metamorphic grade is lower than in the other Franciscan terranes (Blake and Jones, 1981). The eastern contact with the Central terrane is formed by the Coastal belt thrust where rocks of the Central terrane are thrust westward over the younger Coastal terrane (Jones and others, 1978).

Foraminifers from limestone of the Coastal terrane range in age from early Campanian to middle Maestrichtian (Fig. 3).

At Hackettsville (loc. 1), a small outcrop consisting of predominantly red, sheared, aphanitic limestone faulted between pillow basalt and Coastal terrane melange contains a Maestrichtian assemblage attributed to the *Gansserina gansseri* Zone (Fig. 2). In addition to the zonal indicator, species include *Globotuncanita stuarti* and *G. stuartiformis*. Along the Bear River (loc. 2), light-gray, aphanitic, heavily veined and mottled limestone yields a late Campanian age. This limestone also is associated with pillow basalt and melange. Present are species such as *Globotruncana arca*, *G. bulloides*, *G. linneiana*, *G. ventricosa*, and *Globotruncanita stuartiformis*. The *Globotruncana ventricosa* Zone is indicated based on this assemblage and the lack of younger index species. At Parkhurst Ridge (loc. 3), approximately 10 to 12 meters of steeply dipping, synclinally deformed manganiferous, stylolitic, red and light gray limestone gives an early Campanian to Maestrichtian age range (see Harbert and others, this volume). Present are species from the *Globotruncanita elevata* Zone to at least the *Globotruncana falsostuarti* Zone. Species present include *Globotruncana ventricosa*, *Globotruncanella havanensis*, *Globotruncanita stuarti* and *G. elevata*.

A red, sheared, and heavily veined limestone from near Dugans Opening (loc. 4) associated with pillow basalt is late Campanian in age and assigned to the *Globotruncana ventricosa* Zone. Present are *Globotruncana arca*, *G. bulloides*, *G. rosetta*, *G. ventricosa* and *Globotruncanita stuartiformis*.

Central Terrane

The Central terrane is largely melange composed of abundant graywacke and metagraywacke, with large blocks of chert, greenstone, and serpentinite, lesser amounts of red and gray limestone, and inclusions of high-grade blue schist, amphibolite and eclogite. On the east, the older Yolla Bolly terrane structurally overlies the Central terrane along a subhorizontal to gentle eastward dipping fault zone (Blake and Jones, 1974).

Foraminifers from limestone in the Central terrane are older than those from Coastal terrane localities and range from the late Albian (Vraconian) to the early Coniacian (Fig. 3).

Near Harris (loc. 5), small, isolated, outcrops of fine-grained, light-gray, heavily veined limestone with abundant stylolites spaced a centimeter or less apart and thin chert interbeds, range in age from late Turonian to early Coniacian. Foraminifers are assigned to the upper half of the *Marginotruncana schneegansi* Zone that straddles the Turonian-Coniacian boundary (Fig. 2). Species present include *Marginotruncana schneegansi*, *M. sigali*, *M. pseudolinneiana*, *Dicarinella imbricata* and *Hedbergella flandrini*. At Cummings (loc. 6), small outcrops of red, sheared limestone in melange associated with pillow basalt and serpentinite lie just east of the Coastal belt thrust. Poorly preserved foraminifers indicate the *Praeglobotruncana helvetica* Zone of Turonian age. Species include the zone indicator plus *Whiteinella archaeocretacea*, *W. baltica*, *W. aprica* and *Dicarinella hagni*.

The red, stylolitic limestone and interbedded chert exposed in several small, isolated outcrops near Laytonville has attracted geologic interest for years (loc. 7). In addition to the early paleontologic studies described above, a lithologic distinction was made by Bailey and others (1964) between the red limestone from the northern Coast Ranges typified by that from Laytonville, and the white-weathering light to dark gray limestone of the San Francisco peninsula. The red limestone was referred to as the Laytonville type and the gray limestone the Calera-type after the Calera Limestone of Lawson (1902, 1914). Detailed paleontologic study of several outcrops in the Laytonville area identified an age range of late Albian to late Cenomanian (Alvarez and others, 1980). Using the zonation in Figure 2, the current range extends from the *Rotalipora appenninica* Zone to the top of the *R. cushmani* Zone.

Permanente Terrane

The Permanente terrane extends from Pacifica on the San Francisco peninsula south to the San Juan Bautista quadrangle. The terrane is bounded by the Pilarcitos and San Andreas faults in the north and by the San Andreas and parallel eastern faults in the south. A complex system of faults and folds within the southern part of the terrane delineate at least three large structural blocks bounded by major faults or shear zones. The terrane consists primarily of graywacke and mafic volcanic rocks with lesser amounts argillite, limestone, chert, metamorphic rocks and conglomerate. Although limestone constitutes only a

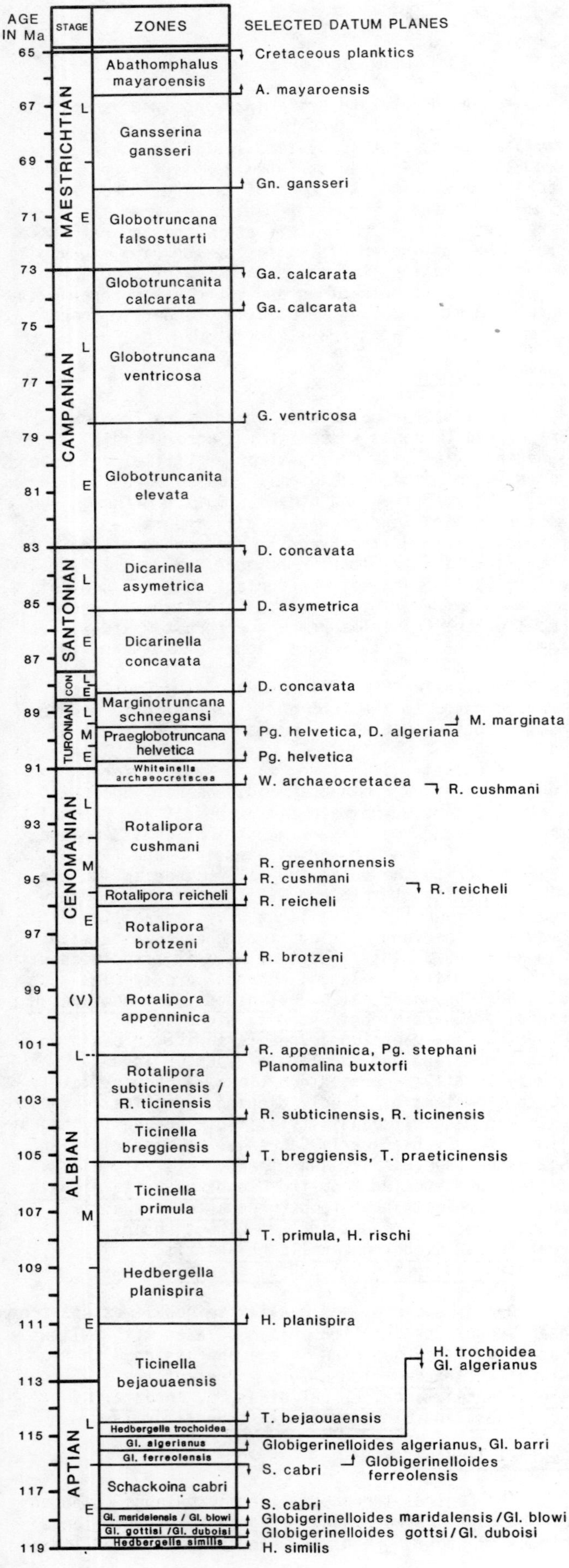

Figure 2. Cretaceous planktic foraminifer zones and selected datum levels. Zonal scheme after Longoria (1974), Sigal (1977), Robaszynski and Caron (1979), and Premoli Silva and Sliter (1981). Geologic time scale after Harland and others (1982). Arrows show first and last occurrences.

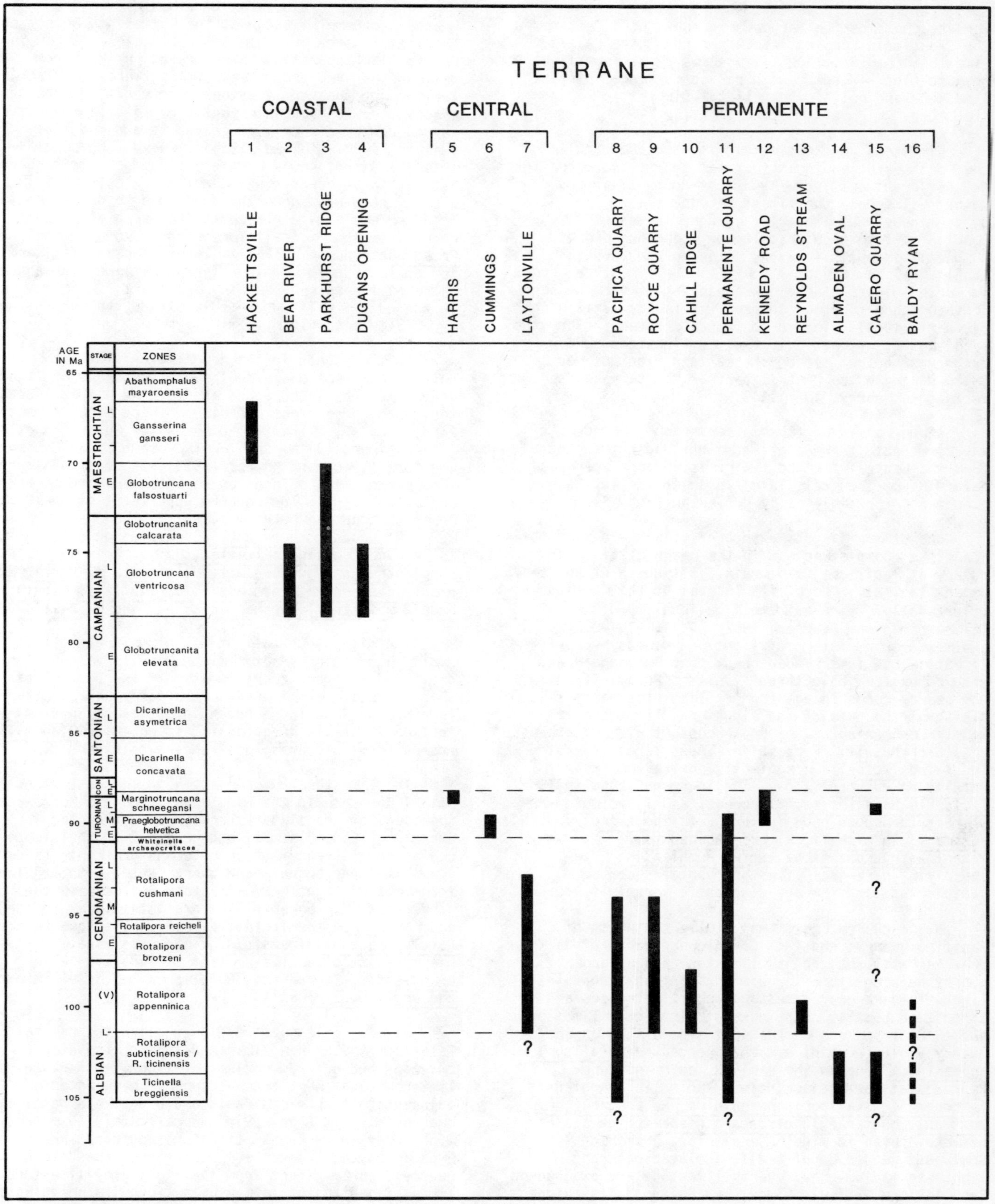

Figure 3. Stratigraphic distribution of limestone localities from Franciscan terranes (Coastal, Central, and Permanente) in northern California. Ages based on planktic foraminifers. Geologic time scale and planktic foraminifer zones as in Figure 2. Locality numbers refer to Figure 1 and Table 1.

small part of the rock assemblage, the Permanente terrane contains the largest exposures of limestone in the Franciscan Complex. Referred to as the Calera Limestone (Lawson, 1902), exposures range from light gray to black in color with rare occurrences of pink limestone and oolitic to pelletal-bioclastic limestone.

Limestone outcrops within the terrane parallel the northwest-trending structural fabric,and occur in two different settings. North of Route 17 at Los Gatos, California the limestone occurs as discontinuous but linearly arranged lenses of imbricated limestone, basalt and graywacke along subparallel faults (Lawson, 1914; Walker, 1950; Dibblee, 1966). South of Route 17 the limestone occurs as series of isolated blocks associated with the Ben Trovato shear zone and subparallel faults (Bailey and Everhart, 1964). The northern occurrences include the large bodies of complexly faulted limestone surrounded by graywacke that lie on the eastern flank of Monte Bello Ridge in the central part of the terrane (see Permanente Quarry, loc. 11).

Known ranges of foraminifers from limestone in the Permanente terrane are somewhat older than those from the Central terrane. At present, the age range extends from the early late Albian to the late Turonian or possibly early Coniacian (Fig. 3).

One of the largest exposures of limestone in the Franciscan Complex occurs in the beach cliffs and quarry at Pacifica, California, just north of Rockaway Beach (loc. 8). The cliffs located at the mouth of Calera Valley represent the type locality of the Calera Limestone (Lawson, 1902; 1914). The quarry exposes over 70 meters of light-gray to dark-gray limestone and interbedded chert that is bounded and sheared by low angle thrust faults. Foraminifers in the quarry range in age from at least the early late Albian at the base of the light-gray limestone to middle Cenomanian at the upper contact with overthrust vesicular basalt. Foraminifer zones range from at least the Ticinella breggiensi Zone to the early Rotalipora cushmani Zone. The lower horizons include Ticinella breggiensis, T. roberti and T. primula whereas the upper beds include Rotalipora brotzeni, R. cushmani, R. micheli, R. greenhornensis, Praeglobotruncana gibba and P. aumalensis. The underlying dark-gray limestone contains common radiolarians and few age-diagnostic foraminifers.

Royce Quarry, less than a mile southeast of Pacifica Quarry, consists of a lower quarry of dark-gray, crystalline, cherty limestone and an upper quarry adjacent to a large water tank that contains light- to dark-gray limestone, chert interbeds and associated basalt. Exposures of light-gray limestone and chert interbeds on the west side of the upper quarry yield formainifers that range the Rotalipora appenninica Zone to the early R. cushmani Zone. Representative species include Ticinella breggiensis, T. roberti, Rotalipora appenninica, R. brotzeni, R. cushmani, and Praeglobotruncana gibba for an age range of late Albian to middle Cenomanian. Further southeast, a series of small, isolated outcrops of light- to medium-gray, cherty limestone are exposed on Cahill Ridge, Fifield Ridge, and Spring Valley Ridge adjacent to the Crystal Springs Reservoir, San Mateo County, California (loc. 10). These localities here are grouped together as they all contain late Albian foraminifers of the Rotalipora appenninica Zone. Species include Ticinella roberti, Rotalipora ticinensis and Planomalina buxtorfi.

The largest exposures of limestone in the Franciscan Complex occur at the Permanente Quarry on the eastern side of Monte Bello Ridge near Cupertino, California (loc. 11). Extensive quarrying since 1900 has exposed over 130 meters of light-gray to black, faulted and sheared limestone with chert interbeds associated with volcanic rocks, radiolarian chert, and surrounded by a matrix of graywacke and mudstone (Walker, 1950; Dibblee, 1966). Due to the complex internal structure, a complete stratigraphic sequence has yet to be identified. Dating of individual blocks, however, shows the age range to be at least from the early late Albian to the Turonian. Zones range from the Ticinella breggiensis Zone to the Praeglobotruncana helvetica Zone. Foraminifers are common to abundant in the light-gray limestone and all the zone indicator species are present. Future work in the quarry will undoubtedly extend the ranges, especially into older ages.

Further south in the New Almaden district, collections were made from several outcrops of light-gray limestone interbedded and faulted with tuffaceous sediments, greenstone and ribbon chert along Kennedy Road (loc. 12). Identified by Bailey and Everhart (1964, figure 13) as the locality sampled by Cushman and Todd (1948) and Kupper (1955), the outcrops range in age from middle Turonian to late Turonian or early Coniacian. Zones include the upper part of the Praeglobotruncana helvetica Zone to the Marginotruncana schneegansi Zone based on species such as Praglobotruncana helvetica, P. gibba, Marginotruncana schneegansi, M. sigali, M. renzi, M. pseudolinneana, M. coronata, Dicarinella imbricata, D. hagni, Whiteinella archaeocretacea, and W. baltica.

Two miles southeast at Reynolds Stream (loc. 13), small outcrops of fractured and heavily veined light-gray limestone and chert located in the Ben Trovato shear zone are attributed to the early half of the late Albian Rotalipora appenninica Zone. Species include Ticinella breggiensis, T. roberti, Planomalina praebuxtorfi and Praeglobotruncana delrioensis.

The limestone outcrop shown on the east flank of Mt. El Sombroso in the Ben Trovato shear zone near the upper end of Guadalupe Reservoir by Bailey and Everhart (1964, Plate 1) and here called the Almaden Oval (loc. 14) consists of outcrops of light- to dark-gray limestone breccia and chert lenses with angular fragments of light-gray limestone enclosed in the darker matrix. Foraminifers are late Albian in age and range from the Ticinella breggiensis Zone to the Rotalipora subticinensis/R. ticinensis Zone. Characteristic species present include Ticinella breggiensis, T. roberti, Rotalipora subticiensis and Hedbergella delrioensis.

Several miles east and just below the Calero Reservoir in the New Almaden district, isolated outcrops and quarry workings of light- to dark-gray limestone and chert interbeds occur between or adjacent to the Calero fault and a southern parallel branching fault (loc. 15). Approximately 10 meters of light-gray to dark-gray limestone and chert are exposed in the quarry face here termed the Calero Quarry. Foraminifers from the light-gray limestone are assigned to the late Albian Ticinella breggiensis Zone on the basis of T. breggiensis, T. roberti, Hedbergella delrioensis and H. planispira. Isolated outcrops of light-gray limestone and chert interbeds 400 meters southeast of Calero Quarry contain foraminifers assigned to the early part of the Marginotruncana schneegansi Zone of Turonian age. The

Table 1. Franciscan limestone location data for samples shown on Figure 1.

No.	Sample No.	Terrane	Latitude/Longitude	Location
1	S84-11 to 12	Coastal	40° 28.43' N 124° 11.13' W	Hackettsville Taylor Peak 7 1/2' Quadrangle S:16, T:1N, R:1W
2	S84-13	Coastal	40° 26.90' N 124° 16.10' W	Bear River Capetown 7 1/2' Quadrangle S:23, T:1N, R:2W
3	MT-76-82-1 to 3, MT-76-82A to D, PRO-4A to D, PRO-5A to C	Coastal	40° 17.42' N 124° 08.80' W	Parkhurst Ridge Buckeye Mountain 7 1/2' Quadrangle S:14, T:2S, R:1W
4	S83-35 to 37	Central	39° 52.82' N 123° 48.72' W	Dugans Opening Piercy 7 1/2' Quadrangle S:U, T:23N, R:18W
5	S84-15 to 20	Central	40° 05.48' N 123° 38.07' W	Harris Harris 7 1/2' Quadrangle S:29, T:4S, R:5E
6	S84-8 to 10	Central	39° 50.07' W 123° 38.15' W	Cummings Leggett 7 1/2' Quadrangle S:16, T:23N, R:16W
7	S84-1 to 7	Central	39° 42.08' N 123° 29.13' W	Laytonville Laytonville 7 1/2' Quadrangle S:1, T:21N, R:15W
8	S84-77 to 80, 83 to 93, 101 to 102	Permanente	37° 36.80' N 122° 29.65' W	Pacifica Quarry Montara Mountains 7 1/2' Quadrangle S:U, T:4S, R:6W
9	S84-68 to 74	Permanente	37° 35.98' N 122° 29.27' W	Royce Quarry Montara Mountain 7 1/2' Quadrangle S:U, T:4S, R:6W
10	S84-103, 105 to 108	Permanente	37° 32.45' N 122° 24.43' W	Cahill Ridge Montara Mountain 7 1/2' Quadrangle S:33, T:4S, R:5W
11	S83-25 to 34, S84-143 to 170	Permanente	37° 19.13' N 122° 06.58' W	Permanente Quarry Cupertino 7 1/2' Quadrangle S:17, T:7S, R:2W
12	S84-52 to 58	Permanente	37° 13.15' N 121° 55.58' W	Kennedy Road Los Gatos 7 1/2' Quadrangle S:U, T:8S, R:1E
13	S84-50 to 51	Permanente	37° 12.03' N 121° 53.87' W	Reynolds Stream Los Gatos 7 1/2' Quadrangle S:U, T:8S, R:1E
14	S84-39 to 42, 44 to 45, 48	Permanente	37° 11.25' N 121° 42.50' W	Almaden Oval Los Gatos 7 1/2' Quadrangle S:U, T:8S, R:1E
15	S84-32 to 35, 38	Permanente	37° 09.80' N 121° 46.32' W	Calero Quarry Santa Teresa Hills 7 1/2' Quadrangle S:U, T:9S, R:2E
16	S84-22 to 31	Permanente	37° 08.90' N 121° 47.42' W	Baldy Ryan Santa Teresa Hills 7 1/2' Quadrangle S:U, T:9S, R:2E

stratigraphic relationship of these outcrops to those at the quarry is unknown. Species present include Marginotruncana schneegansi, M. sigali, Dicarinella hagni, Whitinella brittonensis, W. aprica and W. archaeocretacea.

The oolitic limestone from Baldy Ryan Canyon (loc. 16), formerly Longwall Canyon, is included in this reconnaissance study as it represents the thickest and best preserved example of shallow-water derived material known from the Franciscan Complex. No planktic foraminifers have yet been identified from the sequence, but the presence of agglutinated foraminifers with complex wall structures holds some promise for deriving an age. A description of the limestone and earlier attempts to date the sequence are given by Bailey and others (1964, p. 72; figures 38-40). The geographic location of Baldy Ryan Canyon located in the Ben Trovato shear zone makes an Albian age likely based on the age of simularily located limestone (locs. 13-14).

ORIGIN AND ENVIRONMENTAL SIGNIFICANCE OF FRANCISCAN LIMESTONE

Franciscan limestone records a complex depositional history that began in a variety of oceanic, low-latitude environments primarily as sediment capping ancient oceanic topographic highs such as aseismic ridges, rises, seamounts or plateaus. These localities may have been located on the eastern flank of the Pacific-Farallon ridge, but also may include depositional environments associated with other active spreading ridges and offridge volcanic settings. The interbedded limestone and vesicular basalt sequence on the western flank of Monte Bello Ridge in the Permanente terrane is regarded as seamount deposition. So too are the smaller limestone outcrops at Bear River in the Coastal terrane (Fig. 1, loc. 2), and at Laytonville (loc. 7) and Harris (loc. 5) in the Central terrane. The thicker limestone deposits at Permanente Quarry (loc. 11) and Pacifica Quarry (loc. 8) likewise may represent seamount or perhaps plateau environments. The much smaller deposits of red limestone that occur with pillow basalts, flow breccias and manganiferous beds such as at Hackettsville (Fig. 1, loc. 1) and Parkhurst Ridge (loc. 3) suggest deposition associated with spreading ridge or seamount volcanism and hydrothermal activity.

Accretion of progressively younger oceanic deposits is recorded by the older ages of the limestone in the Central and Permanente terranes and the younger ages in the Coastal terrane (Fig. 3). A similar pattern of age reduction is seen in the distribution of existing oceanic plateaus and seamounts in the western North Pacific that are oriented perpendicular to ancient spreading ridges as from Shatsky Rise to Hess Rise or from Magellan Plateau to the Line Islands (Winterer, 1976; Hilde and others, 1976; Thiede, Vallier and others, 1981).

In the Permanente terrane, an earlier period of volcanic activity associated with a low-oxygen or anoxic depositional environment is recorded by up to 75 m of dark-gray to black, recrystallized, bituminous limestone with recrystallized radiolarians and transported shallow-water biogenic debris interbedded with tuffaceous sediments. Intensification of an oxygen-minimum zone through increased surface productivity and/or the increased influx of organic debris associated with the downslope transport of volcaniclastic material presumably resulted in the accumulation of organic carbon. The dark limestone predates the overlying early late Albian light-gray limestone and is presumed to be early Albian to Aptian in age. The Franciscan anoxic sediments thus correlate with the late Barremian through Albian anoxic event (Schlanger and Jenkyns, 1976; Jenkyns, 1980). Similar Aptian/Barremian anoxic sediments are known from North Pacific oceanic plateaus at Shatsky Rise (Larson, Moberly and others, 1975), Manihiki Plateau (Schlanger, Jackson and others, 1976) and Hess Rise and the Mid-Pacific Mountains (Vallier and others, 1983).

Middle Cretaceous sedimentary and faunal sequences from Hess Rise and the Mid-Pacific Mountains, in particular, are very similar to those from the Permanente terrane and are thus important for drawing conclusions as to the paleogeographic and tectonic history of the respective areas. For example, the period of Albian or Aptian volcanic activity recognized in sediments from the Permanente terrane may correlate on a regional scale with generation of volcanic basement in the eastern Mid-Pacific Mountains (Winter, Ewing and others, 1973; Thiede, Vallier and others, 1981).

Deposition of Franciscan pelagic limestone took place mostly at bathyal water depths between 1000 and 1500 m and above the carbonate compensation depth (CCD) as evidenced from the rare but persistent calcareous and agglutinated benthic foraminifers. In the Permanente terrane, depositional environments deepened from pre-late Albian through Cenomanian but there is as yet no strong evidence for environments shoaler than upper bathyal depths (200 to 500 m). Occurrences of shallow-water biogenic debris and ooids in Albian and older(?) limestone such as at Baldy Ryan Canyon (Fig. 1, loc. 16) that indicate nearby shallow-water environments are regarded as displaced into the bathyal environment. Accumulation rates for the light-gray pelagic Calera Limestone with its matrix of calcareous nannofossils, foraminifers and radiolarians (Garrison and Bailey, 1967; Wachs and Hein, 1974, 1975) are noticeably lower than those from coeval sections from Hess Rise and the Mid-Pacific Mountains. Rates of approximately 3.0 m/m.y. at Pacifica Quarry (Fig. 1, loc. 8) are compared to somewhat less than 10 m/m.y. at Deep Sea Drilling Project Site 463 from the Mid-Pacific Mountains (Thiede, Vallier and others, 1981). The Pacific sites were not adjusted for compaction, however, nor were adjustments made for stylolitization, recrystallization or chertification that may have reduced the section at Pacifica Quarry by as much as 20 percent.

Planktic foraminiferal faunas from Franciscan limestone bear striking resemblance in diversity and species composition to those from low-latitude, tethyan environments in the Pacific Ocean (Douglas, 1973; Caron, 1975; McNulty, 1976; Boersma, 1981; Premoli Silva and Sliter, 1981). Using the faunal and sedimentologic data from western North Pacific plateau sites as a model, deposition of Franciscan pelagic limestone took place beneath or adjacent to the wide biologically productive equatorial divergence. North or south latitudinal limits of approximately 20° are presumed for middle Cretaceous deposition although biogeographic distribution patterns for Cretaceous planktic foraminifers in the North Pacific are as yet poorly known. Seamounts capped by thick limestone deposits probably formed south of the paleoequator and passed beneath the productive zone as part of northward plate migration. Seamounts originating within the productive zone or on the northern margin of the zone would have thinner deposits of limestone.

Paleomagnetic results from Franciscan limestone identified and continue to support the allochthonous nature of the accreted blocks. Data from two limestone localities at Laytonville in the Central terrane (Fig. 1, loc. 7) identified a 17° south latitude site of deposition for the late Albian to Cenomanian sequence (Alvarez and others, 1980). Results from Albian limestone in the Permanente Quarry (Fig. 1, loc. 11) gave a depositional paleolatitude of 21° north (Courtillot and others, 1983). Recent work on the limestone from Parkhurst Ridge (Fig. 1, loc. 3) indicates a depositional environment 22° shallower than the expected inclination calculated for this locality on the North American plate (Harbert and others, this volume).

SUMMARY

Preliminary dating of Franciscan limestone from northern California defines two broad groups of samples; an older group that ranges in age from late Albian to early Coniancian from the Central and Permanente terranes, and a younger group of early Campanian to middle Maestrichtian samples from the Coastal terrane. The age difference between the two groups resulted from the accretion of progressively younger oceanic materials removed from the subducted Farallon plate. Existing data indicates that Franciscan limestore was deposited at bathyal water depths on topographic highs that were located on the eastern flank of the Pacific-Farallon Ridge or other active spreading ridges. Deposition of the thicker sections took place beneath or adjacent to the Cretaceous equatorial productivity zone as the Farallon plate migrated northeastward. Seamounts capped by thick limestone deposits probably were formed south of the paleoequator. The limestone records the subsidence of the oceanic seamount deposits from pre-late Albian through Cenomanian. Dark bituminous limestone probably accumulated in an intensified oxygen-minimum zone during the late Barremian to Albian anoxic event.

ACKNOWLEDGMENTS

J. A. Barron and D. L. Jones are thanked for constructive criticism of the manuscript. Special thanks to I. Premoli Silva who awakened me to the wonders of thin-section identifications through numerous discussions and microscope sessions. I also thank M. L. McGann for aid in field reconnaissance, laboratory preparation and drafting.

REFERENCES CITED

Alvarez, Walter, Arthur, M. A., Fischer, A. G., Lowrie, William, Napoleone, Giovanni, Premoli Silva, Isabella, and Roggenthen, W. M., 1977, Upper Cretaceous-Paleocene magnetic stratigraphy at Gubbio, Italy, 5: Type section for the Late Cretaceous-Paleocene geomagnetic reversal time scale: Geological Society of America Bulletin, v. 88, p. 383-389.

Alvarez, Walter, Kent, D. V., Premoli Silva, Isabella, Schweickert, R. A., and Larson, R. A., 1980, Franciscan complex limestone deposited at 17° south paleolatitude: Geological Society of America Bulletin, v. 91, p. 476-484.

Arthur, M. A., and Fischer, A. G., 1977, Upper Cretaceous-Paleocene magnetic stratigraphy at Gubbio, Italy, 1: Lithostratigraphy and sedimentology: Geological Society of America Bulletin, v. 88, p. 367-389.

Atwater, Tanya, 1970, Implications of plate tectonics for the Cenozoic tectonic evolution of western North America: Geological Society of America Bulletin, v. 81, p. 3513-3535.

Bailey, E. H., and Everhart, D. L., 1964, Geology and quicksilver deposits of the New Almaden District, Santa Clara County, California: U.S. Geological Survey Professional Paper 360, 206 p.

Bailey, E. H., and Irwin, W. P., 1959, K-feldspar content of Jurassic and Cretaceous graywackes of northern coast ranges and Sacramento Valley, California: American Association of Petroleum Geologists Bulletin, v. 43, p. 2797-2809.

Bailey, E. H., Irwin, W. P., and Jones, D. L., 1964, Franciscan and related rocks, and their significance in the geology of western California: California Division of Mines and Geology Bulletin 183, 177 p.

Ben-Avraham, Z., Nur A., Jones, D. L., and Cox, Allan, 1981, Continental accretion: From oceanic plateaus to allochthonous terranes: Science, v. 213, p. 47-54.

Berry, K. D., 1982, New age determinations in Franciscan limestone blocks, northern California [abs.]: American Association of Petroleum Geologists Bulletin, v. 66, no. 10, p. 1685.

Blake, M. C., Jr., and Jayko, A. S., 1983, Geologic map of the Yolla Bolly-Middle Eel Wilderness and adjacent roadless areas: U.S. Geological Survey Miscellaneous Field Studies Map MF-1595-A, Scale 1:62,500.

Blake, M. C., Jr., and Jones, D. L., 1974, Origin of Franciscan melanges in northern California, in Dott, R. H., and Shaver, R. H., eds., Modern and ancient geosynclinal sedimentation: Society of Economic Paleontologists and Mineralogists Special Publication 19, p. 345-357.

_______ 1978, Allochthonous terranes in northern California?--A reinterpretation, in Howell, D. G., and McDougall, K., eds., Mesozoic paleogeography of the western United States: Society of Economic Paleontologists and Mineralogists, Pacific Section, Pacific Coast Paleogeography Symposium 2, p. 397-400.

_______ 1981, The Franciscan assemblage and related rocks in northern California: A reinterpretation, in Ernst, W. G., ed., The geotectonic development of California: Englewood Cliffs, New Jersey, Prentice-Hall, v. 1, p. 307-328.

Blake, M. C., Jr., Howell, D. G., and Jones, D. L., 1982, Preliminary tectonostratigraphic terranes map of California: U.S. Geological Survey Open-File Report 82-593, Scale 1:750,000.

Boersma, Anne, 1981, Cretaceous and early Tertiary foraminifers from Deep Sea Drilling Project Leg 62 sites in the central Pacific, in Thiede, Jorn, Vallier, T. L., and others, Initial Reports of the Deep Sea Drilling Project: Washington, U.S. Government Printing Office, v. 62, p. 377-396.

Caron, Michele, 1975, Late Cretaceous planktonic foraminifera from the northeastern Pacific: Leg 32 of the Deep Sea Drilling Project, in Larson, R. L., Moberly, R., and others, Initial Reports of the Deep Sea Drilling Project: Washington, U.S. Government Printing Office, v. 32, p. 719-724.

Champion, Duane, Gromme, Sherman, and Howell, D. G., 1980, Paleomagnetism of the Cretaceous Pigeon Point Formation and the inferred northward displacement of 2500 km for the Salinian block, California: EOS (American Geophysical Union, Transactions), v. 61, p. 948.

Church, C. C., 1952, Cretaceous Foraminifera from the Franciscan Calera limestone of California: Cushman Foundation for Foraminiferal Research Contributions, v. 3, p. 68-70.

Coney, P. J., 1978, Mesozoic-Cenozoic cordilleran plate tectonics, in Smith, Robert B., and Eaton, Gordon P., eds., Cenozoic tectonics and regional geophysics of the western cordillera: Geological Society of America Memoir 152, p. 33-50.

Coney, P. J., Jones, D. L., and Monger, J. W. H., 1980, Cordilleran suspect terranes: Nature, v. 288, p. 329-333.

Courtillot, V., Feinberg, H., McWilliams, M., and Cox, Allan, 1983, Franciscan complex limestone deposited at 21° north [abs.]: EOS (American Geophysical Union Transactions), v. 64, p. 686.

Cushman, J. A., and Todd, Ruth, 1948, A foraminiferal fauna from the New Almaden District, California: Cushman Foundation for Foraminiferal Research Contributions, v. 24, p. 90-98.

Dibblee, T. W., Jr., 1966, Geologic map and sections of the Palo Alto, 15' quadrangle, Santa Clara and San Mateo Counties, California: California Division of Mines and Geology Map Sheet 8, scale 1:62,500.

Dickinson, W. R., 1970, Relation of andesites, granites, and derivative sandstones to arc-trench tectonics: Reviews of Geophysics and Space Physics, v. 8, p. 813-860.

_______, 1981, Plate tectonics and the continental margin of California, in Ernst, W. G., ed., The geotectonic development of California: Englewood Cliffs, New Jersey, Prentice-Hall, p. 1-18.

Douglas, R. G., 1973, Planktonic foraminiferal biostratigraphy in the central North Pacific Ocean, in Winterer, E. L., Ewing, J. I., and others, Initial Reports of the Deep Sea Drilling Project: Washington, U.S. Government Printing Office, v. 17, p. 673-674.

Engebretson, D. C., Cox, Allan, and Gordon, R., 1981, Relative motions between oceanic and continental plates in the northern Pacific since the early Jurassic [abs.]: EOS (American Geophysical Union Transactions), v. 62, p. 1034.

Engebretson, D. C., and Cox, Allan, 1982, Plate interactions in the northeast Pacific since the Oxfordian [abs.]: EOS (American Geophysical Union Transactions), v. 63, p. 911.

Ernst, W. G., 1970, Tectonic contact between the Franciscan melange and the Great Valley sequence--crustal expression of a late Mesozoic Benioff Zone: Journal of Geophysical Research, v. 75, p. 886-901.

_______ 1984, California blueschists, subduction, and the significance of tectonostratigraphic terranes: Geology, v. 12, p. 436-440.

Garrison, R. E., and Bailey, E. H., 1967, Electron microscopy of limestone in the Franciscan Formation of California: U.S. Geological Survey Professional Paper 575-B, p. B94-B100.

Glaessner, M. F., 1949, Foraminifera of Franciscan (California): American Association of Petroleum Geologists Bulletin , v. 39, p. 2336-2340.

Hamilton, W., 1978, Mesozoic tectonics of the western United States, in Howell, D. G., and McDougall, K., eds., Mesozoic Paleogeography of the western United States: Society of Economic Paleontologists and Mineralogists, Pacific Section, Pacific Coast Paleogeography Symposium 2, p. 33-70.

Harbert, W. P., McLaughlin, R. J., and Sliter, W. V., 1984, Preliminary paleomagnetic and tectonic interpretation of the Parkhurst Ridge Limestone, Coastal Belt Franciscan, nothern California: This Volume.

Harland, W. B., Cox, A. V., Llewellyn, P. G., Pickton, C. A. G., Smith, A. G., and Walters, R., 1982, A geological time scale: New York, Cambridge University Press, 131 p.

Heirtzler, J. R., Dickson, G. O., Herron, E. M., Pitman, W. C. III, and LePichon, Xavier, 1968, Marine magnetic anomalies, geomagnetic field reversals, and motions of the ocean floor and continents: Journal of Geophysical Research, v. 73, p. 2119-2136.

Henderson, L. J., Gordon, R. G., and Engebretson, D. C., 1984, Mesozoic aseismic ridges on the Farallon plate and southward migration of shallow subduction during the Laramide Orogeny: Tectonics, v. 3, no. 2, p. 121-132.

Hilde, T. W. C., Isezaki, Nobuhiro, and Wageman, J. M., 1976, Mesozoic sea-floor spreading in the North Pacific, in Sutton, G. H., Manghnani, M. H., and Moberly, R., The Geophysics of the Pacific Ocean basins and its margin: American Geophysical Union, Geophysical Monograph Series, v. 19, p. 205-226.

Hilde, T. W. C., Uyeda, S., and Kroenke, L., 1977, Evolution of the western Pacific and its margin: Tectonophysics, v. 38, p. 145-165.

Howell, D. G., and Vedder, J. G., 1978, Late Cretaceous paleogeography of the Salinian block, California, in Howell, D. G., and McDougall, K., eds., Mesozoic paleogeography of the western

United States: Society of Economic Paleontologists and Mineralogists, Pacific Section, Pacific Coast Paleogeography Symposium 2, p. 523-534.

Howell, D. G., Crouch, J. K., Greene, H. G., McCulloch, D. S., and Vedder, J. G., 1980, Basin development along the late Mesozoic and Cainozoic California margin: A plate tectonic margin of subduction, oblique subduction and transform tectonics: Special Publication of the International Association of Sedimentologist, no. 4, p. 43-62.

Ingersoll, R. V., 1978, Submarine fan facies of the upper Cretaceous Great Valley sequence, northern and central California: Sedimentary Geology, v. 21, p. 205-230.

_______ 1979, Evolution of the late Cretaceous forearc basin, northern and central California: Geological Society of America Bulletin, v. 90, p. 813-826.

Irwin, W. P., 1957, Franciscan group in coast ranges and its equivalents in Sacramento Valley, California: American Association of Petroleum Geologists Bulletin, v. 41, p. 2284-2297.

_______ 1960, Geologic reconnaissance of the northern coast ranges and Klamath Mountains, California, with a summary of the mineral resources: California Division of Mines and Geology Bulletin 179, 80 p.

_______ 1972, Terranes of the western Paleozoic and Triassic belt in the southern Klamath Mountains, California, in Geological Survey Research 1972: U.S. Geological Survey Professional Paper 800-C, p. C103-C111.

Jenkyns, H. C., 1980, Cretaceous anoxic events: from continents to oceans: Journal of the Geological Society of London, v. 137, p. 171-188.

Jones, D. L., Blake, M. C., Jr., Bailey, E. H., and McLaughlin, R. J., 1978, Distribution and character of upper Mesozoic subduction complexes along the west coast of North America: Tectonophysics, v. 47, p. 207-222.

Kupper, Klaus, 1955, Upper Cretaceous foraminifera from the "Franciscan Series", New Almaden District, California: Cushman Foundation for Foraminiferal Research Contributions, v. 6, p. 112-118.

Lancelot, Yves, 1978, Relations entre evolution sedimentaire et tectonique de la plaque pacifique depuis le cretace inferieur: Societe Geologue de France Memoire 134, 40 p.

Larson, R. L., and Chase, C. G., 1972, Late Mesozoic evolution of the western Pacific Ocean: Geological Society of America Bulletin, v. 83, p. 3627-3643.

Larson, R. L., Moberly, R., and others, 1975, Initial Reports of the Deep Sea Drilling Project: Washington, U.S. Government Printing Office, v. 32, p. 980.

Lawson, A. C., 1902, A geological section of the mddle coast ranges of California: Science, v. 15, p. 415-416.

_______ 1914, Description of the San Francisco district, Tamalpais, San Francisco, Concord, San Mateo, and Hayward quadrangles, California: U.S. Geological Survey Geologic Atlas, Folio 193, 24 p.

Loeblich, A. R., Jr., and Tappan, Helen, 1961, Cretaceous planktonic foraminifera: Part 1--Cenomanian: Micropaleontology, v. 7, p. 257-304.

Longoria, J. F., 1974, Stratigraphic, morphologic and taxonomic studies of Aptian planktonic foraminifera: Revista Espanola de Micropaleontologia, (numero extraordinario, 1974), p. 134.

Lowrie, William, Alvarez, Walter, Premoli Silva, Isabella, and Monechi, Simonetta, 1980, Lower Cretaceous magnetic stratigraphy in Umbrian pelagic carbonate rocks: Geophysical Journal of the Royal Astronomical Society, v. 60, p. 263-281.

McLaughlin, R. J., Kling, S. A., Poore, R. Z., McDougall, Kristin, and Beutner, E. C., 1982, Post-middle Miocene accretion of Franciscan rocks, northwestern California: Geological Society of America Bulletin, v. 93, p. 595-605

McNulty, C. L., 1976, Cretaceous foraminiferal stratigraphy, DSDP Leg 33, Holes 315A, 316, 317A, in Schlanger, S. O., Jackson, E. D., and others, Initial Reports of the Deep Sea Drilling Project: Washington, U.S. Government Printing Office, v. 33, p. 369-381.

Morgan, W. J., 1972, Deep mantle convection plumes and plate motions: American Association of Petroleum Geologists Bulletin, v. 56, p. 203-213.

Pitman, W. C., III, Larson, R. L., and Herron, E. M., 1974, The age of the ocean basins: Geological Society of America Map and Chart Series MC-6, Scale 1:39,000,000 at the equator.

Premoli Silva, Isabella, 1977, Upper Cretaceous-Paleocene magnetic stratigraphy at Gubbio, Italy, 2: Biostratigraphy: Geological Society of America Bulletin, v. 88, p. 371-374.

Premoli Silva, Isabella, and Sliter, W. V., 1981, Cretaceous planktonic foraminifers from the Nauru Basin, Leg 61, Site 462, western equatorial Pacific, in Larson, R. L., and others, Initial Reports of the Deep Sea Drilling Project: Washington, U.S. Government Printing Office, v. 61, p. 423-437.

Robaszynski, Francis, and Caron, Michele, eds., 1979, Atlas, des foraminiferes planctoniques du cretace moyen (mer boreale et tethys): Cahiers de Micropaleontologie, 2 volumes, 366 p.

Robaszynski, Francis, Caron, Michele, Gonzalez Donoso, J. M., and Wonders, A. H., eds., 1983-1984 [i.e., 1984], Atlas of Late Cretaceous globotruncanids: Revue de Micropaleontologie, v. 26, no. 3-4, 305 p.

Schlanger, S. O., and Jenkyns, H. C., 1976, Cretaceous oceanic anoxic events: causes and consequences: Geologie en Mijnbouw, v. 55, no. 3-4, p. 179-184.

Schlanger, S. O., Jackson, E. D., and others, 1976, Initial Reports of the Deep Sea Drilling Project: Washington, U.S. Government Printing Office, v. 33, 973 p.

Sigal, Jacques, 1977, Essai de zonation du Cretace mediterraneen a laide des foraminiferes planctoniques: Geologie, Mediterraneen, v. 4, p. 99-108.

Sliter, W. V., 1984, Cretaceous low-latitude pelagic limestone from the Franciscan complex, California, and the Sixes River terrane, Oregon [abs.]: Geological Society of America, Abstracts with Programs, v. 16, p. 333.

Thalmann, H. E., 1942, Globotruncana in the Franciscan limestone, Santa Clara County, California [abs.]: Geological Society of America Bulletin, v. 53, p. 1838.

_______ 1943, Upper Cretaceous age of the "Franciscan" limestone near Laytonville, Mendocino County, California [abs.]: Geological Society of America Bulletin, v. 54, p. 1827.

Thiede, Jorn, Vallier, T. L., and others, 1981, Initial Reports of the Deep Sea Drilling Project: Washington, U.S. Government Printing Office, v. 62, 1120 p.

Vallier, T. L., Dean, W. E., Rea, D. K., and Thiede, Jorn, 1983, Geologic evolution of Hess Rise, central North Pacific Ocean: Geological Society of America Bulletin, p. 1289-1307.

Wachs, Daniel, and Hein, J. R., 1974, Petrography and diagenesis of Franciscan limestones: Journal of Sedimentary Petrology, v. 44, p. 1217-1231.

_______ 1975, Franciscan limestones and their environments of deposition: Geology, v. 3, p. 29-33.

Walker, G. W., 1950, The Calera Limestone in San Mateo and Santa Clara Counties, California: California Division of Mines and Geology, Special Report 1-B, 8 p.

Winterer, E. L., 1973, Sedimentary facies and plate tectonics of equatorial Pacific: American Association of Petroleum Geologists Bulletin, v. 57, p. 265-282.

_______ 1976, Bathymetry and regional tectonic setting of the Line Islands Chain, in Schlanger, S. O., Jackson, E. D., and others, Initial Reports of the Deep Sea Drilling Project: Washington, U.S. Government Printing Office, v. 33, p. 731-747.

Winterer, E. L., Ewing, J. I., and others, 1973, Initial Reports of the Deep Sea Drilling Project: Washington, U.S. Government Printing Office, v. 17, 930 p.

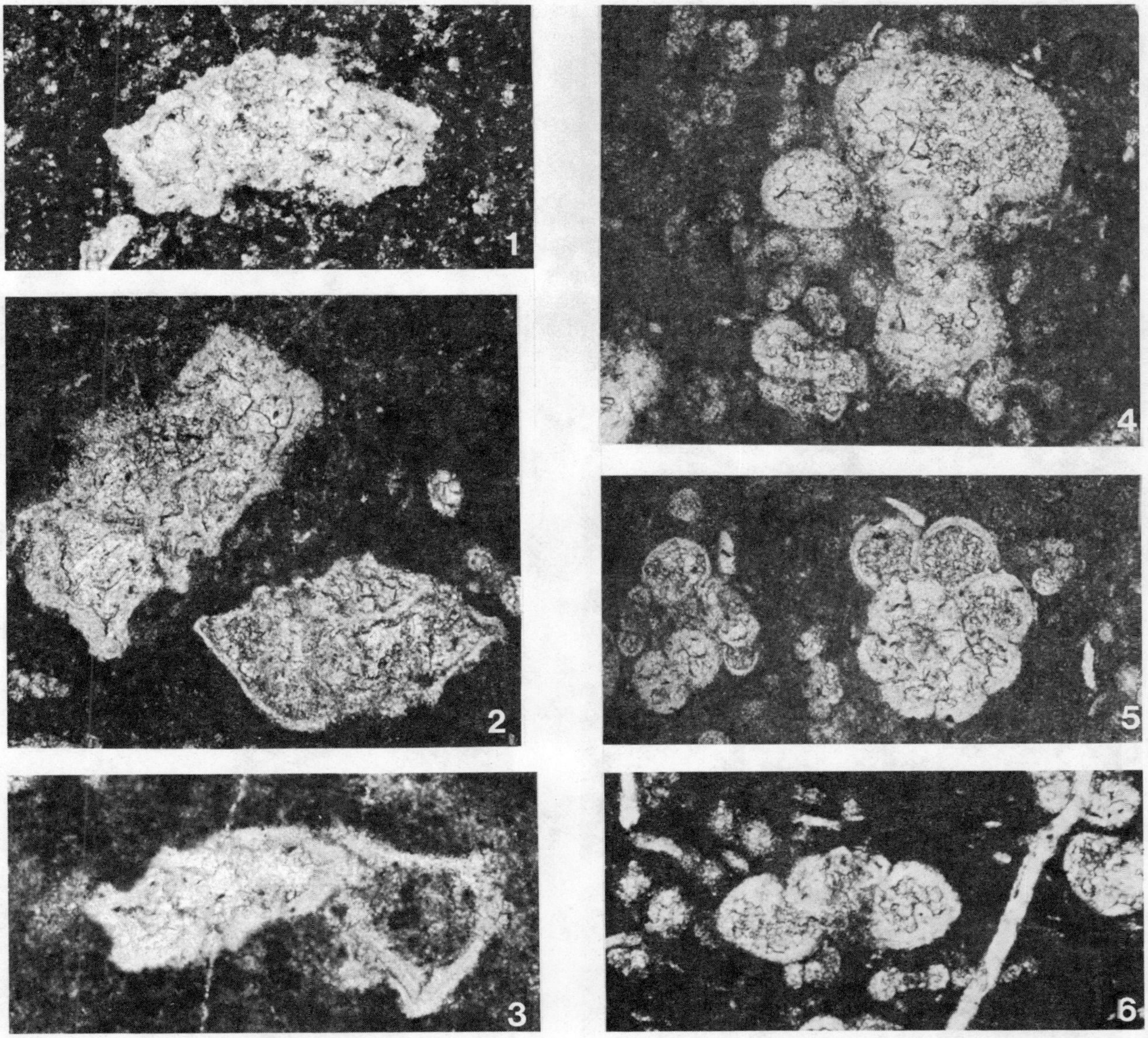

PLATE 1

Figures 1-3 from locality 2 of the Coastal terrane, sample S84-13. All X114.

1. *Globotruncana arca* (Cushman)
2. *Globotruncana linneiana* (d'Orbigny) on left, *Globotruncanita stuartiformis* (de Lapparent) on right.
3. *Globotruncana ventricosa* White.

Figures 4-6 from locality 11 of the Permanente terrane, sample S84-148. All X114.

4. *Ticinella breggiensis* (Gandolfi), peripheral view.
5. *Ticinella roberti* (Gandolfi), spiral view.
6. *Rotalipora subticinensis* (Gandolfi).

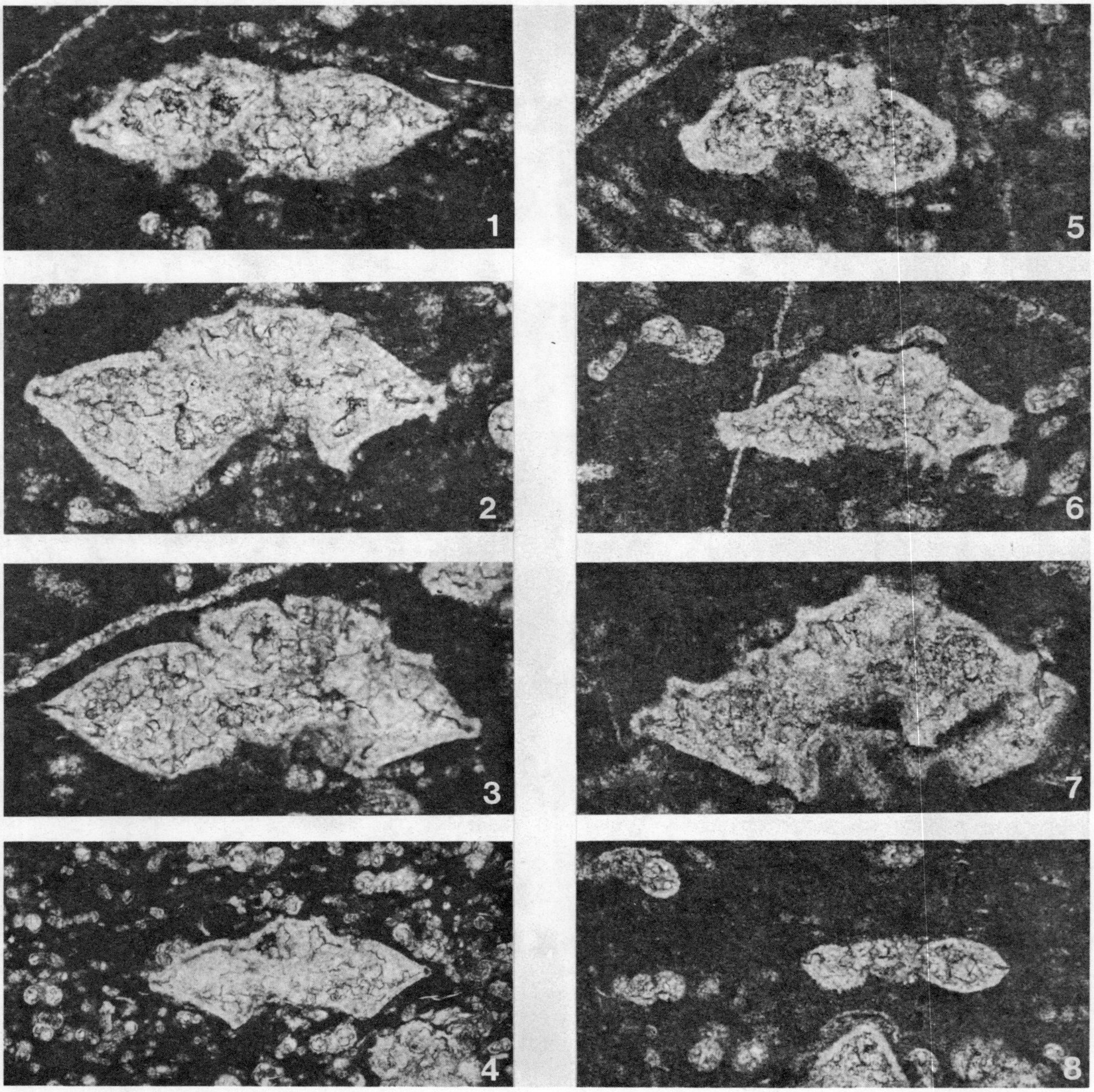

PLATE 2

Figures 1-4 from locality 11 of the Permanente terrane, sample S84-165. Figures 1-3, X114; Figure 4, X57.

1. Rotalipora appenninica (Renz).
2. Rotalipora brotzeni (Sigal).
3. Rotalipora cushmani (Morrow).
4. Rotalipora greenhornensis (Morrow).

Figures 5-8 from locality 5 of the Central terrane, sample S84-15. All X114.

5. Dicarinella imbricata (Mornod).
6. Marginotruncana schneegansi (Sigal).
7. Marginotruncana sigali (Reichel).
8. Hedbergella flandrini Porthault.

TECTONIC WEDGING ASSOCIATED WITH EMPLACEMENT OF THE FRANCISCAN ASSEMBLAGE, CALIFORNIA COAST RANGES

by

C. M. Wentworth, M. C. Blake, Jr., D. L. Jones, A. W. Walter, and M. D. Zoback*
U.S. Geological Survey
345 Middlefield Road, Menlo Park, CA 94025

*now at Department of Geophysics
Stanford University
Stanford, CA 94305

ABSTRACT

Structure along the eastern margin of the California Coast Ranges, where the Franciscan assemblage and Great Valley sequence are juxtaposed, seems to be marked by tectonic wedges. These are allochthonous masses thrust between underlying basement and overlying rocks peeled up by the wedge. Structure inferred from seismic reflection and refraction profiles in the southern Coast Ranges requires that the buried east boundary of Franciscan rocks is the tip of a tectonic wedge. In this interpretation, the Coast Range thrust, long interpreted to be a fossil subduction zone that juxtaposes Franciscan assemblage and Great Valley sequence, is the roof thrust of the Franciscan wedge.

In the northern Coast Ranges, exposed structural relations can best be explained by several imbricate thrust wedges that have been driven eastward over Klamath-Sierran basement. The easternmost wedge tip is evident in reflection profiles. The inferred roof thrusts in the Great Valley sequence presumably merge northward into exposed, northwest trending strike-slip faults alined with the southern Klamath boundary. Eastward movement of the tectonic wedges has produced many tens of kilometers of offset on these tear faults. The Coast Range ophiolite has been thrust eastward far beneath the Great Valley sequence and onto Klamath-Sierran basement. Several terranes of Franciscan rocks have been thrust beneath the Coast Range ophiolite across the same basement. Subsequent east-directed reverse faulting in the late Cenozoic has offset the Coast Range thrust.

INTRODUCTION

For twenty years the structural relation between the Franciscan assemblage and adjacent Great Valley sequence has been considered known. The two units are juxtaposed by a major thrust, the Coast Range thrust, which was inferred to extend eastward to depth through the crust and, in the context of plate tectonics, represented a subduction zone suture. The need for some structural explanation for the adjacence of these distinctly different but largely coeval rocks was explicitly posed by Bailey, Irwin and Jones (1964) in their landmark paper on the Franciscan assemblage. They proposed alternative explanations, of which Porter Irwin's thrust hypothesis has since been generally adopted.

The evidence for the Coast Range thrust comes from structural and stratigraphic relations determined from geologic mapping: in addition to the structural and lithologic contrasts between the two coeval terranes, it includes such things as the fault boundary between them, the presence of the Coast Range ophiolite at the base of the Great Valley sequence, outliers of Great Valley sequence and Coast Range ophiolite within the Franciscan terrane, and the window of Franciscan terrane surrounded by Great Valley sequence in the central Diablo Range.

The view that the Coast Range thrust extends eastward to depth through the crust as a subduction zone suture is less specifically founded. Fundamentally, extending the thrust down through the crust seemed to make sense, particularly in the context of plate tectonics (Hamilton, 1969). In his original hypothesis (Bailey, Irwin and Jones, 1964; fig. 35; and Irwin, 1964), Irwin carried the thrust down along the serpentinite that is typically exposed at the contact between Franciscan and Great Valley sequence and suggested that this serpentinite connected eastward in the subsurface to a mantle ridge beneath the Great Valley. This interpretation was stimulated, in part, by interpretation of magnetic anomalies attributed to the serpentinite (Irwin and Bath, 1962). After the plate tectonic model had been established, much of the serpentinite mapped between the Franciscan and Great Valley sequence was recognized by Bailey, Blake, and Jones (1970) to be Mesozoic oceanic crust and upper mantle (the Coast Range ophiolite) on which the Upper Jurassic part of the Great Valley sequence had been deposited. They placed the Coast Range thrust at the base of the Coast Range ophiolite and described it, in effect, to be a subduction zone suture along which the Franciscan assemblage had been thrust eastward beneath the Coast Range ophiolite and overlying Great Valley sequence.

In late 1981, the U.S. Geological Survey began a project to trace the Coast Range thrust eastward to depth using deep seismic reflection profiling and detailed refraction surveys. Initial emphasis on a profile across the central Diablo Range and northern San Joaquin Valley was expanded by occurrence of the M 6.7 earthquake at Coalinga in May, 1983. Although data processing is not yet complete, present results seem to require major revision of our view of structural relations along the Mesozoic oceanic-continental boundary in central California. Tectonic wedging, rather than subduction, seems to be the dominant tectonic style at the east margin of the Franciscan assemblage (Wentworth, Walter, Zoback, and Blake, 1983).

As field assistants to Ed Bailey in 1959 and 1960 during his mapping of the Skaggs 15-minute quadrangle, Blake and Wentworth helped grapple with the task of subdividing the Franciscan assemblage into useful, mappable rock units and puzzled at the fact that most contacts seemed to be faults. Vast progress in understanding the Franciscan assemblage has been made since then, but geologic mapping, essential though it is,

in Blake, M.C. Jr., ed., 1984, **Franciscan Geology of Northern California:** Pacific Section S.E.P.M., Vol. 43, p. 163-173

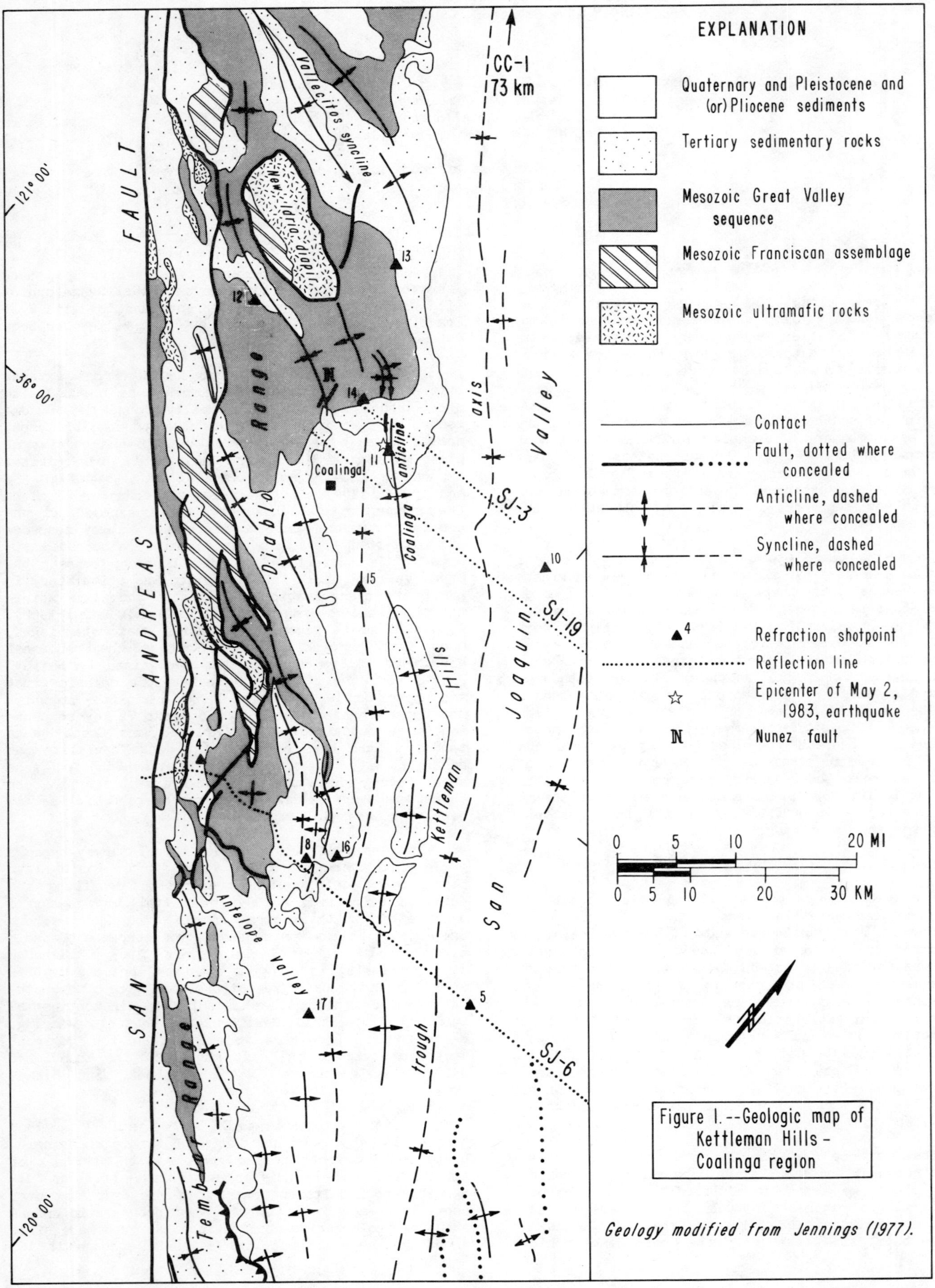

Figure 1.--Geologic map of Kettleman Hills - Coalinga region

has left important ambiguities in subsurface relations. To understand these structural relations, we must explore the subsurface directly. Seismic reflection and detailed refraction studies seem especially effective, as indicated by the emerging results of this study. Were Ed still with us, he would undoubtedly caution that surprises were still to come.

DATA

Seismic reflection profiles extending westward across the western Great Valley to or into the Coast Range have been obtained at four latitudes representing much of the length of the Great Valley (see the Geologic Map of California, Jennings, 1977, and figures 1 and 6 for regional geology and locations of places cited).

An aggregate of 260 km of 24-fold, 6-s commercial VIBROSEIS profile, 160 km recorrelated to 12 s, was purchased by the Geological Survey. (a) One long line (SJ-6) extends westward across the southern San Joaquin Valley, Kettleman South Dome, and the southern Coast Ranges (lat. 35.75° N); (b) two shorter lines (SJ-19 and SJ-3) cross Coalinga anticline (lat. 36.18° and 36.26° N). (c) Further north, a 15-s VIBROSEIS line with nominal fold of 133 was collected under contract using an 800-channel, sign-bit system. This 140-km-long, composite line (CC-1, 2) extends westward across the northern San Joaquin Valley and into the Coast Range at Garzas Creek north of Los Banos (lat. 37.25° N). (d) In the northern Sacramento Valley an aggregate of 160 km of commercial explosion reflection data, 24-fold and 5 s long, was purchased. Two of these lines extend westward to the edge of the Coast Range, S-45 north of Black Butte reservoir and east of Paskenta (lat. 39.86° N) and S-51 near the north end of Sites anticline (lat. 39.58° N).

Seismic refraction profiles, coordinated with the reflection profiles, have been collected and interpreted in at least preliminary form at the two southern crossings. The refraction surveys typically involved 900 kg shots spaced 30 km apart and recorded by 100 portable stations. Interpretation was carried out using two-dimensional ray tracing to match arrival times.

Velocity structure along the reflection lines is determined largely from normal-moveout analysis of reflection data in the simply layered rocks of the Great Valley, and, where possible, from analysis of refraction data in the deeper or more complicated parts of the cross sections. Stratigraphic identities in the cross sections are based on correlation with stratigraphy at the surface and in drill holes, correlation along the reflection record sections, and velocities characteristic of particular units.

Figure 1. Geologic map of the Kettleman Hills-Coalinga region. Locations of seismic reflection lines SJ-3, -6, and -19 are shown relative to the Coalinga-Kettleman Hills anticlinal trend, the uplifted Franciscan and Great Valley sequence exposed in the Diablo Range and the epicenter of the 1983 Coalinga main-shock.

COAST RANGE-GREAT VALLEY BOUNDARY

An abrupt and fundamental structural change occurs across the boundary between the California Coast Ranges and the Great Valley. Beneath the valley, a regional unconformity on crystalline rock dips gently westward at the base of similarly dipping Upper Cretaceous and Cenozoic strata and reaches depths of 5-10 km at the west side of the valley. The thickness of Jurassic and Cretaceous strata beneath the west side of the Great Valley for most of its length ranges from about 1.5 to 4 km. Immediately to the west in the Coast Ranges, the Jurassic and Cretaceous Great Valley sequence stands nearly on edge, dips steeply eastward toward the valley, and has a minimum stratigraphic thickness that ranges from 3 to 8 km in the southern Coast Ranges to more than 15 km in the northern Coast Ranges. The western, basal boundary of this upturned Great Valley sequence abuts the vast Franciscan assemblage, separated in the northern Coast Ranges by as much as 5 km of Coast Range ophiolite. Locally, it can be demonstrated that basal Great Valley sequence of late Jurassic age was deposited on the ophiolite (Bailey, Blake, and Jones, 1970), although in most places the contact is faulted.

The Coast Range-Great Valley boundary is thus marked by abrupt westward thickening and upturning of the Great Valley sequence as a narrow border along the east margin of the Franciscan assemblage. The persistence of this relation along the 600-km-length of this boundary must mark some fundamental underlying cause.

STRUCTURE

The subsurface structure at the Coast Range-Great Valley boundary that the reflection and refraction profiles indicate is fundamentally similar at all four latitudes for which data are available. Cretaceous and Cenozoic strata and the surface of underlying crystalline basement dip gently westward across the Great Valley, broken only locally by faults. Where the sedimentary strata turn upwards at the west side of the valley, the basement surface continues or increases its westward dip. An eastward-thinning wedge of material of intermediate velocity (5.7 km/s) lies between this basement (6.5 km/s) and the upturned sedimentary strata (less than 5 km/s) with its eastern tip located beneath or just east of the Coast Range-Great Valley boundary.

Kettleman South Dome

Reflection line SJ-6 and the coincident refraction line across South Dome at the southeast end of Kettleman Hills was interpreted (fig. 2) in the context of the Coalinga earthquake (Wentworth, Walter, Bartow, and Zoback, 1983). The west-dipping basement drops from a depth of about 9 km to a depth of 15 km while the strata above are repeated and folded by eastward-directed thrusting. The presence of shallow Franciscan rock underlying Great Valley sequence at the west side of the cross section is inferred from (a) nearby surface relations, (b) appropriate depth of reflection and refraction boundaries relative to the nearby exposed 3-km thickness of Great Valley sequence, and (c) a seismic velocity similar to that determined for Franciscan elsewhere in the Diablo Range -- 5.4-5.9 km/s below a depth of about 3 km (Walter and Mooney, 1982).

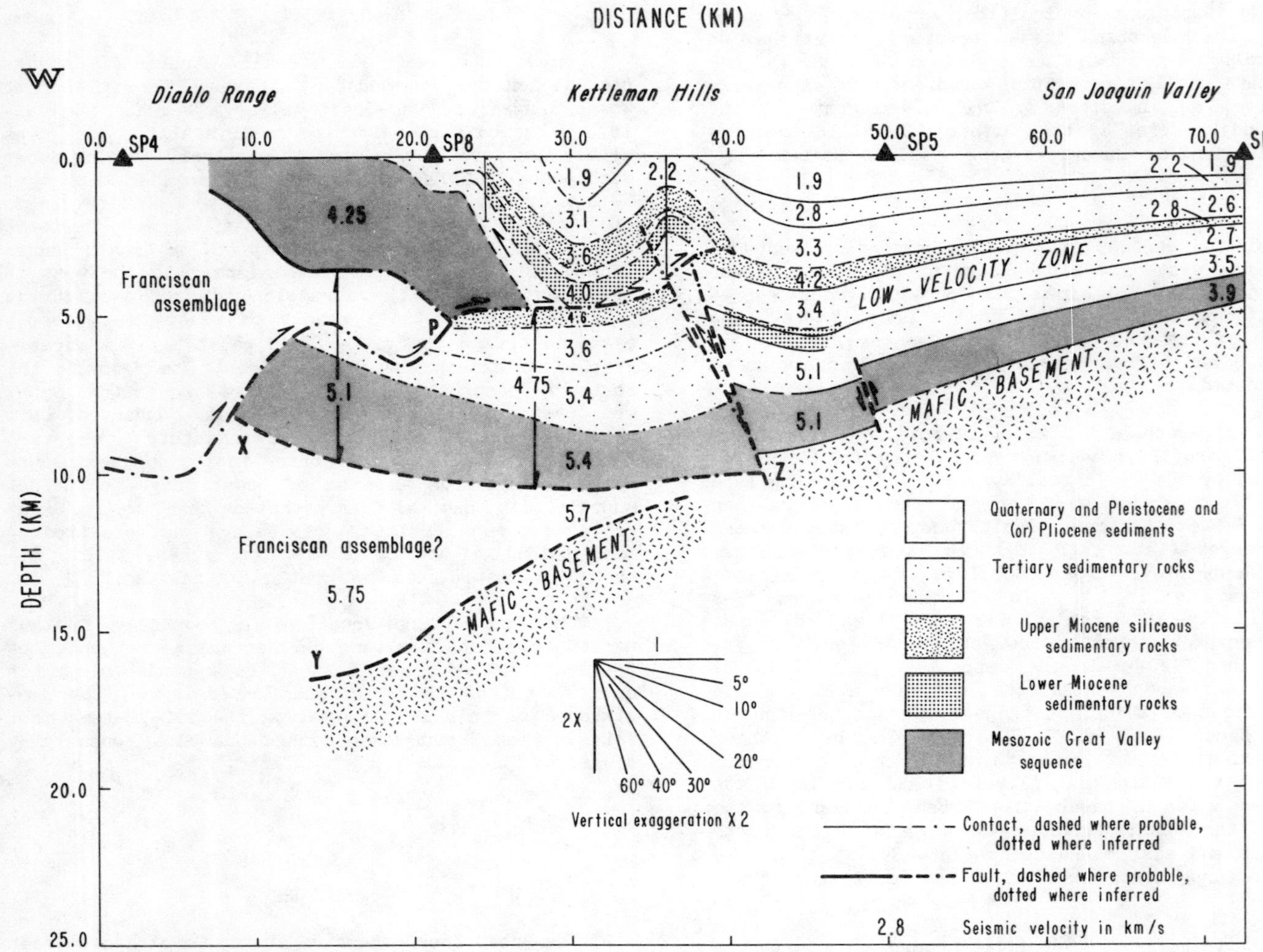

Figure 2. Interpretation of reflection profile SJ-6. Taken from Wentworth, Walter, Bartow, and Zoback (1983). Basement and Franciscan based in part on associated refraction velocity model. Note the two-fold vertical exaggeration.

Beneath and east of this shallow Franciscan mass, a deeper wedge of rock with boundaries that are partially defined by both the SJ-6 reflection and refraction profiles overlies the 6.5 km/s basement. This wedge has a refraction-defined velocity of 5.6-5.9 km/s and is overlain by slower rock. It seems best interpreted as Franciscan rock overlain by Great Valley sequence, from which the shallow Great Valley sequence and Franciscan assemblage have been offset by eastward-directed thrusting.

Reinterpretation now underway may change some details in figure 2, particularly the shape and position of the young, west-dipping thrust. Conflict in the depth of the basement surface beneath the syncline between the reflection and refraction interpretations (12 and 15 km, respectively) has not been resolved (see Wentworth, Walter, Bartow, and Zoback, 1983, figs. 2 and 3).

Coalinga

Reflection lines SJ-19 and SJ-3 and the intervening refraction line extend from the San Joaquin Valley westward across Coalinga anticline, beneath which the 1983 Coalinga earthquake occurred, and onto the west limb of Pleasant Valley syncline (fig. 1). These lines were obtained to help determine the origin of the Coalinga earthquake (Wentworth, Zoback, and Bartow, 1984; Walter, 1984). SJ-3 contains less detail than SJ-19 and serves largely to support continuity of SJ-19 relations along strike, albeit with changing amplitude of Coalinga anticline in the shallow section. Interpretations of reflection line SJ-19 and the associated refraction line are presented in figure 3. The interpretation of SJ-19 below the top of the Upper Cretaceous Panoche Formation is critically dependent on the refraction model.

About 1.5 km of upper Cretaceous strata (Moreno and Panoche Formations) are present above crystalline basement in the San Joaquin Valley 40 km east of Coalinga anticline, compared to an exposed section of Great Valley sequence that is about 8 km thick in the Diablo Range to the west.

The surface of crystalline basement (6.3-6.6 km/s) drops westward to a depth of 14 km beneath the Diablo Range at the west edge of the section, whereas in the same traverse the Cretaceous and Cenozoic strata rise westward through Coalinga anticline and on into exposures in the Diablo Range. The refraction model shows a prominent wedge of 5.7-6.0 km/s material separating the deep basement from slower, overlying strata. In the reflection record, the top of this wedge coincides with a prominent reflection within a

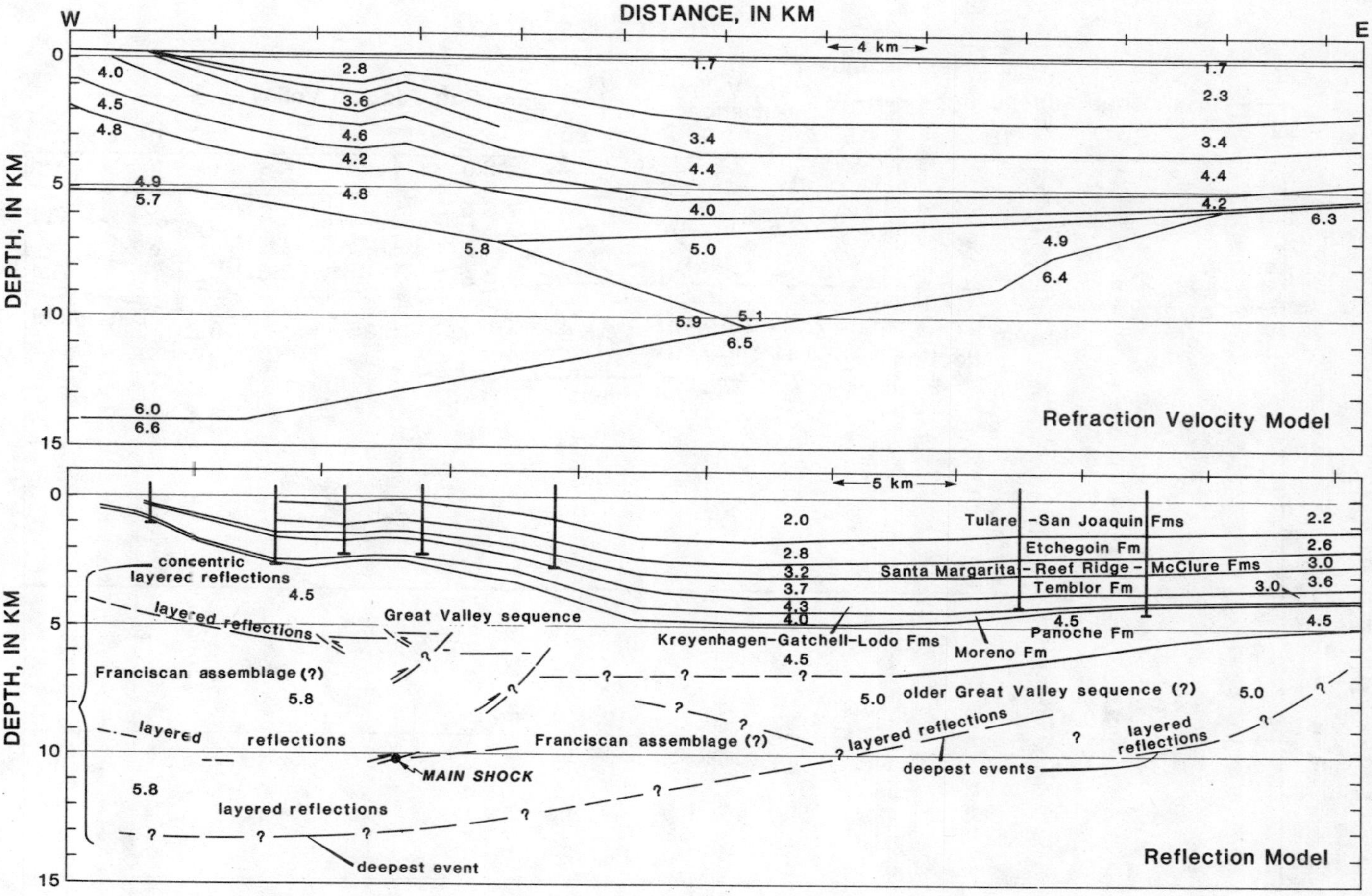

Figure 3. Refraction velocity model and associated reflection interpretation along profile SJ-19. Taken from Wentworth, Zoback, and Bartow (1984). Velocities above 5 km in reflection model are based on reflection interval velocities, those below are taken from refraction model. "Deepest event" marks deepest clear reflection in 12-s reflection record. Coalinga main-shock hypocenter and southwest-dipping focal plane projected onto this W-E section parallel to the anticline axis. Gatchell is a name of local useage.

band of concentric reflections that extends for 3 s below the base of Tertiary strata. The deepest reflections in this band fall near the base of the refraction wedge. Although the regular layering implied by the concentric reflections is unexpected in Franciscan rocks, the seismic velocity of the wedge is similar only to Franciscan rocks amongst those rocks exposed at the surface in the region. Partially serpentinized ultramafic rock could also be involved, but the velocity is too fast to represent Great Valley sequence.

Garzas Creek

Structural interpretation at Garzas Creek (fig. 4) is based principally on the 130-fold reflection profile CC-1 and the preexisting refraction profile along the axis of the Diablo Range (Walter and Mooney, 1982), which crosses the west end of profile CC-1. Refraction profiles have been recorded both along CC-1 and across it in the San Joaquin Valley, but only preliminary interpretations have been made.

The simple, west-dipping basement surface beneath the San Joaquin Valley reaches a depth of about 6 km at the structural trough near the valley margin (km 36, fig. 4), where the Upper Cretaceous section is 4 km thick. Farther west, the section bends upward slightly and extends beneath the leading edge of the Diablo Range (km 24, fig. 4), where steeply east-dipping Great Valley sequence is exposed at the surface. A prism of 5.6-5.8 km/s material, inferred to be Franciscan rocks from the reflection interval velocity, lies between this slightly upturned Great Valley sequence and underlying basement.

The Great Valley sequence exposed at the east edge of the Diablo Range dips steeply eastward and is 8 km thick, which is twice as thick as the Upper Cretaceous section that dips gently eastward beneath the range front (km 14, fig. 4). At the surface, just north of the mouth of Garzas Creek where the reflection line runs, the attitude of the Eocene Kreyenhagen formation defines an abrupt hinge between 60° easterly dips in the range and 10° dips in front of the range. This hinge, faulted or not, must extend down and westward beneath the range to account for the penetration of flat dips beneath the range front in the reflection record.

A structure younger than the Coast Range thrust, the Ortigalita fault, raises Franciscan rocks against Great Valley sequence, probably through eastward-directed thrust or reverse faulting. In a fashion similar to that inferred in the subsurface at Kettleman South Dome (fig. 2), this faulting should repeat the base of the Great Valley sequence, now eroded from the crest of the range.

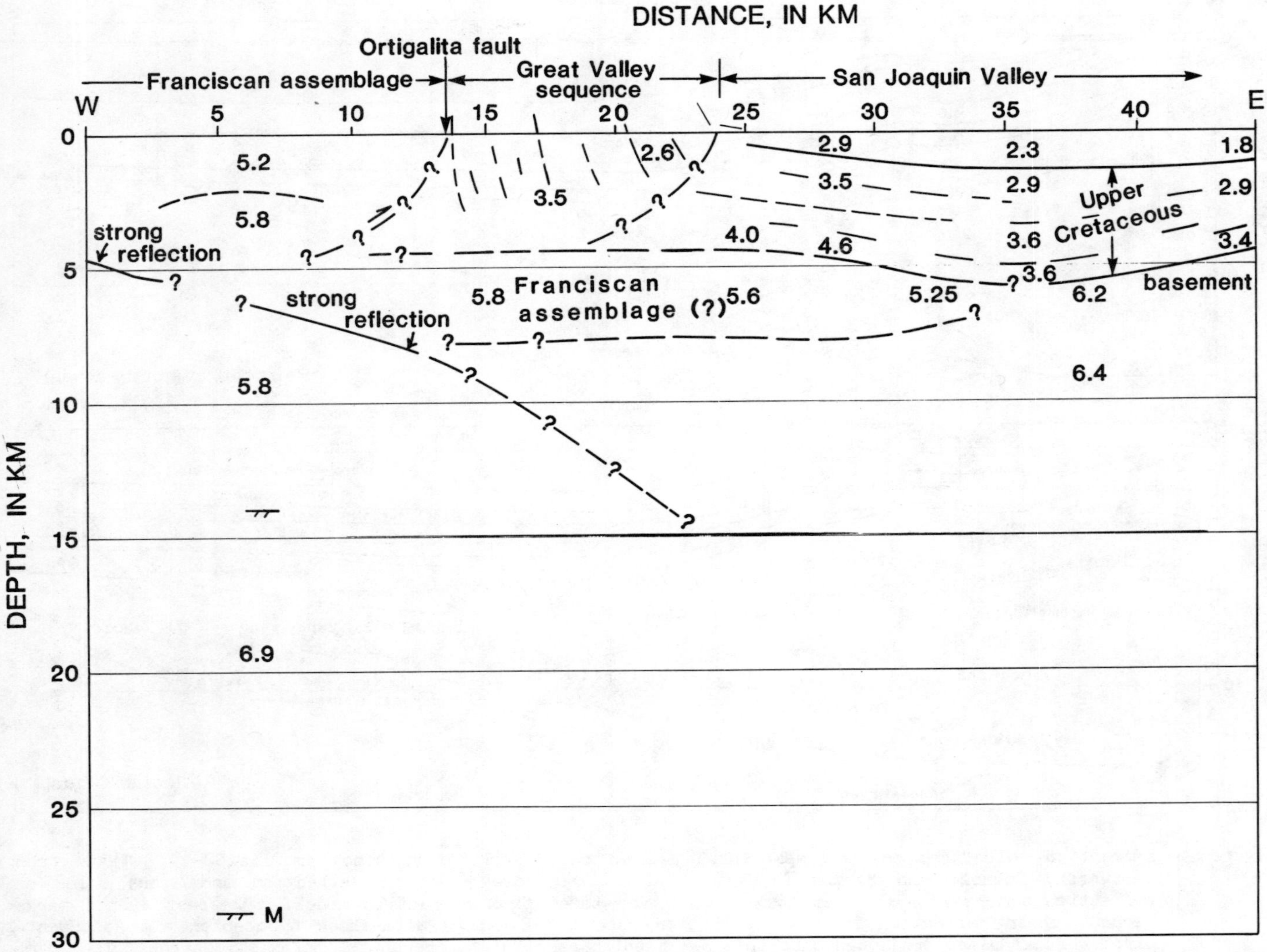

Figure 4. Preliminary interpretation of reflection profile CC-1. Taken from Wentworth, Zoback, and Bartow (1984). Profile lies 140 km northwest of profile SJ-19 along Garzas Creek and highway 140 (see Jennings, 1977). Velocity structure 6 km east of west end of profile taken from Walter and Mooney (1982); pi symbols mark deep boundaries in that velocity model, which crosses this profile at a high angle. Surface attitudes shown at range front (km 24). West-dipping queried faults and steeply east-dipping layering in Great Valley sequence not evident in reflection record.

In the core of the Diablo Range, strong, moderately east-dipping reflections at a depth of 5 to 8 km beneath exposed Franciscan may lie within the thick Franciscan section suggested by the refraction interpretation of Walter and Mooney (1982), or may define the base of an unexpectedly thin Franciscan. Analogy with the deep position of the base of the Franciscan in the SJ-6 and SJ-19 sections (figs. 2 and 3) suggests that the Franciscan rocks extend to a depth of about 15 km here as well. Weak eastward continuation of the east-dipping reflections in the record extend to a depth of at least 15 km beneath the range front. This seems to represent a fault that, because it truncates the Franciscan wedge, must post-date emplacement of that wedge.

Northern Sacramento Valley

The reflection profiles in the northern Sacramento Valley have not yet been carefully studied, but one principal structural relation that they exhibit is striking. Reflection lines S-45 and S-51, which are 30 km apart north-south, both show the same basic features (fig. 5). Near the center of the Sacramento Valley, a well defined set of reflections represents the Great Valley sequence overlain by a thin Cenozoic section. The top of crystalline basement (B, fig. 5) dips gently westward with the concordant overlying strata to a depth of about 5 km, where the Great Valley sequence is about 4 km thick. Beyond this point, the basement reflection continues its westward dip for another 5-8 km (E to B, fig. 5) before the reflection is lost. The reflections from the overlying Mesozoic strata, in contrast, begin to rise westward and within a short distance achieve east dips as steep as 30°.

At the west end of these reflection lines, then, east-dipping reflections from the sedimentary section abut downward against the west-dipping basement reflection at a considerable angle, whereas farther east the two are concordant and subhorizontal. The eastern tip of the downward-truncated reflections (E in fig. 5 and 7) lies about 8 km east of the valley margin. Reflection interval velocities suggest that only Great Valley sequence is involved here, and that a

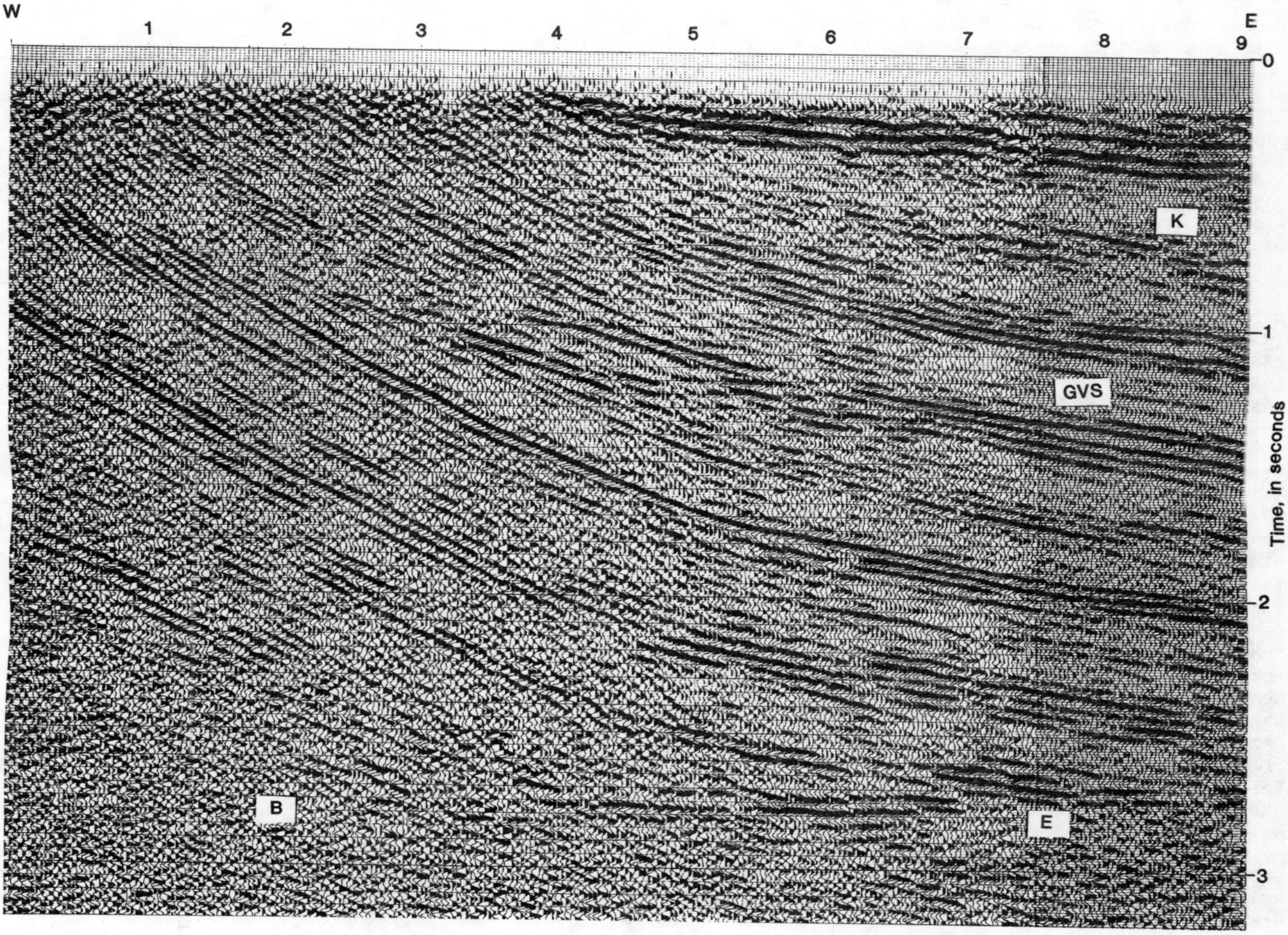

Figure 5. The easternmost tectonic wedge of Great Valley sequence expressed in seismic reflection line S-51 (migrated). The lowermost reflection is inferred to represent the top of crystalline basement (B). K marks the top of the Great Valley sequence (GVS). East-dipping Great Valley sequence butts downward against basement as far east as E, the eastern tip of the wedge; farther east the Great Valley sequence is essentially concordant with the basement surface. Horizontal scale in km.

wedge of Franciscan rock like those in the cross sections to the south must be farther west. Where exposed at the surface to the west of S-45 and S-51 along the front of the Coast Range, the steeply east-dipping Great Valley sequence is at least 15 km thick, several times the 4-km thickness of Mesozoic strata in the valley.

TECTONIC INTERPRETATION

At the boundary between the Coast Range and the San Joaquin Valley, an eastward-thinning wedge of rock of intermediate seismic velocity lies between fast, crystalline basement and slower, overlying Great Valley sequence. This wedge is evident in all three of the cross sections from the San Joaquin Valley. Largely because of its seismic velocity, the wedge is inferred to consist of Franciscan rock. The presence of this wedge at the eastern margin of the Franciscan assemblage is not compatible with the established view that the Franciscan was emplaced there beneath a master thrust that bottomed eastward in the mantle. Instead, the Franciscan wedge appears to overlie the same basement that unconformably bears the Great Valley sequence immediately to the east.

The Franciscan wedge is overlain by Great Valley sequence, but Irwin (1957) long ago put to rest the possibility that this was an unconformable relation by demonstrating that the Franciscan assemblage and Great Valley sequence are coeval. The clear alternative to account for the wedge configuration is that the Franciscan wedge was emplaced tectonically between the Great Valley sequence and its basement. The wedge was thrust eastward onto that basement while it concurrently peeled the Great Valley sequence up from the basement along the Coast Range thrust.

This interpretation is dependent on continuity of the basement surface westward from the unconformity beneath Mesozoic strata in the Great Valley to the base of the Franciscan wedge. Otherwise the basement beneath the Franciscan wedge could be a separate, sub-Franciscan basement. Any change in basement identity would have to occur at the tip of the Franciscan wedge, where the Coast Range thrust reaches basement. On line SJ-6, the conflict between the refraction and reflection depths of the western basement segment beneath the syncline (fig. 2, km 30) might permit a major east-dipping fault to juxtapose different basements (at about km 38). In the refraction model, however, the eastern tip of the 5.7 km/s wedge lies east of this possible break and on top of the eastern basement segment (Wentworth, Walter, Bartow, and Zoback, 1983, fig. 2).

On line SJ-19, the shape of the basement is poorly constrained where the basement steps down to the west near the east side of the cross section (fig. 3). A fault could be inferred here. In this case, however, the wedge tip is at least 10 km farther west where the basement surface appears to dip gently.

On line CC-1 (fig. 4), the exact position of the surface of the eastern basement segment is uncertain in the reflection record and a mismatch in alinement of as much as 0.5 s (1.4 km) is possible between it and the top of basement beneath the wedge to the west. It would be an extraordinary coincidence, however, that two separate basements would end up so closely alined after major shortening across a subduction zone.

The basic structural relation, we conclude, is that the Franciscan wedge is bounded below by a fault along which eastward-directed thrusting across Great Valley basement occurred and is bounded above by a fault along which westward-directed thrusting of peeled-up Great Valley sequence occurred. Neither of these faults, in the simplest case, extends eastward beyond the tip of the wedge; there, the Great Valley sequence still rests unconformably on its original basement. The Coast Range thrust, then, is the roof thrust of the Franciscan wedge and separates the Franciscan rock in the wedge from the peeled-up Great Valley sequence. It extends no further down dip to the east than the tip of the wedge.

This Franciscan wedge seems to characterize the east margin of the Franciscan assemblage along a 200-km length of the boundary between the Franciscan and Great Valley sequence, from reflection line SJ-6 at South Dome to line CC-1 at Garzas Creek. It could well be the fundamental structural relation between the two terranes along the whole length of the Coast Range. The occurrence of similar wedge relations in Great Valley sequence on reflection lines S-45 and S-51 in the northern Sacramento Valley supports this view.

Northern Coast Ranges

At the north end of the Coast Ranges, wedge tectonics help to make sense of what are otherwise very puzzling structural relations. Mapping by Bailey and Jones (1973; see also Jennings, 1977) in the mid 1960's in the Colyear Springs-Paskenta area, at and south of the junction between the Great Valley sequence, Franciscan, and Klamath terranes, documented an extraordinary system of faults. Three principal fault zones (fig. 6) strike southeastward across the exposed, east-dipping Great Valley sequence and mark drastic changes in the thickness, and even the presence, of Upper Jurassic and Lower Cretaceous strata in the Great Valley sequence (Jones, Bailey, and Imlay, 1969; Jones and Bailey, 1973). Despite these effects in the Great Valley sequence, the faults do not offset the Franciscan rocks to the west nor are they evident farther east along strike. In particular, the linear magnetic and gravity anomalies in the Great Valley, which are derived from the basement there, extend across the projected trend of the faults without deviation.

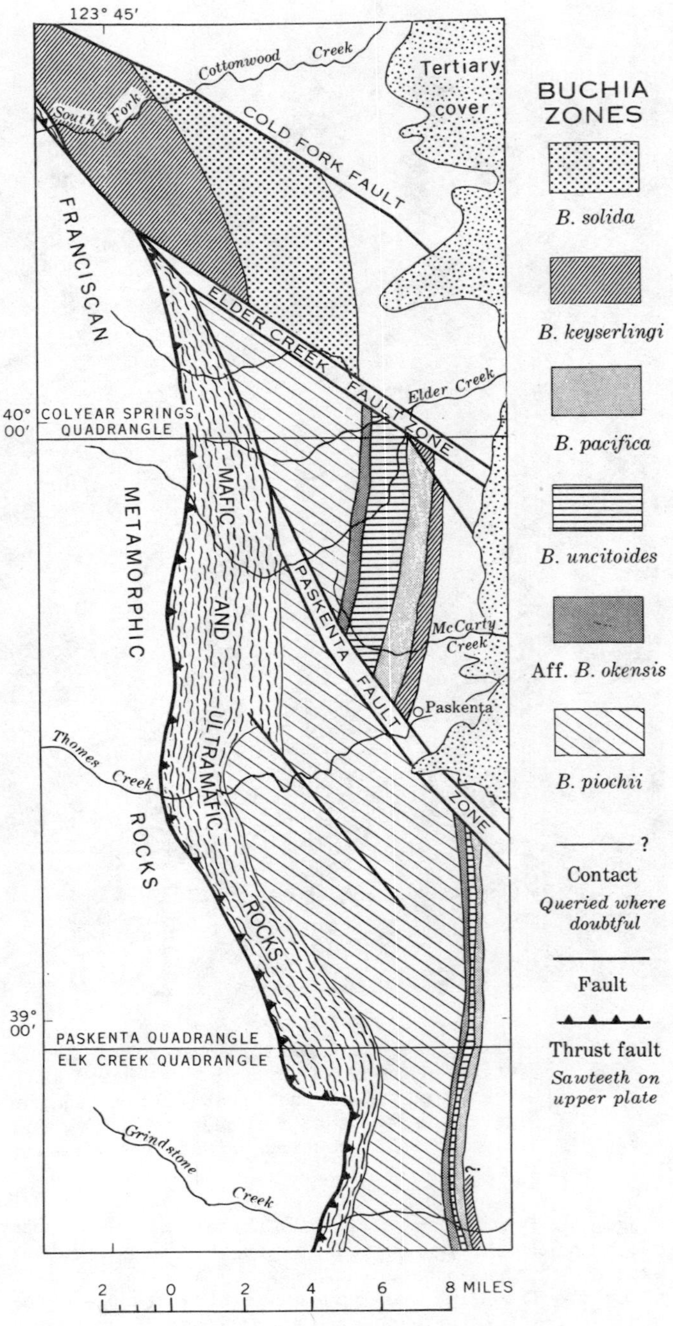

Figure 6. Major fault zones in the Colyear Springs-Paskenta area and the Buchia zones that document large fault offsets. Taken from Jones, Bailey, and Imlay (1969). The complex fault zones are shown only schematically. South of the Paskenta zone, a fault splay bends southward and parallel to strike near the top of the B. piochii zone (Jennings, 1977). The ages of the Buchia zones are: B. solida - late Valanginian; B. keyserlingi - middle Valanginian; B. pacifica - early Valanginian; B. uncitoides - Berriasian; Aff. B. okensis - late Tithonian; B. piochii - middle and late Tithonian.

North of the Cold Fork fault zone, which is alined with the south margin of the Klamath terrane, the Great Valley sequence lies on Klamath basement and is principally younger than Valanginian. To the south between the Cold Fork and Elder Creek zones, there are several kilometers of Valanginian strata, but no Jurassic strata. South of the Elder Creek zone, the Valanginian section is drastically thinner but a thick Upper Jurassic section is present, lying above the Coast Range ophiolite. South of the Paskenta zone the Jurassic section nearly doubles in thickness.

The stratigraphic contrasts require large left-lateral strike slip on the faults. The occurrence of marine-nonmarine transitions in Valanginian strata to the northwest on Klamath basement (Jones and Irwin, 1971) implies a large amount of differential eastward movement of the Valanginian deep-water marine strata south of the Cold Fork zone. Jones and Irwin suggest that the Valanginian shoreline is offset about 100 km. Their shoreline reconstruction may be too simple, but the contrast between paralic and deep water environments in the Valanginian strata requires major offset amounting to at least many tens of kilometers.

A persistent problem with large left slip on these fault zones has been how the faults are rooted to the east. It is reasonable that the faults are limited to the upper plate of the Coast Range thrust and therefore don't offset the Franciscan rocks. But what happens to the faults southeastward along strike? Tectonic wedges offer a reasonable solution. We interpret the southeast-trending faults to be tear faults at the north margin of tectonic wedges in the Great Valley sequence. These wedges moved southeastward, parallel to the tears, and produced imbricate shortening in the Great Valley sequence. The tear faults are inferred to bend southward, parallel to bedding, to become the roof faults of the tectonic wedges (fig. 7). Parts of the Paskenta zone seem to do just this (Jennings, 1977), but for the most part, evidence for such a relation is either buried beneath Cenozoic rocks in the Sacramento Valley or is as yet unrecognized. Reflection lines S-45 and S-51 (fig. 5) show the tip of the easternmost of these wedges.

The driving force for the wedges of Great Valley sequence must come from farther west. Generally eastward thrusting of the Franciscan assemblage, responsible for the Franciscan wedge recognized in the sections farther south, is the obvious candidate. This Franciscan thrusting, in turn, presumably derives from collisional tectonics at a subduction zone farther west.

Faults typically associated with the top of the Coast Range ophiolite seem to mark another wedge boundary between a wedge occupied by the Coast Range ophiolite and the overlying wedge of basal Great Valley sequence. The depositional contact at the top of the ophiolite is only locally preserved, presumably where the thrust lies slightly up in the basal Jurassic strata.

This ophiolite wedge seems to have been thrust eastward a long distance across the underlying basement (fig. 7). As noted by Griscom (1983; and see Blake, Zeitz, and Daniels, 1978), such eastward thrusting above a west-dipping basement could account for (a) the northward termination of the magnetic anomaly produced by the Coast Range ophiolite (terminated at the Elder Creek fault along with the Upper Jurassic strata), (b) the tear faulting in the Great

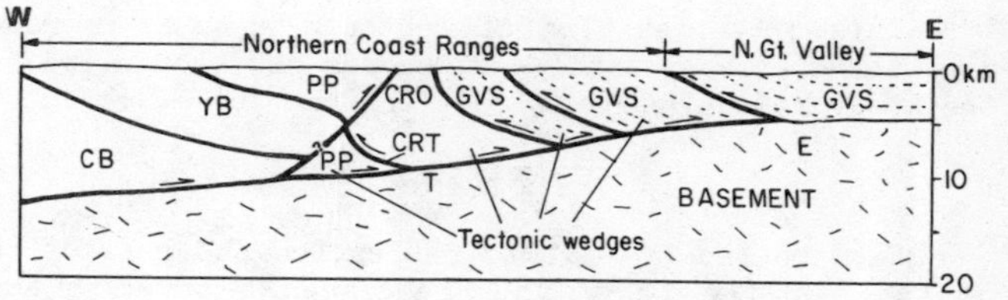

Figure 7. Schematic cross section across Coast Range-Great Valley boundary at latitude 39.88° N. GVS - Great Valley sequence, CRO - Coast Range ophiolite, CRT - Coast Range thrust, PP - Pickett Peak terrane, YB - Yolla Bolly terrane, CB - Central Franciscan belt.

Valley sequence, and (c) the lack of offset in basement rocks farther east.

Thrusting must have been sufficient to carry the Coast Range ophiolite onto basement that is Klamath-Sierran in character, in order to account for small, fault-bounded masses of Klamath rock (Galice(?) metagraywacke and slate together with Rogue(?) meta-andesite) caught between the Coast Range ophiolite and underlying Franciscan assemblage. These anomalous fragments extend southward discontinuously along that contact for at least 55 km beyond the southern Klamath boundary (Blake, Jayko, and McLaughlin, 1984).

These relations imply that a wedge of Franciscan rock, like that inferred along the west margin of the San Joaquin Valley, underlies the Coast Range ophiolite. Its tip is the down dip end of the Coast Range thrust (T, fig. 7). The Franciscan assemblage consists of a number of separate tectonostratigraphic terranes that have had somewhat different histories (Blake, Jayko, and McLaughlin, 1984). The easternmost terranes at the latitude of the cross section in figure 7 are the Pickett Peak and Yolla Bolly terranes of the Eastern belt Franciscan.

These two terranes were juxtaposed, deformed, and metamorphosed to blueschist grade before being emplaced here beneath the Coast Range ophiolite (Jayko, 1984). Neither the Coast Range ophiolite nor the Great Valley sequence have experienced such metamorphic conditions, which presumably developed at greater depth in a subduction zone farther to the west. Post-metamorphic movement that has occurred on the thrust fault separating the two terranes (Jayko, 1984) may be associated with the wedging. The underlying Central belt Franciscan mass (fig. 7) may represent a separate tectonic wedge emplaced beneath the Eastern belt wedges. Subsequently, the main mass of Franciscan rock was offset from the easternmost tip of the Franciscan wedge by eastward-directed reverse faulting, similar to the offset along the Ortigalita fault in the Diablo Range, which probably occurred in late Cenozoic time (Blake, Jayko, and McLaughlin, 1984).

McLaughlin and Ohlin (this vol.) describe relations in the Franciscan east of Clear Lake that imply tectonic wedging at the east margin of the Franciscan. They suggest that the East Clear Lake terrane was thrust eastward onto a west-dipping magnetic basement inferred from a smoothly west-dipping magnetic gradient. Farther east, they show the easternmost Franciscan terrane (Little Indian Valley terrane) to terminate as a wedge that overlies this west-dipping basement and underlies the east-dipping Coast Range thrust.

This interpretation of tectonic wedging is made more believable by similar structure now recognized in the Rocky Mountain thrust belt (Price, 1984).

CONCLUSIONS

Exploration by seismic reflection and refraction of the east margin of the Franciscan terrane in the California Coast Ranges is leading us to revise our view of structural relations between the Franciscan assemblage and Great Valley sequence. The Coast Range thrust now appears to be the roof thrust of a Franciscan tectonic wedge that was driven eastward between a west-dipping basement and the structurally overlying Coast Range ophiolite and Great Valley sequence. The well established view of the Coast Range thrust as a subduction zone suture seems incorrect, although such a relation may have existed far to the west prior to eastward thrusting of the wedge.

Inclusion of tectonic wedging in the history of the Franciscan assemblage and Great Valley sequence allows the following reconstruction. The Franciscan rocks of the northern Coast Ranges were metamorphosed to blueschist grade, presumably in a subduction zone, and then were thrust upward and eastward as a wedge onto Klamath-Sierran basement. As it moved, the Franciscan wedge progressively peeled up and carried before it, in imbricate fashion, a slab of Coast Range ophiolite and several slabs of Great Valley sequence (fig. 7). This thrusting greatly shortened the original distance across the basin in which the Great Valley sequence was deposited.

Fragments of Klamath rock were torn from the basement surface and dragged up along the top of the Franciscan wedge. This requires that the overlying wedge of Coast Range ophiolite continued to slip westward relative to the underlying Franciscan wedge even after the ophiolite had overridden Klamath-Sierran basement.

The absence of major repetition of stratigraphy in the Great Valley sequence requires that the imbricate wedges are progressively older downward and westward within the Great Valley sequence. This seems to indicate that, in its original distribution, the Great Valley sequence involved eastward onlap of younger strata such that the strata decreased progressively in age eastward across the basin. This would permit older, western sections to be thrust beneath younger, eastern sections.

The tectonic wedging of a thicker, western section of Great Valley sequence beneath a thinner, eastern section, which farther east is still unconformably attached to its original basement, may account for the abrupt westward thickening of the Great Valley sequence across the Great Valley-Coast Range boundary. This seems to be the case in the northern Coast Range and may explain similar abrupt westward thickening in the southern Coast Range as well.

The possibility that roof thrusts of tectonic wedges occur within the Great Valley sequence requires further scrutiny of the internal structure of the Great Valley sequence along the east margin of the Coast Range. The apparent unconformity east of Clear Lake, which separates Lower Cretaceous strata involved in the Wilbur Spring antiform from Upper Cretaceous strata that are not (Blake and Jones, 1981, p. 317; Jennings, 1977), could, for example, be such a thrust. Other thrusts should exist, particularly if tectonic wedging does accomplish the abrupt westward thickening of section at the west side of the valley, and some of these should produce at least local repetition of section.

The westward extent of the relatively undeformed, west-dipping basement beneath the Franciscan terrane needs examination. The refraction work in the Diablo Range (Walter and Mooney, 1982; Walter, 1984) suggests that it flattens at a depth of about 15 km and extends at least 20 km west of the eastern front of the Coast Range. At the latitude of Clear Lake, it seems to extend at least 40 km west of the range front (McLaughlin and Ohlin, this vol.).

The abrupt structural change at the west margin of the Great Valley thus seems to be related to tectonic wedge emplacement of the Franciscan assemblage. The emplacement process probably began shortly after Valanginian time in the Early Cretaceous and continued sporadically through the Cretaceous and probably into early Tertiary time. Arrival of each successive Franciscan terrane at the continental margin probably caused further thrusting. Existing thrusts may have moved further and new ones may have formed, both near the east margin of the Franciscan terrane and back within it. Continued transverse compression in late Cenozoic time has caused the cross-cutting east-directed faulting (Ortigalita fault and others), raised the Coast Range topography, and most recently, produced the 1983 Coalinga earthquake.

REFERENCES CITED

Bailey, E. H., and Jones, D. L., 1973, Preliminary lithologic map, Colyear Springs quadrangle, California: U.S. Geological Survey Miscellaneous Field Studies Map MF-516, scale, 1:48,000.

Bailey, E. H., Blake, M. C. Jr., and Jones, D. L., 1970, On-land Mesozoic oceanic crust in Californian Coast Ranges: U.S. Geological Survey Proessional Paper 700-C, p. C70-C81.

Bailey, E. H., Irwin, W. P., and Jones, D. L., 1964, Franciscan and related rocks and their significance in the geology of western California: California Division of Mines and Geology, Bulletin 183, 177 p.

Blake, M. C., Jr., Jayko, A. S., and McLaughlin, R. J., 1984, Tectonostratigraphic terranes of the northern Coast Ranges, California, in Howell, D. G., ed., Tectonostratigraphic terranes of the circumpacific region: Circumpacific Council for Energy and Mineral Resources, Earth Science Series, v. 1, (AAPG Bookstore), in press.

Blake, M. C., Jr., and Jones, D. L., 1981, The Franciscan assemblage and related rocks in northern California: a reinterpretation, in Ernst, W. G., (ed.), The geotectonic development of California: Prentice-Hall (Rubey Volume I), p. 306-328.

Blake, M. C., Jr., Zietz, Isadore, and Daniels, D. L., 1977, Aeromagnetic and generalized geologic map of parts of central California: U.S. Geological Survey Geophysical Investigations Map GP-918, scale 1:1,000,000.

Griscom, Andrew, 1983, The contact between the Great Valley sequence and the Franciscan assemblage, California, from magnetic data (abs.): American Geophysical Union, Transactions, EOS, v. 64, no. 45, p. 867-868.

Hamilton, Warren, 1969, Mesozoic California and the underflow of Pacific mantle: Geological Society of America Bulletin, v. 80, p. 2409-2430.

Irwin, W. P., 1957, Franciscan group in Coast Ranges and its equivalents in Sacramento Valley,

California: American Association of Petroleum Geologists Bulletin, v. 41, no. 10, p. 2284-2297.

-----,1964, Late Mesozoic orogenies in the ultramafic belts of northwestern California and southwestern Oregon: U.S. Geological Survey Professional Paper 501-C, p. C1-C9.

Irwin, W. P., and Bath, G. D., 1962, Magnetic anomalies and ultramafic rock in northern California: U.S. Geological Survey Professional Paper 450-B, p. B65-B67.

Jayko, A. S., 1984, Structure and deformation in the Eastern Franciscan belt, northern California: unpublished Ph.D. dissertation, University of California, Santa Cruz, 204 p.

Jennings, C. W., 1977, Geologic map of California: California Division of Mines and Geology, Geologic Data Map 2, scale 1:750,000.

Jones, D. L., and Bailey, E. H., 1973, Preliminary biostratigraphic map, Colyear Springs quadrangle, California: U.S. Geological Survey Miscellaneous Field Studies Map MF-517.

Jones, D. L, and Irwin, W. P., 1971, Structural implications of an offset early California shoreline in northern California: Geological Society of America Bulletin, v. 82, p. 815,-822.

Jones, D. L., Bailey, E. H., and Imlay, R. W., 1969, Structural and stratigraphic significance of the Buchia zone in the Colyear Springs-Paskanta area, California: U.S. Geological Survey Professional Paper 647-A.

Price, R. A., 1984, The southeastern Canadian Cordillera: thrust faulting, tectonic wedging and delamination of the lithosphere: submitted to Journal of Structural Geology, June, 1984.

Walter, A. W., 1984, Velocity structure near Coalinga, California, in Rymer, M. J., and Ellsworth, W. L. (eds.), Mechanics of the May 2, 1983 Coalinga, California, earthquake: U.S. Geological Survey Open-File Report 84---, in press.

Walter, A. W., and Mooney, W. D., 1982, Crustal structure of the Diablo and Gabilan Ranges, central California: A reinterpretation of existing data: Bulletin of the Seismological Society of America, v. 72, no. 5, p. 1567-1590.

Wentworth, C. M., Walter, A. W., Bartow, J. A., and Zoback, M. D., 1983, Evidence on the tectonic setting of the 1983 Coalinga earthquakes from deep reflection and refraction profiles across the southeastern end of Kettleman Hills, in Bennett, J. H., and Sherburne, R. W., eds., The 1983 Coalinga, California earthquakes: California Division of Mines and Geology Special Publication 66., p. 113-126.

Wentworth, C. M., Walter, A. W., Zoback, M. D., and Blake, M. C., Jr., 1983, Possible obduction of Franciscan assemblage, southeasternmost Diablo Range, California (abs.): American Geophysical Union, Transactions, EOS, v. 64, no. 45, p. 868.

Wentworth, C. M., Zoback, M. D., and Bartow, J. A., 1984, Tectonic setting of the 1983 Coalinga earthquakes from seismic reflection profiles: a progress report, in Rymer, M. J., and Ellsworth, W. L. (eds.), Mechanics of the May 2, 1983 Coalinga, California, earthquake: U.S. Geological Survey Open-File Report 84---, in press.

PALEOMAGNETIC AND TECTONIC INTERPRETATION OF THE PARKHURST RIDGE LIMESTONE, COASTAL BELT FRANCISCAN, NORTHERN CALIFORNIA

W. P. Harbert;
Department of Geophysics,
Stanford University,
Stanford CA 94305

R. J. McLaughlin; W. V. Sliter;
U.S.Geological Survey,
345 Middlefield Rd.,
Menlo Park CA 94025

ABSTRACT

Paleomagnetic results are reported for a deformed manganiferous pelagic limestone of Late Cretaceous (Campanian-Maastrichtian) age in the Coastal Belt of the Franciscan Complex. Sixty-eight oriented samples were collected from a broken sequence of limestone that is steeply folded synclinally providing the basis for an inconclusive fold test. Sedimentary structures within the limestone and the depositional contact with underlying volcanics show that the limestone body is right-side-up. The limestone and underlying volcanics together compose a large block within a zone of melange. Thermal demagnetization of the limestone samples to 600° C define two components of magnetization. The component with the lowest blocking temperature ≈ 250° C was always of normal polarity. This component failed the fold test, its direction before applying a tectonic correction was rotated 23° in a clockwise sense. The second component observed between 250° and 540° C is of reversed and normal polarity and considered the characteristic remanent magnetization. The resultant inclination mean of $I_c=-47°$, $I_{95}=27°$ N=5, 23° shallower than the expected inclination calculated for the North American plate. Eocene dinoflagellates from carbonate concretions in the matrix of the Coastal Belt melange and presence of a post-accretion overlap sequence indicate that the time of accretion of the limestone with the North American plate was approximately late Eocene to late Oligocene. To account for this paleomagnetic result a plate reconstruction is favored whereby the limestone of Parkhurst Ridge is moved as part of the Farallon plate from 70 Ma until it encounters terrigenous deposition from North America at 45 Ma. Northward strike-slip movement after or during accretion may be responsible for mixing of the limestone with Eocene terrigenous sediment. The model assumes deposition of the limestone on or close to the Farallon-Pacific plate spreading center at 70 Ma.

INTRODUCTION

Recent paleomagnetic results from the Franciscan Complex of northern and central California show that some limestone blocks within the Complex were deposited in localities far removed from the North American plate. Study of the Laytonville limestone of Late Cretaceous (Albian to Turonian) age located in the Central Belt of the Franciscan Complex (Alvarez and others 1980) suggested a latitude of deposition of 17° south.

This surprising result requires about 7500 km of relative poleward motion between the Laytonville limestone and the North American plate since Albian time. To test the hypothesis that limestone blocks within the Coastal Belt of the Franciscan Complex might also be allochthonous to North America, we undertook a study of a limestone near Honeydew, California on Parkhurst Ridge (latitude 41° 15' N, longitude 124° 10' W, figure 1). We first present results from a preliminary paleomagnetic study of the limestone of Parkhurst Ridge and then suggest a terrane model to explain the paleomagnetic results based on recently published oceanic plate reconstructions for the Pacific basin (Engebretson 1982, Engebretson and others 1984b). The paleomagnetic data together with the plate model provide insight into the probable tectonic setting of deposition of the limestone.

GEOLOGIC SETTING OF THE LIMESTONE OF PARKHURST RIDGE

The Franciscan Complex of Northern California (figure 1) is divided into several belts that are aligned roughly north 30° west. Parkhurst Ridge lies within the western-most of these belts, referred to as the Coastal Belt. A brief review of the geology of the Franciscan Central and Coastal Belts follows, (see Bailey and others 1964, Blake and Jones 1981, Jones and others 1978, Page 1981 and McLaughlin and others 1982, for a more detailed description).

In the area of this study (figure 1) the Coastal Belt is composed of a melange that consists dominantly of highly sheared arkosic and lithic sandstone and argillite originally deposited in various middle to outer slope, marginal basin and fan settings. Rare basaltic volcanic rocks, pelagic limestone and rare blueschist locally occur as tectonic blocks enclosed in the sheared matrix of sandstone and argillite. The Coastal Belt contains microfossils of Late Cretaceous, Paleocene and Eocene age (Blake and Jones 1978, Evitt and Pierce 1975, Sliter 1984) and is juxtaposed with the Franciscan Central Belt to the east along the Coastal Belt Thrust. To the southwest, the Coastal Belt is juxtaposed with Late Cretaceous through middle Miocene rocks of the Franciscan King Range terrane (McLaughlin and others 1982) along a major southwest-dipping Neogene thrust. The Coastal Belt and Central Belt are both overlain by a marine and non-marine post-accretion overlap sequence of middle Miocene and younger age that fills the east-west on-land extension

in Blake, M.C. Jr., ed., 1984, **Franciscan Geology of Northern California:** Pacific Section S.E.P.M., Vol. 43, p. 175-183

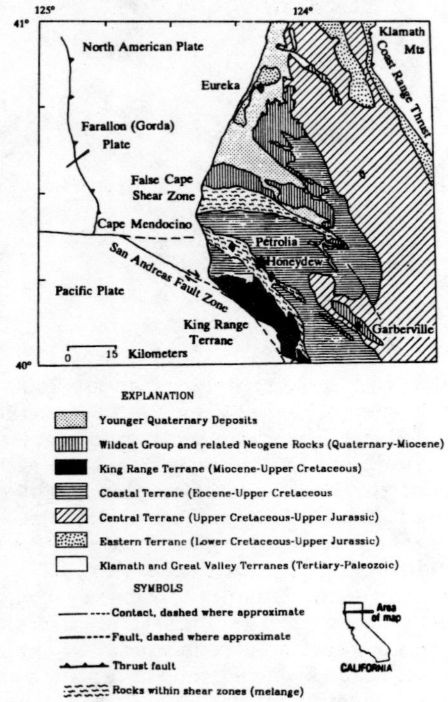

Figure 1

Regional map of Parkhurst sampling locality, Parkhurst Ridge is within the Coastal Belt of the Franciscan Complex, slightly northwest of Honeydew, California.

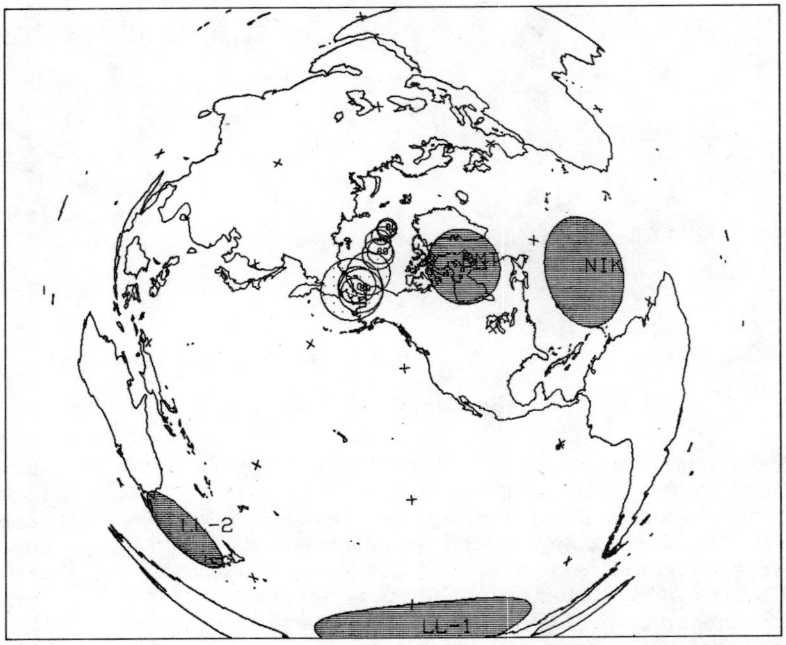

Figure 2

Pole positions of paleomagnetic results from the Central Belt of the Franciscan Complex shown on a Lambert projection. LL-1, LL-2 are poles calculated from Alvarez and others(1981), NIK is calculated from Grommé and Gluskoter (1965), RMT calculated from Saad (1969). Courtillot and others (1983) have suggested the pole positions for LL-1 and LL-2 are incorrect. Note the consistent clockwise rotation of poles away from the Apparent Polar Wander Path for North America (Irving and Irving 1982), which are labelled in 20 Ma intervals.

of Humboldt basin. These depositional and structural relations bracket the timing of accretion of the Coastal Belt and the limestone of Parkhurst Ridge at its present position on the North American plate to between late Eocene and late Oligocene time.

At the latitude of Parkhurst Ridge (41° N lat, 124° W long) the Central Belt consists of melange (figure 1) made up of sheared lithic and arkosic graywackes, shale, chert, and basic volcanic rocks of middle Jurassic to Late Cretaceous age (Bailey and others 1964, Blake and Jones 1978). The limestones of the Franciscan Central Belt and Coastal Belt generally are pink, red, or gray micritic limestones, usually associated with mafic volcanic rocks. The limestones may be massive to thin bedded, and may consist of 60-90% micrite and 10-30% foraminifera (Bailey and others 1964, Wachs and Hein 1975). Some of the limestones in the Central Belt have a micritic matrix consisting entirely of coccoliths (Garrison and Bailey 1967). Based on lithic associations and their foraminiferal assemblages the limestones of both the Central and Coastal Belts are thought to be of low latitude (20° N to 20° S latitude) and intermediate (1,000-2,000 m) water depth oceanic origin (Sliter 1984). Deposition may have occurred in a seamount, plateau or ridge setting (Wachs and Hein 1975, Sliter 1984).

PREVIOUS PALEOMAGNETIC STUDIES FROM THE FRANCISCAN

Previous paleomagnetic studies (figure 2) of the Franciscan have been conducted in the Central Belt. The more reliable of these studies have shown incorporated Late Cretaceous limestone blocks to be allochthonous. Alvarez and others (1980) paleomagnetically sampled two blocks (LL-1, LL-2) of pink limestone near Laytonville California. In their study all samples were of reversed magnetic polarity and no fold test was possible. Alvarez and others (1980) calculated a latitude of deposition of 17°±7° South latitude avoiding the usual polarity ambiguity by showing a paleontologic boundary within the limestone blocks allowed determination of top and bottom. The polarity of the geomagnetic field was also known, as deposition of the limestone occurred during the Cretaceous Normal Polarity Superchron (118-83 Ma, using the time scale of Harland and others (1982)).

Paleomagnetic studies of Grommé and Gluskoter (1965), Saad (1969), and Courtillot and others (1983) (see also Grommé, and Curry and others, in this volume) produced results with uniformly steeper mean inclinations than the Laytonville study. Courtillot and others (1983) studying the 100 to 103 Ma (Albian) aged Calera-type limestones of the Permanente quarry used paleontologic variations within the sampled section to determine top and bottom and assumed a normal po-

Reference	Age	N	K	α_{95}	D_s	I_s	F ΔF	Age(Ref)
Grommé and Gluskoter (1965)	Tithonian-Cenomanian	7	20	14	78	47	19 16	120
Saad (1969)	Upper Creta-ceous	11	34	8	29	68	-1 7	85
Alvarez et al. (1980)	LL-2 Albian-Cenomanian	40	21	11	230	-32	99 10 34# 10	100
	LL-1	12	10	22	183	-23	89 18 43# 18	100

#Northern Hemisphere interpretation

Table 1

Summary of previous paleomagnetic data from the Central Belt Franciscan. N= Number of sites (for Alvarez and others 1980, N refers to the number of samples), κ= Dispersion parameter, α_{95}= Cone of 95% confidence about mean, D_s, I_s= Declination and inclination after bedding correction. F and ΔF= Flattening and error in measuring flattening (Beck 1976). Age (Ref)= Age of reference pole (Irving and Irving 1982) used to calculated F and ΔF. All ΔF terms have been multiplied by .78 as suggested by Demarest (1983).

larity for the geomagnetic field. They reported that deposition of the Permanente quarry limestone had occurred at 21°±6° North latitude. Courtillot and others (1983) suggested the possibility of a polarity error in the study of Alvarez and others (1980) indicating LL-1 and LL-2 were really deposited at 17°±7° North latitude.

Table 1 shows the studied limestones (Courtillot and others 1983, Alvarez and others 1980) and basaltic volcanic rocks (Grommé and Gluskoter 1965) are far travelled. It is interesting to note that the measured declinations (figure 2) are generally rotated in a clockwise manner away from the expected pole positions for the North American plate. As the sampled localities outcrop in melange this consistency is contrary to generally accepted models for melange deformation; which seem to imply random declinations would be observed from studies of separate blocks.

PALEOMAGNETIC SAMPLING AND RESULTS

Sixty-eight oriented cores were collected at Parkhurst Ridge from about 7 meters of red, light grey and light green micritic limestone (figure 1). Five separate outcrops were sampled corresponding to localities 83pr01, 83pr04, 83pr05, 84pr02, and 84pr03. Tectonic corrections used to restore sites to horizontal are shown in table 2. These sampling sites do not represent separate blocks in a melange such as LL-1 and LL-2 in the study of Alvarez and others (1980). All sites are within a generally continuous, structurally coherent synclinally deformed limestone unit. The age of synclinal folding of the limestones is not well constrained.

Plots of magnetic intensity show a smooth decay although at higher thermal demagnetization steps some samples acquired a random component of magnetization during heating effectively limiting their usefulness for tectonic interpretation. The magnetic intensity in the demagnetization step before this component was added was usually on the order of 2×10^{-7} emu. NRM intensities averaged about $J_{nrm}=5.4\times10^{-6}$ emu. Magnetic intensities after cleaning averaged $J_p=5.2\times10^{-7}$ emu.

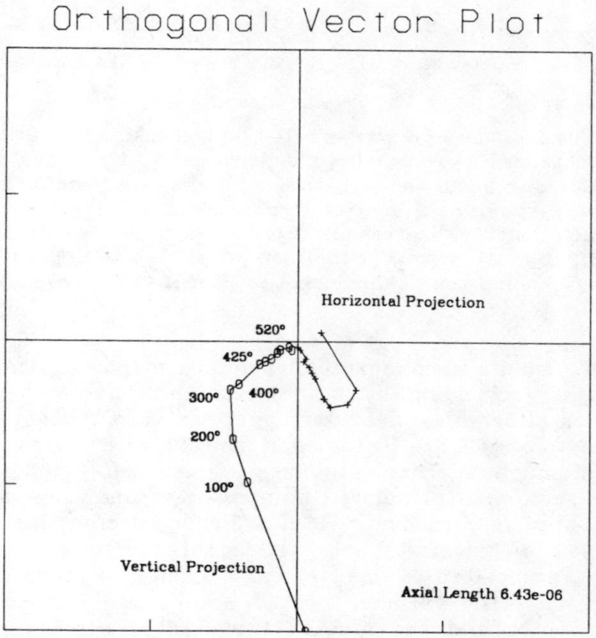

Figure 3

Typical orthogonal vector plot during thermal demagnetization. Note two component behavior. Temperatures along demagnetization path are in degrees Celsius. Open circles represent the magnetic vector in stratigraphic coordinates projected onto a vertical plane.

Locality	Tectonic Correction
83pr01	North 68° East 85° South
83pr04	North 40° West 75° North
83pr05	North 15° West 65° North
84pr02	North 46° West 39° North
84pr03	North 60° West 42° North

Table 2

Structural corrections for Parkhurst Ridge localities.

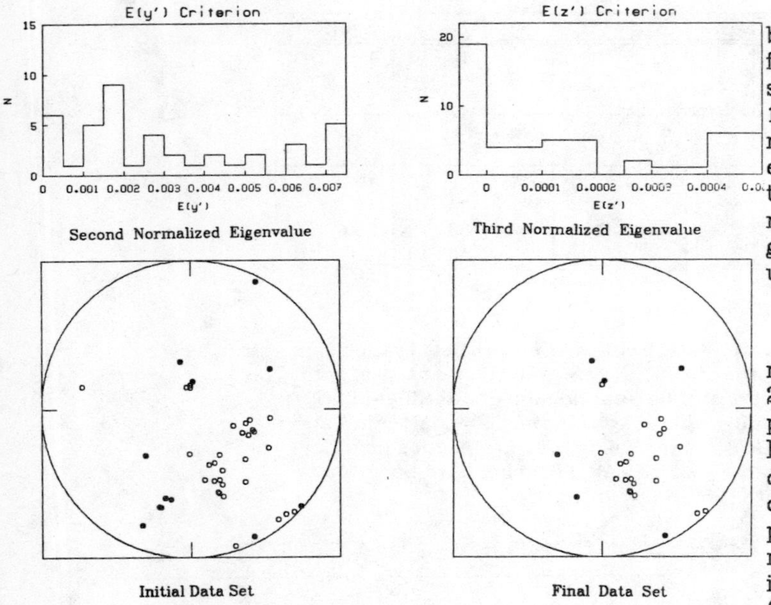

Figure 4

The final data set was selected on the basis of linearity (as measured by normalized eigenvalue) and number of points along linear demagnetization paths. The number of samples (shown here in stratigraphic coordinates) included in each locality decreased, 83pr01 (20 to 14 samples), 84pr02 (5 to 2), 84pr03 (7 to 3), and 83pr04 (4 to 2) 83pr05 (7 to 6), as did the α_{95} for localities 83pr01 (10° to 8°) and 83pr05 (13° to 11°).

To isolate secondary and primary magnetic components each sample was subjected to stepwise thermal or alternating field demagnetization. Orthogonal vector plots (figure 3) for each sample were reviewed and demagnetization data for each was subjected to least-squares (principal component) analysis as presented in Kirschvink (1980). Principal component analysis of trivariate data yields three eigenvectors and associated eigenvalues. Each eigenvalue represents the variance of the data parallel to the eigenvector and the method reviewed in Kirschvink (1980) maximizes the variance parallel to the principal component axes (figure 4). These eigenvalues are a direct measure of the degree to which "line segments" seen on orthogonal vector plots are actually linear. If a perfectly straight array of points was subjected to principal component analysis all variance would be parallel to one eigenvector that would have an associated normalized eigenvalue of 1.00. Two non-zero eigenvalues suggest that variation between points takes place on a plane. Three non-zero values suggest variation with three associated variances. To increase the quality of the final data set normalized eigenvalues (E_y, E_z) associated with each specimen were reviewed. Criteria for final data selection were $E_{y+z}=E_y+E_z \leq .57\%$ and $N_{ov} \geq 3$ (N_{ov}= number of points used to calculate the principal components including the origin). $E_z \leq .02\%$ (E_z= third principal component) and $E_y \leq .55\%$ (E_y= second principal component) were chosen after reviewing figure 4 and noting the weakly bimodal distribution. These criteria assure single component behavior over a reasonable magnetic cleaning range. Twenty-seven samples from five localities were chosen for final analysis (figure 4).

Figure 5 shows that even after application of reliability criteria to each sample the means (and modes) for each locality differ. We generated a preliminary study mean by averaging the Fisherian means (Fisher 1953) for each locality using the inclination-only method of McFadden and Reid (1982), and assumed each locality represented an independent sampling of the geomagnetic field. Assuming each sample represents an independent measurement of the geomagnetic field would underestimate the final study uncertainity about the mean.

A secondary component (table 4) present at thermal demagnetization steps with temperatures below 250°C failed the fold test. Before unfolding this component had a mean Fisherian direction of D_g=23°, I_g=69°. Where D_g, I_g represents the mean magnetic declination and inclination in present day geographic coordinates before tectonic correction. Another component was observed between 250° to 540° C. The mean of this component for 5 localities calculated using inclination-only statistics (McFadden and Reid (1982)) is I_s=−47°, I_{95}=27.°, κ_I=5 N=5. Where I_s is the inclination with respect to stratigraphic coordinates after tectonic correction for tilt (table 3). The inclination mode was found to be within the mode interval enclosing the inclination mean (figure 5). Figure 5 shows these magnetic vectors in stratigraphic and geographic coordinates. Because of the possibility of anomalous declinations caused by inaccurate structural corrections, plunging structure, local block rotations, lack of azimuthal symmetry of directions about the Fisher mean, and the relatively small number of samples from each locality we choose to use the primary direction calculated using inclination-only statistics.

If Fisher statistics (Fisher 1953) and the associated κ are calculated before and after tectonic correction with N=5 this results in a negative (k_g=5.1, k_s=3.1) though inconclusive fold test at 95% confidence (using the criterion of McElhinny (1964)). Figure 5 allows two interpretations. The final magnetic component may represent an overprint or the characteristic remanent magnetization. If this component represents a secondary magnetic overprint it is interesting to note (figure 6) the similarity in the uncorrected directions from the localities of Alvarez and others (1980) and this study. This similarity is disturbing as no correlation would be expected between the uncorrected locality means. This may indicate the failure of a 'megaconglomerate test' between these two study localities. However this interpretation requires no differential rotation or folding between the sites since the time of overprinting. Given the complex geology and separation of the study areas this seems unlikely. A second interpretation is that this magnetic component was acquired during a low latitude (I_g=11°,λ_g=6°) post-folding remagnetization event. The age of folding of the limestone is not well constrained and may have occurred before accretion. This would require greater convergence between the limestone and the North American plate. Such convergence could be provided by the presence of a more rapidly poleward moving oceanic plate offshore of California during the Late Cretaceous-lower Tertiary as suggested by Page and Engebretson (1984a). We adopt the latter interpretation which is that the magnetic component of −47° inclination is of primary origin acquired during or short-

Statistic	Age	N	I_o	λ_o	I_e	λ_e	d^* d_{km}	Age (Ref)
McFadden and Reid (1982)	Maastrichtian-Campanian	5	-46.6	28	68	51	23 2553	70
Fisher (1953)	Maastrichtian-Campanian	5	-62.8	44	68	51	7 753	70

Table 3

Summary of preliminary results from Parkhurst Ridge. N refers to the number of sites, I_o is the observed primary component inclination, λ_o is the observed paleolatitude. The APW path of Irving and Irving (1982) was used to calculated I_e the expected cratonal inclination and λ_e the cratonal expected paleolatitude, using their 70 Ma reference pole. d^* is the observed displacement (λ_e-λ_o) and d_{km} is the observed displacement in kilometers. The inclination-only mean is prefered to the Fisherian average.

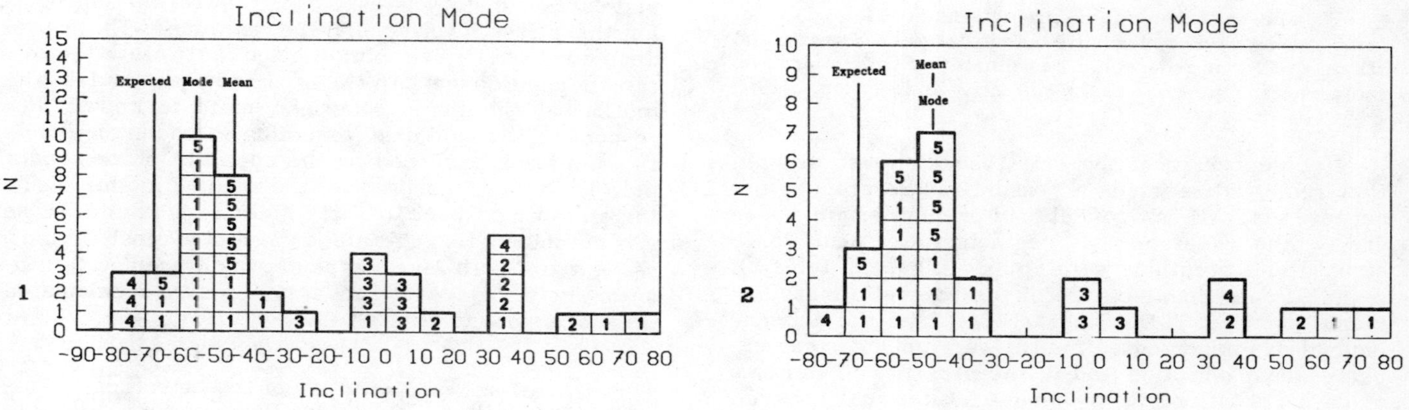

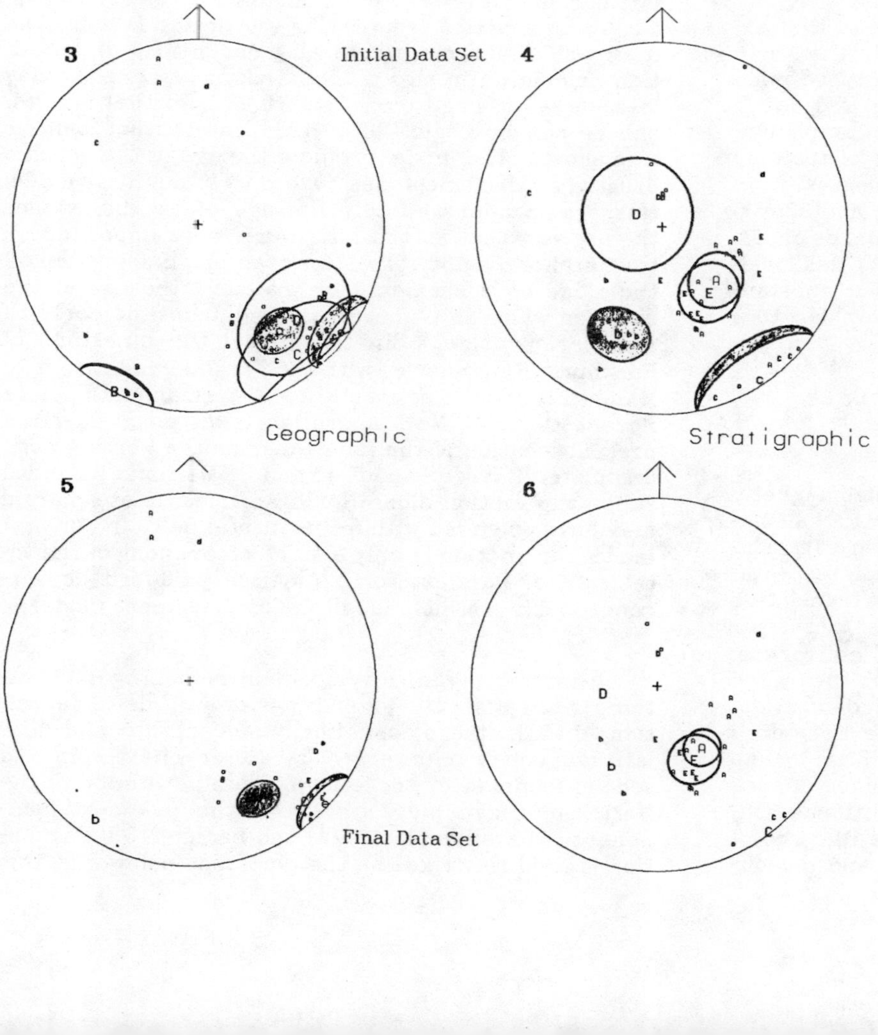

Figure 5

Sample inclinations before and after application of reliability criteria are shown in plots 1 and 2 on which the expected, mode and mean inclinations are noted. 1 refers to samples from locality 83pr01, 2 samples from 84pr02, 3 samples from 84pr03, 4 samples from 83pr04 and 5 samples from 83pr05. Plots 3,4 and 5,6 show individual samples and associated uncertainities on equal area sterographic projections in geographic and stratigraphic coordinates before and after application of reliability criterion. Small upper case letters represent reversely magnetized samples, lower case letters normally magnetized samples. Large letters on these plots refer to the locality Fisherian mean where A=83pr01 (symbols A, a), B=84pr02 (symbols B, b), C=84pr03 (C, c), D=83pr04 (D, d) and E=83pr05 (E, e). Darkened ovals represent lower hemisphere α_{95} confidence intervals In the final data set for localities 84pr02, 84pr03, and 83pr04 Fisherian confidence intervals were not calculated. Because of the small number of samples only the Fisherian means are shown for these localities.

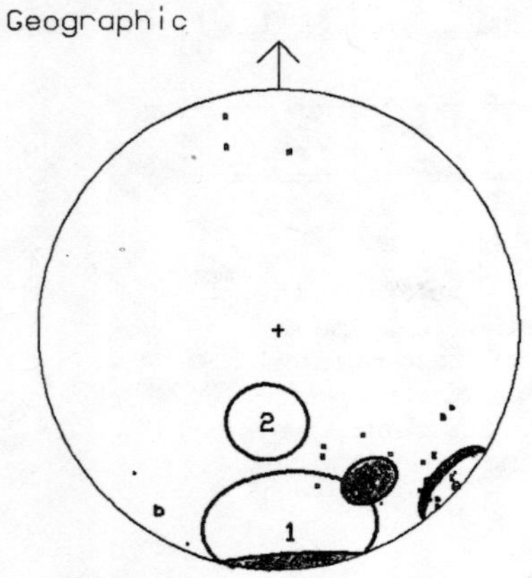

Figure 6

Equal area stereographic projection of the final data set from Parkhurst Ridge and LL-1 and LL-2 from Alvarez and others (1980) in geographic coordinates. Symbols are the same used in figure 5 note the similarity in direction.

ly after deposition of the limestone. However we note that remagnetization is a possibility. Several factors suggest that this component is of primary origin. These are, a, the coherence of the inclinations after large structural corrections were applied (figure 5, table 2), b, single component nature of the magnetic measurements above 250° C, c, observation that the primary inclinations uncorrected for tilt were not near any expected field direction and d, the presence of samples of both polarities within individual sites and between sites in the final data set.

This result is 23° (table 3) shallower than that expected had the limestone been deposited on the North American plate and implies 2550 km of relative poleward convergence between the limestone and North American plate since Late Cretaceous (Campanian-Maastrichtian). Figure 5 shows the danger of tectonic interpretation based on paleomagnetic results from one or two outcrops. Both localities 83pr01 and 83pr05 yield stable samples and their respective means agree. Locality 84pr03 also yields stable samples. If this locality had been the only accessible outcrop or substantially larger than the others this could have led to a very different tectonic interpretation.

INTERPRETATION OF PALEOMAGNETIC DATA AND PLATE TECTONIC MODEL

The preliminary Parkhurst Ridge paleomagnetic data suggests the limestone was deposited at a paleolatitude of 28°. High rates of required poleward translation suggest that a Southern Hemisphere interpretation of latitude is unlikely (78° of poleward translation in approximately 40 Ma). Therefore we interpret deposition to have occurred in the Northern Hemisphere. Two models are presented which satisfy the mean and the lower acceptable range of latitude of deposition. The large associated uncertainity about the mean does not allow rejection of the possibility of little or no movement with respect to North America since deposition. However the mean result and additional constraints posed by the age of the limestone, the age of its accretion to the North American plate and paleontologic evidence provide the foundation for our terrane model. Of the plates in the present Pacific basin at the time of deposition of the limestone (Kula, Farallon, and Pacific, (Woods and Davies 1982)), Farallon plate motions (Engebretson 1982) most easily satisfy all these constraints. Farallon-North America motions between 70-45 Ma also provide an appealing explanation for the Laramide orogeny (Engebretson and others 1984a, Page and Engebretson 1984). Kanter (1983) also used Farallon-North America relative motions to model Cretaceous and lower Tertiary paleomagnetic results from central Salinia and the Point Arena terrane.

The method of calculation of a terrane trajectory is explained fully in Engebretson (1982) and Engebretson and others (1984c) but is briefly reviewed. Once the age and location of accretion are defined and a plate travel history is assumed, relative motion poles for the plates involved may be combined. These produce a set of relative motion poles that rotate the terrane in accordance with the assumed plate model. The motion of the North American plate is known with respect to the spin axis (assuming a geocentric dipole this motion is recorded by the apparent polar wander path for North America) and the relative motion of the terrane with respect to North America is known. These are combined to generate a predicted position with respect to North America along which predicted magnetic inclination and declination may be calculated (Engebretson and others 1984c).

Figure 7 shows the motion of the limestone of Parkhurst Ridge across the Pacific Basin in a North America fixed reference frame. The solid lines represent 10° latitude lines drawn around the 70 Ma reference pole for North America from Irving and Irving (1982). The dark solid line shows the mean paleolatitude of deposition calculated in this study. Ticks show present day latitude and longitude. The 70 Ma Northern hemisphere paleontologic (Sliter 1984) low-latitude band is also shown. The model predicts the age of the oceanic crust that the limestones rode on. In our first model this age is approximately the age of the limestones themselves, that is, the limestones were deposited on the Farallon-Pacific ridge, (for example, if deposited off the ridge on a seamount or plateau , the age of the crust on which the limestones were deposited could be considerably older). In this model the limestone of Parkhurst Ridge moves with the Farallon plate from 70 Ma until it encounters the North American plate in late Eocene time (40 Ma). Accretion takes place near the present location of the limestones on the North American plate. Between about 40 and 28 Ma some poleward strike-slip motion along North American plate margin may have occurred within the Coastal Belt. In our first model this motion is only a small component of the total rate of Farallon-North America poleward convergence and is not included in the displacement model.

A second terrane displacement model (figure 7) attempts to satisfy the lower range of latitude of deposition that is paleomagnetically acceptable and low-latitude constraints posed by Sliter (1984). In the second model after accretion of the limestone with the North American plate at 45 Ma the poleward component of Farallon (45-10 Ma) and Pacific (10-0 Ma) motion is used to strike-slip the limestone poleward. This

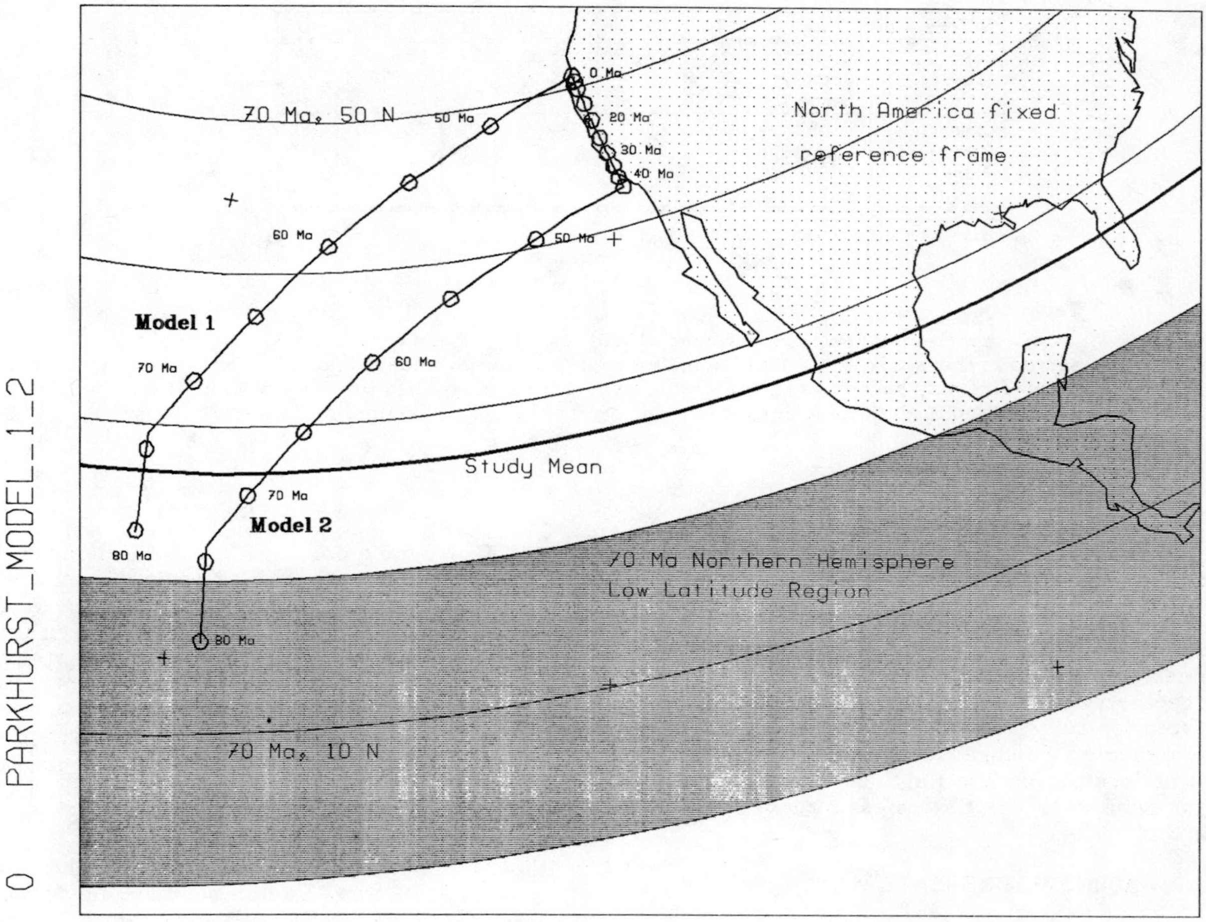

Figure 7

Terrane trajectory calculated using the Farallon-North America poles of Engebretson (1982) in a North America fixed reference frame. The study mean calculated from the final data set is shown as a dark latitudinal band. Open circles represent positions of the limestone at 10 Ma intervals along two possible terrane trajectories. Note the longitudinal as well as latitudinal displacement as the limestone move away from the Farallon-Pacific spreading center and accrete to the North American plate. The shaded band represents the 70 Ma interval between 20° N and the 70 Ma equator calculated about the 70 Ma reference pole of Irving and Irving (1982). 30° latitude and longitude tick marks are shown in present-day coordinates for reference. Model 2 nearly satisfies the constraint of low latitude deposition (between 20° North and 20° South) and differs from model 1 by using the poleward component of Farallon (45-10 Ma) and Pacific (10-0 Ma) motion after accretion of the limestones to North America.

defines the maximum amount of poleward translation possible given the Farallon-North American relative motion poles of Engebretson (1982). This relative convergence is ≈ 30°, or roughly 3300 km, which represents a paleolatitude of deposition of ≈ 20° N. This result satisfies the low-latitude requirement of both the lower bound of the paleomagnetic data (<28° N) and the paleontologic data (20° N). Figure 7 shows that both paleomagnetic and paleontologic data suggest the limestone of Parkhurst Ridge is allochthonous to the North American plate and the observed paleolatitudes of deposition fit well into the context of Farallon-North America plate motions as presented in Engebretson (1982). Further paleomagnetic sampling is required to produce constraints needed to choose between these terrane models.

SUMMARY

The paleomagnetism of Parkhurst Ridge limestones consists of two components. The characteristic remanent magnetism with $I=-47°$, which is 23° shallow from the expected inclination for the North American plate. The data may be modelled with Farallon-North America relative motions (Engebretson 1982) by assuming the limestones were accreted by approximately late Eocene or Oligocene time. A displacement model constructed around the parameters of observed latitude of deposition, time of accretion and location of accretion, fits the observed paleomagnetic data and satisfies paleontologic constraints. Together the paleomagnetic data and terrane model suggest the limestones are allochthonous and were deposited on the Farallon plate. This model assumes the limestones

Reference	Age	N	Lat N	Long E	D	I	λ	φ
Saad (1969)	Upper Creta-ceous	11	37.5	238.5	29	68	65	286
Overprint, Parkhurst Ridge	Post-accretion (40-28 Ma)	14*	41	236	23	69	71	282

*Averaged over 14 samples.

Table 4

Comparison of the result of Saad (1969) and the secondary component observed in this study. Note the calculated pole positions are nearly indistinguishable. This may reflect a regional low temperature remagnetization event which affected both Parkhurst Ridge (lat 41° North) and Red Mountain (lat 37.5° North).

of Parkhurst Ridge had a depositional site on or near the Farallon-Pacific spreading center.

The second magnetic component was acquired after the limestone was accreted and deformed into roughly the geometry seen today. The overprint direction, before rotation to horizontal is oversteepened relative to any expected Paleogene direction calculated for the present location on the North American plate (figure 4) but similar to the CRM measured by Saad (1969) (table 3).

ACKNOWLEDGEMENTS

We thank Dave Engebretson for insight regarding relative plate motions and the use of rotation poles from Pacific basin plate models. Computer programs written by Michel Debiche allowed calculation of a accurate terrane trajectory and were kindly made available to us. Ben Page, Ray Wells, Leah Frei, and John Tarduno read preliminary versions of this paper, their excellent and helpful suggestions greatly improved the final version. Special thanks to Allan Cox for insightful discussions concerning adequate sampling of the geomagnetic field and calculation of paleomagnetic study uncertainities.

REFERENCES CITED

Alvarez, W., Kent, D. V., Silva, I. P., Schweickert, R. A., and Larson, R. L., 1980, Franciscan-Complex limestone deposited at 17° south paleolatitude: Geological Society of America Bulletin, v. 91, p. 476-484.

Bailey, E. H., Irwin, W. P., and Jones, D. L., 1964, Franciscan and related rocks and their significance in the geology of western California: California Division of Mines and Geology Bulletin 183, 177p.

Beck, M. E. Jr., 1976, Discordant paleomagnetic pole positions as evidence of regional shear in the western Cordillera of North America: American Journal of Science, v. 276, p. 694-712

Blake, M. C. Jr. , and Jones, D. L., 1978, Allochthonous terranes in northern California?--A reinterpretation: in Howell, D. G., and McDougall K. eds., Mesozoic paleogeography of the western United States: Society of Economic Paleontologists and Mineralogists, Pacific Section, Pacific Coast Paleogeography Symposium 2, p. 397-400.

Blake, M. C. Jr., and Jones, D. L., 1981, The Franciscan assemblage and related rocks in northern California: A reinterpretation: in Ernst, W. G., ed., The Geotectonic development of California: Rubey, v. 1, p. 330-417.

Courtillot, V., Feinberg, H., McWilliams, M., and Cox A., 1983, Franciscan Complex Limestone deposited at 21° North: EOS, Transactions, American Geophysical Union, v. 64, no. 45, p. 686.

Demarest, H. H., Jr., 1983, Error analysis for the determination of tectonic rotation from paleomagnetic data: Journal of Geophysical Research, v. 88, no. 5, p. 4321-4328.

Engebretson, D. C., 1982, Relative Motions between Oceanic and Continental plates in the Pacific Basin: Ph. D. thesis, Stanford University, 211 p.

Engebretson, D. C., Cox, A., and Thompson, G. A., 1984a, Correlation of plate motions with continental tectonics: Laramide and Basin-Range: Tectonics, v. 3, p. 115-119.

Engebretson, D. C., Gordon, R. G., and Cox. A., 1984b, Relative motions between oceanic and continental plates in the pacific basin: in preparation.

Engebretson, D. C., Debiche, M. G., and Cox, A., 1984c, Plate motions, paleomagnetism and terranes: in preparation.

Evitt, W. R., and Pierce, S. T., 1975, Early Tertiary ages from the Coastal Belt of the Franciscan Complex, northern California: Geology, v. 3, no. 8, p. 433-436.

Fisher, R. A., 1953, Dispersion on a Sphere: Proceeding of the Royal Society of London, Series A. Mathematical and Physical Sciences, v. 217, p. 295-305.

Garrison, R. E., and Bailey, E. H., 1967, Electron microscopy of limestones in the Franciscan formation of California: United States Geological Survey Professional Paper 575-B, p B94-B100.

Grommé, C. S., and Gluskoter, H. J., 1965, Remanent magnetization of spilite and diabase in the Franciscan formation, western Marin county, California: Journal of Geology, v. 73, p. 74-94.

Harland, W. B., Cox, A. V., Llewellyn, P. G., Pickton, C. A. G., Smith, A. G., and Walters, R., 1982, A geologic time scale: Cambridge University Press, 131 pp.

Irving, E., and Irving, G. A., 1982, Apparent polar wander paths Carboniferous through Cenozoic and the assembly of Gondwana: Geophysical Surveys, V. 5, p. 141-188.

Jones, D. L., Blake Jr., M. C., Bailey, E. H., and McLaughlin, R. J., 1978, Distribution and character of upper Mesozoic subduction Complexes along the west coast of North America: Tectonophysics, v. 47, p. 207-222.

Kanter, L. R., 1983, Paleomagnetic constraints on the movement history of Salinia: Ph. D. thesis, Stanford University, 156 p.

Kirschvink, J. L., 1980, The least-squares line and plane and the analysis of paleomagnetic data: Geophysical Journal of the Royal Astronomical Society, v. 62, p. 699-718.

McElhinny, M. W., 1964, Statistical significance of the fold test in palaeomagnetism: Geophysical Journal, v. 8, no. 3, p. 338-340.

McFadden P. L., and Reid. A. B., 1982, Analysis of palaeomagnetic inclination data: Geophysical Journal of the Royal Astronomical Society, v. 69, p. 307-319

McLaughlin, R. J., Kling, S. A., Poore, R. Z., McDougall, K., and Beutner, E. C., 1982, Post-middle Miocene accretion of Franciscan rocks, northwestern California: Geological Society of America Bulletin, v. 93, p. 595-605.

Page, B. M., 1981, The southern Coast Ranges: in Ernst, W. G., ed. The Geotectonic development of California: Rubey, v. 1, p. 330-417.

Page, B. M., and Engebretson, D. C., 1984, Correlation between the geologic record and computed plate motions for central California: Tectonics, v. 3, p 133-155.

Saad, A. H., 1969, Paleomagnetism of Franciscan ultramafic rocks from Red Mountain, California: Journal of Geophysical Research, v. 74, no. 27, p. 6567-6578.

Sliter, W. V., 1984, Cretaceous low latitudes from late Cretaceous limestones from the Franciscan Assemblage, California, and Sixes River terrane, Oregon: in, Geological Society of America, Abstracts with Program, 80th Annual Cordilleran Meeting, p 333.

Wachs, D., and Hein, J. R., 1975, Franciscan limestones and their environments of deposition: Geology, v. 3, p. 29-33.

Woods, M. T., and Davies, G. F., 1982, Late Cretaceous genesis of the Kula Plate: Earth and Planetary Science Letters, v. 58, p. 161-166.

FRANCISCAN COMPLEX GEOLOGY OF THE CRESCENT CITY AREA, NORTHERN CALIFORNIA

K. R. Aalto
Department of Geology
Humboldt State University
Arcata, California 95521

J. M. Murphy
Division of Geological and Geophysical Surveys
Department of Natural Resources
State of Alaska
794 University Avenue - Basement
Fairbanks, Alaska 99701

ABSTRACT

Largely overturned, northwest-striking, northeast-dipping units of broken formation and sheared olistostrome are exposed along abrasion platforms and sea cliffs. Units grade in character from Mutti and Ricci Lucchi (1972) mixed facies B, C, and D, to F, to mixed A, B and E from east to west, upwards within the section. Sandstones are litharenitic. Uncommon pebble to boulder conglomerates have the followng mean clast composition: chert-8%, clastic sedimentary-37%, tonalite-13%, porphyritic volcanic-24% and non-porphyritic volcanic-18%.

Facies F olistostromes contain olistoliths up to hundreds of meters across composed of chert, brecciated tonalite, serpentinite, greenstone, limestone and graywacke despersed in a shaley matrix. Olistoliths are commonly discoidal and oriented subparallel to a poorly to well developed anastomosing matrix foliation, or irregular in shape and randomly oriented where foliation is lacking. At Point Saint George a 25 m thick olistostrome is in depositional contact with bounding turbidites.

The deformation history of all units includes soft-sediment folding associated with olistostrome emplacement, followed by pervasive layer parallel brittle and ductile extension causing widespread boudinage and olistolith form modification, followed by compression resulting in thrust faulting and folding of all units. These deformation events can be related to a regional plate tectonic model for the development of the Franciscan accretionary wedge.

INTRODUCTION

The deformation history of subduction complexes and origin of associated melanges are controversial. Effects of both compressional and extensional deformation have been described within the Franciscan subduction complex and ascribed to events that occurred during its initial formation (Hsu 1967, 1968; Maxwell 1974; Cowan 1974, 1982; Aalto 1981). The origin of Franciscan melanges has been ascribed to both sedimentary (Aalto 1982; Maxwell 1974; Cowan, 1978) and tectonic (Cloos 1982; Korsch 1982) mixing processes. The purpose of this paper is to provide a preliminary description of stratigraphy, lithology and structure of an olistostrome within the Franciscan Complex and relate them to regional tectonic history. This olistostrome is well exposed on abrasion platforms and seacliffs between Crescent City and Point Saint George (Fig. 1, 2). The reader is referred to Aalto (1982) for a more detailed review of the Franciscan geology of northernmost California and description of the Lighthouse Island (Fig. 2) stratigraphic section.

REGIONAL GEOLOGIC SETTING

The Franciscan Complex of the Crescent City area is part of the Central Belt of Blake and Jones (1974), consisting of intercalated melanges and broken formations (terminology of Hsu 1974) ranging in age from Late Jurassic through Cretaceous. Aalto (1982) has subdivided Franciscan rocks of northernmost California into an eastern belt of solely broken formation and western belt of melange with subordinate broken formation (Fig. 1), both characterized by unfoliated graywacke (textural zone 1 of Blake et al., 1967). The study area falls within the western belt. Fossils found between Trinidad and the Klamath River indicate an age range of Tithonian-Cenomanian for Franciscan rocks of this region (Aalto et al., 1983; Harden et al., 1981).

The eastern limit of Franciscan Complex exposure is defined by the Coast Range thrust fault (Fig. 1), along which older Jurassic rocks of the Klamath Mountains plate have overrider the Franciscan by some tens of kilometers (Blake and Jones 1981). Other major thrust faults portrayed in figure 1 juxtapose semischistose and schistose (textural zones 2 and 3) Franciscan rocks over those of textural zone 1. All of these thrust faults are marked by a 5-100 m thick tectonic melange composed chiefly of graywacke, greenstone and chert blocks dispersed in argillaceous gouge (Aalto 1982, 1983; Monsen and Aalto 1980). Drag folding beneath these faults is confined to narrow zones beneath this gouge. There is no evidence that the multiple deformation and metamorphic events that affected upper plate rocks genetically relate to deformation of lower plate rocks beyond this zone of drag folding. Blake and Jones (1983) and Jayko (1983) provide evidence constraining motion on the Coast Range thrust and faults which appear correlative to the other thrusts of figure 1 to the Cenomanian-Paleocene (92-60 Ma). The Grogan fault (Fig. 1) is characterized by a narrow tectonic melange zone containing blocks of diverse lithology dispersed in argillite with vertical folds and foliation (Aalto 1983). It is thought to have large right lateral displacement (Monsen and Aalto 1980; Kelsey and Hagans 1982)

in Blake, M.C. Jr., ed., 1984, **Franciscan Geology of Northern California:** Pacific Section S.E.P.M., Vol. 43, p. 185-201

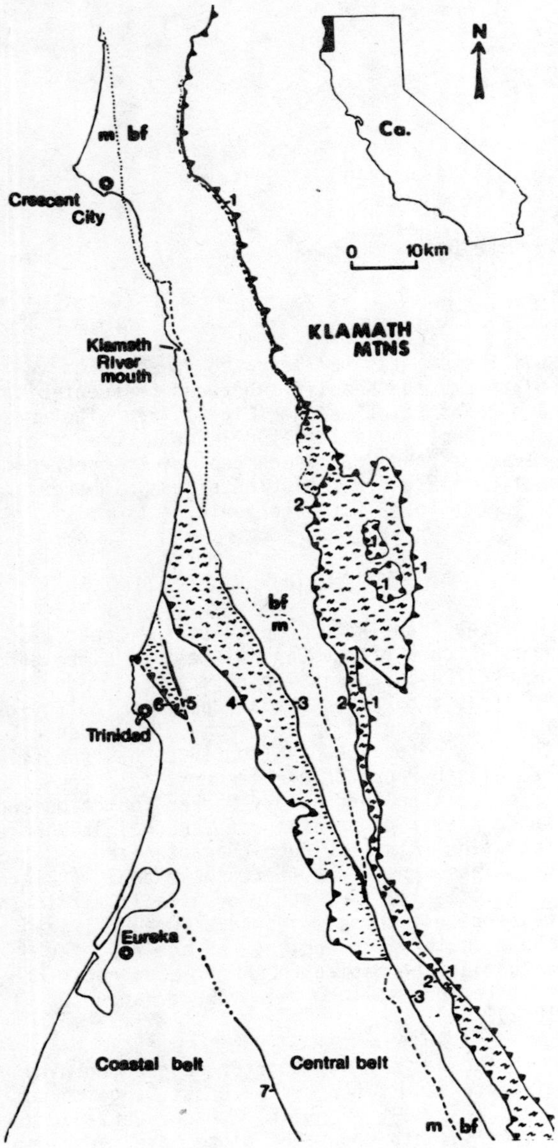

Figure 1: Location map showing distribution of textural zones and generalized regional geology. The numbered faults aeneralized regional geology. The numbered faults are: (1) the Coast Range thrust fault (or South Fork fault) which constitutes the eastern boundary of the Coast Ranges geological province and of Franciscan Complex basement rock, (2) the Redwood Mountain fault (or South Fork Mountain fault), (3) the Grogan fault, (4) the Bald Mountain fault, (5) the Big Lagoon fault, (6) the Patricks Point fault, and (7) the Coastal Belt fault (Freshwater fault of the Eel River basin) which separates Coastal Belt and Central Belt Franciscan terranes. Textural zone 1 Franciscan basement - no pattern, zone 2 - dashes, zone 3 - wavy dashes, undifferentiated zone 2/3 - mixed pattern. Central belt zone 1 rocks are further subdivided into an eastern belt of chiefly broken formation (labeled bf), and a western belt of intercalated melange and subordinate broken formation units (labeled m). Geology generalized from published and unpublished mapping of K. R. Aalto, P. Cashman, S. M. Cashman, H. M. Kelsey, D. Hardiin, T. Stevens, S. Morrison and G. D. Harper and the Weed and Redding sheets of the Geologic Map of California. Only basement geology and faults bounding major Franciscan terranes are shown.

and may be equivalent to a Cenozoic lateral fault described by Jayko (1983) to the south.

Late Cenozoic deformation is profound throughout the northern California Coast Ranges. Carver and others (1982, 1983) describe thrust faults between Eureka and Trinidad that place Franciscan rocks over Pleistocene terrace deposits and involve several kilometers of crustal shortening. Kelsey (1982, 1983) and Kelsey and Cashman (1983) describes strike-slip faults that offset Pleistocene terraces between Trinidad and the Klamath River. Both styles of faulting produce fractures and well-defined shear zones among Franciscan rocks, but cannot account for extensive Franciscan folding because of their shallow crustal level. North of the Klamath River, Late Cenozoic deformation is less severe and Neogene to Holocene formations have been uplifted and tilted, but not extensively faulted (Aalto and Harper 1982; Davenport 1982). An excellent summary of Late Cenozoic deformation is provided in Harden and others (1982).

SEDIMENTOLOGY

Introduction:

Franciscan Complex sedimentary rocks of the Crescent City area are predominantly submarine sediment gravity flow deposits and may be described using standard facies terminology of Mutti and Ricci Lucchi (1978). In aerial distribution (Fig. 2, 3) chaotic deposits of facies F and massive to graded sandstone turbidites of facies B and C are most abundant. Facies A conglomeratic sandstone and facies D and E shales with sandstone interbeds are subordinant in abundance. No significant compositional differences exist between shales and sandstones of the different facies. Deformation features are discussed in a subsequent section.

Bedding features are graphically described using the model of Aalto (1976):

Division	Description
I	Massive sandstone or inversely graded pebbly sandstone.
II	Pebbly sandstone or conglomerage with or without imbrication.
III	Normally graded conglomerate or pebbly sandstone.
IV	Diffusely laminated and/or cross stratified pebbly sandstone.
V	Massive sandstone.
VI	Laminated sandstone with dish and pillar structures.

Following the newly proposed model of Lowe (1983), divisions I-II-III correspond to R_2-R_3, and IV-V-VI to S_1-S_2-S_3. Refer to Lowe for a thorough discussion of their genesis.

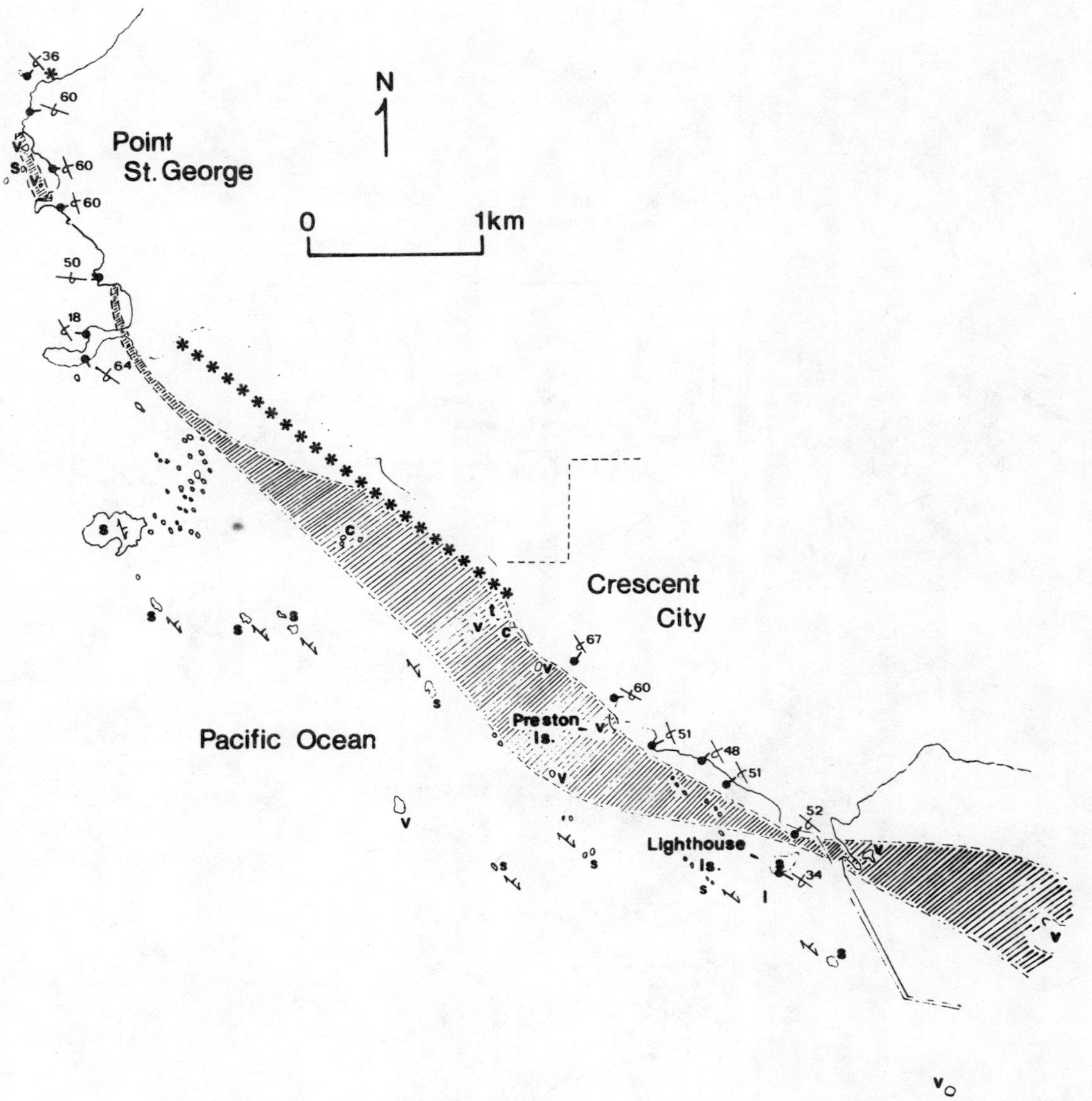

Figure 2: Geologic map of the Crescent city area. Only Franciscan Complex geology is shown. Franciscan rocks are exposed in seacliffs, offshore stacks and on abrasion platforms. Uplifted marine terrace deposits and the Plio-Pleistocene Saint George Formation (Back 1959) underlie inland areas. The Franciscan-St. George Formation unconformity is exposed in seacliffs east of the line of asterisks to the northeast of Crescent City and at northernmost Point Saint George. The main olistostrome outcrop area is cross-hatched. Attitudes are given for locations indicated with dots. Composition of larger olistoliths is given as follows: C - chert, V - greenstone and T - tonalite. Graywacke composition of stacks visited by boat is indicated with an "S". The barbed strike and dip symbols depict bedding orientation as observed on air photos and field-checked by boat. Facing of offshore beds is unknown, but most likely beds are overturned. Geologic mapping by J. M. Murphy (1982) at Point Saint George and K. R. Aalto elsewhere.

Facies A:

Conglomeratic sandstones and conglomerates exist at Lighthouse Island (Fig. 2) and in fault-bounded patches and measured sections at Point Saint George (Fig. 3). The southern coarse conglomerate exposure at Point Saint George is 10 m thick and comprised of an overturned coarsening upward sequence. Beds are commonly graded, 0.75 to 2 m thick and show basal scour structures. The northern conglomerate is massive and highly sheared. Clasts are subrounded to well rounded, poorly sorted and range from medium pebble to boulder size, with a maximum exposed long axis length of 0.75 m. Three field counts of approximately 100 clasts each show an average composition of chert - 8%, clastic sedimentary - 37%, tonalitic plutonic - 13%, porphyritic volcanic - 24%, and non-porphyritic volcanic - 18%. Volcanic clasts include a variety of tuffs and microlitic, felsitic and lathwork textural types of Dickinson (1970).

Coarse pebbly sandstones are present in the measured section of figure 4 and on Lighthouse

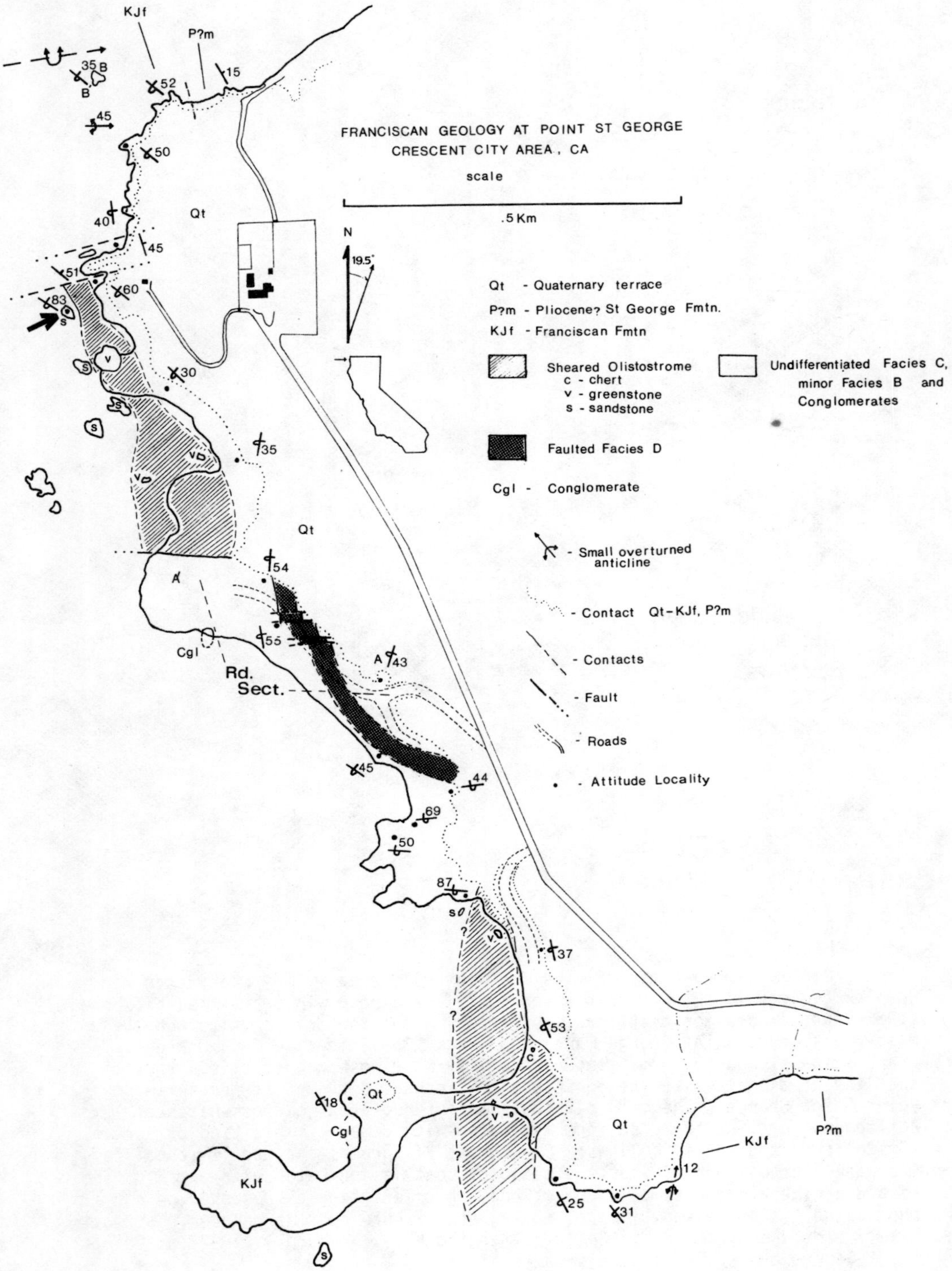

Figure 3: Geologic map of Point Saint George. Franciscan beds are largely overturned and are unconformably overlain by Quaternary terrace deposits and the Plio-Pleistocene Saint George Formation. Sheared olistostromes are well exposed in the north, less so in the south. Note stepping of facies D beds along high angle faults in the central portion of the map. Geologic map from Murphy (1982).

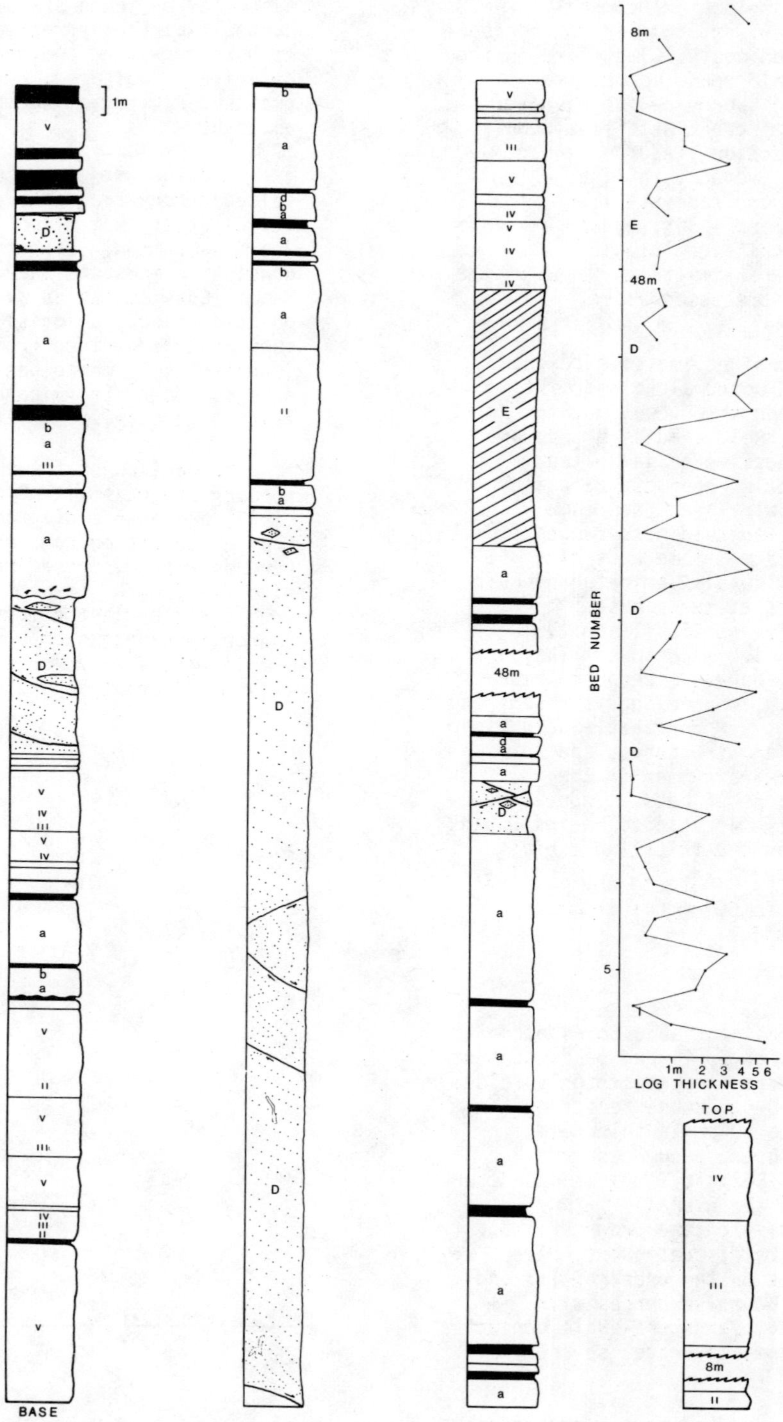

Figure 4: Stratigraphy of the road section (A-A') at Point Saint George. Bouma sequence divisions within beds are shown by lower case letters; Roman numerals depict bedding features among pebbly sandstones according to the Aalto (1976) model. Mutti and Ricci Lucchi (1978) Facies D and E are labeled with upper case letters. Black depicts thick shale and shale ripup clasts within beds. Facies D units are shown faulted, folded and contain sandstone phacoids. Facies E beds are not as extensively deformed. The graph portrays thicknesses of successive beds, from base to top. Breaks in graphs occur where facies D beds are contorted, thin facies E beds occur, or the section is covered. Section location given in figure 3. Figure from Murphy (1982).

Island (Aalto 1982). Beds range in thickness from a decimeter to 6 m, but are commonly 1 - 2 m. Less scouring is evident than in the coarser conglomerates described above. Clasts are in the granule to medium pebble range, poorly to moderately sorted, and subrounded to rounded. Two 100 point counts of granule to fine pebble size clasts in thin section yield an average composition of quartz (monocrystalline, stretched metamorphic and recrystallized metamorphic) - 11%, chert - 36%, mudstone - 28%, volcanic - 23% and tonalitic plutonic - 2%. Volcanic lithologies are similar to those of the coarser conglomerates discussed above.

Facies B and C:

Fine to coarse-grained, massive to graded, gray sandstones are exposed along seacliffs below Crescent City and throughout much of Point Saint George (Fig. 4, 8e). Facies B beds are commonly massive or massive to laminated, normally graded in their uppermost part and overlain by a thin shale layer (sequence V-VI-T_{ae} or S_3-T_{te}). Bed thickness ranges from 1 - 6 m, averaging 1 - 2 m. Some beds have massive, graded or diffusely laminated granule to small pebble layers at their bases corresponding to divisions II, III or IV overlain by thick massive sandstone. They are assigned to facies B, however, because these coarse layers are thin, channeling is absent and massive sand divisions are dominant. Shale ripup clasts, dish and pillar structures, and convoluted lamination are common in the laminated upper portions of beds. Groove and load casts exist at two locations. Facies C beds are commonly less than 1 m thick and exhibit classic Bouma sequences: T_{abce}, T_{abe} and T_{ae}, with T_e divisions up to 50 cm but commonly several centimeters thick.

Facies D and E:

Shales with subordinant sandstones occur at Point Saint George (Fig. 3, 4) and on Lighthouse Island. Both facies are in places highly folded and faulted. Facies D sandstone beds range from several centimeters to 50 cm in thickness, exhibit base-cut-out Bouma sequences and uncommon flute casts. Pelitic T_e divisions occur up to 45 cm thick and the overall sandstone-shale ratio is on the order of 1:5. Facies E beds are quite discontinuous, have sandstone shale ratios on the order of 1:1 and exhibit base-cut-out Bouma sequences with extensive ripple drift. Sandstone-shale contacts are sharp. Bed thickness averages several centimeters.

Sandstone Composition and Sediment Provenance:

Sandstones are commonly litharenites or feldspathic litharenites (Fig. 5). Based upon six counts of 300 grains per thin section the following Dickinson and Suzcek (1979) mean framework modes have been determined: Q - 44%, Q_m - 21%, Q_p - 23%, F = P - 11%, L_t - 69%, L - 46%, L_s - 11%, and L_v - 35%.

Most monocrystalline quartz grains exhibit undulose extinction and some have deformation bands and lamellae. Some monocrystalline quartz grains have the clarity, idiomorphism, fractures and embayments which suggest a volcanic origin. All varieties of polycrystalline quartz described by Young (1976) exist. Chert is commonly partially recrystallized. Plagioclase is both untwinned and sericitized or twinned and relatively unaltered. Pumpellyite exists in some grains. Albite and sodic oligoclase are most abundant.

All four major textural categories of volcanic rock fragments described by Dickinson (1970) exist, but microlitic and felsitic grains are most common. Lithic sedimentary clasts consist of mudstone and very fine sandstone. Matrix constitutes an average of 20% of samples point-counted. Chloritic orthomatrix and pseudomatrix derived from mudstone grains (counted as L_s where possible) are common. Accessory detrital minerals include pigeonite, zoisite, augite, muscovite and sphene.

Comparison of framework modes plotted on ternary diagrams with Dickinson and Suzcek (1979) provenance diagrams suggests derivation from a subduction complex source terrane, perhaps with minor sediment contribution directly from a somewhat dissected arc. Conglomerate clast composition fits such a source terrane, as does data for sandstones and conglomerates of a large area of the northern California coast Ranges (Fig. 5; Aalto 1982).

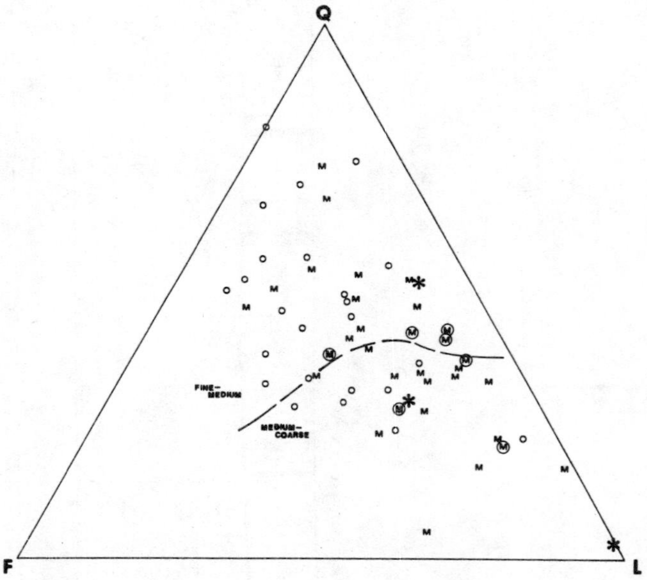

Figure 5: Franciscan sandstone composition, northern Humboldt and Del Norte Counties. Following Folk (1968): Q=mono- and polycrystalline quartz (excluding chert), F=feldspar and granitic rock fragments, and L=unstable lithic fragments plus chert. Data points for the western melange belt of figure 1 indicated with M, the eastern broken formation belt with circles. Data points for the Crescent City area are encircled. The dashed broken line approximately divides finer from coarser-grained samples. Asterisks represent textural zone 2 sandstones from the coastal metasedimentary rock exposure immediately north of Trinidad. Data from counts of 300 grains per thin section.

Figure 6: Submarine slump deposit showing progressive development of chaotic fabric at Point Saint George, bedding is overturned and dips to the left. From left to right, up within the bed, slump folds become indistinct, the fabric becomes random and graywacke clast roundness increases. The top of the bed is off the photo to the right. Scale marked in decimeters.

Figure 7: Olistostrome at Preston Island, Crescent City. Slightly foliated diamictite with blocks of graywacke (s) and greenstone (g) dispersed in an argillaceous matrix (a). Note the irregular shapes and random orientation of clasts.

Facies F:

Olistostromes and associated slump-folded turbidite beds are extensively exposed along the coast and offshore in the Crescent City - Point Saint George area (Fig. 1). Two types of deposits exist: diamictites with and without an anastomosing matrix foliation.

Diamictites lacking matrix foliation are exposed in the northern part of Point Saint George, in the second cove north of the fault truncating the main olistostrome exposure (Fig. 3) and in seacliffs due east of Preston Island (Fig. 2). In both localities diamictites are composed chiefly of angular to rounded clasts of graywacke, with very minor chert and greenstone, dispersed in a sand-silt matrix. In the northern exposure (Fig. 6), upwards within a bedded diamictite unit, thin-bedded T_{abe} and T_{abcde} turbidites become slump-folded, folds become irregular and detached and basal portions of beds appear as isolated clasts. Clast roundness increases upwards within the bed. These changes are ascribed to downslope motion, soft sediment deformation and the progressive homogenization of a thin-bedded turbidite sequence. The presence of foreign greenstone and chert clasts in the upper portion of the bed suggests that disruption of the turbidites was initiated by the overriding of the sequence by a subaqueous mudflow which produced a mixture of autocthonous and allocthonous diamictite components. The same relationships exist east of Preston Island, but are less well exposed. The presence of the major olistostrome unit immediately to the west of slump-folded sand-silt matrix diamictites suggests a genetic relationship between the two. These diamictites attest to unstable submarine slope conditions, which resulted in slope failure during the initial stages of the main olistostrome emplacement.

The main olistostrome portrayed in the regional maps (Fig. 1, 2) is composed of olistoliths of diverse lithology dispersed in a shaley matrix with pervasive anastomosing foliation, the origin of which will be discussed in a later section. The overall matrix/olistolith ratio, estimated by area at outcrop and in outcrop photographs, varies from 5:1 to 1:1, with an average of 2 - 3:1. Where olistoliths are highly fragmented and dispersed, the distinction between olistolith and matrix becomes somewhat arbitrary and is set approximately at a 1 cm short axis dimension in planar view. In the few areas where matrix foliation is absent or poorly developed, olistoliths are equant and angular to subangular (Fig. 7). Post depositional form modification of olistoliths, discussed ahead, has resulted in rounding, development of a disc form, and reorientation parallel to matrix foliation. Olistoliths range in size from centimeters to tens of meters in maximum exposed dimension.

The exposure of the main coastal olistostrome is well constrained by onshore and offshore mapping (Fig. 2). Contact relationships between the olistostrome and underlying, overturned turbidites are well exposed at Point Saint George (Fig. 3, 8a, 8b; Murphy and Aalto 1982, 1983). Here the olistostrome is approximately 25 m thick and in depositional contact with facies B and C turbidites both to the east and west. To the east the chaotic unit rests upon a 6 m thick V-T_{bcde} (or S_3 - T_{t-e}) facies B sandstone bed. The T_e division of this bed in some areas is terminated by a calcite-filled fracture beneath the scaly matrix of the olistostrome, but in other places grades into this matrix. The matrix foliation passes at a shallow angle into bedding into the T_e division, is defracted to a shallower angle in the T_{cd} divisions and fades out within the T_b division of the bed. Similar, but less well preserved contacts of the same olistostrome are seen to the west at this location and on the east side of the southernmost Point Saint George olistostrome exposure (Fig. 3). The olistostrome is truncated by faults along strike, but is bounded conformably by turbidites on both sides where outcrop exists.

Olistolith Composition:

Olistoliths are chiefly composed of sandstone with less abundant greenstone and trace amounts of radiolarian chert, limestone, serpentinite, phyllite and tonalite. Composition of large olistoliths is given on maps (Fig. 1, 2, 3). Sandstone composition is similar to that of turbidite sandstones discussed above, although more feldspathic. Two point counts of 300 grains give average framework modes: Q - 41%, Q_m - 21%, Q_p - 20%, F = P - 26%, L_t - 53%, L - 33%, L_s - 2%, and L_v - 31%.

Greenstone includes metabasalt, dacite porphyry, quartz keratophyre and tuff with highly variable proportions of crystals, shards and lithic fragments. Some tuffaceous olistoliths contain highly indurated black argillite, intermixed complexly with tuff in wispy convoluted intrusions, similar to blocks described by Aalto (1981) to the south. Tonalite contains granophyric quartz and albite (An_8) in subequal proportions and trace amounts of apatite, sphene, zircon and chloritized amphibole. Scattered keratophyre dikes intrude the tonalite and both are thoroughly brecciated at thin section scale. All olistoliths could logically be derived by recycling of a subduction complex. Tuffs may have been deposited with pelagic mud on an oceanic plate prior to its subduction accretion.

Shale Composition:

Cloos (1983a, 1983b) suggests that the role of tectonic mixing in melange formation may be assessed by X-ray diffraction study of shaley melange matrix and shales of adjacent turbidites. Tectonic mixing during subduction accretion would result in a higher pressure mineral assemblage within melange than in muds deposited upon the trench slope. True olistostromes, even if derived from diapirs or thrust plates of tectonic melange that pierce the trench slope sediment cover, should contain a fair amount of seafloor mud in their matrix derived by mixing during gravity flow. To evaluate this hypothesis and whether the Crescent City - Point Saint George olisostrome might be a sill-like intrusion of tectonic melange as suggested by Cloos (1983b), X-ray studies were undertaken on olistostrome matrix and shales and ripup clasts in bounding turbidites (Aalto, unpublished data; Humboldt

TABLE 1: SOME SHALE COMPOSITION DATA DETERMINED ACCORDING TO TECHNIQUES DESCRIBED BY CLOOS (1983a)*

Parameter	Range (r), mean ($\bar{x}$) and standard deviation (s) for: OLISTOSTROME MATRIX (n=8)	TURBIDITE SHALE (n=7)
Illite crystallinity index	r= 3.6 - 6.4 $\bar{x}$= 5.15 s= 1.04	r= 1.6 - 6.6 $\bar{x}$= 4.17 s= 1.78
Mica b_o lattice spacing	r= 9.006 - 9.054 $\bar{x}$= 9.021 s= 0.017	r= 9.002 - 9.027 $\bar{x}$= 9.015 s= 0.008

*Data from Peggy Guare, Humboldt State University B.Sc. thesis in progress.

State X-ray diffraction class, 1983; Peggy Guare, senior thesis in progress).

Table 1 presents data for samples analyzed. Procedures suggested by Cloos (1983a) were utilized: powder analysis of samples run untreated, glycolated, heated to 550°C and treated with concentrated hydrochloric acid. All samples contain iron-rich chlorite, quartz, sodic plagioclase and white mica. Kaolinite exists in some samples of both melange matrix and turbidite shale. Mixed-layer clays are absent. Based upon absence or poor low intensity of the 2.80 Å 2M mica peak, despite the high intensity of other mica peaks, most mica is of the 1Md variety. The illite crystallinity index (Kubler 1969) and the b_o lattice spacing of the mica overlap for melange matrix and turbidite shales. The b_o spacing is related to directly to increasing phengitic component (Sassi and Scoleri 1974), and determined from the position of the 1.505 Å peak. Composition of samples thus far studied suggests that both olistostrome and turbidites have undergone deformation somewhat less intense than that of "high pressure" shales and melange matrix of the Central Belt Franciscan rocks to the south (Cloos 1983). No differences exist which would indicate a separate deformation history of olistostrome matrix versus shale in bounding turbidites.

Paleocurrent Indicators:

Current indicators among thickly bedded turbidites at Point Saint George are uncommon due to pervasive shearing. Three sets of groove casts trend between N34°W and N76°W when

....

Figure 8: (A) Sheared olistostrome unit at Point Saint George. The unit is approximately 25 m thick and contains lenticular blocks of graywacke (labeled "s"), greenstone, chert and limestone dispersed in an argillaceous matrix. The incompetence of this matrix accounts for the beveling of the unit. More competent, overturned facies B turbidites (stratigraphic top to left) are exposed bounding, and in depositional contact with, the olistostrome unit. Note the thickening and coarsening upward sequence atthe right. Exposure location indicated on figure 3 by a large arrow in the northern part of the map.

(B) Olistostrome contact with turbidites. Closeup photo of the depositional contact between the sheared olistostrome unit of figure 8A and an overturned, underlying turbidite bed. Note how the olistostrome matrix foliation (parallel to half-meter scale) passes into the T_{cde} divisions of the turbidite bed, indicating that the development of the foliation is tectonic, not compactional. Greenstone clasts (labeled V) and sandstone (S) are stretched parallel to foliation.

(C) Olistostrome at Preston Island. Olistoliths of chert (c) and sandstone (s) are oriented and elongated parallel to foliation. The matrix foliation is developed between shear fractures A and B. Scale is marked in decimeters.

(D) Overturned syncline with extensive disruption of the shaley divisions of graded beds. Arrows show up direction in beds. Outcrop located adjacent to that of figure 6. Proximity to slump folded turbidites suggests that this fold may be a slump fold; however, it is bounded by faults and could be fault-related. Scale is 50 cm.

(E) Graded bed with small scale extensional shear fractures in the laminated divisions. The bed contains divisions $S_2S_3T_{t-e}$ of Lowe (1983). Scale marked in decimeters.

A

B

C

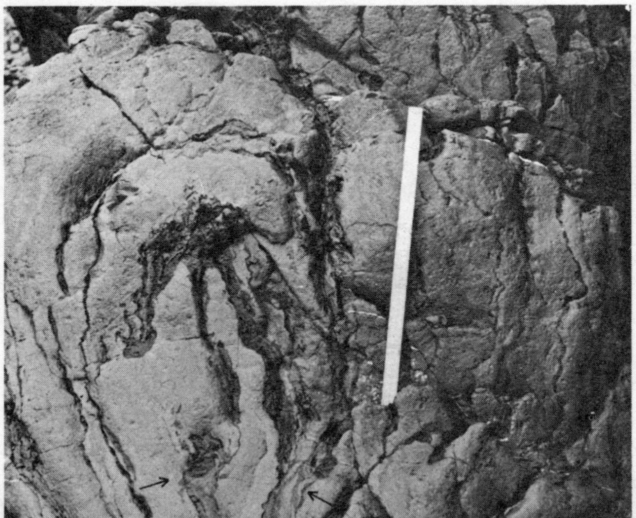

«D

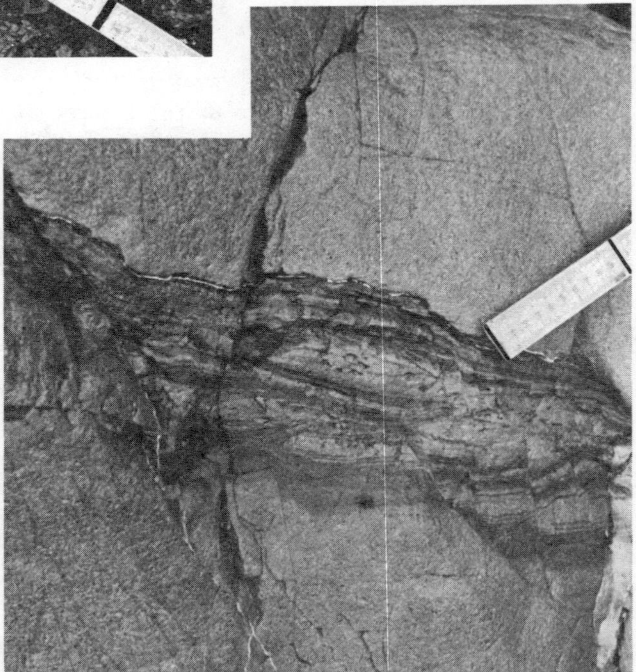

E»

restored to horizontal. One measurement of parting lineation trends N27°W. One set of flutes casts indicates flow towards N83°W. Westerly flow is consistent with sediment derivation from the Klamath - Sierran arc, suggested by Aalto (1976, 1982). Complex folding among thinly bedded turbidites precludes their use in paleocurrent analysis. The accuracy of paleocurrent analysis is limited by possible post-accumulation rotation of Franciscan rocks (M. C. Blake, Jr., pers. comm., 1980).

Facies Associations and Depositional Setting:

Aalto (1982) presented data demonstrating that turbidites associated with melanges along the northernmost California coast are predominantly inner- to mid-submarine fan deposits. Extreme deformation has obscured much of the evidence vital for detailed facies reconstruction. Measured sections are too short to assess stratigraphic trends. The prevalence of facies F, B, C and A deposits, listed in order of decreasing abundance, and dearth of facies D and G deposits suggests an inner- to mid-fan setting. Facies E deposits may represent over-levee spills from high-energy sediment gravity flows on inner regions of submarine fans.

Progressing from east to west, upwards within the stratigraphic sections in both Point Saint George and Crescent City coastal exposures, the facies succession generally passes from mixed B, C and D, to F, to mixed A, B and E. This suggests progradation of an inner fan over a mid fan, chiefly by emplacement of a large olistostrome. Melange units farther offshore (Fig. 1) may reflect similar events.

STRUCTURE

Introduction:

Intercalated olistostrome and broken formation units of the Crescent City area have undergone at least three periods of pre-Quaternary deformation. Structures preserved within some olistoliths yield information on deformation events which occurred prior to their incorporation into olistostomes. These have been reviewed by Aalto (1981) and will be summarized in the section concerning tectonic interpretation.

Pliocene to Holocene deformation involved uplift and tilting but little faulting and folding. These events may have produced joints within Franciscan rocks, but complex folds, fractures and foliation discussed below are apparently pre-Pliocene in age. Because an attempt will later be made to relate local deformation to regional tectonic history, local events will be referred to sequentially as deformations D_x, D_y and D_z. Interpretation of direction of tectonic transport for Franciscan structures is limited by possible post-Franciscan rotation of the region (M. C. Blake, Jr., pers. comm., 1980).

Deformation D_x:

Emplacement of olistostromes, gravity sliding of bedded turbidites and soft-sediment intrusion produced D_x slump folds. Because of subsequent deformation, these are clearly recognizable at only one locality (Fig. 6) but were likely extensively developed. Their existence is best confirmed where they are associated genetically with submarine mass flow deposits. Elsewhere folds are associated with faults. Although some may be slump folds subjected to later faulting, this origin cannot be confirmed (Fig. 8D).

Deformation D_y:

This deformation records an extensional realm which affected Franciscan Complex textural zone 1 broken formation and melange throughout northernmost California, and possibly the entire Franciscan Complex (Cowan 1982). Features observed in outcrop are precisely those described and interpreted by Cowan (1982) in the southern California Coast Ranges. These include ductile necking, pinch and swell and boudinage of thinly bedded sandstones and olistoliths (Fig. 8c) and distortion of thinly bedded sandstones into irregular convolutions.

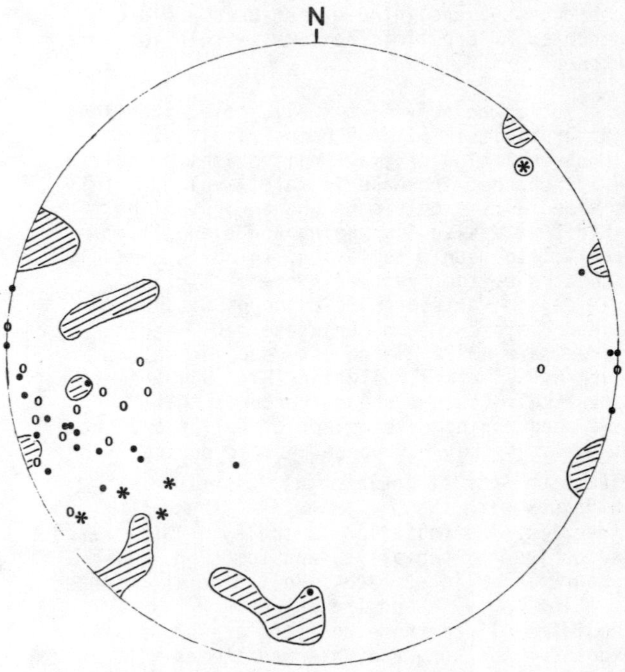

Figure 9: Clast orientation and matrix fabric data for the Point Saint George sheared olistostrome. Data presented for the outcrop shown in figures 8A and 8B. Depicted are poles to planes containing the intermediate and long axes of disc-shaped olistoliths (dots), poles to matrix foliation (circles) and poles to overturned (asterisks) and upright (encircled asterisk) beds of bounding turbidites adjacent to the exposure. Concentrations of greater than 7% per 1% area of poles to joints within the olistostrome and depicted by the pattern (n=28).

Effects of brittle extension include formation of chaotic breccia zones within necks of sandstone boudins and both large and small scale shear fracture systems, which are commonly oriented at 30° - 60° to bedding. Movement on large scale conjugate fractures spaced meters apart among thickly bedded sandstones causes wedging of beds. With fracture separation and shale intrusion lozenge-shaped sandstone olistoliths can form. Ellipsoidal ribbon chert and sandstone olistoliths can be modified to a lozenge shape by this process.

Motion on small scale extensional fractures spaced centimeters apart results in synthetic and antithetic normal offset of laminated divisions of thick turbidite beds, in many cases without offset of bed soles and coarser-grained basal divisions. Among thinly bedded turbidites motion on and separation of small scale fractures is accompanied by ductile intrusion of shale. This results in normal offset, rotation and isolation of miniature sandstone lozenges. This can occur within a thinly bedded sandstone-shale interval bounded by thickly bedded sandstones which remain undeformed. Thus a stratigraphic section can thin solely by extension within select horizons, with extension occurring parallel to bedding. Sandstone dikes both crosscut and intrude fractures and are truncated by parallel fractures within the same outcrop.

Motion on pervasive small scale extensional fractures within olistostromes results in microbrecciation of small olistoliths, resulting in an apparent increase in matrix volume. This is expecially significant where olistoliths exhibit extensive flattening and elongation parallel to matrix foliation, which is commonly truncated by the fracture systems (Fig. 8c).

In some olistostrome outcrops discrete fracture systems form conjugate sets with an approximate 60° angle of intersection which contains S_y matrix foliation (Fig. 9). These are very likely Coulomb shear fractures that developed during development of foliation. At these sites they have been rotated during extension so that their obtuse bisector coincides with the D_y principle stress direction. The foliation is scaley in appearance and defined by pervasive, anastomosing, subparallel slip surfaces. Foliation is oriented at a low angle to bedding wherever turbidite-olistostrome contacts are clearly exposed. The long and intermediate axes (d_L and d_I) of olistoliths are contained within the S_y plane (Fig. 9), although foliation may flow around larger olistoliths. Olistoliths commonly are coated with a hard argillite rind and may have slickensided surfaces, suggesting a history of rotation during foliation development.

Figure 10 presents form data according to olistolith composition for two localities. Where foliation is poorly developed, olistoliths are more angular and spherical (Fig. 7). Development of foliation is coincident with flattening of olistoliths, resulting in discs oriented parallel to foliation for all lithologies. Necking, pinch and swell and boudinage accompany olistolith flattening.

The relative importance of ductile versus brittle behavior in the deformation of broken formation and melange inclusions has been assessed differntly by Cowan (1978, 1982) and Korsch (1982). Korsch believes that melange inclusions originate chiefly by movement on shear fracture systems which disrupted intercalated greenstone, chert and turbidites. Cowan believes ductile and brittle deformation were largely contemporaneous and affected intercalated olistostromes and turbidites, which is apparently what occurred in the Crescent City area.

Although many olistoliths are truncated, and thus form-modified, by discrete fractures, many others exhibit no fragmentation and solely ductile shortening perpendicular to foliation. This flattening, the intrusion of matrix into discrete fractures in olistoliths and the geometric relationship of larger scale fractures and foliation support the Cowan (1982) model. Contemporanity of ductile and brittle behavior is as well indicated by the contrasting behavior of sandy and shaley layers of turbidite sequences discussed above.

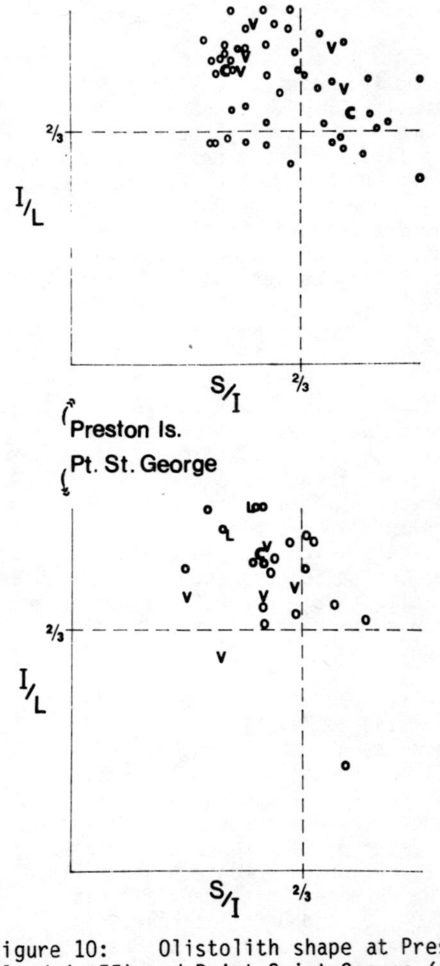

Figure 10: Olistolith shape at Preston Island (n=55) and Point Saint George (n=21). Shape of olistoliths of sandstone (o), limestone (l), chert (c) and greenstone (V) is depicted on Zingg diagrams. The long (L), intermediate (I) and short (S) olistolith axes were measured directly on extracted clasts with long axes exceeding 5 cm.

Deformation D_z:

Olistostrome and broken formation units are folded and truncated by both low and high angle faults. D_z is best defined by large scale folds which have resulted in the overturning of most Franciscan rocks (Fig. 11) and small-scale folds clearly associated with low angle faults.

Large scale folds are visible on abrasion platforms and in aerial photographs at the northern extreme of Point Saint George (Fig. 3; Murphy 1982). Axial surfaces of assymetric, overturned anticlines and synclines among facies B and C turbidites commonly dip northwest and axes plunge steeply to moderately north-eastward. Folds are open and concentric with curvatures ranging from 3 - 30 m, averaging 5 m. Some offshore folds appear to be isoclinal.

Small scale folds occur within facies D and E deposits and are associated with faults. Folds are tight with acute interlimb angles and planar limbs. Most axial surfaces strike north to northwest and axes plunge at low to moderate angles east to southeast. They are genetically associated with east-dipping faults, which truncate bedding at a low angle. Should bedding regionally be restored to upright horizontal, such faults would be east-dipping thrusts with fold vergence indicating upper plate transport to the west.

High angle faults and many fractures cannot be clearly related to any one particular deformation event. Minor fractures are pervasive and, based upon 134 measurements at Point Saint George, show random distribution.

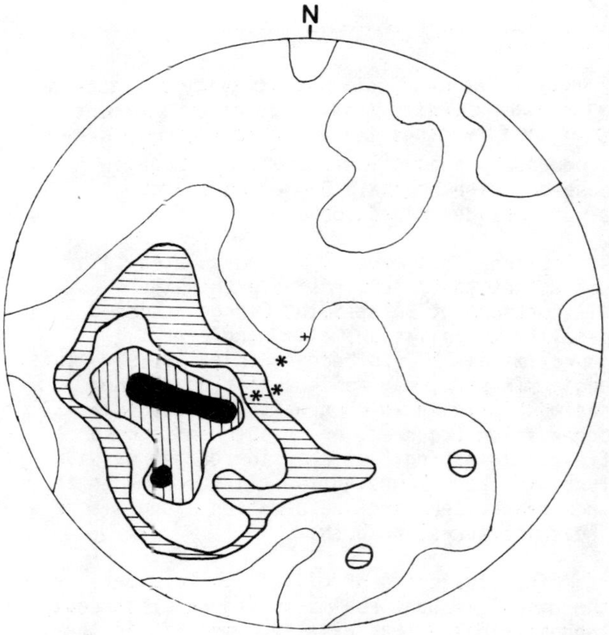

Figure 11: Structural data for the Crescent City - Point Saint George area. Ninety poles to bedding, of which 82 represent overturned beds, are depicted by contours of 0, 2, 4, 6, and 8% per 1% area. The asterisks show poles to bedding of the overlying Saint George Formation.

TECTONIC INTERPRETATION

Introduction:

Table 2 provides a tectonic interpretation of pre-Cenomanian (pre-Coast Range thrust) deformation events that affected textural zone 1 Franciscan Complex melange and broken formation in northernmost California. these events are distinct in style and effect from those affecting foliated Franciscan rocks in upper plates of thrust faults protrayed in figure 1. Evidence compiled to develop this table was gathered throughout northernmost California and critical localities are listed. D_{3-5} correspond to D_{xyz} at Crescent City. The sedimentary tectonic model for the Crescent City area (Fig.12) is based chiefly upon that of Moore and Karig (1976).

Deformation D_1:

Aalto (1981) describes melange terraces near Trinidad and the Klamath River mouth composed chiefly of blocks of pillow basalt, autoclastic volcanic breccia, hyaloclastite, ribbon chert and textural zone 1 and 2 graywacke both faulted together and dispersed in a highly indurated tuff-argillite matrix. The tuff and argillite are complexly intermixed and appear to have undergone ductile deformation prior to the development of foliation and brittle shear fractures. Contacts between tuff and argillite are convoluted, and delicate intrusions of one lithology into the other are common, as well as intrusions into fractures within melange blocks. These features are offset by fractures and along foliation.

Aalto hypothesized that tuffs were deposited with pelagic muds on an oceanic plate prior to its descent into a trench. With continued subduction the tuff-pelagic mud was ductilly deformed, became intermixed with other oceanic plate components and trench fill sediment, and was subjected to dehydration and brittle deformation. Tectonic melanges thus generated became indurated and were uplifted within the subduction complex during D_2 deformation along thrust faults, and possibly by diapirism (Cloos 1982; Hussong and Fryer 1983). Olistoliths derived from tectonic melange thus exposed on the trench slope are present throughout northernmost California and exist at Point Saint George (Murphy 1982). They constitute melange within melange and are characterized by lithologic mixtures, foliations and fracture systems unrelated to those of the encompassing matrix.

Deformation D_2:

Elevation of D_1 tectonic melange and oceanic crustal slabs is thought to occur by thrust faulting and diapirism. Faults deform trench slope sediment cover, generate trench slope basins and deform deposits within these basins (Karig and Sharman 1975; Moore and Karig 1976; Moore and Lundberg 1983). With respect to D_1 tectonic melange olistoliths at Point Saint George, such faulting and related folding would be "pre-olistostrome" and thus not evident. Aalto, Court and Engebretson (1983) describe a fold and bounding faults within a pillow

TABLE 2: DEFORMATION EVENTS RELATED TO FORMATION OF THE FRANCISCAN SUBDUCTION CONPLEX OF NORTHERNMOST CALIFORNIA AND THEIR TECTONIC INTERPRETATION

D_1: Tectonic melange formation by mixing of oceanic plate components and trench sediments during subduction accretion by underplating. Primary ductile deformation producing tuff-argillite intrusions and complex folds; secondary brittle deformation producing shear fractures. Examples: Klamath Coast and Elk Head terranes and some olistoliths described by Aalto (1981).

D_2: Elevation of D_1 tectonic melanges and oceanic slabs by faulting and diapirism within an evolving accretionary prism. Discrete northeast-dipping thrust faults and fault-related large and small-scale west-verging folds developed within D_1 tectonic melanges, oceanic slabs and trench-slope basin turbidite fill. Examples: Bounding faults and folds within the mouth-of-the-Klamath chert-basalt slab described by Aalto, Court and Engebretson (1983).

D_3: Gravity transport on the trench slope associated with the emplacement of olistostromes, derived in part from D_1 melanges and slabs exposed by D_2 events. Soft-sediment ductile deformation producing slump folds and sediment intrusions. Examples: F_x slump folds in Point Saint George unfoliated facies F deposits.

D_4: Layer parallel coaxial extension within thick sequences of trench-slope sediments and D_3 mass transport deposits occurring during burial. Ductile to brittle extensional deformation producing pinch and swell, necking, boudinage, granulation, foliation and large- and small-scale extensional shear fractures among bedded turbidities and olistostromes. Examples: Crescent City area D_y structures.

D_5: Continued compression producing thrust faults and folds within an evolving accretionary prism. Equivalent to and coeval with D_2, but affecting different parts of the accretionary prism. Oceanic rocks which have experienced D_{1-4}, intermixed with trench-slope sediments which have experienced D_{3-4}, may again be subjected to D_2 style deformation. Examples: Crescent City area D_z structures and Trinidad area folds described by Aalto (1981).

basalt-chert slab near the Klamath River mouth which may have developed during this event.

Deformation D_3:

Gravity transport on the trench slope may result in the emplacement of olistostromes in the trench and in trench slope basins on the active lower trench slope (Karig et al. 1980). Some olistostromes might contain olistoliths derived from D_1 tectonic melanges and oceanic slabs exposed on the trench slope by D_2 events. Overriding of turbidites by olistostromes and mass failure within thick turbidite sequences produce F_3 folds, which are equivalent to F_x folds at Point Saint George. Olistostromes containing oceanic olistoliths derived from D_1 melanges and upfaulted oceanic slabs are uncommon on modern trench slopes (Hamilton 1979; Karig 1979; Karig et al. 1980). They exist in some Pliocene subduction complex olistostromes (Page and Suppe 1981), however, and should not be discounted within the Franciscan Complex.

Deformation D_4:

Cowan (1982) presents a model for deformation of trench slope sediments which accounts for axially symmetric bedding parallel extension observed in Franciscan melanges and turbidites in the central California Coast Ranges. The model involves gravity collapse within a superstructure composed of sediment gravity flow deposits, which rest upon basement produced by subduction accretion. Such a basement might contain D_1 melanges and D_2 structures described above.

Extensional event D_y is postulated to be equivalent to D_4. Extension within the olistostrome at Point Saint George occurred parallel to foliation, at an angle of approximately 20° to bedding (Fig. 8b, 9). This angle, in part, may reflect tilting of the sediment pile in which gravity-induced triaxial compression occurred, or an aberrant stress field originating from rotation of the massive turbidite beds bounding the olistostrome as they underwent extensional deformation along discrete shear fractures, or both.

The clustering of olistolith form data in the discoid-equant realms of figure **10** strongly suggests that D_y was axially symmetric in the plane of foliation and thus, following the logic of Cowan (1982), resulted from pure shear. This places the deformation event in the upper portion of the trench slope superstructure of Cowan (1982).

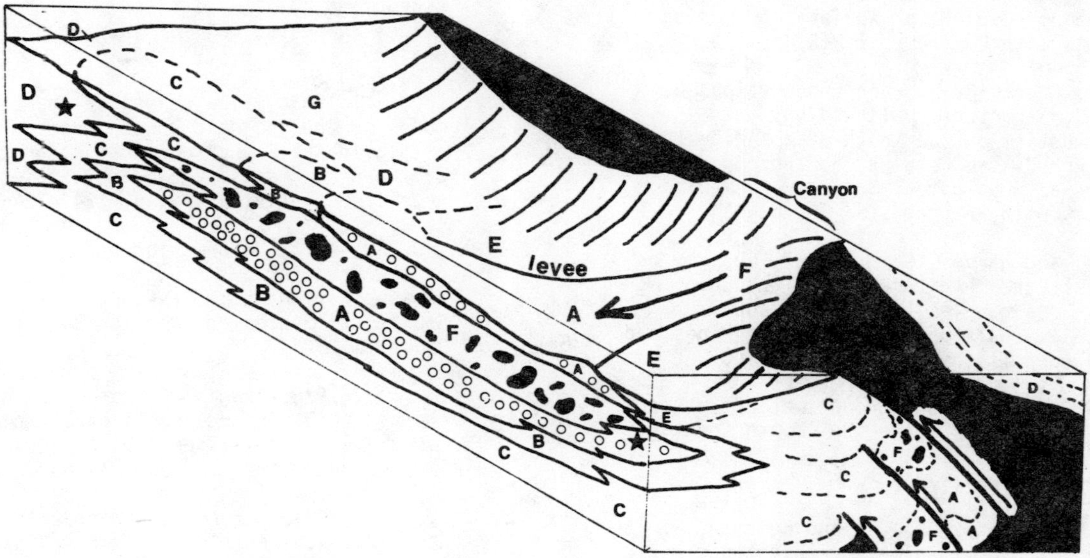

Figure 12: Sedimentary tectonic model for the origin of the Crescent City - Point Saint George Franciscan Complex. The block diagram depicts an actively developing, submarine trench slope basin. Facies are labeled according to Mutti and Ricci Lucchi (1972). Tectonic melange of the accretionary prism is shown in black. Black blobs in facies F deposits represent olistoliths derived from slumping of upfaulted or diapirically uplifted tectonic melange in the vicinity of the submarine canyon. An arrow is drawn in the canyon axis depicting transport directionfor sediment gravity flows. Thrust faults and related folds are shown schematically on the front of the block.

<u>Deformation D_5:</u>

The Crescent City - Point Saint George olistostrome and bounding broken formation are folded and faulted by deformation D_z discussed above. This deformation clearly postdates D_4 = D_y and logically corresponds to the faulting and folding of trench slope cover and basin fill that occurs as an accretionary prism grows (Lucas and Moore 1983). Such deformation exerts a continued but diminishing effect on basin fill as basins are elevated on the trench slope. Lundberg (1983) describes a transition from a compressional to an extensional realm progressing up the inner trench slope, based upon study of small scale deformation features preserved in DSDP cores of trench slope cover. With burial, perhaps, sediments which have suffered extension can be compressed and tectonically kneaded into the accretionary substrate, thus mimicking D_4 - D_5.

Late Mesozoic and Cenozoic deformation has undoubtedly altered original structure of the Thithonian-Cenomanian Franciscan subduction complex. However, the regional northwest strike and northeast dip of most bedding, faults subparallel to bedding and axial surfaces of folds may logically be attributed to the geometry of plate convegence during the development of the subduction complex.

ACKNOWLEDGEMENTS

John Longshore helped considerably with X-ray diffraction studies. Camilia Armstrong typed the manuscript. Rolf Aalto and Ben Garlick salvaged the manuscript from a faulty disk. This article is condensed from a field trip guidebook prepared for the Penrose Conference on Structural Styles and Deformation Fabrics of Accretionary Complexes held in Eureka, CA, 30 April-4 May, 1984. The Humboldt State University Foundation provided support for offshore study (Grant #109-83). Study of the Crescent City area was originally funded by the Division of Earth Sciences of the National Science Foundation (Grant #EAR 7809957).

REFERENCES CITED

Aalto, K. R., 1976, Sedimentology of a melange: Franciscan of Trinidad, California: Journal of Sedimentary Petrology, v. 46, p. 913-929.

............, 1981, Multistage melange formation in the Franciscan Complex, northernmost California: Geology, v. 9, p. 602-607.

............, 1982, The Franciscan Complex of northernmost California: sedimentation and tectonics: <u>in Trench-forearc Geology</u>, J. K. Leggett, ed., <u>Geological Society of London</u> Special Publication 10, p. 419-432.

............, 1983, Franciscan Complex geology of the Pilot Creek quadrangle, northern California: Geological Society of America Abstracts with Programs, v. 15, p. 275.

............, Court, B. L., and Engebretson, D. C., 1983, Structure, paleontology and paleomagnetism of the Mouth-of-the-Klamath Chert, Franciscan Complex, northern California: Geological Society of America Abstracts with Programs, v. 15, p. 314.

............, and Harper, G. D., 1982, Geology of the Coast Ranges in the Klamath and Ship Mountain quadrangles, Del Norte County, California: California Division of Mines and Geology open file map, scale 1:62,500.

Back, W., 1957, Geology and ground water feature of the Smith River plain, Del Norte County, California: U.S. Geological Survey Water Supply paper 1254, 76 p.

Blake, M. C., Jr., Irwin, W. P. and Coleman, R. G., 1967, Upside down metamorphic zonation, blueschist facies, along a regional thrust in California: U.S. Geological Survery Professional Paper 515, p. 1-9.

............, and Jones, D. L., 1974, Origin of Franciscan melanges in northern California, in Dott, R. H., Jr., and Shaver, R. H., eds., Modern and ancient geosynclinal sedimentation: Society of Economic Paleontologists and Mineralogists Special Publication 19, p. 345-357.

............, 1981, The Franciscan assemblage and related rocks in northern California: A reinterpretation, in Ernst, W. G., ed., The geotectonic development of California: Englewood Cliffs, N. J., Prentice-Hall, p. 306-328.

............, 1983, the Late Mesozoic Coast Range orogeny and subsequent collision events in northern California and southwestern Oregon: Geological Society of America Abstracts with Programs, v. 15, p. 295.

Carver, G. A., Stephens, T. A., and Young, J. C., 1982, Quaternary reverse and thrust faults, Mad River fault zone: in Harden, D. R., Marron, D. C. and MacDonald, A., convenors: Late Cenozoic history and forest geomorphology of Humboldt County, California: Friends of the Pleistocene Pacific Cell field trip guidebook, p. 93-99.

............, 1983, Quaternary thrust and reverse faulting on the Mad River fault zone, coastal northern California: Geological Society of America Abstracts with Programs, v. 15, p. 316.

Cloos, M., 1982, Flow melanges: numerical modeling and geologic constraints on their origin in the Franciscan subduction complex, California: Geological Society of America Bulletin, v. 93, p. 330-345.

............, 1983a, Comparative study of melange matrix and metashales from the Franciscan subduction complex with the basal Great Valley Sequence, California: Journal of Geology, v. 91, p. 291-306.

............, 1983b, Flow melanges: numberical modeling and geologic constraints on their origin in the Franciscan subduction complex, California: Reply: Geological Society of America Bulletin, v. 94, p. 1243-1244.

Cowan, D. S., 1974, Deformation and metamorphism of the Franciscan subduction zone complex northwest of Pacheco Pass, California: Geological Society of America Bulletin, v. 85, p. 1623-34.

............, 1978, Origin of blueschist-bearing chaotic rocks in the Franciscan Complex, San Simeon, California: Geological Society of America Bulletin, v. 89, p. 1415-23.

............, 1982, Deformation of partly dewatered sediments near Piedras Blancas Point, California: in Leggett, J. K., ed., Trench-forearc Geology: Geological Society of London Special Publication 10, p. 439-457.

Davenport, C. W., 1982, Geology and geomorphic features related to landsliding, Hiouchi 7.5' quadrangle, California: California Division of Mines and Geology 1:24,000 preliminary map.

Dickinson, W. R., 1970, Interpreting detrital modes of graywacke and arkose: Journal of Sedimentary Petrology, v. 40, p. 695-707.

.... and Suzcek, C. A., 1979, Plate tectonics and sandstone compositions:

American Association of Petroleum Geologists Bulletin, v. 63, p. 2164-2182.

Hamilton, W., 1979, Tectonics of the Indonesian region: U.S. Geological Survey Professional Paper 1078, 345 p.

Harden, D. R., Marron, D. C. and MacDonald, A., convenors, 1982, Late Cenozoic history and forest geomorphology of Humboldt County, California: Friends of the Pleistocene Pacific Cell field trip guidebook, 247 p.

............, Kelsey, H. M. Morrison, S. D. and Stephens, T. A., 1981, Geologic map of the Redwood Creek basin: U.S. Geological open file map 81-496, scale 1:62,500.

Hsu, K. J., 1968, Principles of melanges and their bearing on the Franciscan-Knoxville paradox: Geological Society of America bulletin, v. 79, p. 1063-74.

............, 1974, Melanges and olistostromes: in Dott, R. H., Jr. and Shaver, R. H., eds., Modern and ancient geosynclinal sedimentation: Special Publications of the Society of Economic Palentologists and Mineralogists 19, p. 321-33.

Hussong, D. M. and Fryer, P., 1983, Forearc tectonics and morphology of the northern Mariana arc: EOS, v. 64, p. 828-829.

Jayko, A., 1983, Deformation in low-grade blueschists of the eastern Franciscan belt, northern California: Geological Society of America Abstracts with Programs, v. 15, p. 313.

Korsch, R. J., 1982, Structure of Franciscan Complex in the Stanley Mountain window, southern Coast Ranges, California: American Journal of Science, v. 282, p. 1406-1437.

Kelsey, H. M., 1982, Structure and tectonics of Plio-Pleistocene sediments exposed within and adjacent to Prairie Creek State Park, Humboldt County, north coastal California: in Harden, D. R., Marron, D. C. and Macdonald, A., convenors: Late Cenozoic history and forest geomorphology of Humboldt County, California: Friends of the pleistocene Pacific Cell field trip guidebook, p. 130-150.

............, and Allwardt, A. A., 1983, Evidence for a major Quaternary fault zone in the central belt Franciscan melange, northern California: Geological Society of America Abstracts with Programs, v. 15, p. 316.

............, and Cashman, S. M., 1983, Wrench faulting in northern California and its tectonic implications: Tectonics, v. 2, p. 565-576.

............, and Hagans, D. K., 1982, Major right-lateral faulting in the Franciscan assemblage of northern California in late Tertiary time: Geology, v. 10, p. 387-391.

Karig, D. E. 1980, Material transport within accretionary prisms and the "knocker" problem: Journal of Geology, v. 88, p. 27-39.

............, and Sharman, G. F. 1975, Subduction and accretion in trenches: Geological Society of America Bulletin, v. 86, p. 377-389.

............, Moore, G. F., Curray, J. R. and Lawrence, M. B., 1980, Morphology and shallow structure of the lower trench slope off Nias Island, Sunda Arc: in Hayes, D. E., ed., The Tectonic and Geologic Evolution of southeast Asian Seas and Islands: American Geophysical Union Geophysical Monograph 23, p. 179-208.

Kubler, B., 1969, Crystallinity of illite. Detection of metamorphism in some frontal parts of the Alps: Fortschr. der Mineralogie, v. 47, p. 39-40.

Lowe, D. R., 1982, Sediment gravity flows II: depositional models with special refernce to deposits of high-density turbidity currents: Journal of Sedimentary Petrology, v. 52, p. 279-298.

Lucas, S. E. and Moore, J. C., 1983, Cataclasis and stratal disruption in trench and trench slope deposits off southern Mexico, DSDP Leg 66: EOS, v. 64, p. 866.

Lundberg, N., 1983, Deformation styles in active margins: upper slope extension and lower slope shortening seen in DSDP cores: EOS, v. 64, p. 833.

Maxwell, J. C., 1974, Anatomy of an orogen: Geological Society of America Bulletin, v. 85, p. 1195-1204.

Monsen, S. A., and Aalto, K. R., 1980, Petrology, structure and regional tectonics: South Fork Mountain Schist of Pine Ridge Summit, northern California: Geological society of America Bulletin, v. 91, p. 369-373.

Moore, G. F., and Karig, D. E., 1976, Development of sedimentary basins on the lower trench slope: Geology, v. 4, p. 693-697.

Murphy, J. M., 1982, Franciscan Complex geology at Point Saint George, California: Humboldt State University B. Sc. thesis, 64 p.

............, and Aalto, K. R., 1982, Franciscan complex geology at Point Saint George and Crescent City, California: Geological Society of America Abstracts with Programs, v. 14, p. 219.

............, and Aalto, K. R., 1983, Flow melange: numerical modeling and geologic constraints on their origin in the Franciscan subduction complex, California: Discussion: Geological Society of America Bulletin, v. 94, p. 1241-1243.

Mutti, E. and Ricci Lucchi, F., 1978, Turbidities of the northern Appennines: Introduction to facies analysis: International Geology Review, v. 20, p. 125-166.

Page, B. M., and Suppe, J., 1981, The Pliocene Lichi melange of Taiwan: Its plate tectonic and olistostromal origin: American Journal of Science, v. 281, p. 193-227.

Sassi, F. P., and Scoleri, A., 1974, The b_o value of the potassic white micas as a barometric indicator in low-grade metamorphism of pelitic schists: Contributions to Mineralogy and Petrology, v. 45, p. 143-152.

THE LASSIC OUTLIER, AN OUTLIER OF COAST RANGE OPHIOLITE, NORTHERN CALIFORNIA

Terry A. Kaplan
United States Department of Agriculture
United States Forest Service
Mendocino National Forest
P.O. Box 1019
Corning, CA 96021

ABSTRACT

The Lassic outlier, the most northerly outlier of Coast Range ophiolite in California is a remnant of the upper plate of the Coast Range thrust, which rests on Franciscan rocks. The rocks of the outlier consist of Coast Range ophiolite and strata of the Great Valley sequence, but the original layered sequence of the upper plate has been disrupted to such a degree that the terrane resembles an ophiolitic melange.

The sedimentary rocks exposed on Black Lassic peak are correlative with the Stony Creek petrofacies of the Great Valley sequence. This sequence is widely exposed approximately 100 km to the southeast and is similar in general appearance, degree of alteration, and detrital modes.

Mafic rocks of the Coast Range ophiolite consist mainly of basalt and diabase with minor amounts of gabbro. This sequence of mafic rocks has been detached from the ultramafic rocks by thrust faulting. The ultramafic rocks consist predominantly of serpentinite and harzburgite which have undergone intense shearing. The ultramafic rocks are generally the lowest unit of the ophiolite, and are separated from the underlying metasedimentary rocks of the Franciscan Complex by the Coast Range thrust.

Franciscan metasedimentary rocks range from low textural zone 2 away from the thrust fault to textural zone 3 near the thrust fault. The textural zone 3 Franciscan rocks are quartz mica schist that contain minerals indicative of the blueschist facies.

Late Jurassic radiolarians were extracted from chert associated with the opholitic rocks. Late Jurassic or Early Cretaceous radiolarians were extracted from chert associated with the Franciscan Complex. Conglomerate pebbles from the Great Valley sequence at Black Lassic and Franciscan conglomerates yielded Triassic radiolarians.

INTRODUCTION

The study area includes one of the largest exposures of ultramafic rock within the northern Coast Ranges. It is located in Ranges 5 and 6 East; Townships 1 and 2 South, Humboldt Base and Meridian, and is within the boundaries of Six Rivers National Forest. It is covered by the Blocksburg Southeast (7 1/2') quadrangle. The nearest town is Mad River to the north, which is on state highway 36, approximately 120 miles west of Red Bluff and 80 miles east of Eureka. Mad River is approximately 15 miles north of the study area via Van Duzen River Road and U.S. Forest Service Road 1S07, a high quality gravel road which transects the study area. Figure 1 shows the regional geologic setting and the location of the Lassic outlier.

The Lassic ultramafic body (fig. 2) is exposed at altitudes ranging from about 4,500 feet to nearly 6,000 feet in the vicinity of three peaks on a broad ridge. The northernmost peak is Black Lassic, elevation 5,916 feet; the western peak is Mount Lassic, elevation 5,876 feet, and the eastern peak is Red Lassic, elevation 5,898 feet. The Lassic peaks are approximately 400 feet above the general crest of the ridge. From Black Lassic, the uniform summit level of South Fork Mountain can be seen to the east. This summit level is considered to be one of the best developed portions of the Klamath Mountain Peneplain (Irwin, 1966; Diller, 1902). To the west toward the coast, the mountain ranges are generally lower.

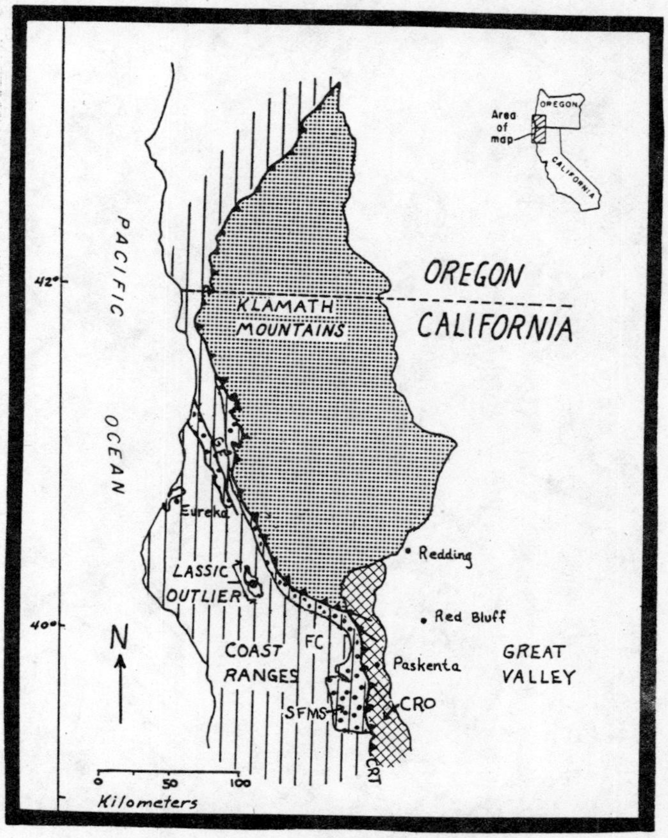

Figure 1. Location map of northwestern California-southwestern Oregon showing regional geologic setting and study area, the Lassic outlier. Geologic map after Irwin, 1977: SFMS=South Fork Mountain Schist; CRO-Coast Range ophiolite and associated sedimentary rocks of the Great Valley sequence; CRT=Coast Range thrust; GF=Grogan fault; FC=Franciscan Complex.

in Blake, M.C. Jr., ed., 1984, **Franciscan Geology of Northern California:** Pacific Section S.E.P.M., Vol. 43, p. 203-219

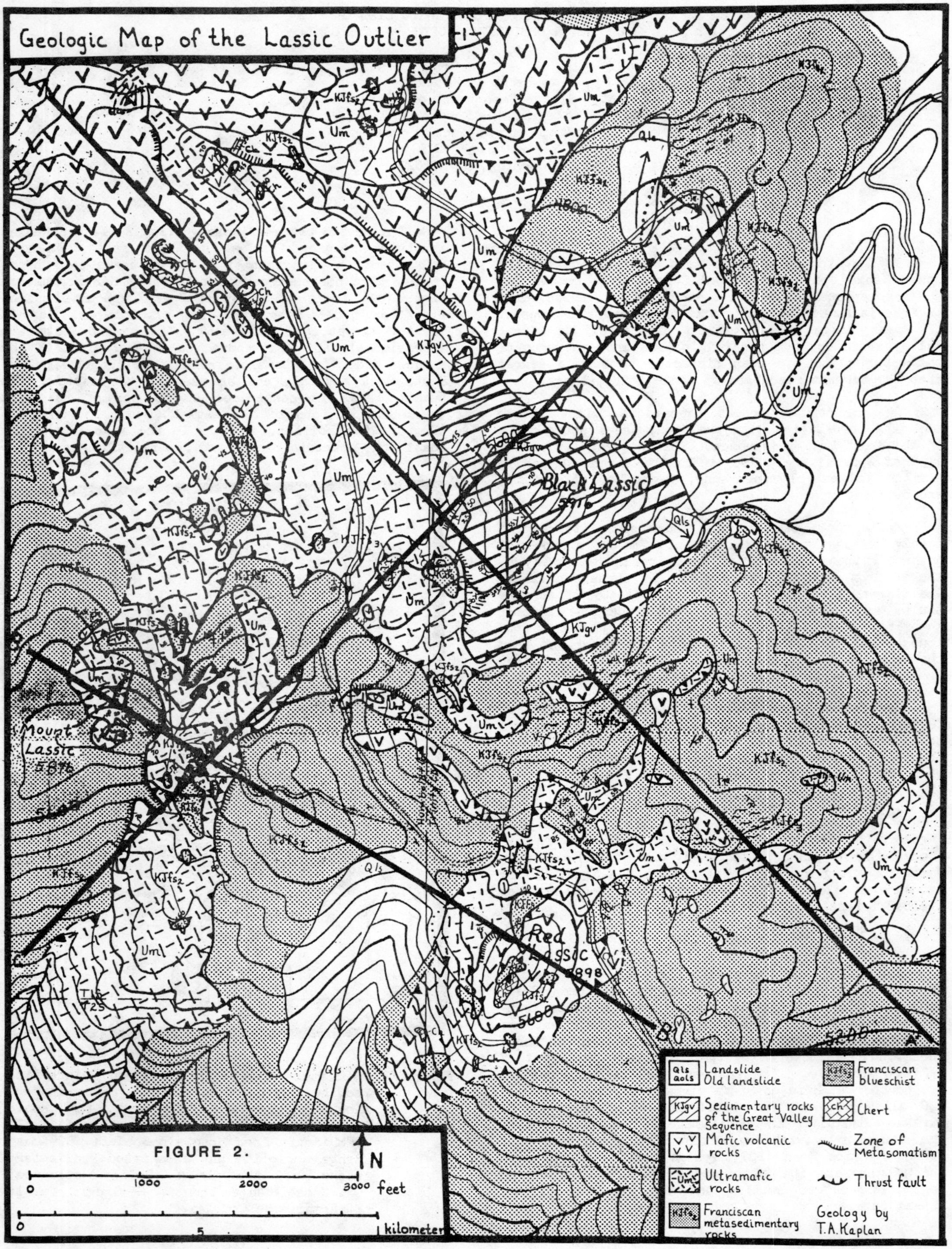

Geologic Map of the Lassic Outlier
Black Lassic
Mount Lassic
5876
Red Lassic
5898
Qls aols Landslide Old landslide
KJgv Sedimentary rocks of the Great Valley Sequence
Mafic volcanic rocks
Um Ultramafic rocks
KJfs2 Franciscan metasedimentary rocks
KJfs3 Franciscan blueschist
Ch Chert
Zone of Metasomatism
Thrust fault
Geology by T.A. Kaplan
N
0 1000 2000 3000 feet
0 5 kilometer

FIGURE 2.

The Lassic ultramafic body was first shown on a geologic map by Irwin in 1960. He later used a panorama of the Lassic area to illustrate the uniform summit level of South Fork Mountain, and briefly described the terrain between South Fork Mountain and Mount Lassic (Irwin, 1966). In August of 1964, Irwin and R. G. Coleman (oral communication) spent a few days in the Lassic area mapping the geology and collecting samples of the metasomatic reaction zone between the serpentine and overlying sedimentary rocks. None of this work was published. In 1977, Irwin used a picture of this metasomatic zone to illustrate a tectonic contact, and later a geologic structure section across the Klamath Mountains and Coast Ranges was drawn through Black Lassic (Irwin and Dennis, 1979).

The Lassic ultramafic body was mapped on a regional scale (1:62,500) by Kelsey and Allwardt (1975) during a study of the van Duzen River Basin. They designate Black Lassic peak as Jurassic and Cretaceous Franciscan sedimentary rocks.

Prior to this study, the most extensive work in this area was accomplished by Hagans (1978) as a senior thesis project at Humboldt State University. In his study Hagans states that Black Lassic was formed as a result of Cenozoic (post Franciscan) reverse faulting which placed unmetamorphosed rocks of the Franciscan Complex structurally above dismembered ophiolite melange.

In 1982, Kelsey and Hagens changed their opinion as to the formational affiliation of the sedimentary rocks on Black Lassic and postulated that the rocks of the Lassic outlier were transported northward by right-lateral faulting.

LITHOLOGY

Coast Range Ophiolite

Highly serpentinized and sheared ultramafic rocks account for about 80 percent of the ophiolitic rocks in the field area. The ultramafic sheet consists mainly of sheared serpentinite and bastitic harzburgite. These rocks form the lowest unit of the ophiolite and are separated from underlying metasedimentary rocks of the Franciscan complex by a regional thrust fault. The ultramafic rocks are overlain by a mafic complex consisting predominantly of basaltic and diabasic volcanics with subordinate gabbroic rocks. The mafic complex has been detached from the ultramfic sheet by thrusting.

Ultramafic Rocks

About one-third of the ultramafic body is serpentinized and sheared while about two-thirds of the body is blocky and massive. The sheared serpentinite commonly occurs as a highly polished and slickensided rock of a pale green to brownish-red color. This variety of serpentinite is usually found near tectonic contacts and forms smooth broad ridges that are subject to landsliding.

The more massive olive green to brownish-red variety of serpentinite forms resistant knobby outcrops. Although the rocks have undergone almost total serpentinization, the blocky outcrops are studded with relict pyroxene crystals. This bastitic texture accounts for the warty appearance in outcrop. Veins of highly slickensided "fish scale" serpentine occur along shears and fractures in the more resistant and massive serpentinite. These slip veins indicate that some movement has taken place during or after emplacement. Serpentinite of a rare brecciated variety is associated with mafic volcanic rocks (sample #47-79). The breccia usually occurs as isolated pods poking through mafic volcanics. The serpentinite breccia is pale green in color and composed of a completely serpentinized, brecciated harzburgite with relict pryroxenes readily visible. Reddish-brown to yellow serpentine replacement of olivine can occasionally be seen among the pale green pyroxene crystals. Relict textures indicate that in these rocks brecciation preceded serpentinization. Table 1 shows sample localities noted in the text and located on figure 3.

In general, the serpentinite crops out along ridge crests forming topographic highs supported by pods of volcanic rock. These pods crop out at the nose of the ridges and protect the serpentinite from eroison by forming a resistant "retaining wall." Soil mantle is always very thin and gives way to a disaggregated rock which is characteristically reddish-brown in color.

Metasomatism of Serpentinite and Adjacent Rocks

A metasomatic contact relationship between ultramafic rock and country rock exists throughout the Lassic outlier. Metasomatic alteration zones between serpentinite and rocks with a higher silica content are common to many ultramafic bodies. This alteration zone results from an innerchange of ions along contacts which results in an alteration product called rodingite. Rodingites are composed predominatly of calcium aluminum silicate and occur as a normal by-product of serpentinization (Coleman, 1977). The time and formation of serpentinization is not known, but according to Barnes and O'Neil (1969) can occur at the present time under "conditions approaching mean annual temperatures of the earth's surface". Metasomatic reaction zones develop when ultramafic rocks are tectonically emplaced within the earth's crust and serpentinization usually occurs with tectonism (Coleman, 1967). It has been suggested by Coleman (1980, oral communication) that the rodingites of the Lassics are some of the best exposures in the world.

Mafic Complex

The remainder of the ophiolitic rocks in the field area are composed of basaltic and diabasic volcanics with minor gabbro and igneous breccia distributed randomly throughout the volcanic rocks. Associated with fine-grained mafic rocks is red radiolarian chert. In many instances, the chert is found in depositional contact with the mafic volcanic rocks.

Pillow basalt is the most easily recognized member of the mafic complex and crops out on or near all of the three major peaks. Red Lassic is almost completely composed of well-developed pillow basalt. The basalt makes up steep towers at the top of Red Lassic and weathers to a characteristic reddish-purple. Soil development is almost nonexistent on Red Lassic, due to steep slopes, but around its base, on the more stable slopes, a reddish-brown soil mantle has developed on the basalt. Associated with the pillow basalt of Red Lassic is a reddish-purple radiolarian chert. The contact relationship between the chert, and the basalt on Red Lassic is unclear.

Mt. Lassic, although composed mostly of serpentinite and Franciscan metasedimentary rock, has two major exposures of volcanic rock situated near the top. These rocks are fine-grained and range in color from light orange-brown to green-brown on weathered surfaces. Their color is dependent on the amount of metasomatic alteration these rocks have undergone. A vague pillow shape can still be recognized even though these bodies have undergone extensive weathering, metasomatic alteration, and are covered with lichen.

Black Lassic is flanked on the north and west sides by pillow basalt. On the west flank, the volcanic rocks are clearly in depositional contact with sedimentary rocks of the Great Valley sequence exposed on Black Lassic. The depositional nature of the contact is suggested by continuity of structural trends and intercalation of volcanic and sedimentary rocks (figs. 4a,b). On the northern flank, the volcanic rocks are thrust over the sedimentary rocks of Black Lassic. Here the bedding in the sedimentary rocks is truncated by the overriding slab of volcanic rock.

The overthrust volcanic rocks on the Black Lassic form an elongate zone that extends across the northwestern portion of the study area and account for the bulk of the mafic rocks that make up the ophiolite. Most of the outcrops are resistant knobby piles of rubble with lichen growths on highly weathered surfaces. Outcrops are generally massive, fine-grained, and gray-to-black in color. Some of these rocks are metasomatically altered and are light salmon-pink to light greenish-brown in color. Some outcrops exhibit vague pillow structures, but others have no obvious structural form.

Sparsely intermingled with the fine-grained volcanic rocks are randomly distributed pods of coarse-grained gabbroic rock. These rocks constitute no more than two percent of the total mafic complex. The gabbroic rocks are distinguished from the basalts and diabases by their coarser texture, salt-and-pepper appearance, and smooth rounded form.

Associated with the northern zone of mafic rocks is an unusual igneous breccia. The breccia crops out as resistant knobby piles of welded rubble much like the fine-grained volcanic rocks. The breccia is distributed irregularly thoughout the mafic zone. It is composed of unsorted angular fragments that form a tightly fitted, clast-supported framework in a matrix of microbreccia that is cemented with iron oxide. Microscopic study indicates that only igneous clasts are present. Diabasic and basaltic clasts are the most abundant, but other clast types include gabbro, plagiogranite, palagonitic glass, and pyroxenite. The breccia is relatively unaltered; however, sparse patches of prehnite and chlorite exist.

The breccia is similar, if not identical, to breccia that crops out on Crow Foot Point near Paskenta, California (Kaplan, 1983; Kelsey and Hagans, 1982). The breccia at Crow Foot Point is overlain stratigraphically by the lowest beds of the Great Valley sequence (Betrucci, 1983). Several miles north of Crow Foot Point, along the South Fork of Elder Creek the mafic igneous breccia of the Coast Range ophiolite is depositionally overlain by sediments of the basal Great Valley sequence (Hopson, Mattinson, and Pessagno, 1981; Blake and Jones, 1977; Bailey and others, 1970). The breccia has been interpreted to be a submarine talus or scree deposit formed on the ocean floor, possibly near an active sea floor fault (Bertucci, 1983; Hopson and others, 1981).

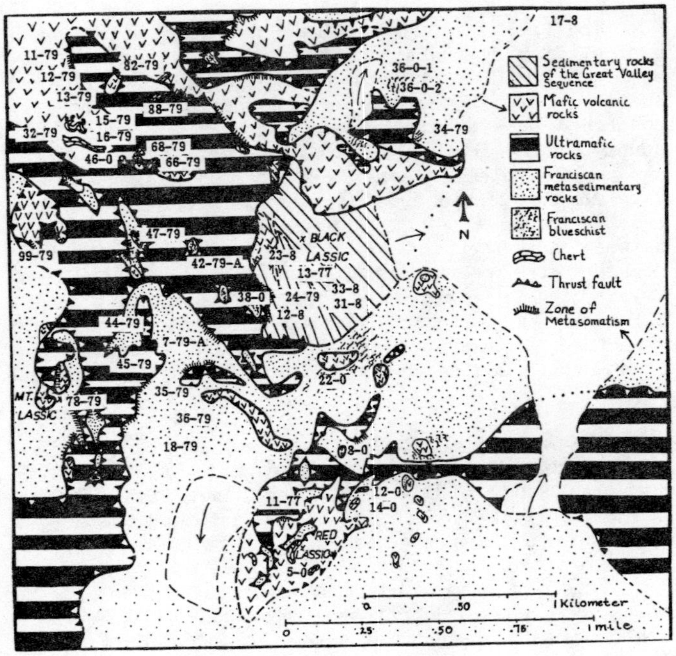

Figure 3. Geologic map of the Lassic outlier showing sample localities noted in the text and listed on table 1.

Sample Number	Sandstone	Conglomerate	Metasandstone	Metaconglomerate	Blueschist	Volcanic Breccia	Ultramafic Breccia	Bedded Chert	Chert Float	Locality	
5-0								X		SW1/4, SE1/4Sec. 31,	T1S, R6E.
8-0						X				NW1/4, SE1/4Sec. 31,	T1S, R6E.
12-0								X		SE1/4, SE1/4Sec. 3,	T1S, R6E.
14-0								X		SE1/4, SE1/4Sec. 3,	T1S, R6E.
22-0					X					SW1/4, NE1/4Sec. 3,	T1S, R6E.
36-0-1					X					NE1/4, SE1/4Sec. 30,	T1S, R6E.
36-0-2					X					NE1/4, SE1/4Sec. 30,	T1S, R6E.
38-0					X					NE1/4, NW1/4Sec. 3,	T1S, R6E.
46-0								X		NE1/4, SE1/4Sec. 25,	T1S, R5E.
17-8								X		NE1/4, NE1/4Sec. 25,	T1S, R6E.
12-8	X									SE1/4, NW1/4Sec. 3,	T1S, R6E.
23-8	X									NE1/4, NW1/4Sec. 3,	T1S, R6E.
31-8	X									SW1/4, NE1/4Sec. 3,	T1S, R6E.
33-8	X									NW1/4, NE1/4Sec. 3,	TIS, R6E.
11-77									X	SE1/4, SW1/4Sec. 3,	T1S, R6E.
13-77		X								NW1/4, NE1/4Sec. 3,	T1S, R6E.
7-79-A			X							SW1/4, NW1/4Sec. 3,	T1S, R6E.
11-79						X				SW1/4, NE1/4Sec. 25,	T1S, R5E.
12-79									X	SE1/4, NE1/4Sec. 25,	T1S, R5E.
13-79								X		SE1/4, NE1/4Sec. 25,	T1S, R5E.
15-79								X		NE1/4, SE1/4Sec. 25,	T1S, R5E.
16-79								X		NE1/4, SE1/4Sec. 25,	T1S, R5E.
18-79			X							NW1/4, SW1/4Sec. 3,	T1S, R6E.
24-79	X	X								SE1/4, NW1/4Sec. 3,	T1S, R6E.
32-79								X		NE1/4, SE1/4Sec. 29,	T1S, R5E.
34-79					X					SE1/4, SE1/4Sec. 30,	T1S, R6E.
35-79			X							SW1/4, NW1/4Sec. 3,	T1S, R6E.
36-79				X						NW1/4, SW1/4Sec. 3,	T1S, R6E.
42-79-A					X					NW1/4, NW1/4Sec. 3,	T1S, R6E.
44-79			X							SE1/4, NE1/4Sec. 36,	T1S, R5E.
45-79		X								SE1/4, NE1/4Sec. 36,	T1S, R5E.
47-79							X			NW1/4, NW1/4Sec. 3,	T1S, R6E.
66-79								X		SW1/4, SW1/4Sec. 30,	T1S, R6E.
68-79								X		SW1/4, SW1/4Sec. 30,	T1S, R6E.
78-79			X							SE1/4, NE1/4Sec. 36,	T1S, R5E.
82-79						X				NW1/4, SW1/4Sec. 30,	T1S, R6E.
88-79								X		NW1/4, SW1/4Sec. 30,	T1S, R6E.
99-79				X						NE1/4, NE1/4Sec. 36,	T1S, R5E.

Table 1. Table of sample localities shown on figure 3.

Figure 4a. View northeast toward Black Lassic, altitude 5,916 feet, center of photograph, composed predominately of dark shale that is correlative with the Stony Creek petrofacies of the Great Valley sequence.

Figure 4b. View north towards resistant beds of volcanic rock (V) in depositional contact with dark shaley rock of the Great Valley sequence (GV) on the west slope of Black Lassic. The barren rounded hill in the middle distance, lower left of photograph, is sheared serpentinite (Um).

Chert

Chert comprises less than one percent of the rocks in the study area. Most of the chert is red-purple to red-brown, although green chert is also present. Milky altered chert is also quite common and occurs in a variety of colors. Chert is generally associated with volcanic rocks but also crops out in the Franciscan metasedimentary rocks and the ultramafic body.

The best preserved chert is associated with the pillow basalts of Red Lassic and the volcanic body in the northwestern quarter of the study area. Contact relations between the chert on Red Lassic and the pillow basalt is unclear; however, it is assumed that these two rock types are genetically related. This assumption is based on a manganiferous stain and a similar degree of alteration affecting both rock types. On Red Lassic, the chert crops out as resistant blocks with color banding that varies from pinkish-white to purplish-red. The banding is parallel to bedding planes, many of which have been broken and folded.

A few diagnostic radiolarians were extracted from the chert exposed on Red Lassic. These radiolarians include: Ristola sp. with a range of upper Bathonian to upper Tithonian (C. D. Blome, oral commun., 1983); an undescribed Nassellariina with a range of Oxfordian to Tithonian (C.D. Blome, oral commun., 1983); Hsuum sp aff. H mclaughlini with a persistance in the Tithonian which may extend into later stages (Pessagno and Blome, 1984); Hsuum obispoensis with a range of upper Kimmeridgian to upper Valanginian (Pessagno, 1977); and Parvicingula vera with a range of upper Bathonian to Tithonian (Pessagno and Whalem, 1982) (fig. 5). The ranges of these radiolarians overlap in the Tithonian indicating deposition during this time (fig. 7).

The chert cropping out in the northwest portion of the study area is clearly in depositional contact with the surrounding volcanic rocks. At the southernmost exposure, rhythmically bedded chert is interbedded with fine-grained volcanic rocks. Individual beds are discontinuous near the contact with the volcanic rock but have more continuity away form this contact (fig. 8). Proceeding north perpendicular to strike, the chert beds become chaotic with no bedding structures, and isolated beds, broken folds, and chunks of red chert are incorporated in a melange of purple and green volcanic rocks.

Proceeding east, isolated bodies of chert and volcanic rock crop out as knockers in the serpentinite terrane. These knockers are probably remnants of the volcanic body just mentioned and at one time may have been present above the serpentinite.

The outcrops of chert in the northwest portion of the study area were extensively sampled for radiolarians. Radiolarians from this area include: Crucella sp., Hsuum sp., and Parvicingula sp. More diagnostic radiolarians from this area include: an undescribed Nassellasriina which, at present, appears to be restricted to the Upper Jurassic and is under present study (C.D. Blome, oral commun., 1983); Eucyrtidium ptyctum with a range of Kimmeridigian to upper Tithonian (Riedel and Sanfilippo, 1974); Pantanellium riedeli with a range of Tithonian to Berriasian (C.D. Bome, oral commun., 1983); Tripocyclia blakei, with a range of Tithonian (Pessagno, 1977); and Acanthocircus variabilis (Squinabol) with a range of Tithonian (C.D. Blome, oral commun., 1983). These radiolarians, like those associated with the chert exposed on Red Lassic indicate deposition during the Tithonian. Figure 6 displays the various radiolarians and figure 7 shows their ranges.

Volcanic knockers, generally associated with the serpentinite body, are found with highly altered and milky chert interbeds. The chert interbeds are highly variable in color due to intense alteration that is present in both the volcanic rocks as well as the chert. Although this chert was sampled for radiolarians, diagnostic forms were not observed.

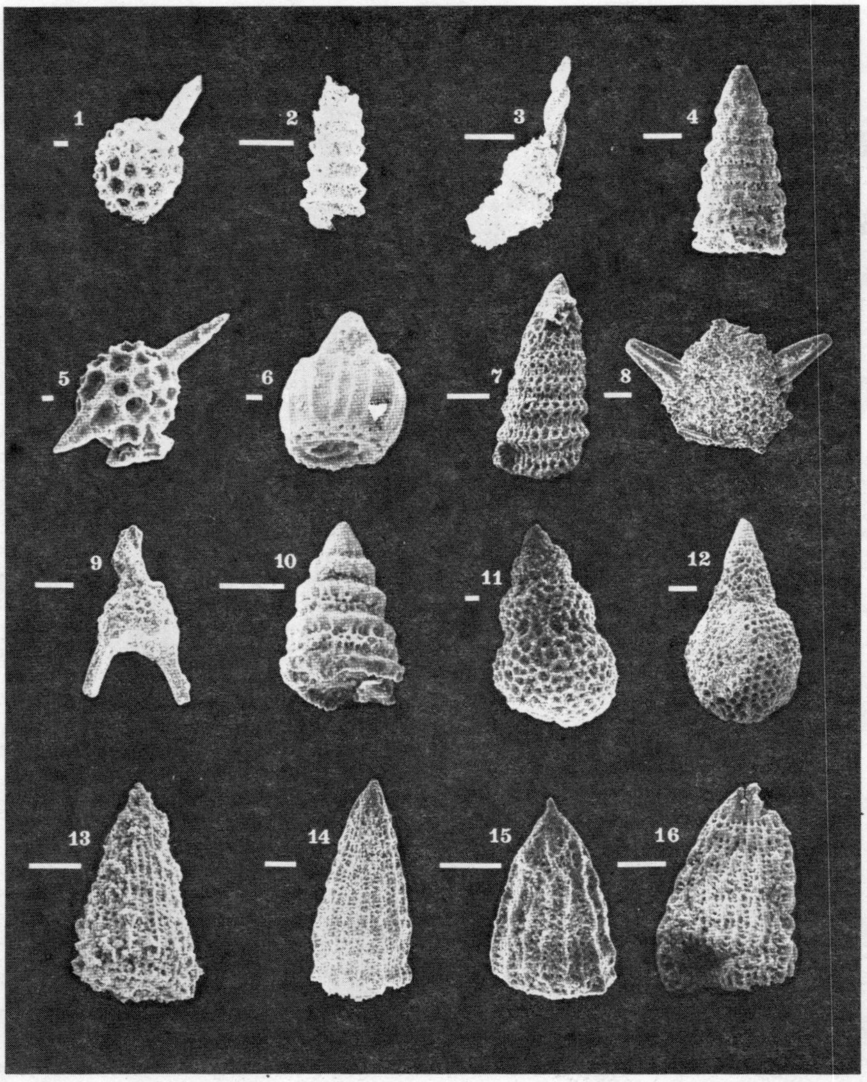

Figure 5. Photomicrographs of radiolarians from the Lassics area. White scale bars represent lengths indicated. Table 1 shows sample localities noted in this figure and located on figure 3. 1: Capnodoce sp., Upper Triassic (upper Karnian to Norian) (Blome, 1983); from chert pebble conglomerate on Black Lassic, sample 13-77; scale bar = 10 μm. 2: Triassocampe sp., Middle (Ladinian) to Upper Triassic (mid Norian); from chert pebble conglomerate on Black Lassic, sample 13-77; scale bar = 50 μm. 3: Sara sp., Upper Triassic (Karnian to Norian); from chert pebble conglomerate on Black Lassic, sample 13-77; scale bar = 50 μm. 4: Corum sp., Upper Triassic (upper Karnian to Norian); form chert pebble conglomerate on Black Lassic, sample 13-77; scale bar = 50 μm. 5: Pantanellium riedeli, Upper Jurassic (Tithonian) to Lower Cretaceous (Berriasian) (C. D. Blome, oral commun., 1983); NW 1/4 of the study area, sample 15-79, scale bar = 10 μm. 6: Eucyrtidium ptyctum, Upper Jurassic (Kimmeridigian to upper Tithonian) (Riedel and Sanfilippo, 1974); NW 1/4 of the study area, sample 13-79; scale bar = 10 μm. 7: Ristola sp., Middle Jurassic (upper Bathonian) to Upper Jurassic (upper Tithonian); Red Lassic, sample 11-77; scale bar = 25 μm. 8: Tripocyclia blakei. Upper Jurassic (Tithonian) (Pessagno, 1977); NW 1/4 of the study area, sample 13-79, scale bar = 50 μm. 9: Napora sp., Lower Jurassic to Upper Cretaceous; Franciscan terrane, sample 17-78; scale bar = 50 μm. 10: Undescribed Nassellarina, Upper Jurassic (Tithonian?) (C. D. Blome, oral commun., 1983); NW 1/4 of the study area, sample 66-79; scale bar = 50 μm. 11, 12: Undescribed Nassellariina; 11, Red Lassic, sample 11-77; 12, NW 1/4 of the study area, sample 66-79; scale bar = 10 μm, 50 μm, respectively. 13: Hsuum sp., Lower Jurassic (lower Pliensbachian) to Lower Cretaceous (upper Valanginian); Franciscan terrane, sample 17-78; scale bar = 50 μm. 14: Hsuum sp. aff. H. mclaughlini, Upper Jurassic (Tithonian?) (Pessagno and Blome, 1983); Red Lassic, sample 11-77; scale bar = 50 μm. 15: Hsuum obispoensis Upper Jurassic (upper Kimmeridgian-lower Tithonian) to Lower Cretaceous (upper Valanginian) (Pessagno, 1977); Red Lassic, sample 11-77; scale bar = 50 μm. 16; Hsuum sp., Lower Jurassic (lower Pliensbachian) to Lower Cretaceous (upper Valanginian); NW 1/4 of the study area, sample 88-79; scale bar = 50 μm.

Figure 6. Photomicrographs of radiolarians from the Lassics area. White scale bars represent lengths indicated. Table shows sample localities noted in this figure and located on figure 3. 1: Crucella sp., Jurassic to Cretaceous; from the NW 1/4 of the study area, sample 13-79; scale bar = 100 μm. 2, 3: Acanthocircus variabilis (Squinabol) (C. D. Blome, oral commun., 1983); NW 1/4 of the study area, sample 88-79; scale bar = 50 μm. 4, 7, 8: Parvicingula vera, Middle Jurassic (upper Bathonian) to Upper Jurassic (Tithonian) (Pessagno and Whalem, 1982); Red Lassic, sample 11-77; scale bar = 50 μm, 50 μm, 100 μm, respectively. 5, 6: Parvicingula sp., Middle Jurassic (lower Bajocian) to Lower Cretacaeous (upper Valanginian); 5, Red Lassic, sample 11-77, 6, NW 1/4 of the study area, sample 66-79; scale bar = 25 μm, 100 m, respectively. 9, 10, 11, 12: Archaeodictyomitra sp., Middle Jurassic (lower Bajocian) to Upper Cretaceous (Campanian); NW 1/4 of the study area, sample 88-79, 66-79, 13-79, 13-79, respectively; scale bar = 50 μm. 13, 14, 15: Mirifusus sp., Upper Jurassic (Kimmeridgian) to Lower Cretaceous (upper Valanginian); NW 1/4 of the study area, sample 66-79, 88-79, 66-79, respectively; scale bar = 100 μm.

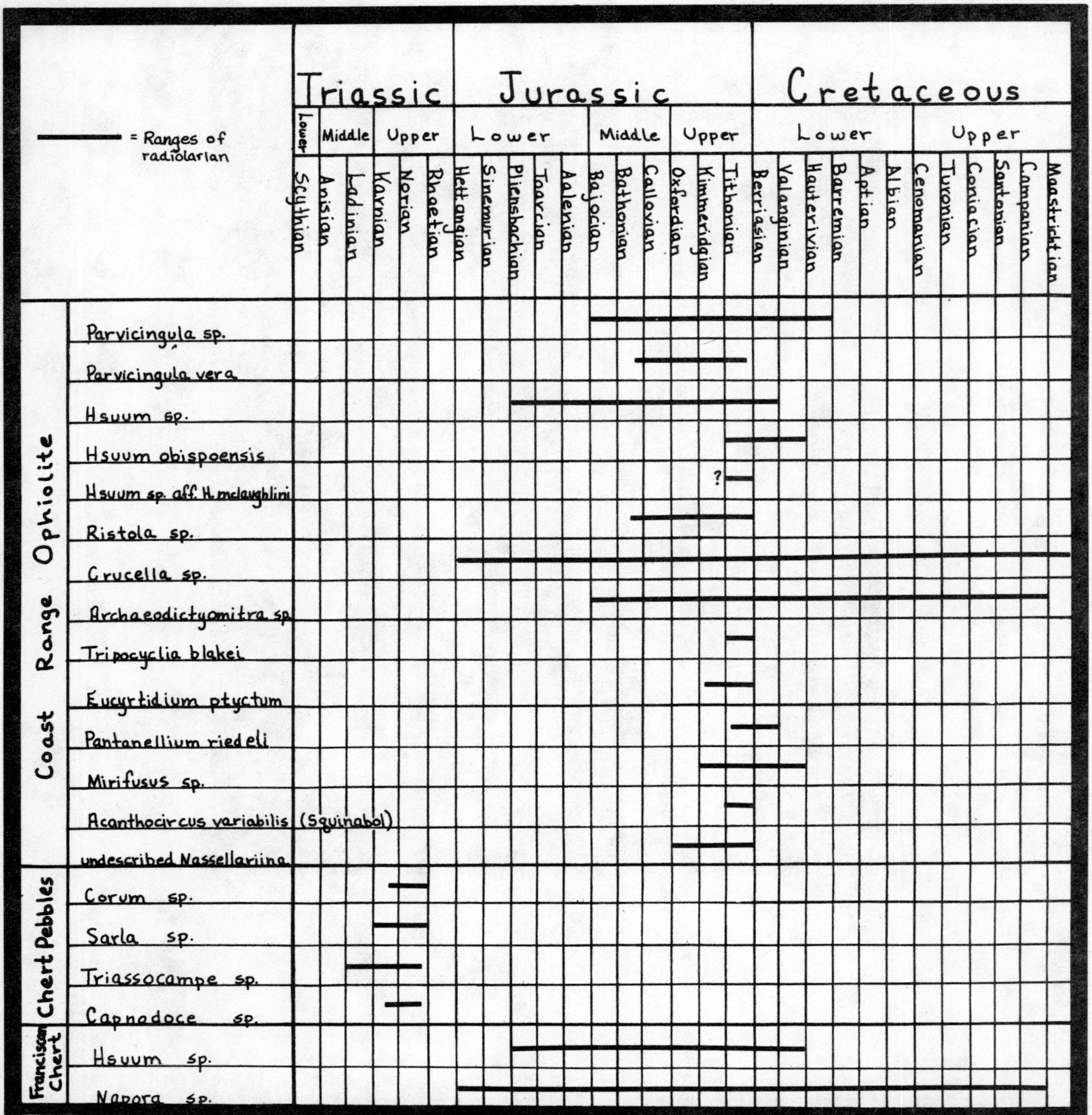

FIGURE 7. Ranges of radiolarians extracted from chert and chert pebble conglomerate of the Lassics area.

Figure 8. Red chert (ch) interbedded with green volcanic rocks (v) of the Coast Range ophiolite in the northwest portion of the study area. This outcrop yielded radiolarians of Late Jurassic to Early Cretaceous age (samples 15-79 and 16-79). A more specific age of Tithonian was obtained from float further up-section (sample 13-79).

Great Valley Sequence

The sedimentary rocks of Black Lassic are clearly distinct from the Franciscan sedimentary deposits, and have detrital modes correlative with those of the Stony Creek petrofacies of the Great Valley sequence (fig. 9). Unlike the Franciscan metagraywackes in the area, the Great Valley sandstones commonly contain glauconite pellets and biotite grains. Great Valley sequence rocks have detrital textures with little evidence of shearing or exensive recrystallization. There were no blueschist minerals or clasts found in the Great Valley sequence sedimentary rocks as were found in the Franciscan metasediments.

Black Lassic peak is the only place in the field area where sedimentary rocks of the Great Valley sequnence crop out. Sedimentary strata of Black Lassic have been metamorphosed to the zeolite and prehnite-pumpellyite facies but in the field appear unmetamorphosed. Secondary pyrite found as cubes as well as undefined pseudomorphs occurs throughout the sediments. Calcite occurs as fills of veins and voids within sedimentary rocks and as zones of mineralization that mark fault planes. Here the calcite acts as a matrix in which angular clasts of black shale are embedded. The white zones of calcite cut the Great Valley sequence sedimentary rocks of Black Lassic peak and contrast visibly with the dark sedimentary host rock.

The sedimentary rocks exposed on Black Lassic, like those at Stony Creek, consist predominantly of a fissile, dark greeish-gray to black, thinly bedded to laminated, locally concretionary, siltsone and mudstone. These rocks comprise about 70 percent of the sedimentary sequence. Within the laminated siltstone and mudstone exist fine- to coarse-grained sandstone lenses and beds as well as lenses of conglomerate. The sandstone and conglomerate comprise about 30 percent of the unit. Altered pillow lavas occur in a matrix of shaley Great Valley sequence sedimentary rock near the base of Black Lassic.

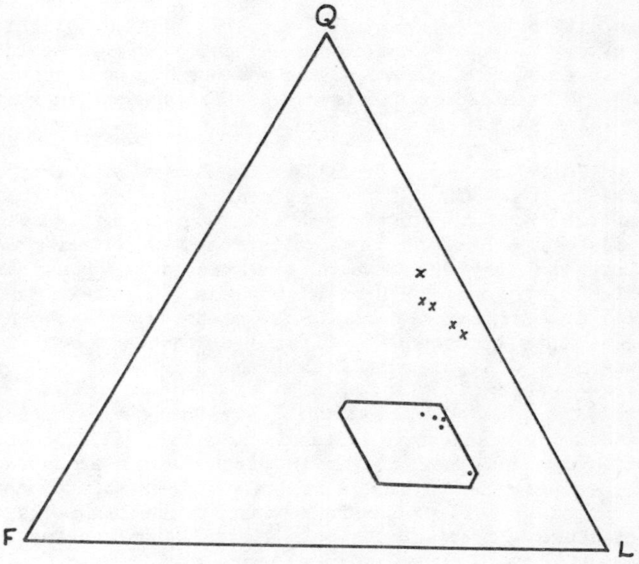

Figure 9. Triangular percentage plot showing the volumetric Q-F-L percentages of quartzone grains (Q), feldspar grains (F), and lithic fragments (L) for the Great Valley sandstone (·) versus Franciscan sandstone (x) of the Lassic outlier. Correlative Stony Creek petrofacies of the Great Valley sequence plot in the diamond (data from Ingersoll, Rich, and Dickinson, 1977).

Shale and Sandstone

Shales at Black Lassic can be seen to have plastically flowed around more coherent lenses of rock. Concretions are common and when processed for microfossils, yielded pyrite growths which appeared to have replaced radiolarians.

Sandstone beds and lenses are characterized by their dark green color, high percentages of volcanic clasts, and low quartz content. Some beds are of uniform thickness (fig. 10) while others pinch and swell; however all maintain the same general structural attitude. Generally, these rocks are well indurated, fine- to coarse-grained nonporous dirty sandstone which breaks into angular blocky chunks. With the exception of bedding, sedimentary structures are not obvious. Calcite is commonly found filling voids or forming veins. Quartz also occurs as vein material. Pyrite crystals are common and weathered pyrite stains the surface of some rocks.

Microscopically, the sandstones of Black Lassic are microbreccias rich in volcanic detritus. The sandstones are classified as lithic graywackes bordering on lithic arenites. Grains are extremely angular with a lack of abrasion and sorting, and occur in an argillaceous matrix. In most cases, grains are highly compacted but not tectonically deformed.

In many instances, it is hard to tell grains from matrix due to the size gradation between the two. The matrix has undergone patches of recrystallization to a green-yellow chloritic material. Zeolite and prehnite-pumpellyite minerals can be observed. Calcite also occurs as matrix, but its occurrence is very sporadic and is not present in all rocks sampled.

The sandstones of Black Lassic are composed of approximately 55 percent to 75 precent lithic clasts. Clastic serpentinite and an occasional schist clast accounts for most of the metamorphic rock detritus. Plutonic detritus is rare but is present in very small amounts.

Shale grains of probable intraformational origin make up the bulk of the sedimentary rock detritus. The remainder of clastic sedimentary rock detritus is sandstone. Some of these clasts appear intraformational as they are almost identical to the rocks in which they form. Others have almost no matrix, are high in quartz and feldspar fragments, and have very few lithic components. Clastic carbonate occurs in extremely small amounts.

The most abundant lithic detritus are basaltic, diabasic or antesitic volcanic clasts. The second most abundant rock clasts in Black Lassic sandstone are quartz, chert, and quartzite. The least abundant clasts in the Black Lassic sandstones are plagioclase feldspar. The plagioclase is albite to oligoclase in composition. Other mineralogic constituents that appear to be of primary (detrital) origin include biotite altering to chlorite and oxides. Garnet, zircon, sphene, and spinel of euhedral habit occur as accessory grains.

A few shelly fossils were deposited in the sandstone as clastic debris. These are broken and can only roughly be identified as echinoderm plates (Rodney Watkins, oral commun., 1980). Glauconite occurs sparsely distributed throughout the sandstones. Occasionally a brown reaction rim of limonite is present. Chert clasts high in argillaceous detritus have the outline of radiolarians preserved within the clast. Some radiolarians have well defined shapes, others occur as recrystallized "blotches," and a few are replaced by carbonate.

Conglomerate

The conglomerates of Black Lassic crop out as resistant lenticular pods. These pods are concentrated on the southwest side of the ridgetop forming Black Lassic peak and have no apparent bedding features. The conglomerates are surrounded on all sides by ductile black shale that "flows" around the resistant pods.

The clasts that make up the conglomerate are chiefly chert and volcanic rock detritus with subordinate clasts of shale, graywacke, quartz, and feldspar. The matrix surrounding these clasts ranges from 10 percent to 30 percent of the rock. Radiolarians are common in most chert clasts and can be seen with the aid of a hand lens. Volcanic rock detritus is the primary constituent in some of conglomerate; while in others it is only of secondary abundance.

Shale clasts in the conglomerate are probably of intraformational origin as they are very angular, ductile, and are similar in appearance to the shale beds exposed on Black Lassic. They have been interpreted as rip-up clasts deposited with other constituents during sediment gravity flows into a low energy marine basin (Hagans, 1978). Graywacke clasts are another minor constituent of the conglomerates. Quartz and feldspar clasts are both subangular in form and are generally confined to the matrix.

Figure 10. Tithonian sandstone and shale beds of the Great Valley sequence on Black Lassic.

Radiolarians and conodonts of extraformational origin have been extracted from the conglomerates. These organic remains are found primarily in the chert clasts. Although many of the conglomerates yielded radiolarians, only a few samples were diagnostic as to the age of the chert. None of the conodonts extracted were diagnostic. The age of the chert, based on the presence of conodont fragments and radiolarians Corum sp., Sarla sp., Triassocampe sp., and Capnodoce sp. (Blome, 1983), appears to be Late Triassic (Karnian to Norian) (D.L. Jones, oral commun., 1978; B.L. Murchey, oral commun., 1981; C.D. Blome, oral commun., 1983).

Metamorphic Minerals

Basal rocks of the Great Valley sequence typically have metamorphic minerals common to the zeolite and/or prehnite-pumpellyite facies (Baily and Jones, 1973; Dickinson and others, 1969). Alteration minerals indicative of the zeolite facies include chloritization of biotite and replacement of plagioclase by albite and laumontite (Dickinson and others, 1969).

These critical phases have been observed in Great Valley sequence graywackes exposed on Black Lassic. Chlortization and albitization of matrix and detrital grains occur within these sandstones. Other metamorphic minerals observed in these rocks include: laumontite, pumpellyite, prehnite, analcite, sphene, epidote, and minor muscovite. Metamorphic mineral assemblages in Great Valley sequence rocks exposed on Black Lassic are indicative of two metamorphic facies, the zeolite facies and the prehnite-pumpellyite facies. The zeolite facies is typically underlain by the prehnite-pumpellyite facies and there is substantial overlap in the range of occurrence of both facies (Hyndman, 1972; Winkler, 1974; Blatt and others, 1972). Because the rocks of Black Lassic contain laumontite + pumpellyite + prehnite + epidote, it is best to assume that the rocks were metamorphosed near the boundary of the two facies. Temperatures and pressures for minerals in Great Valley sequence rocks exposed on Black Lassic indicate that these rocks were subjected to temperatures between 250 to 300°C and pressures between 2.5 and 2 kb during burial metamorphism (Kaplan, 1983).

Comparison with Stony Creek Sediments

The general megascopic field appearance of both Stony Creek sediments and Black Lassic sediments strongly suggest a correlation of the two petrofacies. Contact relations, typified by intertonguing of shales with pillow lavas and breccias at the base of both sedimentary sequences, are another feature suggestive of this correlation. The most diagnostic of all similarities is the agreement of petrographic characteristics. The content of mica, mostly biotite, is an extremely diagnostic parameter for the recognition of petrologic intervals in the Great Valley sequence (Dickinson and Rich, 1972). Sedimentary rocks of both Stony Creek and Black Lassic have an unusually low percentage of biotite. Comparison of detrital modes show sandstones of both localities distinctive in their low percentage of quartz, high percentage of lithics, low ratio of plagioclase to total feldspar, and high ratio of volcanic lithics to total lithics. Both sedimentary outcrops have also undergone similar diagenetic alteration histories. In both localities, albitization of plagioclase and chloritization of biotite are common features. Similarly, both outcrops have been subjected to burial metamorphism resulting in mineral assemblages common to the zeolite and prehnite-pumpellyite facies. Table 2 summarizes mineral composition of the Great Valley sequence exposed on Black Lassic.

Table 2. Mineral composition of sandstone of the Great Valley sequence

Sample Number	12-8	23-8	31-8	33-8	24-79
% Matrix	13	16	7	7	15
$CaCO_3$	x	x		x	
Chlorite	x	x	x	x	x
Mud	x	x	x	x	x
Grain Size (mm)	.1 to 1	.1 to 8	.2 to 10	.1 to 2	.1 to 1
Detrital Modes %					
Q	23	10	23	21	22
F	16	10	14	12	10
L (total Lithics)	48	64	56	60	53
Lv (volcanic)	42	51	46	57	38
Lm+p (meta. & plut.)	5	10	3	3	3
Ls (sedimentary)	1	3	7	0	12
Other Detrital Const. < .1 percent					
Biotite	x		x	x	x
Orthoclase	x				x
Hematite	x	x	x	x	x
Ilmenite			x		
Chromite	x	x	x	x	x
Sphene	x		x		x
Magnetite	x	x	x	x	x
Zircon	x	x	x	x	x
Garnet	x				x
Authigenic Const.					
Dolomite	x	x			x
Pyrite	x	x	x	x	x
Glauconite	x	x	x		x
Quartz	x	x	x	x	x
Albite	x	x	x	x	x
Fossils	x				x
Radiolarians in Clasts	x		x		x
Metamorphic Min.					
Albite	x	x	x	x	x
Pumpellyite	x	x	x	x	x
Laumontite	x	x	x	x	x
Epidote	x	x	x	x	x
Sphene			x		
Muscovite	x		x	x	x
Analcite (Pseudo)			x		
Vein Minerals					
Calcite		x			x
Albite + Quartz	x			x	

Franciscan Complex

The Franciscan Complex underlies and completely surrounds the Lassic ultramafic body and its associated volcanic and sedimentary rocks. These rocks include metasandstone, metashale, metaconglomerate, metachert, greenstone, and schist rich in lawsonite ± glaucophane. There are no textural zone 1 Franciscan rocks exposed in the study area. Most of the Franciscan graywackes are of textural zone 2 or transitional between zones 2 and 3. There are at least four locations where textural zone 3 rocks crop out.

Franciscan Metaclastic Rocks

Metagraywacke is the dominant rock type in the Franciscan terrane. This rock type grades into a semischist where relict grains (mostly quartz) are visible with the aid of a hand lens. These rocks are well-indurated, grayish-white to bluish-gray on fresh surfaces, and buff on weathered surfaces. Interbedded with the graywackes are conglomerate and shale beds. The shale beds are typically schistose due to their fine grain size, while the conglomerates have little if any schistosity or cleavage developed. The shale is a fissile, dark black to gray rock which weathers buff. This rock is highly crenulated and breaks into paper thin sheets and needles. The conglomerates form extremely resistant rocks. They are found as massive discontinuous beds, some of which may be tectonic inclusions.

The majority of detritus forming the conglomerates are lithic and quartzose clasts. Feldspar detritus is rare in the coarser-grained conglomerates, but is more abundant in the finer-grained varieties. Lithic clasts include shale, graywacke, coarse-grained and fine-grained volcanics. Conglomerates can best be observed in the Franciscan terrane around Mt. Lassic and through windows in the ultramafic body to the north.

Fine-grained conglomerates to fine-grained sandstones were studied microscopically. Evidence of cataclasis was viewed in all varieties of clastic sedimentary rock studied. Cataclasis is confined chiefly to grain boundaries, which are highly sutured, sheared, and broken, and to the matrix which has undergone extensive recrystallization to chlorite, white mica, stilpnomelane, and pumpellyite. These rocks show much evidence of compaction and are fractured within and across grains. Calcium carbonate can be observed filling voids and possibly replacing matrix material. Other mineralogic constituents include hematite, magnetite, leucoxene, and zircon. Dolomite is present in chert clasts and reveals a characteristic rhombic habit, which tends to degrade and become granular as grain size decreases and cataclasis increases. In the finer-grained sandstones, dolomite is not present.

Alteration minerals include quartz, albite, pumpellyite, chlorite, sphene, zircon, white mica, and stilpnomelane. Lawsonite is present in very small amounts as minute tabular crystals with low birefringence and higher relief.

As clast size decreases, the size of lawsonite grains increases. Epidote is also present in some conglomerates and is generally found in aggregates around sphene crystals. As cataclasis increases and grain size decreases, the epidote disappears.

Some of the conglomerates in the Franciscan terrane were processed for radiolarians. A few of these yielded radiolarians of Triassic age (Karnian to Norian) (C.D. Blome, oral commun., 1983). Franciscan metasandstones are highly compacted and show evidence of shearing and cataclasis. The matrix of the Franciscan metasandstone has been totally recrystallized to chlorite, white mica, and stilpnomelane. In sandstones approaching textural zone 3, quartz also becomes a component of the matrix. These minerals reflect the presence of an original clay matrix.

Quartz occurs in the Franciscan metasandstones as primary quartz clasts, quartzite, and chert fragments, as well as matrix in the more cataclastic samples. Metasandstones generally contain albitic plagioclase with very little potassium feldspar, most of which occurs within lithic volcanic fragments. The lithic volcanic detritus ranges in composition from basaltic or diabasic to andesitic. Volcanic clasts are the least resistant and are the first to be affected by metamorphism. Metamorphic rock detritus consists mostly of serpentinite and schistose clasts. Glucophane was not observed in schist clasts. Plutonic rock clasts are sparse. Sedimentary rock detritus consists mainly of graywacke and shale clasts. These fragments show ductile deformation and are generally bent, compacted, and sheared. These clasts become smeared between more competent grains and form part of the matrix with increasing textural zone.

Other constituents include hematite, leucoxene, and pyrite. Rounded sphene and zircon are present as detrital grains. Dolomite forms euhedral crystals in chert clasts, which become aggregates and finally disappear with increasing deformation.

Vein minerals include albite + quartz, chlorite, and potassium beldspar. These observations apply mainly to rocks of low to high textural zone 2.

The Franciscan metasandstones contain alteration minerals common to Franciscan semischists (textural zone 2) and schistose (textural zone 3) rocks. Exceptions are the presence of glaucophane in higher textural zone rocks and the presence of epidote in the coarser grained, lower textural zone sandstones. Primary alteration minerals include albite, pumpellyite, chlorite, lawsonite, sphene, zircon, white mica, and stilpnomelane. Table 3 summarizes the mineral composition of Franciscan metagraywacke in the study area and table 4 shows the composition of schistose metagraywacke.

Other Constituents

Other constituents of the Franciscan Complex include isolated volcanic and chert bodies that crop out within the Franciscan metasedimentary terrane. Their occurrence is sporadic and no structural trends were observed.

The chert bodies were assigned to this terrane by the lack of association with any other rock type. They do not appear continuous within the Franciscan terrane but are surrounded by a matrix of Franciscan metasedimentary rock. Radiolarians extracted from these chert bodies were too recrystallized to be diagnostic; however, one chert locality outside the northern boundary of the study area found interbedded with volcanic greenstone yielded Napora sp., Hsuum sp., and Pantanellium sp. radiolarians of Late Jurassic to Early Cretaceous age (B.L. Murchey, oral commun., 1981).

Table 3. Mineral composition of Franciscan Metagraywacke

	Conglomerate		Coarse Sandstone		Fine Sandstone		
Sample #	36-79	99-79	18-79	44-79	7-79-A	35-79	78-79
% Matrix	3	2	4	5	10	6	6
$CaCO_3$	x		x				x
Phyllosilicates	x	x	x	x	x	x	x
Grain Size (mm)	5 to 20	5 to 10	.1 to 2	.1 to 2	.5 to .1	.5 to .2	.5 to 1
Detrital Modes							
Q	28	62	40	41	43	44	50
F	4	4	4	6	7	7	6
L	63	30	52	48	40	43	38
Lv	55	11	29	24	23	14	21
Lm+p	5	12	8	15	14	22	13
Ls	3	7	15	9	3	7	4
Other Const.	<2%						
Chessboard alb.	x				x		
Hematite		x	x	x	x	x	x
Magnetite	x	x	x	x	x	x	x
Ilmenite/Leucox	x	x	x	x	x	x	x
Pyrite			x	x			x
Sphene				x			
Zircon	x		x	x			
Garnet							
Dolomite	x	x	x	x			
Vein Minerals							
Calcite							
Qtz+Albite	x	x	x	x	x	x	x
Other	x			x	x		
Alteration Min.							
Albite	x	x	x	x	x	x	x
Pumpellyite	x	x	x	x	x	x	x
Chlorite	x	x	x	x	x	x	x
Lawsonite		x	x		x		x
Epidote	x		x				
Sphene	x	x	x	x	x	x	x
Zircon		x		x			
White Mica	x	x	x	x	x	x	x
Stilpnomelane	x	x	x	x	x	x	x
Zoisite			x				
Amphibole		x					
Pyroxene							
Textural Zone	-	-	Tz2-	Tz2-	Tz2	Tz2	Tz2

Table 4. Mineral composition of Franciscan Schistose Metagraywacke

Sample #	34-79	36-0-2	42-79-A	38-0	22-0	36-0-1
%Relict grains	20	15	0	2	0	0
Quartz	x	x				
Feldspar	x	x		x		
Garnet	x					
Zircon	x					
%Metamorphics						
Albite + Qtz.	50	67	56	55	55	32
Muscovite	7	2	7	2		
Chlorite	5	2	10	5		5
Sphene	3	3	5	2	2	
Pumpellyite	5	5	9	5	2	
Stilpnomelane	5	3	3	12	3	
Lawsonite	5	3	3	3	3	3
Trem./Act.			6	12	15	
Glaucophane			1	2	20	
Zircon	t					
Ilmenite/Leucox.	t	t				
Magnetite	t	t				
Hematite	t	t	t		t	
Carbonate						60
Vein Minerals						
Quartz & Albite		x				
Albite				x		
Textural Zones	Tz2+	Tz3-	Tz3	Tz3	Tz3	Tz3

t = trace

STRUCTURE

The Franciscan Complex forms the lowermost unit in the study area (figs. 2 and 11). This unit underlies and completely surrounds the exposure of the Lassic ultramafic body and its associated sedimentary rocks.

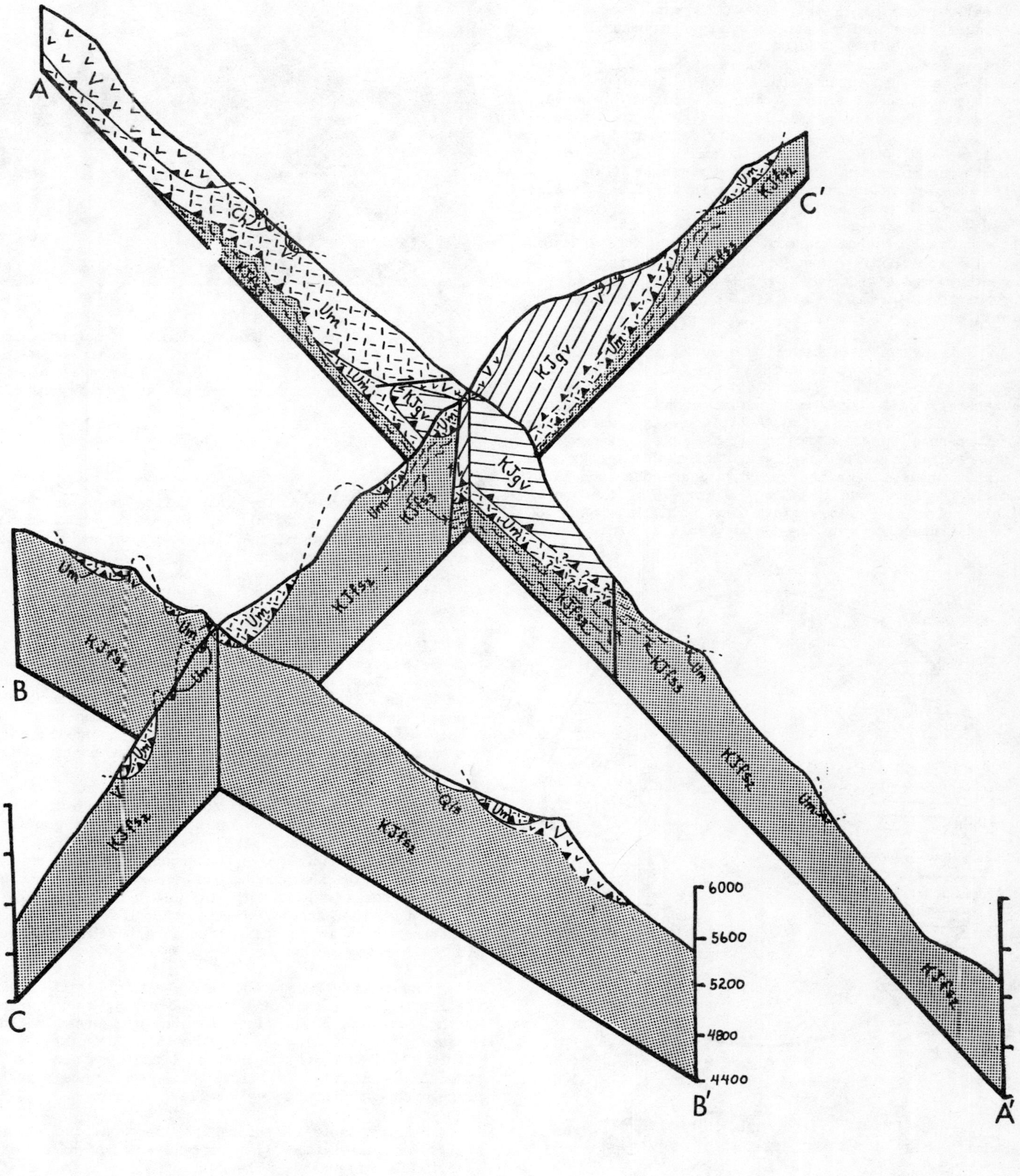

FIGURE 11. Fence Diagram of the Lassic Outlier.

In most places the Franciscan terrane superficially resembles a melange with a matrix of sheared metagraywacke and metashale having structural attitudes that differ radically over short distances. Greenstone, metachert, metagraywacke, metaconglomerate, and blueschist appear to be incorporated in this matrix as tectonic blocks or knockers. However, pole to bedding plots suggest that the exposed Franciscan rocks are a relatively coherent but highly folded unit with fold axes trending roughly north-south and plunging steeply to the south (fig. 12). Evidence of this can be observed in the field looking east from Mount Lassic (fig. 13). Here, Franciscan strata are exposed beneath sheared serpentinite with the contact forming the nose of a fold. Fold axes can be traced through other exposures of Franciscan sedimentary rocks that are seen through windows in the overlying ultramafic sheet along the crest of the antiform.

The contact between the Lassic ultramafic body and the underlying Franciscan Complex is the Coast Range thrust fault. The fault zone is marked by an extremely well developed zone of metasomatic rodingite. This fault is also typified by upside-down metamorphic zonation in the Franciscan rocks of the lower plate. The overlying ultramafic rocks are highly sheared near the thrust and are characteristically reddish-brown in color. Wicks (1977) interprets this color change to result from a higher degree of serpentinization near the thrust fault contact.

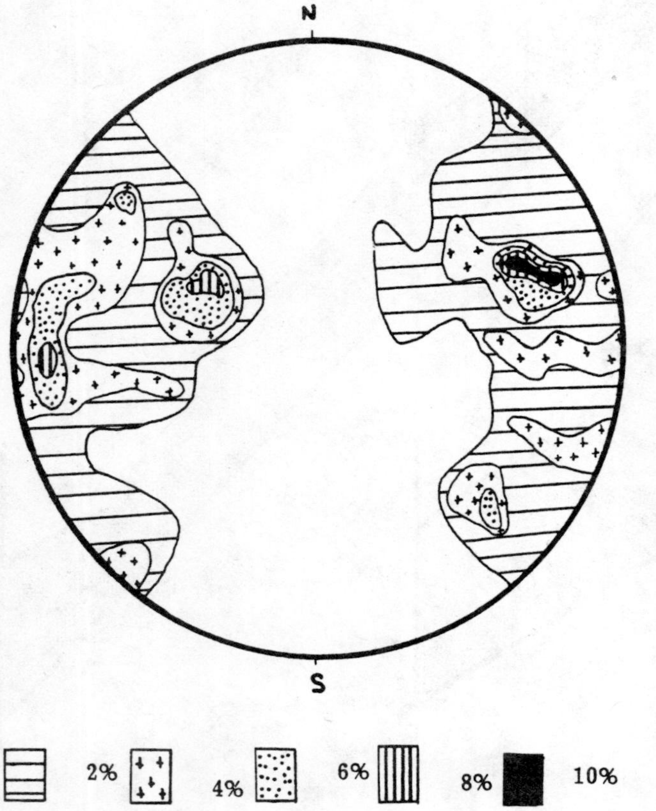

Figure 12. Pole to bedding plot of Franciscan metasedimentary strata (50 points). Plot indicates fold axes trend roughly north-south. Patterned areas represent percent of total points per 1% area of the hemisphere.

Figure 13. View east toward window in serpentinite (Sp) exposing underlying Franciscan metasedimentary rocks (F). Fold axes plunge 55° to the north as indicated in figure 12.

Shearing within the ultramafic rocks decreases away from the thrust fault and the rocks grade into a serpentinized harzburgite which is blocky and massive, with relict textures that are visible in hand specimen. There is also a color change from the light green sheared variety of serpentinite to the dark olive green, massive bastitic harzburgite.

Associated with the ultramafic body are mafic volcanic rocks and Great Valley sequence sedimentary rocks. These rocks have been separated from the underlying ultramafic rocks by faults that probably formed during regional thrusting. Shearing near the upper boundary of the ultramafic rocks is present but is minor compared to that near the lower contact with the regional Coast Range thrust fault.

The volcanic rocks on Mount Lassic are inferred to be thrust beneath serpentinite. At this contact, there is a fairly well developed metasomatic reaction rind on the volcanic rocks that probably formed during thrusting. On the northern flank of Black Lassic, volcanic rocks are in thrust fault contact with the sedimentary sequence. The contact is clearly not conformable as it abruptly truncates bedding features in the sedimentary rock.

The contact between the strata of the Great Valley sequence rock and the underlying rock varies with rock type. On Black Lassic, the Great Valley sequence has a lack of sedimentary structures and strikes uniformly to the northeast and dips steeply to the northwest (fig. 14). The thickness of the beds is irregular and pinching and swelling is common.

The contact relationships between the Great Valley sequence and the volcanic rocks have been previously discussed. The contact between the ultramafic sheet and the strata of the Great Valley sequence is a thrust fault which is marked by an extremely well developed zone of metasomatized shale composed predominantly of prehnite and pectolite (fig. 15). It should be noted that the ultramafic rocks have undergone only slight shearing at this contact as their warty nature and bastitic texture are still

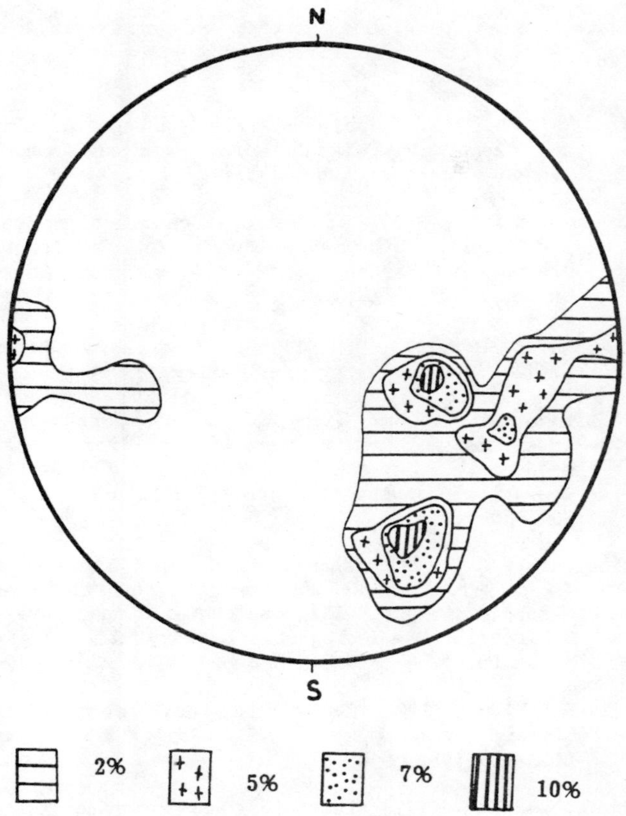

Figure 14. Pole to bedding plot of sedimentary rocks exposed on Black Lassic Peak (20 points). Plot indicates beds strike northeast and dip steeply to the northwest. Patterned areas represent percent of total points per 1% areaof the hemisphere.

Figure 15. Wall of resistant metasomatized shale along contact between ultramafic rocks and Great Valley sequence of Black Lassic. Metasomatized shale is predominately prehnite and pectolite.

present. This suggests that thrusting and shearing along the contact between the ultramafic rocks and the sedimentary rocks of the Great Valley sequence was not as intense as the regional thrusting responsible for shearing the serpentinite near the Franciscan contact.

On the southeast flank of Black Lassic peak, sedimentary rocks of the Great Valley sequence are in fault contact with Franciscan sedimentary rocks of high textural zone 2. The contact occurs in a gully with Franciscan metasedimentary rocks vegetated with live oak on a high linear ridge to the south and the black barren unmetamorphosed rocks of Black Lassic to the north. The abrupt nature of the contact and the structural sequence of rocks in the study area strongly suggest this contact is a fault. Thus, rocks of the Great Valley sequence exposed on Black Lassic peak have moved down relative to the metasedimentary rocks of the Francsican Complex. This, plus the resistent volcanic rocks cropping out on the opposite flank of Black Lassic peak is believed responsible for the preservation of the Great Valley sediments of the peak.

SUMMARY

This study indicates that the upper plate rocks of the Lassic area are part of the Coast Range ophiolite and associated Great Valley sedimentary rocks that are more extensively exposed approximately 100 km to the southeast near Paskenta. The succession of rock units observed in the Lassic area is the same as exposures of Coast Range ophiolite along the west side of the Sacramento Valley.

The Great Valley sedimentary rocks exposed on Black Lassic have detrital modes that correlate with the Stony Creek petrofacies of the Great Valley sequence. The sedimentary strata of Black Lassic have undergone similar alteration as those exposures of Upper Jurassic-Lower Cretaceous Great Valley rocks seen elsewhere. The sedimentary rocks of Black Lassic are clearly distinct from Franciscan sedimentary strata structurally beneath the ophiolitic rocks.

Franciscan sedimentary rocks in the study area differ from Great Valley sedimentary rocks in composition, texture, and degree of metamorphism. The Franciscan sedimentary rocks are devoid of biotite and glauconite; these are common constituents of the Great Valley sandstones exposed on Black Lassic. In the Lassic area the Franciscan sedimentary strata contain less lithic clasts and more quartzose clasts than the Great Valley sedimentary rocks. Texturally, Franciscan sandstones range form low textural zone 2 to textural zone 3. Great Valley sequence sandstones which have undergone slight alteration to the zeolite or prehnite-pumpellyite facies are confined to Black Lassic peak. The Franciscan rocks in the study area have mineral assemblages common to high pressure-low temperature metamorphism. The Franciscan sedimentary rocks of textural zone 3 have undergone metamorphism to a lawsonite-bearing quartz-mica schist indicative of the blueschist facies.

ACKNOWLEGEMENTS

I wish to thank all of those individuals who helped me during the course of this project, and especially W. Thomas Henry, my devoted husband, who assisted in mapping the study area, reviewed the manuscript, and was always present to give moral support. Special thanks also goes to W. Porter Irwin (U.S. Geological Survey, Menlo Park, CA) who served on the thesis committee, provided technical support, reviewed the many drafts and helped me through the rough times. Also, I wish to thank Dr. John Williams (Department of Geology, San Jose State University, San Jose, CA) and Dr. Scott Creely (Department of Geology, San Jose State University, San Jose, CA) who served on the thesis committee and reviewed the manuscsript. Thanks to Dr. Ken Aalto (Department of Geology, Humboldt State University, Arcata, CA) and Clark Blake (U.S. Geological Survey, Menlo Park, CA) who reviewed the manuscript and made helpful suggestions. Thanks to David L. Jones (U.S. Geological Survey Menlo Park, CA), Charles D. Blome (U.S. Geological Survey, Menlo Park, CA), and Benita L. Murchey (U.S. Geological Survey, Menlo Park, CA) who gave their time to identify the radiolarians and made helpful suggestions on how to prepare and present the data. Special thanks go to Mary Hathaway for taking scanning electron micrographs of the radiolarians. I also wish to thank the faculty of the Department of Geology and Physical Science, Chico State University, Chico, CA, who helped with this project, especially Dr. Vern McMath who reviewed the geologic map and structure sections, Dr. Rodney Watkins who identified fossils other than radiolarians, and Dr. William Guyton who gave me access to the department's many facilities.

REFERENCES CITED

Bailey, E. H., and Jones, D. L., 1973, Metamorphic facies indicated by vein minerals in basal beds of the Great Valley sequence, northern California: Journal of Research, U.S. Geological Survey, v. 1, no. 4, p. 383-385.

Bailey, E. H., Blake, M. C., Jr. and Jones, D. L., 1970, On-Land Mesozoic Oceanic Crust in California Coast Ranges: U.S. Geological Survey Professional Paper 700-C, p. C70-C81.

Bailey, E. H., Irwin, W. P., and Jones, D. L., 1964, Franciscan and related rocks and their significance in the geology of western California: California Division of Mines and Geology Bulletin 183, 177p.

Barnes, Ivan, and O'Neil, J. R., 1969, The Relationship between Fluids in Some Fresh Alpine-Type Ultramafics and Possible Modern Serpentinization, Western United States Geological Society of America Bulletin, v. 80, p. 1947-1960.

Bertucci, P. F., 1983, Petrology and Provenance of the Stony Creek Formation, Northwestern Sacramento Valley, California, in Bertucci, P. F., and Ingersoll, R. Y., eds., Guidebook to the Stony Creek Formation, Great Valley group, Sacramento Valley California: Pacific Section Society of Economic Paleontologists and Mineralogists, p. 1-16.

Blake, M. C. Jr., and Jones, D. L., 1977, Plate tectonic history of the Yolla Bolly junction, northern California: Fieldtrip Guidebook, 73rd annual meeting of the Cordilleran Section, Geological Society of America, Sacramento, California.

Blake, M. C., Jr., Irwin, W. P., and Coleman, R. G., 1969, Blueschist-facies metamorphism related to regional thrust faulting: Tectonophysics, v. 8, p. 237-246.

Blatt, Harvey, Middleton, Gerard, and Murray, Raymond, 1972, Origin of Sedimentary Rocks: New Jersey, Prentice-Hall, Inc. 634 p.

Blome, C. D., 1983, Upper Triassic Capnuchosphaenidae and Capnodocinae (Radiolaria) from east-central Oregon: Micropaleontology, v. 29, p. 11-49.

Coleman, R. G., 1967, Low-Temperature Reaction Zones and Alpine Ultramafic Rocks of California, Oregon, and Washington: U.S. Geological Survey Bulletin 1247, 49 p.

Coleman, R. G., 1977, Ophiolites: Minerals and Rocks, v. 12: New York, Springer-Verlag, 220p.

Dickinson, W. R., and Rich, E. L., 1972, Petrologic intervals and petrofacies in the Great Valley sequence, Sacramento Valley, California: Geological Society of America Bulletin, v. 83, p. 3007-3024.

Dickinson, W. R., Ojakangas, R. W., and Stewart, R. J., 1969, Burial metamorphism of the late Mesozoic Great Valley sequence, Cache Creek, California: Geological Society of America Bulletin, v. 80, p. 519-526.

Diller, J. S., 1902, Topographic devlopment of the Klamath Mountains: U.S. Geological Survey Bulletin 196, 69 p.

Ernst, W. G., 1977, Tectonics and Prograde Versus Retrograde P-T Trajectories of High Pressure Metamorphic Belts, in High Pressure-Low Temperature Metamorphism of the Oceanic and Continental crust in the Western Alps: Italian Society of Mineralogy and Petrology, Special Session, Sept. 1976, Genova, Italy, p. 33-62.

Hagans, D. K., 1978, Geology and structure of the Lassics area, Southern Humboldt and Trinity counties, California: Humboldt State University, Arcata, California, Unpublished Senior Thesis, 58 p.

Hopson, C. A., Mattinson, J. M., and Pessagno, E. A., Jr., 1981, Coast Range ophiolite, Western California, in Ernst, W. G., ed., The Geotectonic Development of California: Rubey Volume 1, New Jersey, Prentice-Hall, p. 419-510.

Hyndman, D. W., 1972, Petrology of Igneous and Metamorphic Rocks: New York, McGraw-Hill, 533 p.

Ingersoll, R. V., Rich, E. I., and Dickinson, W. R., 1977, Great Valley sequence, Sacramento Valley: Fieldtrip Guidebook, 73 Annual meeting of the Cordilleran Section, Geological Society of America, Sacramento, California.

Irwin, W. P., 1960, Geologic reconnaissance of the Northern Coast Ranges and Klamath Mountains, California, with a summary of the mineral resources: California Divsison of Mines Bulletin 179, 80 p.

-----,1964, Late Messozoic orogenies in the ultramafic belts of northwestern California and southwest Oregon: U.S. Geological Survey Professional Paper 501-C, p. C1-C9.

-----,1966, Geology of the Klamath Mountain Providence, in Bailey, E. H., ed., Geology of Northern California: California Division of Mines and Geology Bulletin, v. 190, p. 19-38.

-----,1977, Ophiolitic terranes of California, Oregon, and Nevada, in Coleman, R. G., and Irwin, W. P., eds., North American Ophiolites: Oregon State Department of Geology and Mineral Industries, Bulletin 95, p. 75-92.

Irwin, W. P., and Dennis, M. D., 1979, Geologic structure section across southern Klamath Mountains, Coast Ranges, and seaward of Point Delgada, California: Geologic Society of America Map and Chart Series, MC-28D, Scale 1:2,500,000.

Kaplan, T. A., 1983, The Lassic Outlier an Outlier of Coast Range Ophiolite: San Jose State University, San Jose, California: Unpublished M.S. thesis, 96 p.

Kelsey, H. M., and Allwardt, A. O., 1975, Geologic map of the Van Duzen River Basin, in Van Duzen River Basin Environmental Atlas: State of California Resources Agency, Department of Water Resources, Northern District.

Kelsey, H. M., and Hagans, D. K., 1982, Major right-lateral faulting in the Franciscan assemblage of northern California in late Tertiary time: Geology, v. 10, p. 387-391.

Miyashiro, Akiho, 1973, Metamorphism and Metamorphic Belts: New York, John Wiley and Sons, 92 p.

Pessasgno, E. A., Jr., 1977, Upper Jurassic Radiolaria and Radiolarian Biostratigraphy of the California Coast Ranges: Micropaleontology, v. 23, no. 1, p. 56-113.

Pessagno, E. A., Jr., and Blome, C. D., 1984, A revised Radiolarian Zonation for the Upper Jurassic of Western North America: Bulletin of American Paleotology, in press.

Pessagno, E. A., Jr., and Whalem, P. A., 1982, Lower and middle Jurassic Radiolaria (multicyrtid Nassellariina) from California, east-central Oregon, and the Queen Charlotte Islands, B. C.: Micropaleontology, v. 28, no. 2, p. 111-169.

Riedel, W. R., and Sanfilippo, A., 1974, Radiolaria from the southern Indian Ocean, Deep Sea Drilling Project Leg 26, in Davies, T. A., and others, initial reports of the Deep Sea Drilling Project, volume XXVI, Washsington D.C.: U.S. Government Printing Office, p. 1-1129.

Turner, F. J., 1981, Metamorphic Petrology, Mineralogical, Field, and Tectonic Aspects, second edition: New York, NcGraw-Hill Book Co., 524 p.

Wicks, F. J., and Whittaker, E. J. W., 1977, Serpentine Textures and Serpentinization: Canadian Mineralogist, v. 15, p. 459-488.

Winkler, H. G. F., 1974, Petrogenisis of Metamorphic Rocks, Third Edition: New York, Springer-Verlag, 320 p.

TECTONOSTRATIGRAPHIC FRAMEWORK OF THE GEYSERS-CLEAR LAKE REGION, CALIFORNIA

Robert J. McLaughlin and Henry N. Ohlin
U.S. Geological Survey
345 Middlefield Road
Menlo Park, CA 94025

ABSTRACT

Our studies of the Franciscan assemblage, the Coast Range ophiolite, and the lower part of the Great Valley sequence in the Geysers-Clear Lake area suggest a complex history of late Mesozoic and Paleogene accretion. The dismembered tectonostratigraphic units involved in this accretion are, proceeding southwestward across strike: the Grizzly Creek melange, which consists of intensely deformed rocks of the lower part of the Great Valley sequence and the Coast Range ophiolite, the Little Indian Valley terrane, the East Clear Lake (=Yolla Bolly? terrane) terrane and several terranes belonging to the Franciscan Central belt. Sediment of the Little Indian Valley terrane was deposited in an outer-slope or basin-plain setting during the Tithonian to Valanginian, concurrently with deposition of strata in the Grizzly Creek unit on the Coast Range ophiolite. Contemporaneously and continuing at least into late Early Cretaceous (Aptian-Albian) time, arc-derived sedimentary materials and tuffaceous radiolarian chert of the East Clear Lake terrane were deposited in an extensional oceanic setting on a basaltic substrate. The sandstone and chert were also intruded locally by mafic igneous rocks. In the 40-50-m.y. interval between the Valanginian and Turonian stages: (1) The Little Indian Valley terrane was folded and metamorphosed to pumpellyite-aragonite grade; (2) the East Clear Lake terrane was folded and metamorphosed to lawsonite-pumpellyite grade; and (3) concurrently with generation of the Grizzly Creek melange and westward movement of rocks in the upper plate of the Coast Range thrust, the Little Indian Valley and East Clear Lake terranes were juxtaposed and obducted eastward together onto a west-dipping mafic basement that is now at a depth of at least 6.8 km. Outboard of the East Clear Lake terrane in Late Cretaceous and Paleogene time, a large component of northward translation accompanied east-west convergence between North America and the offshore oceanic plate(s). This resulted in the incorporation of large slabs of the previously accreted Coast Range ophiolite, Great Valley sequence, and East Clear Lake terrane into the wide zone of pervasive shear between North America and the oceanic plate. The rocks in this zone of oblique convergence are now collectively referred to as the Franciscan Central belt. Also incorporated into the Central belt were such far-travelled exotic rocks as the Calera- and Laytonville-type limestones, high-grade blueschist, amphibolite, and a dismembered oceanic terrane of basalt, overlain by Tithonian to Cenomanian chert and arc- or continent-derived sandstone and conglomerate (The Geysers terrane).

INTRODUCTION

A voluminous amount of new data bearing on the Franciscan assemblage has accumulated since the concepts of plate tectonics were first applied to the Franciscan assemblage more than 15 years ago. Recent paleomagnetic studies strongly suggest that many terranes of Franciscan rocks have been tectonically translated northward thousands of kilometers from their original sites of deposition relative to North America (Alvarez and others, 1980; Gromme, this volume; Harbert and others, this volume). It is possible that some parts of the Franciscan assemblage may never have been associated with a trench environment or subduction zone, and that transform faulting or rifting mechanisms may have been at least as significant as subduction in the process of accretion. The Franciscan is now thought of as a collage of accreted tectonostratigraphic units of various origins, many of which are displaced large distances from their original depositional sites (Blake and others, 1982). The processes involved in their displacement and accretion are presently the subjects of intense interest. As early as 1964, Edgar Bailey and his colleagues (Bailey and others, 1964), considered large-scale strike-slip and oblique rifting as important alternative hypotheses to explain the distribution of the Great Valley sequence and Franciscan assemblage. Their models differ from presently conceived models involving right-slip and oblique-rifting mainly in scale, but the concept of large-scale northward translations is certainly not new to the Franciscan.

The regional geology of the Mayacmas Mountains and Clear Lake areas was earlier studied by Bailey (1946), Yates and Hilpert (1946), Brice (1953), and McNitt, (1968). These areas were also cited by Bailey and others, (1964) and Bailey and Blake (1969) in their recognition of the regional thrust relation between the Franciscan assemblage and Great Valley sequence. Cenozoic thrusting complexities in the Lower Lake area were recognized by Swe and Dickinson (1970).

In the early 1970's, the U.S. Geological Survey embarked on extensive studies in the Mayacmas Mountains and Clear Lake region as part of its geothermal research program. Our detailed studies conducted under this program made use of extensive geophysical studies and geothermal well data, and emphasized the structure of the Franciscan assemblage and Great Valley sequence, and their petrological, chemical, and age relations (McLaughlin and Stanley 1976; McLaughlin 1978, McLaughlin, 1981, McLaughlin and Pessagno, 1978).

The purpose of this paper is to describe major tectonostratigraphic terranes of the Franciscan assemblage and the lower part of the Great Valley sequence in the Mayacmas Mountains and Clear Lake areas of northern California. Several of these tectonostratigraphic terranes were previously unrecognized. The area of this study (figs. 1 and 2) extends from Wilbur Springs, at the west side of the Great Valley province, westward to the Maacama fault zone. The area encompasses several tectonostratigraphic units of the Franciscan assemblage, whose faulted boundaries generally strike northwest-southeast. Also included are strongly- to mildly-disrupted rocks in the lower part of the Great Valley sequence. On the following pages we describe the age, lithologic and sedimentologic

in Blake, M.C. Jr., ed., 1984, **Franciscan Geology of Northern California:** Pacific Section S.E.P.M., Vol. 43, p. 221-254

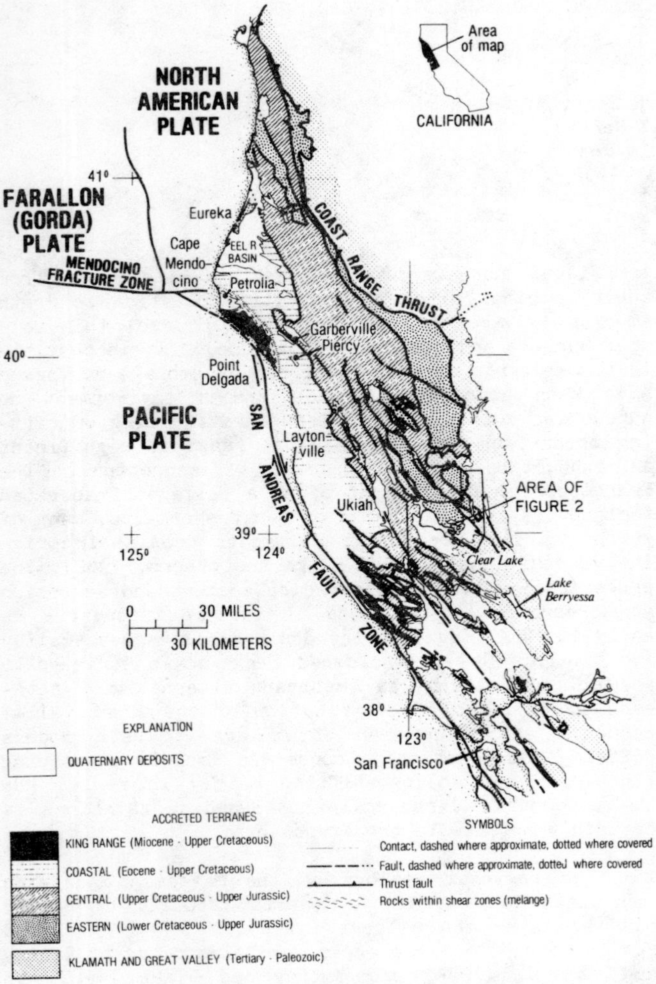

Figure 1. Generalized geologic map of belts and terranes of the Franciscan assemblage in northern and central California, showing location of the Geysers-Clear Lake region.

relations, the structure and metamorphism, and the bounding relations for each terrane. We also present petrographic and chemical data pertinent to interpretation of the sedimentologic and metamorphic histories of the terranes. The overall contrasts between the terranes, along with paleomagnetic data presented elsewhere in this volume show that: (1) there are systematic compositional variations from northeast to southwest within the Franciscan assemblage; (2) from northeast to southwest, metamorphism and deformation accompanying accretion of the Franciscan terranes occurred at about 120-92 m.y., 92-90 m.y., and 90-55 m.y.; and (3) detailed lithic associations of rocks of similar age, differences in metamorphism, and the probable large-scale translations and rotations indicated by structural relations and by paleomagnetic data, preclude stratigraphic correlations and paleogeographic reconstructions of the Franciscan terranes. Exceptions are some slabs and blocks shown to be part of the Central belt, whose field, petrographic, and stratigraphic relations indicate probable original stratigraphic continuity with the Coast Range ophiolite and Great Valley sequence, or with other previously accreted Franciscan terranes that are now in fault contact to the east.

TECTONOSTRATIGRAPHIC TERRANE FRAMEWORK

Bailey and others (1964) first recognized that the northern Coast Ranges east of the San Andreas fault consist mainly of two approximately coeval units now separated by a great regional fault. Bailey and others (1970) further interpreted this fault to be a thrust, and referred to it as the Coast Range thrust. Highly deformed and metamorphosed rocks in the lower plate of the Coast Range thrust were assigned to the Franciscan assemblage. The upper plate of the Coast Range thrust was found to consist of a fragmented ophiolite complex of Jurassic age (the Coast Range ophiolite), overlain depositionally by moderately deformed marine sedimentary rocks of the Great Valley sequence of Late Jurassic, Cretaceous, and Paleogene age. (Bailey and others, 1970; Hopson and others, 1981). The Great Valley sequence has been interpreted as fore-arc basin deposits derived from submarine fan complexes, and eroded Mesozoic island arc terranes of the Klamath mountains and Sierras to the north and east (Dickinson and Seely, 1979; Ingersoll, 1978).

In the area of this study, the Franciscan assemblage is separated from the overlying Coast Range ophiolite and the Great Valley sequence by two major thrusts (fig. 2 C). The lower thrust is the Coast Range thrust (of Bailey and others, 1970), which places a terrane consisting of highly deformed ophiolitic rocks and strata of the Great Valley sequence over the Franciscan assemblage. The upper thrust is the Stony Creek thrust (Brown, 1964), which places a second terrane of discontinuous Coast Range ophiolite remnants, depositionally overlain by intact Tithonian to Valanginian and younger strata of the Great Valley sequence, over the highly deformed rocks beneath the Stony Creek thrust.

In most places, however, the upper plate of the Stony Creek thrust consists of strata of the Great Valley sequence.

Coast Range Ophiolite and Great Valley Sequence

The stratigraphic and structural relationships of the Coast Range ophiolite and Great Valley sequence are here compared for three areas in the Geysers-Clear Lake region. These areas are separated into three discreet structural blocks by major, steep-dipping, northwest-trending faults of the San Andreas fault system. In each of these fault blocks, the stratigraphic and structural relationships of the Coast Range ophiolite and Great Valley sequence with the overlying, underlying, and adjacent rocks differ slightly to significantly. The steep fault zones which bound these structural blocks --the Bartlett Springs, Konocti Harbor Bay, and Maacama (fig. 2 A)-- are major Neogene strike-slip faults, suggesting that direct stratigraphic and structural correlations of units between the fault blocks are unwarranted.

Area between Wilbur Springs and the Bartlett Springs fault zone

In this area and to the north and south along the west side of the Sacramento Valley, the Coast Range ophiolite consists of sheared and serpentinized peridotite that locally is intruded by isolated or discontinous bodies of gabbro and diabase, and locally overlain by basalt flows, breccias, and tuffs. No intact, lower ophiolite section including ultramafic through cumulate ultramafic and mafic rocks is preserved in this area. Only rare isolated blocks or slabs of these rocks occur, mainly within the Grizzly

Creek melange (to be discussed below). The ophiolitic rocks occur largely between the Coast Range and Stony Creek thrusts (fig. 2 C). Locally, however, large masses of pillowed basalt flows, flow breccia, and tuff are present in the upper plate of the Stony Creek thrust. In addition to dismemberment of the ophiolitic rocks themselves, inclusions or diapirs of metaigneous and metasedimentary rocks, probably derived from the underlying Franciscan assemblage, are interleaved with, or enclosed by serpentinized ultramafic rocks in some areas.

K-Ar ages of 140 to 141 m.y. (earliest Cretaceous according to the time scale of Harland and others (1982)) are reported for amphiboles from gabbroic inclusions (possibly dismembered dikes) within the sheared ultramafic part of the Coast Range ophiolite in the Wilbur Springs area (Maxwell, 1974). Hopson and others (1981) questioned these ages, suggesting that the dated amphibole was a product of uralitization associated with serpentinization. Elsewhere, U-Pb dates on zircons from plagiogranitic rocks in noncumulate gabbro sections at numerous other Coast Range ophiolite localities have yielded ages ranging from 153 to 170 m.y. (Mattinson, 1980; Hopson and others, 1981), suggesting a Bathonian to Kimmeridgian age range. Red radiolarian chert between vesicular basalt flows in the upper part of the ophiolite section near Wilbur Springs contains radiolarians of Late Jurassic (early Tithonian) age (subzone 2B of Pessagno, 1977a). At a nearby locality, radiolarians from chert near the faulted base of the terrigenous section overlying the ophiolite are also of early Tithonian age (subzone 2B and Zone 3 of Pessagno, 1977a). Thus, with the exception of some intrusive activity that could have continued into the Early Cretaceous, based on the controversial radiometric dates of Maxwell (1974), the Coast Range ophiolite in the Wilbur Springs area appears to be no younger than Late Jurassic (earliest Tithonian).

The base of the Great Valley sequence in the Wilbur Springs area is mapped at the first appearance of terrigenous sedimentary rocks that directly overlie the igneous section of the Coast Range ophiolite. Tuffaceous cherts in the upper part of the igneous section are generally included as part of the ophiolite. The contact between the ophiolite and the Great Valley section can be observed as locally depositional, but it is faulted in most places. Terrigenous strata at the base of the Great Valley sequence vary from tuffaceous or cherty black mudstone and sandstone composed predominantly of basalt detritus, to coarse sedimentary breccia. The breccia typically is composed of basalt, diabase, gabbro and minor plagiogranite detritus, derived from the underlying ophiolite. At some localites the clasts are well rounded, indicating their sedimentary origin. The coarse breccias and conglomerates of angular to rounded ophiolite detritus associated with basaltic sandstones are widespread from Paskenta in the north, to Lake Berryessa to the south (Blake and others, 1982).

The breccias may have accumulated at the base of fault scarps on the sea-floor (McLaughlin and Pessagno, 1978; Hopson and others, 1981). McLaughlin and others (in press) proposed that these breccias record evidence of uplift in Tithonian time along the west side of the structural basin now filled by the Great Valley sequence. Uplift along the west margin of the proto-Great Valley basin may have accompanied pull-apart extension or oblique rifting of the oceanic basement, similar to that associated with opening of the Gulf of California (McLaughlin and others, in press).

The basal beds of the Great Valley sequence grade upward into thin-bedded black to green turbiditic mudstones and siltstones of Late Jurassic (Tithonian) through Early Cretaceous age, representing a basin plain and/or slope depositional setting (figs. 3 B and 3 C). The mudstones and siltstones in places contain abundant carbonate concretions and lenses. Thick lenses of pebble to cobble conglomerate with abundant well-rounded clasts of quartzite, vein quartz, and chert are also present locally. Lenses of thick-bedded to medium bedded, fine-to coarse-grained lithic sandstone of Late Jurassic (Tithonian) and Early Cretaceous (Valanginian) age (fig. 3 B), representing a channelized middle-fan or slope lithofacies, interfinger with the mudstone and siltstone (McLaughlin and others, in press).

Two thin sections of these sandstones from the Neocomian section of the Great Valley sequence near Wilbur Springs (figs. 2 A, 2 C) which were point counted for this study (fig. 4 A) contain conspicous non-quartzose rock fragments. Based on recalculation of quartz + chert (Q) + total feldspar (F) + total lithic rock fragments less chert and shale chips (L) to 100 percent, rock fragments compose about 30 percent of the framework grains (fig. 4 A). When chert is grouped with the total rock fragments, in one of these samples intermediate and felsitic volcanic rock and chert detritus compose approximately equal percentages (36%), or 72 percent of the total rock fragments. This sandstone locality is in the Valanginian part of the Great Valley section west of Wilbur Springs (figs. 2 A, 2 C). The other point-counted sandstone, collected from a highly faulted, less precisely dated part of the section south of Wilbur Springs, is composed of only 3 percent chert and 82 percent intermediate and felsitic volcanic rock fragments, which together make up 85 percent of the total rock fragments. Apparently, chert and volcanic rock detritus together make up much of the lithic detritus in this part of the Great Valley section. Chert detritus is particularly conspicuous in the Valanginian sandstone, which is consistant with the findings of Seiders and Blome (this volume) for conglomerate detritus of the lower part of the Great Valley sequence.

Detrital serpentinite is present at several stratigraphic horizons in the Neocomian section of the Great Valley sequence in the Wilbur Springs area (Carlson, 1981). Detailed study of these sedimentary serpentinites show that they vary from thin-bedded, fine-grained turbidites, to coarse debris-flow breccias. Carlson also describes clasts of other ophiolitic lithologies that occur in the sedimentary serpentinite, including gabbro cumulates and basalt. Rare clasts of garnet amphibolite, blueschist, greenschist, and metachert were also found. Carlson has interpreted these serpentinitic debris-flows to have resulted from subsea diapirism of upper mantle and lower crustal rocks during Neocomian time.

The Upper Jurassic through Lower Cretaceous section of the Great Valley sequence is well dated paleontologically (McLaughlin and others, in press). Numerous mollusks (mostly Buchia) date most of the lower part of the terrigenous section (fig. 2 C). In addition, reef-like lenses of limestone containing Peregrinella whitneyi (Gabb) are associated with detrital serpentinites that overlie the Coast Range ophiolite in the core of the Wilbur Springs antiform. These brachiopods are indicative of a Hauterivian or younger age for the detrital serpentinites. However, the part of the section containing

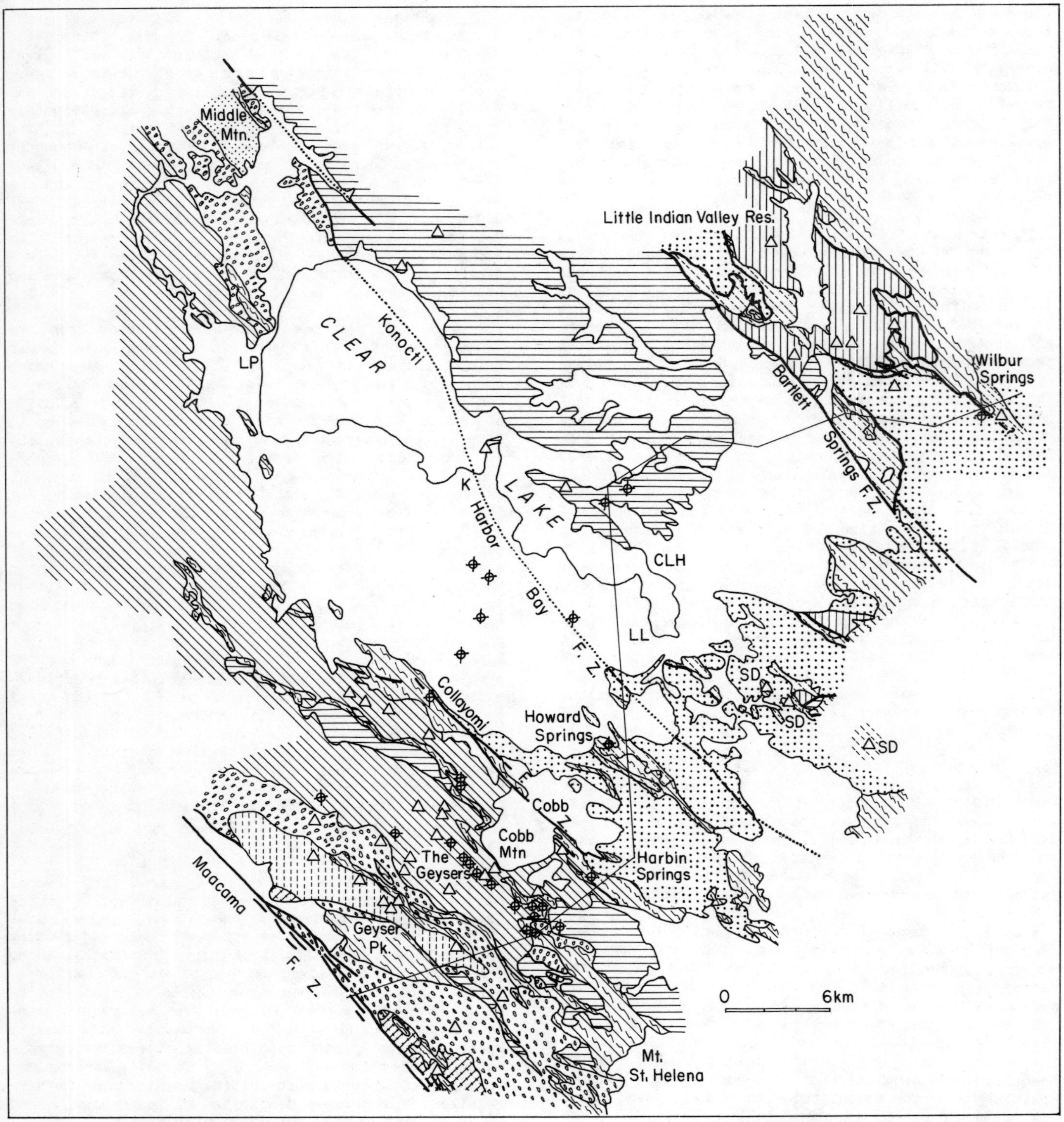

Figure 2. A. Geologic map of the Geysers-Clear Lake area, compiled from McLaughlin (1978), McLaughlin and others (in press), and McLaughlin and Ohlin (unpublished data), showing distribution of tectonostratigraphic terranes, locations of deep geothermal wells used to interpret subsurface geology, locations of point-counted sandstone samples in each terrane, and the location of line of structure section shown in Figure 2B.

Symbols

- Geothermal well location
- Microfossil locality
- Megafossil locality
- Point count locality, numbers ① through ⑤ refer to conglomerate localities
- Radiometric age locality
- Location of measured section
- Line of structure section

ROCK UNIT DESCRIPTIONS

Sedimentary and igneous rocks (Neogene)--Includes Clear lake and Sonoma Volcanics, Cache Formation, Kelseyville Formation, and alluvial deposits north of Clear lake, near Cobb, Middletown, and southeast of Lower Lake and Middletow-.

Great Valley sequence and Paleogene rocks (Eocene to Late Jurassic)--Northeast of the Bartlett Springs fault zone includes highly folded Late Jurassic (Tithonian) through lower Cretaceous mudstone and sandstone, in addition to overlying less deformed Upper Cretaceous (Cenomanian and younger) strata. Southwest of the Bartlett Springs fault zone the unit also includes massive arkosic to quartzose conglomeratic sandstone and silty mudstone of Paleocene and Eocene age, locally channelled into the underlying Great Valley sequence strata.

Coast Range ophiolite (Late and Middle Jurassic)--Includes partly to completely serpentinized ultramfic rocks, gabbro, diabase, basalt flows, and tuffaceous chert, assembled in various states of tectonic disruption.

Grizzly Creek melange (late Early Cretaceous)--Includes penetratively sheared elements of the Coast Range ophiolite and Late Jurassic and Early Cretaceous Great Valley sequence, deformed into a melange sometime between the Hauterivian and Cenomanian.

FRANCISCAN ASSEMBLAGE

Little Indian Valley terrane (Early Cretaceous to Late Jurassic)--

East Clear Lake terrane (Late Cretaceous to Late Jurassic)--Where differentiated (Figure 2C), includes:

Metasandstone and metaargillite

Metachert, tuffacaeous

Metaigneous rocks, largely basalt, and basaltic flow breccia

Terranes of the Franciscan Central belt

The Geysers (early Late Cretaceous to Middle Jurassic)--Where differentiated (Figure 2D), includes:

Broken sandstone and shale and melange, undifferentiated

Conglomerate

Chert

Basaltic rocks (largely flow basalt and flow breccias, but including minor mafic intrusive rocks and gabbro

Metaserpentinite (composed of compact intergrown actinolite, tremolite, antigorite, and chlorite)

Devils Den Canyon (Late Cretaceous?)--

Pine Flat Road (age undetermined)--

Undifferentiated (Mayacmas) melange (Eocene? to Late Jurassic)--

Middle Mountain (Paleocene to Early Cretaceous)--

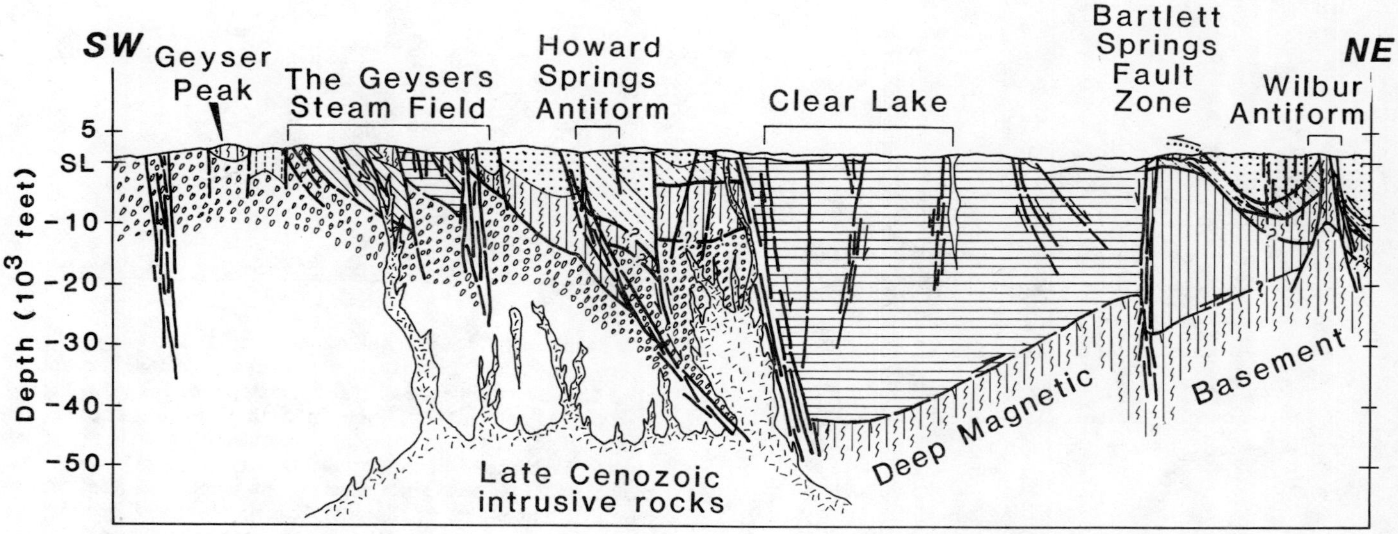

Figure 2 B. Structure section through area shown in Figure 2A., based on surface geology, geothermal drill hole data, and aeromagnetic interpretations of the area. A deep magnetic (probably ophiolitic) basement shown beneath the Franciscan terranes northeast of Clear lake is based upon interpretation of aeromagnetic anomalies shown in Figure 8.

these brachiopods is overlain along the southwest limb of the Wilbur Springs antiform by older, Tithonian and Valanginian strata, indicating that a major thrust fault was present prior to folding of the section into the Wilbur Springs antiform. This folded thrust may be related to the same system of thrusts as the Stony Creek thrust, better exposed further to the southwest.

Area between the Bartlett Springs and Konocti Harbor Bay fault zones

The ophiolite section and Upper Jurassic and Lower Cretaceous Great Valley sequence strata between the Bartlett Springs and Konocti Harbor Bay fault zones are nearly identical in structural style and lithic association to equivalent rocks northeast of the Bartlett Springs fault zone. The Upper Cretaceous rocks are overlain in this area by thin-bedded, silty, greenish gray mudstone and micaceous, thin- to thick-bedded wacke sandstone, with locally abundant carbonate concretions. These strata contain palynomorphs and sparse megafossils of middle Cretaceous through Campanian age, and in contrast to the area east of the Bartlett Springs fault zone, the Upper Cretaceous

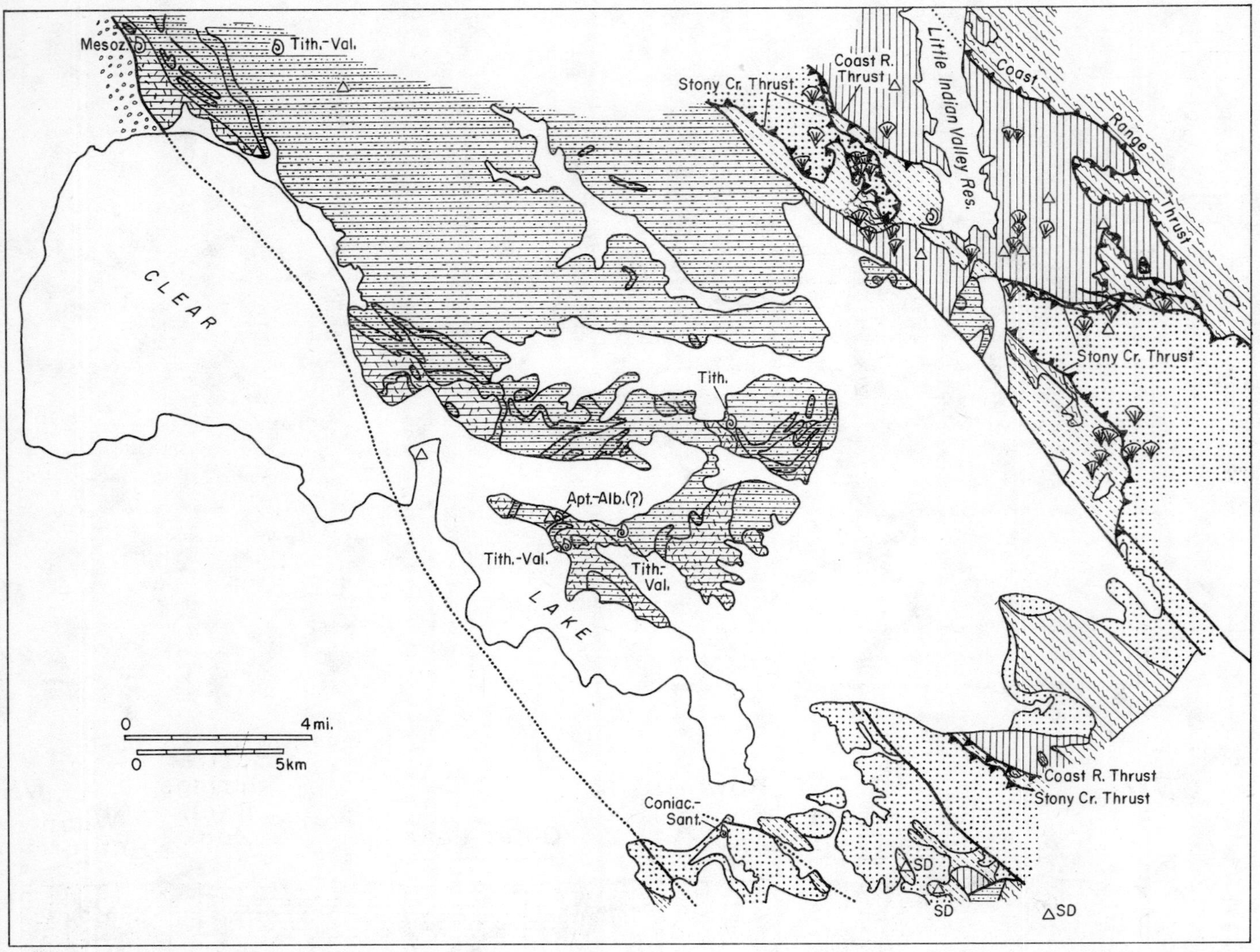

Figure 2. C. Close-up of Franciscan terranes northeast of Clear Lake, showing distribution of folded tuffaceous metachert and metaigneous rocks of the East Clear Lake terrane, point-counted sandstone localities, microfossil and megafossil (mostly invertebrate) localities. Three point-counted sandstone localities from Swe and Dickinson (1970) are designated with trianglular symbol and subscript SD.

rocks are complexly imbricated along high- and low-angle, northwest-trending faults (Swe and Dickinson, 1970). Thick-bedded to massive arkosic and quartzose sandstone, micaceous mudstone and siltstone of Paleocene and Eocene age depositionally overlie the Cretaceous section of the Great Valley sequence. The massive sandstones are conspicuously channeled into the underlying Cretaceous rocks and locally are cobbly and cemented at the base by reddish brown hematitic iron oxide or are hydrothermally altered. Swe and Dickinson (1970) interpreted the ferruginous cementation and hydrothermal alteration near the base of the sandstone section as indicative of thrusting. Instead, the ferruginous cement and hydrothermal alteration are probably the result of ground water reacting with labile minerals in the permeable basal sandstones and possibly in-part are affected by the elevated geothermal gradients around numerous late Tertiary and Quaternary volcanic vents in the area.

Area between the Konocti Harbor Bay and Maacama fault zones

Although the Coast Range ophiolite and Great Valley sequence immediately northeast and southwest of the Konocti Harbor Bay fault zone are largely lithologically equivalent, pre-Late Cretaceous strata and underlying ophiolitic rocks southwest of the structural high near Howard Springs (fig. 2 A) are distinctly different from equivalent rocks to the northeast. Southwest of Howard Springs, these ophiolitic rocks (figs. 3 F and G) and the overlying Great Valley sequence are present in three fault-bounded sections that were either thrust above the melange of the Franciscan Central belt, or were incorporated into it as large blocks during formation of the melange (McLaughlin and Pessagno, 1978). These fragmented ophiolite sections differ significantly from those northeast of Howard Springs in being more intact and complete, although at the same time each of the sections southwest of Howard Springs also show distinct contrasts with each other (fig. 5), due to tectonic fragmentation and attenuation. A composite sequence of the three partial sections (fig. 5 B) begins with a basal sheet up to 1200 m thick, of sheared harzburgite with a minor layer of cumulate dunite in the upper part. This lower ultramafic section is locally pierced by, or includes rocks faulted in from the underlying Franciscan Central belt. The ultramafic rocks are overlain by up to 640 m of cumulate gabbro

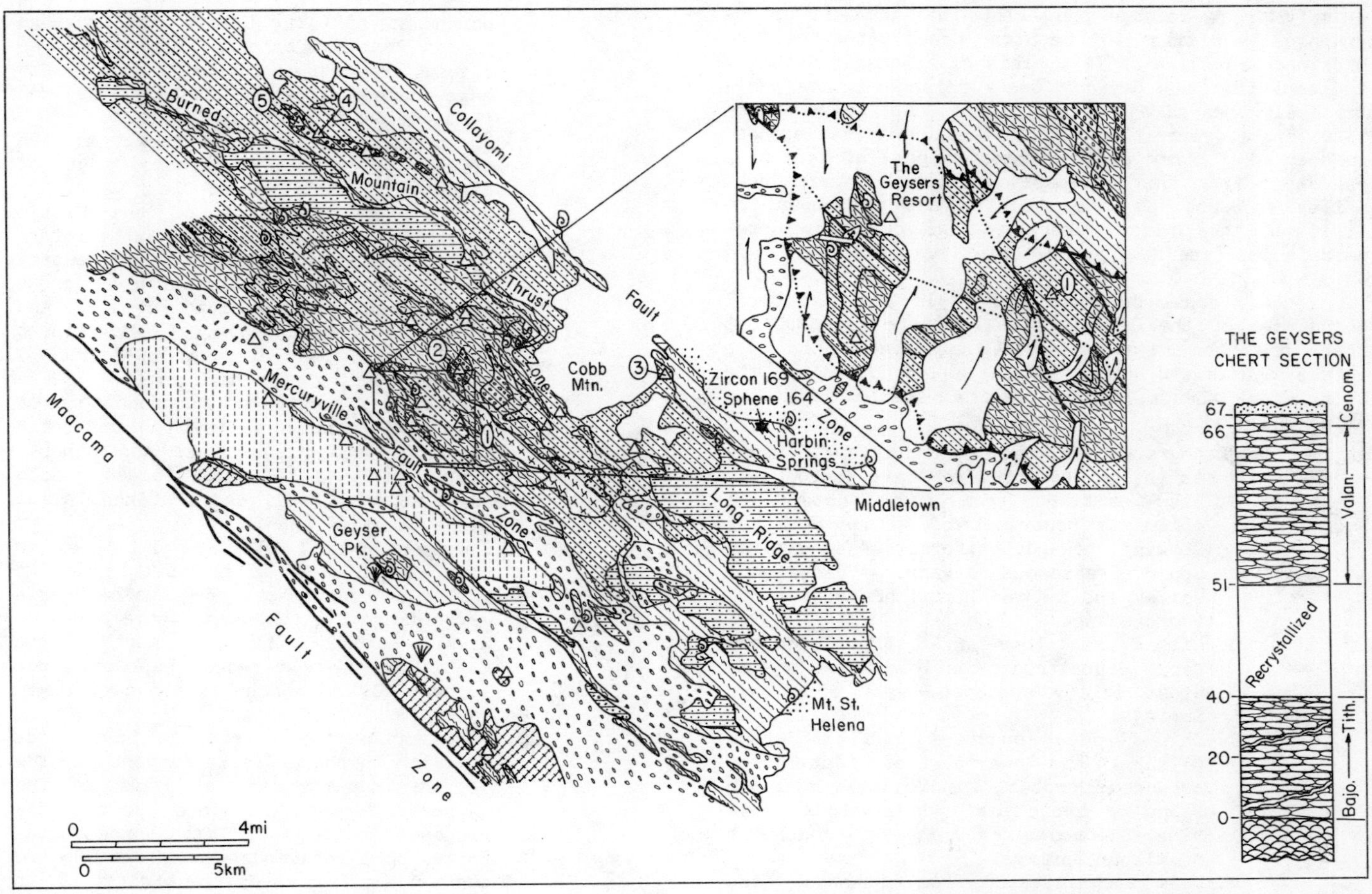

Figure 2 D. Close-up of Franciscan terranes in southwest part of the area, showing distribution of chert, basaltic igneous rocks and broken sandstone and shale in the Geysers terrane, locations of microfossil samples (radiolaria, palynomorphs, and dinoflagellates), molluskan fossils, and the location of plagiogranite dated by U-Pb dating technique by J. M. Mattinson, from the ophiolite of Harbin Springs. Also shown are exploded map view and measured section of the Geysers chert, adapted from McLaughlin, 1978, McLaughlin and Pessagno, 1978, and Murchey, this volume.

(Harbin Springs), succeeded upward by about 460 m of sheeted diabase sills (Geyser Peak), and 490 m of basalt flows (Geyser Peak) locally containing interbedded tuffaceous chert with radiolarians of early Late Jurassic (Oxfordian or Kimmeridgean) age (Pessagno, 1977a and Pessagno and others, in press). Hornblende plagiogranite pegmatites that intrude the upper part of the Harbin Springs gabbro section yielded a concordant U/Pb zircon age of 169 m.y., and a concordant U/Pb sphene age of 164 m.y. (J.M. Mattinson, written commun., 1981) which suggest a late Middle Jurassic (perhaps Callovian) age for the mafic plutonic section of the Coast Range ophiolite in that area.

A sedimentary breccia that may be as thick as 300 m, composed of clasts of the Coast Range ophiolite, overlies all three of these ophiolite sections. Whereas this ophiolitic breccia rests on gabbro at Harbin Springs and Mount St. Helena, it overlies pillow basalt and flow breccia at Geyser Peak (McLaughlin and Pessagno, 1978). Locally, this breccia is intruded by basalt dikes, (Mount St. Helena). The ophiolitic breccia marks the base of the Great Valley terrigenous section in this area, and grades upward into basaltic sandstone and mudstone turbidites. Mudstone which overlies the ophiolitic breccia at Mount St. Helena contains a Late Jurassic radiolarian fauna (early Tithonian Zone 3 of Pessagno, 1977). At Geyser Peak, lithologically similar mudstones and interbedded basaltic sandstone turbidites overlie the ophiolitic breccia, and contain Early Cretaceous dinoflagellates and a belemnite of Early Cretaceous age.

Where present, other Tithonian-through-Valanginian strata of the Great Valley sequence in the area southwest of the Konocti Harbor Bay fault zone consist of interbedded mudstone and lithic wacke sandstone with carbonate concretions that contain sparse mollusks (mostly Buchia), palynomorphs, and Radiolaria. A thin zone of grayish green, fine-grained silicic tuff directly overlies the ophiolite breccia at Harbin Springs. Other 2- to 27-meter-thick tuffaceous intervals occur at about 250 and 325 meters above the ophiolite breccia in the Tithonian through Valanginian part of the Harbin Springs section.

Grizzly Creek Melange Terrane

Rocks between the Coast Range and Stony Creek thrusts consist of an extensive melange that is well exposed along Highway 20 and along Grizzly Creek southwest of Wilbur Springs. This melange also occurs along the crest of the Howard Springs antiform and discontinuously southeast of Clear Lake (figs. 1, 2).

The rocks of the Grizzly Creek disrupted terrane are identical in overall lithology and age to the Coast Range ophiolite and the intact, lower part of the Great Valley sequence. Triangular plots of the framework-grain composition of sandstone from the

Grizzly Creek melange show that this unit is petrographically similar to the Stony Creek petrofacies of Dickinson and Rich (1972). This relationship strongly suggests that the Grizzly Creek melange is the penetratively sheared equivalent of the Coast Range ophiolite and lowermost part of the Great Valley sequence. Two other point-counted samples of Neocomian sandstone from the previously discussed intact Great Valley sequence of this study do not plot near the field of the Grizzly Creek melange or Stony Creek petrofacies (see fig. 4 and table 1).

The internal deformational style and structural boundaries of the Grizzly Creek melange suggest that there has been major transport relative to the more intact equivalent rocks in the upper plate of the Stony Creek thrust, making original spatial relationships uncertain.

Figure 3. Photographs of tectonostratigraphic terranes in The Geysers-Clear Lake region:

A. View eastward from the Coast Range ophiolite northwest of Wilbur Springs, looking toward uniformly east-dipping Upper Cretaceous strata. Sacramento Valley and Sutter Buttes are visible in far distance.
B. Lithic sandstone and silty mudstone slope deposits in the lower part of the Great Valley sequence, west of Wilbur Springs.
C. Thin bedded mudstone and sandstone flysch in the lower part of Great Valley sequence, probably indicative of lower slope or basin plain depositional setting, exposed along Highway 20 southwest of Wilbur Springs.
D. View southeastward from southwest side of Little Indian Valley at Coast Range thrust contact between sheared ultramafic rocks (upper plate) and the Little Indian Valley terrane (lower plate). View is down axis of Wilbur Springs antiform.
E. Partially serpentinized peridotite of the Coast Range ophiolite in a thick sheet exposed northwest of Cobb, along the Collayomi fault zone, and within the Central belt of the Franciscan assemblage.
F. View northeastward at the Harbin Springs ophiolite remnant (Hopson and others, 1981) from vantage point in the Central belt of the Franciscan assemblage.
G. View southeastward at the Geyser Peak ophiolite remnant (Hopson and others, 1980), where it locally lies in thrust contact upon thin melange unit and the Devils Den Canyon terrane.
H. View southeastward from southwest of Little Indian Valley, looking at intact Tithonian and Valanginian strata of the Great Valley sequence in upper plate of the Stony Creek thrust, overlying the Grizzly Creek melange terrane. Melange matrix in this area is largely sheared serpentinite and minor argillite, with blocks of gabbro, basalt and diabase.
I. Broken strata of the Grizzly Creek melange terrane, northeast of Clear Lake along Highway 20, showing extension and boudinage of carbonate beds that are up to .3 m thick.
J. Highly disrupted rocks of the Grizzly Creek melange terrane along Highway 20 northeast of Clear Lake. Here melange matrix consists principally of penetratively sheared argillite. Large lozenge-shaped block is diabase, derived from the Coast Range ophiolite.
K. Large block of pillowed basalt in the Grizzly Creek melange terrane, south of Highway 20, northeast of Clear Lake.
L. Black slaty argillite of the Little Indian Valley terrane, with silty interbeds having fine-scale sedimentary structures (ripple laminations, cross laminations, grading, load features, and convolute structures). A prominent axial plane slaty cleavage (S-2 or S-3) crosses bedding at a high angle.
M. Phyllitic fine-grained metasandstone of the Little Indian Valley terrane (high textural zone 2 of Blake and others, 1967). Note isoclinally folded quartz segregations (F-2), and prominent axial plane cleavage (S-3).
N. Prominent pencil cleavage developed in slaty argillites of the Little Indian Valley terrane. Note specimen of *Buchia keyserlingi* (?) (Lahusen) above pen.
O. Highly folded tuffaceous chert of the East Clear Lake terrane, exposed along Highway 20, on northeast shore of Clear Lake.
P. View northwestward from the north side of Big Sulphur Creek Canyon in The Geysers steam field, at rocks of The Geysers terrane. Note prominently exposed contact of metaserpentinite, thrust upon metasandstone of The Geysers terrane. The metaserpentinite is in turn overlain along another thrust by basaltic pillow flows and flow breccias.
Q. Basaltic pillow flows of The Geysers terrane.
R. Lower part of the chert section of The Geysers terrane exposed along The Geysers - Healdsburg road. The steep-dipping basal contact of chert upon basalt flows lies approximately 2 m to the right of person in photograph.
S. Typical exposure of thick-bedded high textural zone 1 metasandstone of The Geysers terrane.
T. Interbed of slightly stretched, matrix-supported polymict pebble conglomerate in low textural zone 2 metasandstone of The Geysers terrane, exposed in Big Sulphur Creek near The Geyser's resort. Stratigraphic top is to the right.
U. Penetratively sheared unit of pebbly to bouldery mudstone of The Geysers terrane. Large 2 m-long clast is serpentinized peridotite, which is also a prominent component of the pebble and cobble-sized clast population (conglomerate locality 3, Fig. 2D).
V. Fractured, well-bedded metasandstone of the Devils Den Canyon terrane. Note thinning-upward sandstone cycles. Top is to right.
W. Prominent upright fold in thick-bedded K-feldspar-bearing sandstone of the Pine Flat Road terrane.
X. Typical exposures of melange of the Mayacmas Mountains in area southwest of The Geysers terrane.

A
B
C
Coast
Range
thrust
D
Coast
Range
thrust
fault
zone
E
Neogene Clear Lake Volcanics
Boggs Mtn
Great Valley Sequence
Diabase and diabase breccia
Gabbro
Sheared
ultramafic
rock
F

Fig. 3 (A–F)

G

H

I

J

K

L

Fig. 3 (G–L)

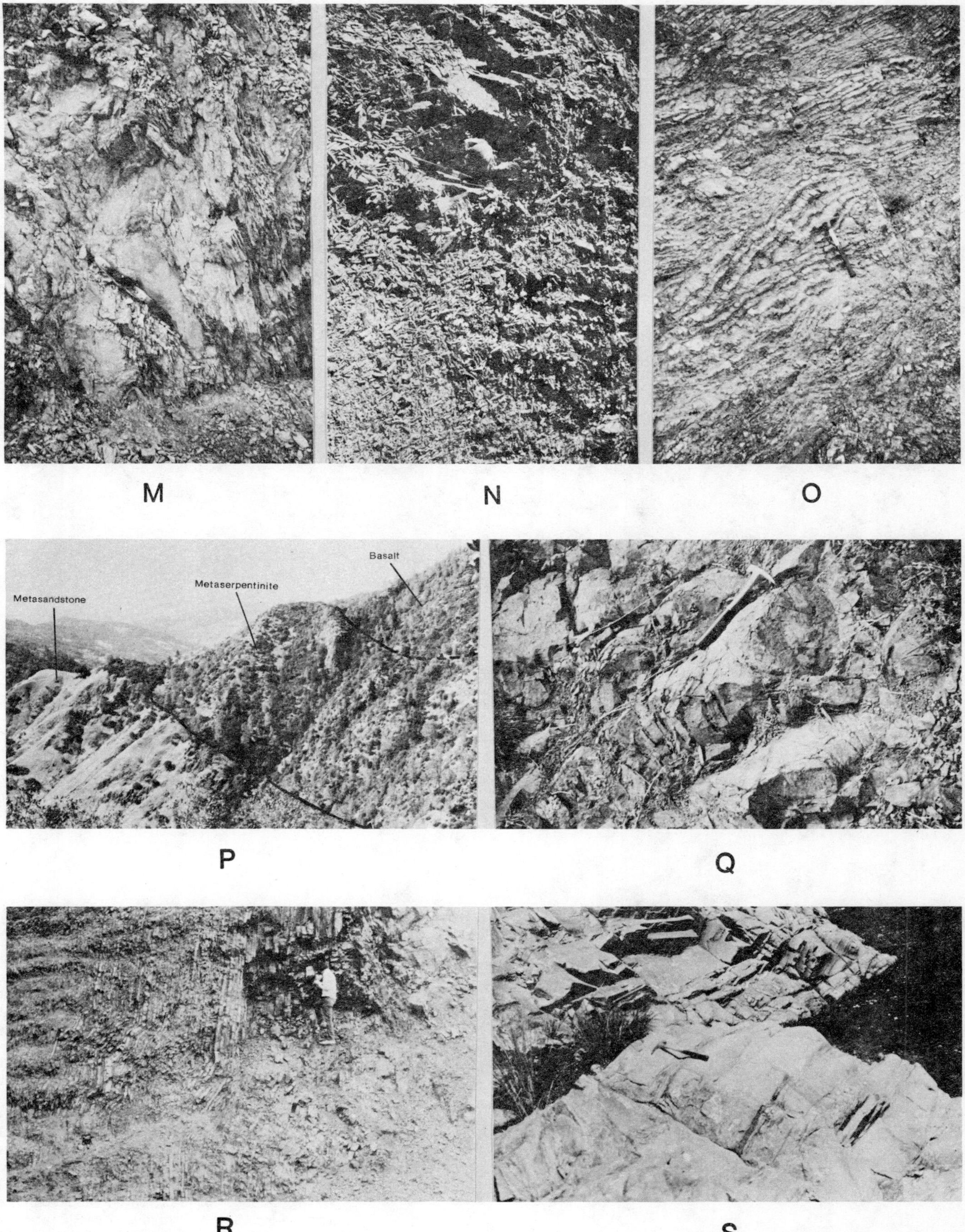

Fig. 3 (M–S)

T

U

V

W

X

Fig. 3 (T–X)

Figure 4. Ternary plots of framework grain composition of sandstone from tectonostratigraphic terranes in The Geysers-Clear Lake area. Symbols for ternary poles are: total monocrystalline quartz + chert (Q+c); total feldspar (F); total rock fragments, less chert and intraformational shale (L). A. Plot for the Little Indian Valley terrane, Grizzly Creek melange terrane, and lower part of the Great Valley sequence. B. Plot for the East Clear Lake terrane. C. Plots for terranes (Devils Den Canyon, Pine Flat Road, Marin Headlands, and The Geysers terranes) of the Central belt of the Franciscan assemblage. D. Comparative plot showing fields of all the Franciscan terranes (Devils Den Canyon, Pine Flat Road, Marin Headlands, and The Geysers terranes) and the Grizzly Creek melange terrane.

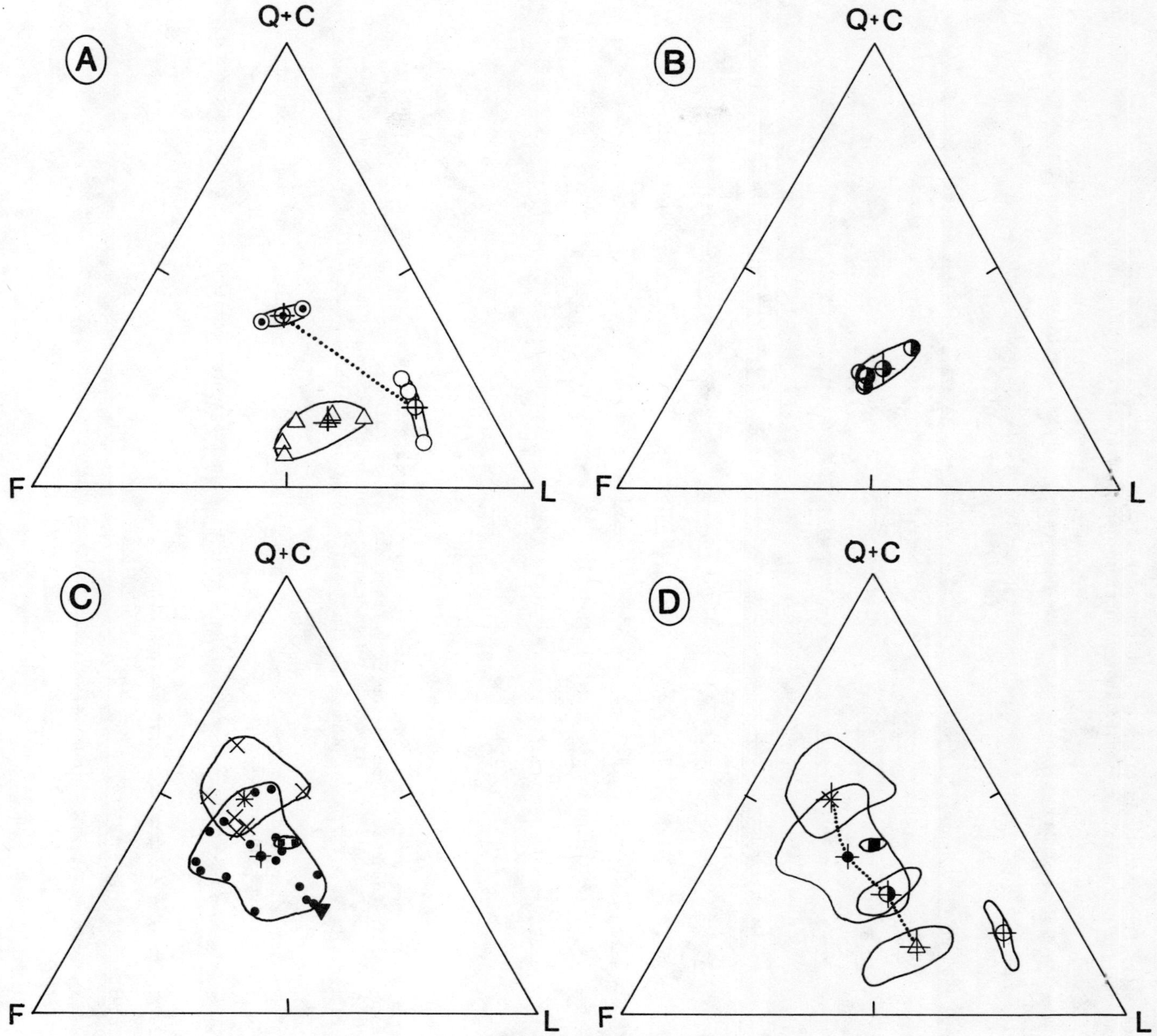

EXPLANATION

× Devils Den (n=6)

• The Geysers (n=16)

▪ Pine Flat Road (n=2)

◑ East Clear Lake (n=4)

△ Little Indian Valley (n=5)

◉ Great Valley Sequence (Late Jurassic and Early Cretaceous of Wilbur Springs area)(n=2)

○ Great Valley Sequence–Grizzly Creek melange (Data from Swe and Dickenson, 1970) (n=3)

▼ Marin Headlands

+ Indicates sample mean

Table 1. Some framework grain parameters from sandstones of Franciscan terranes in the Geysers-Clear Lake area. Notes below indicate the framework grain populations used to determine parameter percentages.

Framework Grain Parameters 1/		Grizzly Creek melange (n=3)		Little Indian Valley (n=6)		East Clear Lake (n=4)		The Geysers (excluding conglomerates) (n=17)		Devils Den Canyon (n=6)		Pine Flat Road (n=2)	
		Range%	Mean%	Range%	Mean%	Range%	Mean%	Range%	Mean%	Range%	Mean%	Range%	Mean%
Qp	2/	5.7-10.0	8.3	1-13	7.3	2-13	9.3	15-47	31	29-71	51	37-36	36.5
Qm	3/	10.1-25.6	19.3	4-22	10.5	14-26	21.5	16-34	25	18-41	31	21-22	21.5
Ft	3/	13.9-16.8	14.9	25-47	36	25-40	35.8	27-51	37	21-41	34	29-32	30.5
Vs	2/	not determined	not determined	39-64	53.7	53-65	57.3	2-36	17.5	9-21	16.7	7-27	17
Vm	2/	not determined	not determined	20-51	33	20-42	30.5	9-70	34.6	5-32	13.7	15-44	29.5
Lv	2/	42.9-71	61.3	83-90	86.7	80-97	87.8	20-84	52.8	18-53	30.3	42-51	46.5
Lt	3/	54.8-65.8	60.3	34-42	38.5	44-58	48.5	13-44	26.6	9-20	14.3	29.32	30.5
Qm/Ft	4/	.60-1.84	1.3	.09-.85	.33	.35-1.04	.64	.39-1.25	.68	.63-1.4	.91	.69-.72	.71
Qp/Lv	5/	.08-.22	.15	.01-.15	.065	.02-.16	.11	.26-1.33	.68	.54-3.71	2.1	.11-.88	.80
Vm/Vs	6/	-----	----	.31-1.31	.66	.31-.76	.54	.35-9.0	2.6	.36-1.5	.80	.56-6.3	3.43

Notes:

1/ Symbols for monocrystalline mineral grains are as follows: Ft = total feldspar, almost entirely plagioclase, except for Pine Pine Flat Road terrane; Qm = monocrystalline quartz. Symbols for rock fragments are as follows: Qp = polycrystalline quartz (chert); Lt = total lithic fragments, excluding Qp; Vs = silicic and intermediate volcanic fragments; Vm = mafic volcanic fragments; Lv = total volcanic rock fragments (= Vs + Vm); Ls = other rock fragments, excluding Qp, which remain after subtraction of Lv from Lt.

2/ Qp and Lv (=V_s + Vm) percentages here are determined from the summation of Qp + Lv (=V_s + Vm) fragments + other rock fragments (Ls), recalculated to 100 percent. These values apply to Figure 6.

3/ Qm, Ft, and Lt percentages here are determined from the summation of Qm + Ft + Lt + Qp fragments, recalculated to 100 percent. These values apply to Figure 4.

4/ Ratio of monocrystalline quartzose grains (Qm) to total feldspar grains (Ft).

5/ Ratio of chert grains (Qp) to volcanic rock fragments (Lv).

6/ Ratio of mafic volcanic rock fragments (Vm) to silicic and/or intermediate volcanic rock fragments (Vs).

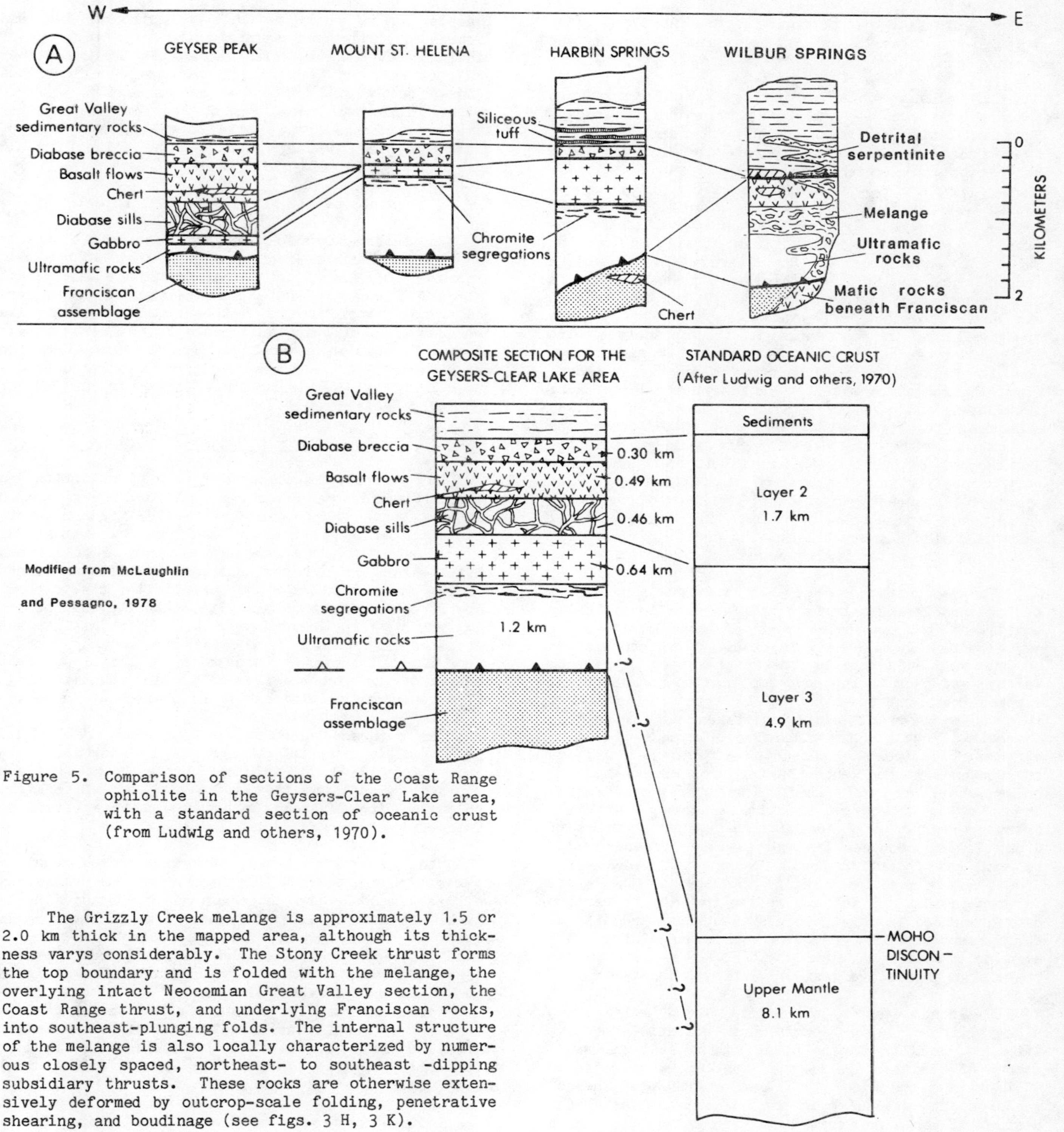

Figure 5. Comparison of sections of the Coast Range ophiolite in the Geysers-Clear Lake area, with a standard section of oceanic crust (from Ludwig and others, 1970).

The Grizzly Creek melange is approximately 1.5 or 2.0 km thick in the mapped area, although its thickness varys considerably. The Stony Creek thrust forms the top boundary and is folded with the melange, the overlying intact Neocomian Great Valley section, the Coast Range thrust, and underlying Franciscan rocks, into southeast-plunging folds. The internal structure of the melange is also locally characterized by numerous closely spaced, northeast- to southeast -dipping subsidiary thrusts. These rocks are otherwise extensively deformed by outcrop-scale folding, penetrative shearing, and boudinage (see figs. 3 H, 3 K).

The melange generally is composed of plagiogranitic, mafic, ultramafic, and pelagic rocks of oceanic derivation, mixed with terrigenous sedimentary rocks. Southwest of Little Indian Valley, the Grizzly Creek melange overlies Franciscan rocks along the faulted southwest limb of the Wilbur antiform and here the melange consists dominantly of blocks of ophiolitic rocks (gabbro cumulates and non-cumulates, diabase, basalt, and mafic breccia) and less abundant Buchia-bearing mudstone, sandstone, and conglomerate, engulfed in a matrix of penetratively sheared serpentinite (fig. 3H). West of Walker Ridge and along Highway 20, however, the Grizzly Creek melange is dominated by a penetratively sheared matrix of thin-bedded mudstone and siltstone with rhythmic interbeds of Buchia-bearing calcareous mudstone that are less than 1 meter thick. This matrix contains the same ophiolitic suite of blocks, which here range from boulder size to slabs and lenticular masses more than a kilometer long (figs. 3 I, 3K). Another variation of melange character is in the Walker Ridge area northwest of Wilbur Springs, where the melange consists almost entirely of partially sheared and serpentinized peridotite with minor discontinous bodies of gabbro, diabase, or remnants of basalt flows (see discussion of the Coast Range ophiolite in Wilbur Springs area).

The oldest rocks of the Grizzly Creek melange terrane are probably the blocks of ophiolitic rocks. This conclusion is based on the interpretation that these ophiolitic rocks are correlative with the Coast Range ophiolite, which has been dated radiometrically in various parts of the Coast Ranges (Mattinson, 1980; Hopson and others, 1981) as Late and Middle Jurassic (150-170 m.y.). Numerous Tithonian molluscan fossils, largely Buchia, that occur locally within the calcareous mudstones of the Grizzly Creek melange terrane (McLaughlin and others, in press), indicate that penetrative deformation that formed the melange was post-Tithonian. If, as we suggest, the Grizzly Creek melange is continuous in the subsurface with sedimentary serpentinites near Wilbur Springs that contain Hauterivian brachiopods, then the melange-forming deformation was post-Hauterivian. Deposits of Cenomanian and younger strata of the Great Valley sequence, that are less severely folded than the Neocomian part of the Great Valley section, and that are thought to overlie the Neocomian rocks unconformably (see earlier discussion of the Great Valley sequence), apparently post-date formation of the Coast Range and Stony Creek thrusts, the Grizzly Creek melange, and subsequent folding of the Neocomian section of the Great Valley sequence.

The recognition of the Grizzly Creek melange as a tectonostratigraphic terrane that is stratigraphically equivalent to the lower part of the Great Valley sequence may be extremely important to interpretation of the deep structure of the northwestern Sacramento Valley, and thus it may have implication for petroleum exploration in that area. The structural style of this unit makes it easily mistaken for Francsican assemblage rocks in the subsurface, which can lead to erroneous conclusions concerning structural level and the potential for encountering hydrocarbon-bearing rocks with further drilling. The major west-directed thrusting or detachment faulting associated with the Grizzly Creek melange suggests that major sections of Great Valley sequence having possible hydrocarbon or natural gas potential could be repeated beneath the Stony Creek and Coast Range thrusts to the east. Interpretations of seismic reflection and refraction data presented in this volume by Wentworth and others, further suggest that Great Valley sequence strata may even be repeated at some greater depth, beneath the Franciscan assemblage west of Wilbur Springs.

The sandstones of the Grizzly Creek melange are highly lithic, with significant volcanic detritus (fig. 6), similar to the Great Valley sequence sandstones of the same age in the Stony Creek petrofacies of Dickinson and Rich (1972). Some areas of sheared sandstone and argillite in the Lower Lake area southeast of Clear Lake that were mapped as part of a melange of the Franciscan assemblage (tectonic breccia of Swe and Dickinson, 1970) have been reassigned in this study to the Grizzly Creek melange. The reassignment of these rocks to the Grizzly Creek melange is based 1) on our mapping of their structural position above the Coast Range ophiolite and below intact strata of the Great Valley sequence; 2) the presence of blocks of ophiolitic rocks and strata commonly containing Buchias; and 3) the absence of metamorphic minerals of blueschist-grade, or of rocks with any textural reconstitution or slaty axial-plane cleavage aside from penetrative shear. Framework modes of sandstones from this area southeast of Clear Lake (sample suite from tectonic breccia of Swe and Dickinson, 1970; and sample suite G-3, table 4 of Dickinson and others, 1982) show that the sandstones are poor in chert, but rich in volcanic rock detritus (figs. 4 and 6, and table 1). Seiders and Blome (this volume) cite two localities of conglomerate in this same area of the Great Valley sequence southeast of Clear Lake, which they report to be Valanginian in age, and rich in volcanic rock detritus.

Terranes of the Franciscan Eastern Belt

Tectonostratigraphic terranes assignable to the Franciscan assemblage crop out southwest of the main exposures of the Coast Range ophiolite and Great Valley sequence. The two easternmost of these terranes, the Little Indian Valley and East Clear Lake terranes (fig. 2), are assignable to the Eastern belt of the Franciscan assemblage (Irwin, 1960, Berkland and others, 1972), and apparently were accreted to the lower plate of the Coast Range thrust by the end of the Early Cretaceous.

Little Indian Valley Terrane

This terrane is composed mainly of argillite and minor metasandstone (figs. 3 L, 3 N) which range in textural grade (Blake and others, 1967) from slightly reconstituted argillite and metasandstone (textural zone 1), to thin, thrust-bounded sheets of slate and phyllite (low to high textural zone 2). Textural zone 1 rocks contain chlorite and pumpellyite; the textural zone 2 rocks contain pumpellyite, chlorite, and possibly very rare incipient lawsonite. The argillite and metasandstone commonly are calcareous and may contain minor shell fragments. Aragonite is locally present as part of the carbonate assemblage and indicates high P/T metamorphic conditions of the blueschist facies.

The metasandstones of this terrane are arkosic and typically contain conspicuous intermediate volcanic (andesitic) and plutonic rock fragments, and felsite, in addition to clastic biotite, white mica, and minor fibrous green to brown amphibole (figs. 4, 6; table 1).

Minor blocks or lenses of metamorphosed basaltic to keratophyric tuffs, flow rocks, and dike rocks, as well as limestone and metachert occur sporadically in the Little Indian Valley terrane. Some of these rocks occur as large, rare, flattened clasts or lenses within intact argillite and metasandstone; others occur as sheared clasts in a unit of melange within the Little Indian Valley terrane.

Molluskan fossils, chiefly Buchia, a few gastropods, and a partial ammonite from the Little Indian Valley terrane indicate a Late Jurassic (middle Tithonian) through Early Cretaceous (Valanginian) depositional age for these rocks.

Sedimentary rocks of the Little Indian Valley terrane probably were deposited in a slope and/or basin plain setting. The outboard, slope-related aspect of this unit is suggested by its predominent fine-grained, flysch-like character and fine-scale sedimentary structures such as ripple cross-laminae, graded beds, and slump and load features. The more sandstone-rich lithofacies probably represents a channelized part of these slope or basin plain deposits.

Our point count data show that sandstones of the Little Indian Valley terrane are highly lithic (fig. 4), and that most of the lithic component consists of

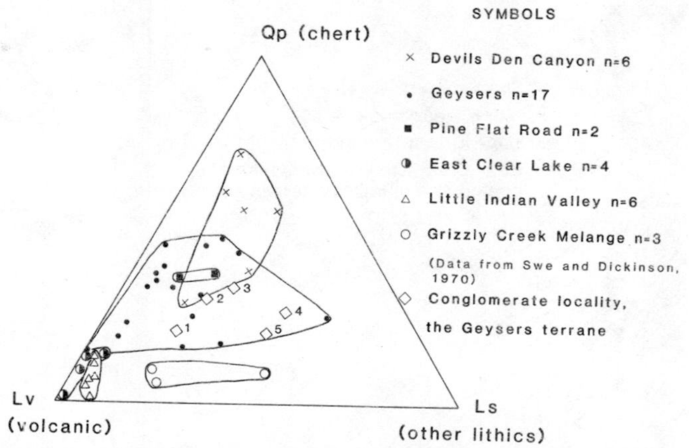

Figure 6. Ternary plot illustrating relationship between chert (Qp), total volcanic rock fragments (Lv), and other rock fragments (Ls). Clast counts from conglomerate localities discussed in text and located on Figure 2 are included. The plot demonstrates distinct increases in cherty detritus at the expense of volcanic rock fragments from northeast to southwest.

volcanic rock fragments (fig. 6). Thus, these rocks are similar (but not identical) compositionally to the Stony Creek petrofacies (Dickinson and Rich, 1972) and to the Grizzly Creek melange terrane of the Great Valley sequence. On an AFM compositional plot (fig. 7 B), sandstones of the Little Indian Valley terrane follow an iron-enrichment trend, varying in composition mostly along the line from the Fe-oxides pole, that is perpendicular to the triangle side connecting the alkali and MgO poles. Another ternary plot (fig. 7C), with total silica at one pole, alumina at another, and iron oxides, MgO, TiO2, P_2O_5, and MnO at the third (fig. 7 C) is comparable to the Q+C,F,L framework grain plot (fig. 4). This plot shows that the compositional field of Little Indian Valley terrane sandstones is somewhat higher in the combined ferromagnesian, MnO, P_2O_5 and TiO_2 component than the sandstone fields of the other Franciscan terranes, which probably is due to the high percentage of volcanic rock fragments in the Little Indian Valley terrane sandstones.

Similar Franciscan assemblage argillite and metasandstone have been reported near Goat Mountain (Suppe and Foland, 1978), approximately 15 km northwest of the report area. Suppe and Foland also reported the presence of Buchias of Tithonian and Valanginian age in argillites of the Goat Mountain area, although they considered the argillites to be part of a melange.

At least two fold sets associated with metamorphism are discernible in rocks of the Little Indian Valley terrane (see figs. 3 L, 3 N). A prominent slaty parting (S-2) at a low angle to bedding (S-1) defines the axial plane of an early fold generation (F-1) in metasandstone and argilite reconstituted to textural zone 2 (Blake and others, 1967). The early S-2 cleavage is folded into a second set of upright to recumbent, generally northeast or southeast-plunging isoclinal folds (F-2), with a prominent NW-SE trending axial plane cleavage (S-3) (fig. 3M). The F-2 folds are truncated by the Coast Range thrust and no axial plane cleavage is recognized in the upper plate of the thrust.

The metamorphosed Little Indian Valley terrane is further folded along with the Coast Range and Stony Creek thrusts, and the overlying pre-Cenomanian rocks of the Great Valley sequence, into large-scale, broad to tightly appressed folds trending N 30°-45° W and plunging E to SE, such as the major antiform associated with the Wilbur Springs area.

The cogenetic relationship of F-1 and F-2 folding with textural metamorphism of the paleontologically well-dated Little Indian Valley rocks, indicates that the folding and metamorphism post-dated Valanginian time. Truncation of the S-3 axial plane cleavage by the Coast Range thrust, furthermore, indicates that this folding and accompanying metamorphism pre-dated emplacement of the Little Indian Valley terrane beneath the Coast Range and Stony Creek thrusts.

Coeval and probably younger (Tithonian to Hauterivian) rocks of the Great Valley sequence in the upper plate of the Coast Range thrust were also involved in complex shearing and folding in post-Valanginian time. This deformation included the folding and west-directed thrusting or detachment faulting associated with the Grizzly Creek melange between the Coast Range and Stony Creek thrusts. It also included the later folding of the Coast Range and Stony Creek thrusts together with their associated Neocomian sections, into E- and SE-plunging folds. This deformation did not affect supposed unconformably overlying Cenomanian and younger strata of the Great Valley sequence (Rich, 1971; Blake and Jones, 1981).

Thus, it appears that final emplacement of the Coast Range and Stony Creek thrusts above the Little Indian Valley terrane occurred sometime following its metamorphism in post-Valanginian time. We interpret the depositional and structural relations in the structurally overlying coeval rocks of the Great Valley sequence to indicate that this emplacement occurred largely prior to, or during the Cenomanian (about 90-92 m.y. ago) and after the Valanginian (about 131 m.y. ago), based on the time scale of Harland and others (1982).

East Clear Lake Terrane

Southwest of the Little Indian Valley terrane lies the East Clear Lake terrane which consists predominently of metasandstone with abundant lenses of metamorphosed tuffaceous chert (fig. 3 O); locally, vitric tuff and basalt are prominent constituents. Rocks of the East Clear Lake terrane are juxtaposed with the Little Indian Valley terrane along the Bartlett Springs fault zone; a major zone of high-angle, probably right-slip faulting of the San Andreas fault system (McLaughlin and others, in press). The East Clear Lake terrane does not appear east of the

Figure 7. Chemistry of sandstones and igneous rocks in The Geysers-Clear Lake area. Rapid rock chemical analyses from which data used in these plots are taken is from McLaughlin (work in progress, 1984) A. Ternary AFM plot of igneous rocks. B. Ternary AFM plot of terrane sandstones. C. Ternary plot of Aluminum, Silica, and FeO $+Fe_2O_3$ + MnO $+TiO_2$ + P_2O_5+ MnO. D. Plot of percent P_2O_5 v.s. percent TiO_2 in basaltic rocks, after Ridley et. al. (1974). E. Plot of SiO_2 v.s. Na_2O+ K_2O. composition of basalts. Compositional fields for alkali and tholeiitic basalts are from MacDonald and Katsura (1964).

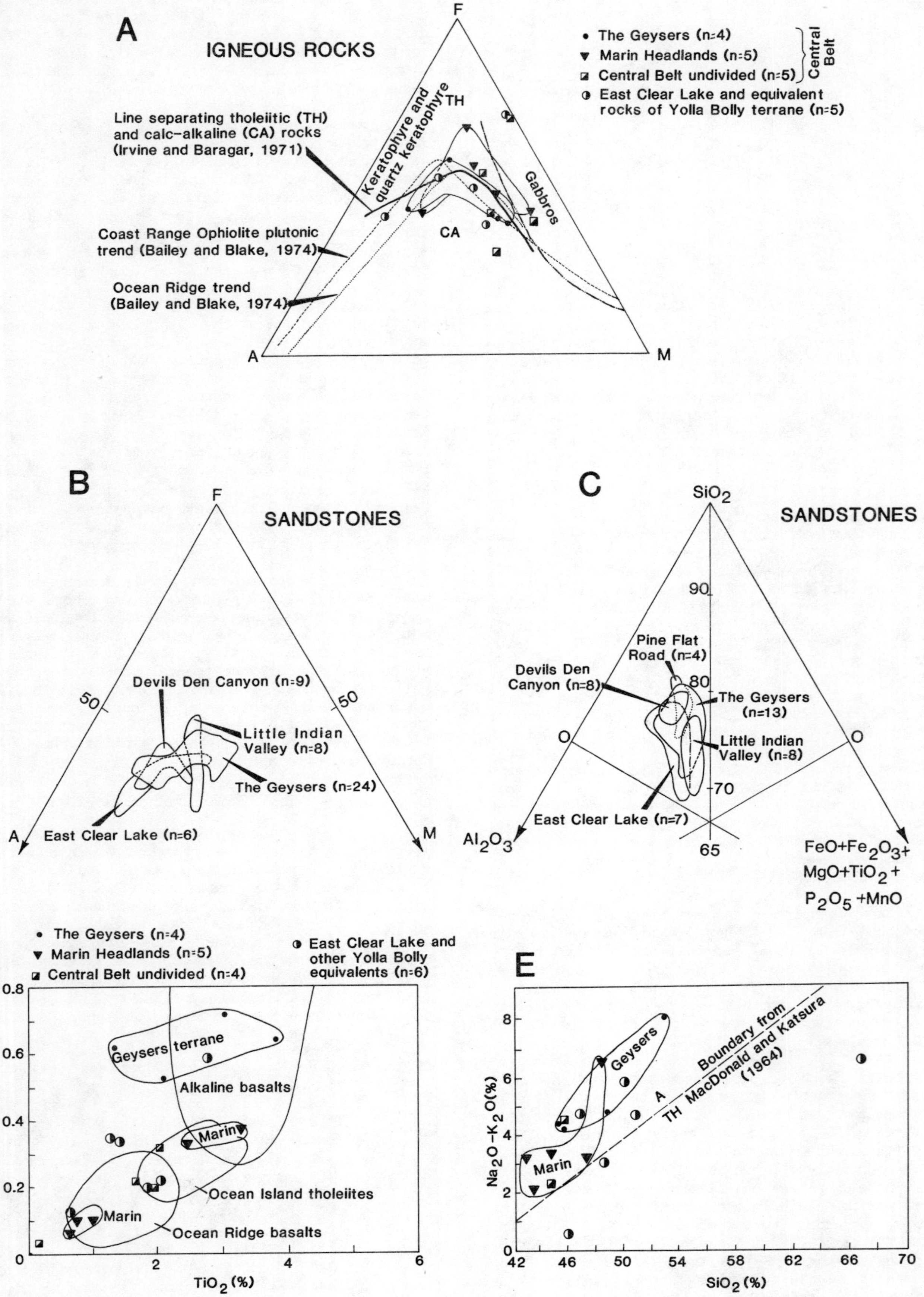
A
IGNEOUS ROCKS
F
A
M
TH
CA
Keratophyre and quartz Keratophyre
Gabbros
Line separating tholeiitic (TH) and calc-alkaline (CA) rocks (Irvine and Baragar, 1971)
Coast Range Ophiolite plutonic trend (Bailey and Blake, 1974)
Ocean Ridge trend (Bailey and Blake, 1974)
The Geysers (n=4)
Marin Headlands (n=5)
Central Belt undivided (n=5)
Central Belt
East Clear Lake and equivalent rocks of Yolla Bolly terrane (n=5)
B
SANDSTONES
F
A
M
50
50
Devils Den Canyon (n=9)
Little Indian Valley (n=8)
The Geysers (n=24)
East Clear Lake (n=6)
C
SANDSTONES
SiO2
90
80
70
65
O
O
Pine Flat Road (n=4)
Devils Den Canyon (n=8)
The Geysers (n=13)
Little Indian Valley (n=8)
East Clear Lake (n=7)
Al2O3
FeO+Fe2O3+MgO+TiO2+P2O5+MnO
The Geysers (n=4)
Marin Headlands (n=5)
Central Belt undivided (n=4)
East Clear Lake and other Yolla Bolly equivalents (n=6)
D
P2O5 (%)
TiO2 (%)
Geysers terrane
Alkaline basalts
Marin
Ocean Island tholeiites
Marin
Ocean Ridge basalts
E
Na2O-K2O(%)
SiO2 (%)
Geysers
Marin
A
TH
Boundary from MacDonald and Katsura (1964)

Bartlett Springs fault zone in the area of this investigation, but extends westward, underlying the entire area between the Bartlett Springs fault zone and the northeast side of Clear Lake, including the south arm of the lake (fig. 2).

The metasandstone of this unit is reconstituted to textural zone 2 of Blake and others (1967), and in contrast to the metasandstones of the Little Indian Valley terrane, commonly contains lawsonite, in addition to pumpellyite, chlorite, and phengitic white mica. The tuffaceous metacherts locally display conspicuous concentrations of fine-grained, dark blue sodic amphibole ("crocidolite") alined along bedding. The metacherts are generally recrystallized, but relict deformed radiolarians are commonly visible. The metabasaltic rocks consist of pillowed flows, flow breccias, and aquagene tuffs, and locally contain blue amphibole, lawsonite, and pumpellyite, and rare aegirine or jadeitic pyroxene.

Locally the East Clear Lake terrane is a melange that contains large blocks of blue amphibole- and white mica-bearing metaigneous rocks, in addition to blocks of less metamorphosed mafic rocks, metachert, and metatuff, dispersed in a matrix of sheared lawsonitic metasandstone.

The East Clear Lake terrane is lithologically similar to the Yolla Bolly terrane of Blake and others (1981) and/or the "Yolla Bolly formation" (in part) of Worrall (1981), despite conspicuously more metabasaltic rocks than typical Yolla Bolly terrane rocks described by Blake and others (1981). The southwest boundary of the East Clear Lake terrane is the steep northwest-trending (probably late Cenozoic) Konocti Harbor Bay fault zone alined with the northeast sides of Middle Mountain and Clear Lake, that projects southeastward beneath the Clear Lake Volcanics and juxtaposes rocks of the Franciscan Central belt, Coast Range ophiolite, and Great Valley sequence against the East Clear Lake rocks. Southwest of this major fault boundary, large slabs of the East Clear Lake terrane greater than 10 km long are imbricated into the Central belt of the Franciscan assemblage along low- to high-angle, east-dipping faults. Locally, slices and slabs of the Coast Range ophiolite have been imbricated into the Central belt melange along with the slabs of the East Clear Lake terrane.

Radiolarians recovered from the tuffaceous metacherts from between the north and south arms of Clear Lake (fig. 2C) indicate a Late Jurassic (Tithonian) through Early Cretaceous (probably Valanginian through Albian) age. A similar Tithonian through Albian(?) fauna was found in the Elk Creek area north of Lake Pillsbury and west of Hull Mountain, 32 km north of Clear Lake (McLaughlin and Ohlin, unpublished data). This range in age implies pelagic sedimentation for this part of the East Clear Lake terrane that was maintained for as much as 40 m.y. In the Hull Mountain area, molluscan fossils of Cenomanian age were reported by Blake and Jones (1974) in argillite associated with metasandstone and tuffaceous metacherts of that area.

The Yolla Bolly terrane of the Hull Mountain area and of the Ortigalita Peak area, 200 km south of Clear Lake (Blake and others, 1982), have associated with them mafic dikes and sills that clearly intrude the interbedded metasandstone and metacherts (Layman, 1977; Mattinson and Echeverria, 1980) and were metamorphosed to blueschist grade following this intrusion. The metachert and metasandstone of the Ortigalita Peak area were folded prior to intrusion of the mafic rocks (Mattinson and Echeverria, 1980). An intrusive gabbro at Ortigalita Peak was dated radiometrically at 95 m.y. (late Cenomanian) by U-Pb methods, and the subsequent blueschist metamorphism of the gabbro and intruded metasandstone was dated at about 92 m.y. by Mattinson and Echeverria (1980). The late Cenomanian age of intrusion and blueschist metamorphism seems to constrain the minimum age of deposition of the Yolla Bolly terrane (and very probably, the East Clear Lake terrane).

Compositionally, sandstones of the East Clear Lake terrane are typically arkosic to lithic, with conspicuous micro-porphyritic intermediate volcanic detritus (figs. 4 and 6). The sandstone field has a distinct alkali trend that overlaps, but is more alkalic and more restricted (tighter) than, the sandstone field of the Franciscan Central belt to the southwest; on an AFM triangular plot (fig. 7 B) the East Clear Lake rocks trend along a line perpendicular to the line between $FeO+Fe_2O_3$ and MgO from the alkali pole.

The high percentage of intermediate volcanic rock detritus in metasandstones of the East Clear Lake terrane, and their alkalic composition (figs. 4, 6, and 7) suggest to us that the metasandstones are arc-derived. The igneous rocks of this terrane are mafic to alkalic, and it has been shown that their correlatives in other areas locally intrude tuffaceous cherts that are interbedded with the terrigenous sedimentary rocks (Layman, 1977; Mattinson and Echeverria, 1980). These relations suggest a depositional setting which underwent extension, and a complex mechanism of sediment dispersal, whereby terrigenous sediment was repeatedly introduced into the pelagic environment represented by the tuffaceous cherts. We suggest that the East Clear Lake terrane may represent an extensional oceanic setting close to an arc, such as a rifted or leaky transform margin (a similar setting was proposed by Blake and others, 1981). However, a marginal basin or even a complex series of lower slope-basins are also plausible depositional environments. A unique determination of depositional environment probably is not possible, due to the extensive metamorphic and tectonic overprint of the East Clear Lake terrane rocks, and a lack of detailed and systematic sedimentologic data.

We recognize at least three fold sets in the East Clear Lake terrane. At outcrop scale, the earliest discernable folds are evident locally in metacherts where primary bedding (S-1) is folded into isoclinal and recumbent folds (F-1) with axial planes (S-2) that now trend N46°E to N75°W. The F-1 folds are refolded by a more evident asymetric isoclinal fold set (F-2), defined by NE- to NW-trending axial planes (S-3) that generally cut steeply across F-1 (fig. 3 O). A third set of N- to NW-trending, large wavelength folds (F-3) are suggested by vergence of S-3 axial planes and changes in the plunge directions of the F-2 folds.

At map-scale (fig. 2 C), the metachert units are folded into early large-wavelength folds that are upright to recumbent, or disharmonic. These folds have approximately N20°- to 40°W-trending axial traces, and are superimposed by secondary folds probably corresponding to F-2 above. Later, broad, open south-southeast- to south-southwest trending warps are defined by broad deflections in the NW-trending fold axes.

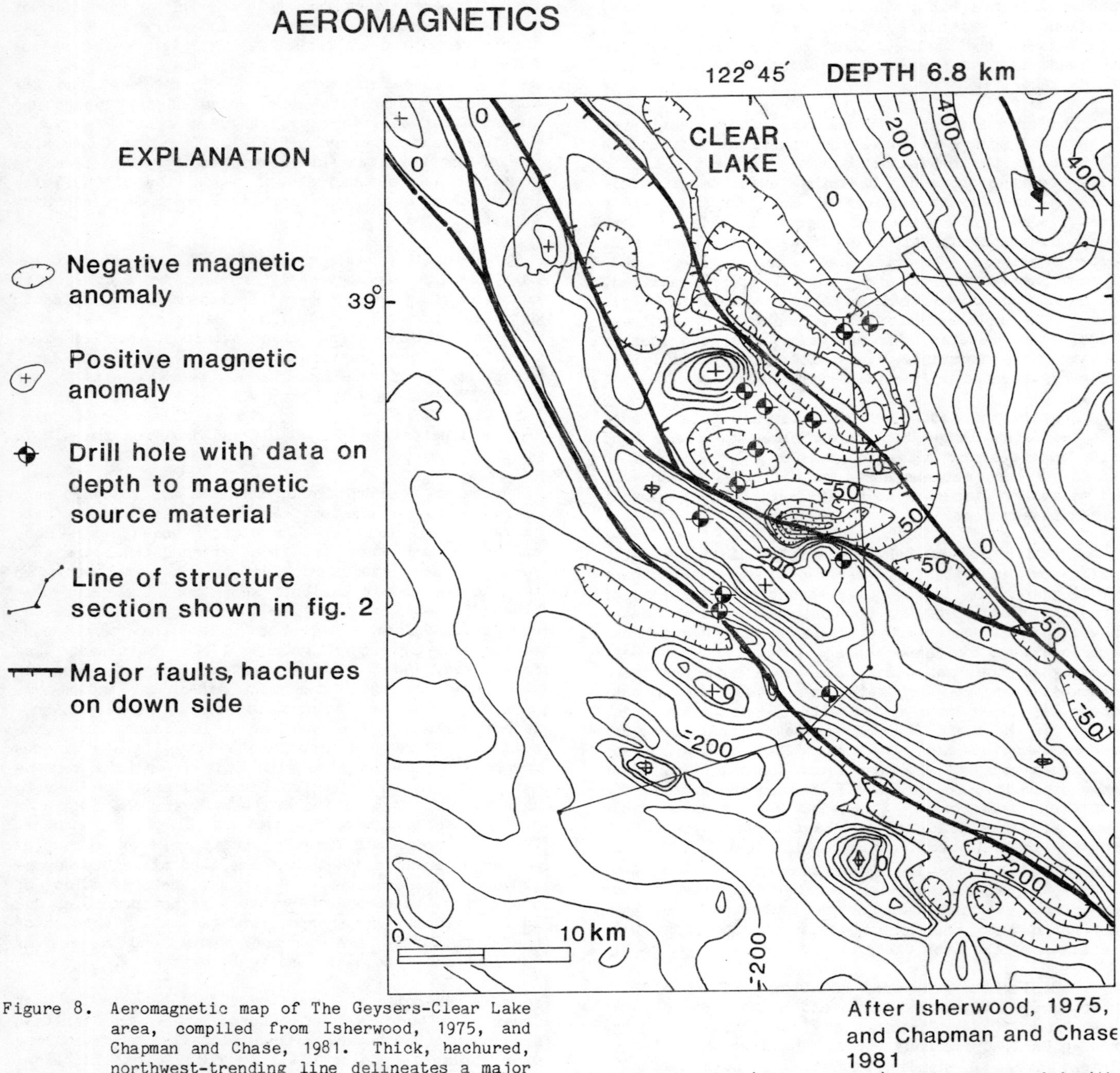

Figure 8. Aeromagnetic map of The Geysers-Clear Lake area, compiled from Isherwood, 1975, and Chapman and Chase, 1981. Thick, hachured, northwest-trending line delineates a major fault boundary. Intact, uniformly southwest-dipping magnetic rocks, at least 6.8 km beneath the East Clear Lake and Little Indian Valley terranes are northeast of this fault. Magnetic anomalies southwest of this fault are near or at the surface, and represent isolated fault-bounded masses within the Central belt of the Franciscan assemblage.

The depositional age of the East Clear Lake terrane and its equivalents elsewhere in the Coast Ranges, and the metamorphic age determination from the Diablo Range equivalent (Mattinson and Echeverria, 1980) imply that the folding of these rocks (F-1 and F-2) was cogenetic with, or preceeded blueschist metamorphism (see discussion, previous pages). These data, then, suggest that metamorphic folding occurred in the middle Cretaceous, probably about late Cenomanian time (92 m.y. ago), and corresponded with the timing of deformation of the Little Indian Valley terrane and rocks in the upper plate of the Coast Range thrust. The timing of the later, broad, open, non-metamorphic folding is not known.

Mafic substrate beneath the Little Indian Valleyand East Clear Lake terranes

Aeromagnetic maps of the northern Coast Ranges and adjacent Sacramento Valley indicate the presence of a major, deep, unexposed magnetic body, presumed to be basalt or ophiolite, beneath the eastern Coast Ranges and Sacramento Valley (Griscom, 1983). Modeling of the major positive magnetic anomaly associated with this magnetic body (Isherwood, 1976) determined that the top of the magnetic body is at a depth of about 6.8 km just east of Clear Lake (fig. 8). The aeromagnetic data indicate that the top surface of the presumed mafic rock body rises eastward and merges with the east-dipping Coast Range ophiolite exposed at the surface in this area. The magnetic data further indicate that the mafic body dips uniformly at a shallow to moderate angle southwestward beneath Clear Lake (fig. 8 and fig. 2 B), and that it underlies both the Little Indian Valley and East Clear Lake terranes. In contrast to the complexly folded, overlying Little Indian Valley and East Clear Lake rocks, however, the uniform magnetic contours indicate that the top surface of the mafic unit has been little deformed, except locally, where offset by major Cenozoic faults of the San Andreas fault system. We interpret the contact between this mafic basement and the overlying Franciscan terranes to be a west-dipping thrust, probably associated with obductive east-directed accretion of the Little Indian Valley and East Clear Lake terranes, concurrent with westward thrusting along the Franciscan assemblage-Coast Range ophiolite boundary (Coast Range and Stony Creek thrusts) (fig. 2 B). The constraints on timing of deposition, metamorphism, and emplacement of the Little Indian Valley and East Clear Lake terranes imply that obductive accretion of these Franciscan rocks probably occurred in late Cenomanian and early Turonian time, concurrently with deformation between the Coast Range and Stony Creek thrusts.

Terranes of the Franciscan Central Belt

A major northwest-trending, steeply-dipping fault zone of probable late Cenozoic age, separates the East Clear Lake terrane on the northeast, from the Franciscan Central belt on the southwest. This steep fault boundary is largely covered by Cenozoic volcanic rocks south of Clear Lake, but aeromagnetic data show that the fault coincides with the Konocti Harbor Bay fault zone, which cuts the volcanic cover in this area (Hearn and others, 1981). The aeromagnetic pattern of the Central belt southwest of this fault (fig. 8) is characterized by numerous shallow magnetic highs and lows together with major elongate anomalies that delineate masses of serpentinite and mafic rocks of the Coast Range ophiolite within the Central belt. All these magnetic rocks are near or at the surface. This contrasts strongly with the deep, uniformly west-dipping magnetic source of the regional anomaly that characterizes the area at depth beneath the Franciscan northeast of the Konocti Harbor Bay fault zone (fig. 8).

The Central belt of the Franciscan assemblage (first defined by Irwin, 1960, and later by Berkland and others. 1972) is composed of a collage of slab-like or intact terranes (McLaughlin and others, 1981; Blake and others, 1982) engulfed in highly sheared and comminuted rocks containing blocks of various lithologies including blueschist and eclogite. The Central belt is typical of what commonly is referred to as "Franciscan melange", and the sheared and comminuted rocks engulfing the slab-like, intact terranes are referred to as "melange matrix" (Cloos, 1983). The matrix rocks of the Central belt usually consist dominantly of argillite that is of a metamorphic grade ranging from pumpellyite to lawsonite (Cloos, 1982, 1983). In some areas, however, these matrix rocks may be dominantly sandstone. Argillic matrix from the Central belt has been shown to contain radiolarians and molluscan fossils of Tithonian through Valanginian age (Blake and Jones, 1974). Microfossils and megafossils from many of the various slab-like bodies enclosed by this matrix are considerably younger than this, however, and ophiolitic rocks and blocks (knockers) of high grade blueschist, eclogite, and amphibolite in the melange are apparently much older.

We have restricted this study to the area northeast of the Maacama fault zone, which bounds the map area on the southwest. Hence, our study of the Franciscan Central belt does not include its entire width. It does include the northeastern two-thirds of the belt exposed in the steep, deeply dissected central Mayacmas Mountains, where considerable structural relief is visible. Furthermore, in The Geysers steam field, lithologic data from numerous geothermal wells extending to depths below 3 km, including cuttings, well logs, and a few core sections, were made available to us by geothermal operators in the area (see fig. 2 A). Many of the rocks from the subsurface of The Geysers steam field were thermally metamorphosed during late Cenozoic intrusion of silicic magmas beneath the area (McLaughlin and others, 1983b).

From northeast to southwest within this area, we have delineated several distinctive slab-like terranes bounded by thrusts, which generally dip at a low to moderate angle northeastward and which are interlayered with units of melange matrix. Some of these terranes can be correlated with their equivalent terranes elsewhere within the Franciscan or along the west side of the Great Valley in northern and central California. Also described in detail is a major unit of melange along the southwest side of the Mayacmas Mountains. Other units of melange to the northeast, which are interlayered with the more intact terranes, differ somewhat in their block contents. A significant block component may be derived locally, from adjacent intact terranes, while other blocks are obviously derived from more exotic sources unrelatable to known source rocks in the region. In addition, it is not clear whether all the melanges of the Central belt constitute a single thrust-imbricated melange unit that shows east to west facies variation, or whether each melange unit represents a distinct tectonostratigraphic pulse of melange generation. With these constraints in mind, the various terranes of the Central belt in The Geysers-Clear Lake region are described.

Middle Mountain Terrane

Northwest of Clear Lake, near the east side of the Central belt, is the southern part of an elongate slab of sandstone and mudstone. This, the Middle Mountain slab, is bounded by northwest-trending, steep-dipping faults; the strata are deformed by folding, metamorphosed to laumontite-grade, and contain a molluscan fauna, a microfauna and a microflora of Early Cretaceous (Berriasian) through Paleocene (Ynezian) age (Berkland, 1969). The slab is bounded along its steep northeast and southwest faulted sides by narrow belts of serpentinite that are probably correlative with part of the Coast Range ophiolite, also exposed along the west side of the Sacramento Valley.

The Middle Mountain slab was previously interpreted as a folded klippe of the Great Valley sequence and Coast Range ophiolite (Berkland, 1969; 1973), thrust westward onto the Franciscan Central belt in Paleogene time. This westward thrust model, however, is incompatible with the emplacement to the east of the Great Valley sequence and Coast Range ophiolite over the Little Indian Valley and East Clear Lake terranes in Cenomanian or Turonian time (see previous discussions of these terranes). The Middle Mountain rocks are a part of, or overlie melange of the Central belt. The Central belt could not have completed forming earlier than earliest Tertiary time at this latitude, based on the youngest ages of fossils in blocks incorporated into the melange. The youngest rocks of the Middle Mountain terrane show that this terrane was not juxtaposed with the Central belt earlier than late Paleocene time (about 55 m.y. ago). The emplacement of the Middle Mountain terrane, therefore, post-dated the major pulse of Coast Range thrusting at the west side of the Sacramento Valley. Furthermore, the steep-dipping, northwest-trending bounding faults, the overall elongate northwest trend, and the tailing-out of lenticular terminations of the Middle Mountain terrane to the northwest and southeast, are not suggestive of thrusting as the mechanism for incorporation of the Middle Mountain into the Central belt. The overall map pattern and constraints on timing of this incorporation are more compatible with the large-scale right-shear associated with the Central belt (to be further discussed in a later section). An alternative to the thrust hypothesis, that more adequately explains incorporation of the Middle Mountain terrane into the Central belt, is that the Middle Mountain rocks were torn away from the west margin of the Great Valley from somewhere to the south, and incorporated into the Central belt as a result of the right-lateral component of convergence between the Central belt and the California margin (represented by the Great Valley sequence and structurally underlying Coast Range ophiolite and rocks of the eastern terranes of the Franciscan assemblage) in Paleogene time.

"Pomo beds" of Berkland (1972) or
Van Arsdale terrane of Blake and others (1982)

Berkland (1972, 1973) has documented a terrane of Franciscan rocks along the west side of Middle Mountain, but north of the area of Fig. 2A. These rocks, which dip steeply eastward beneath the Middle Mountain terrane, were referred to by Berkland as the "Pomo beds", but were later renamed the Van Arsdale terrane by Blake and others (1982). The Van Arsdale terrane is fault bounded, and the lower part of the sequence consists of 1200m of alkali pillow basalt, with minor diabase and gabbro dikes and intercalated radiolarian chert. Berkland reported Late Jurassic (Tithonian) radiolarians from the radiolarian chert. The basalt flows are overlain depositionally by discontinuous, calcite-cemented basalt-pebble conglomerate and massive sandstone, overlain by brown mudstone with abundant carbonate concretions. The conglomerate, sandstone, and shale contain molluscan fossils and palynomorphs of Late Cretaceous (Maestrichtian) age.

The Van Arsdale terrane was not recognized in the area of this study. However, the basalt flows and interbedded Tithonian chert in the lower part of the Van Arsdale terrane are on-trend with cherts and basalts of The Geysers terrane, to be discussed on the following pages (see fig. 2).

The Geysers Terrane

As mapped in the area of this report, The Geysers terrane is a composite of structurally imbricated thrust slabs of similar lithologic affinity, interleaved with fragments of other dismembered terranes of the Franciscan Central belt. The Geysers terrane is characterized by a distinct section of basalt flows, breccias and tuffs, with intercalated lenses of radiolarian chert; that in turn is overlain by lithic sandstone with local coaly laminae, polymict conglomerate, and melange (figs. 3 P - 3 T).

The Geysers terrane is broken into several thrust-bounded slabs. Typically, these slabs are bounded above and below by northeast-dipping thrusts discontinuously lined with serpentinite. Most of the serpentinite along these thrusts is traceable southeastward along trend into major masses of the dismembered Coast Range ophiolite (McLaughlin and Pessagno, 1978), suggesting that the ophiolite was sheared-out and attenuated in a northwest-direction within the thrust zone. One slab of The Geysers terrane (fig. 2D) is bounded above by a thrust (probably the Coast Range thrust, in part) lined with the Coast Range ophiolite that is itself truncated along its northeast side by the steep northwest-trending Collayomi fault zone (fig. 3 E). The lower bounding thrust fault, the Burned Mountain thrust zone (McLaughlin, 1981) is traceable for at least 40 km southeastward from northwest of Highland Springs, through The Geysers area and into a zone of steep reverse faults near the Great Western Mercury mine northwest of Mount St. Helena (figs. 2 A and 2 D).

Metamorphosed serpentinite

A northwest-trending, elongate, fault-bounded serpentinite unit crops out for about 10 km along the northeast side of, and in Big Sulphur Creek canyon. This distinctive unit of metamorphosed serpentinite is composed dominantly of actinolite and/or tremolite ± chlorite ± antigorite; a composition distinctly different from serpentinite associated with the Coast Range ophiolite (McLaughlin, 1981), whose mineral assemblage generally is lizardite ± chrysotile ± talc. This actinolitic serpentinite is indicative of metamorphism to the greenschist facies, probably to temperatures above 300°C.

Well data and the surface geology in The Geysers steam field demonstrate that this serpentinite unit underlies the basalt of The Geysers terrane, that it dips northeastward, and at the same time is tilted gently to the southeast, concordant with the regional plunge (fig. 3 P). Although this unit is fault-bounded above and below, its close association with structurally overlying basalt and chert suggest to us that it represents part of the oceanic crust that originally was beneath the basalt and chert of The Geysers terrane. As such, we assign the metamorphosed serpentinite to The Geysers terrane. The metamorphism of this serpentinite to mineral assemblages of the greenschist-facies indicates a higher-temperature history than for the Coast Range ophiolite, perhaps related to emplacement from a deeper level of the crust or mantle.

Basaltic rocks

Basaltic rocks of The Geysers terrane including pillowed and non-pillowed flows and diabase feeders, typically are distinctly alkalic, plotting well above the trend of the Coast Range ophiolite on a P_2O_5 v.s. TiO_2 diagram (fig. 7 D), and well within the alkali field of MacDonald and Katsura (1964) on an Na_2O+K_2O v.s. SiO_2 variation diagram (fig. 7 E). On an AFM plot (fig. 7 A), The Geysers terrane falls on either side of, but along the trend line of Irvine and Barager (1971), separating tholeiitic from calc-alkalic compositional fields.

This apparently distinct alkalic tendency may reflect low-grade metamorphism and alteration, although we note that more highly metamorphosed volcanic rocks of the East Clear Lake (=Yolla Bolly? terrane) terrane and equivalent components in the Central belt, plot in the ocean ridge, tholeiitic, and alkali fields on the P_2O_5 v.s. TiO_2 diagram; and in both alkali and tholeiitic fields on the Na_2O+K_2O v.s. SiO_2 variation diagram. We would expect these more highly metamorphosed volcanic rocks to have experienced more thorough metasomatic alteration, although they apparently did not . We also note that variation plots of basalts from the Marin Headlands of similar age to the south (Karl; Wahrhaftig; Murchey and Jones; this volume) range from ocean ridge basalt to alkali basalt (figs. 7 A, D, E), are distinctly alkalic, but low in silica (Na_2O+K_2O v.s. SiO_2, fig. 7 E), or distinctly iron enriched (tholeiitic) (AFM plot, fig. 7 A). Thus the basalts of the Marin Headlands terrane tend to be lower in silica and P_2O_5, and more iron-enriched than basalts from The Geysers terrane. We will discuss this further in a following section.

Chert of The Geysers terrane

Perhaps the most significant unit of The Geysers terrane is a 67 m thick section of radiolarian chert of Jurassic, Early Cretaceous and early Late Cretaceous age that depositionally overlies basalt (figs. 2 and 3 R). Earlier, based on radiolarian biostratigraphy, we described this chert section (McLaughlin and Pessagno, 1978), and noted that the upper 16.3 m, of Valanginian and Cenomanian age, is a condensation of approximately 41 m.y. It appears from new data (Murchey and Jones, this volume) that the lower part of this chert sequence is of Bajocian (Middle Jurassic) age, indicating that a total of 70-75 my is condensed into the approximately 67 m of Middle Jurassic to Late Cretaceous pelagic section. These data yield a short term pelagic sediment accumulation rate (for the Valanginian) of .4 cm/1000 yr. or a longer term (Middle Jurassic through early Late Cretaceous) rate of about .1 cm/1000 yr. This is comparable to modern day pelagic sedimentation rates of .2 - .4 cm/1000 yr. (McLaughlin and Pessagno, 1978).

Murchey and others (1983), and Murchey (this volume), describe a nearly identical section of radiolarian chert in the Marin headlands area north of San Francisco, of comparable thickness, that overlies basaltic volcanic rocks, and is in turn overlain by lithic sandstone, similar to the setting of the chert in The Geysers terrane.

In spite of the similarity in age and thickness of chert sections in The Geysers and Marin Headlands terranes, Karl (this volume) has shown that there are distinct geochemical differences between chert sequences in The Geysers and Marin Headlands terranes. In particular, the cherts of the Marin Headlands are more depleted in aluminum relative to iron and manganese, and overlap with, but tend to be somewhat lower in titanium than those from The Geysers terrane. The chert of the Geysers terrane is also distinctly enriched in potassium relative to iron and magnesium compared to the Marin Headlands section. These differences are interpreted to indicate a mid-ocean ridge environment of deposition for the chert of the Marin Headlands terrane, and a more in-board, deep-water setting for the chert of The Geysers, close enough to a continental margin or arc to contain admixed terrigenous detritus.

Sandstones and conglomerates

We have studied the petrography and chemistry of sandstones of The Geysers terrane in some detail, and we also have attempted to characterize variations in the clast types of several polymict and monomict conglomerate lenses in The Geysers terrane (figs. 4, 6, and 7).

Where stratigraphic relations are least obscured by faulting and poor exposures, a prominent section of thick bedded lithic sandstone stratigraphically overlies the chert of The Geysers terrane. The primary contact of the sandstone with the underlying chert is depositional, although in many areas it is now modified by hydrothermal alteration and shearing. These sandstones locally contain thin, ripple-drift, cross laminae along which are concentrations of coaly carbonaceous material or plant debris. Graded bedding in these sandstones is rare, although it is found locally at the tops of some beds. Figure 4 and Table 1 illustrate the framework grain composition of sandstones overlying the chert of The Geysers terrane. The composition field (Fig. 4C) also includes sandstones which do not directly overlie the chert, but are intercalated with the basaltic rocks and chert or are not directly in contact with them, and thus may be stratigraphically lower or higher than sandstones which depositionally overlie the chert of The Geysers terrane. The composition of the sandstone of The Geysers terrane is variable, ranging from arkosic to lithic (fig. 4 C). Lithic fragments consist mainly of chert and mafic to silicic volcanic rocks (fig. 6). Shale fragments of intraformational origin were not counted, but are generally abundant in sandstones throughout all the Franciscan terranes.

Potash feldspar is conspicuously absent from the sandstone of The Geysers terrane, and chert detritus is distinctly more abundant near the southwest margin of the terrane than to the north-northeast, which is true for the whole Franciscan assemblage of this region (figs. 4 and 6, and table 1). This characteristic indicates that these Franciscan terranes are oriented such that pelagic detritus is concentrated and possibly eroded from sources to the west or southwest. This direction probably does not represent true pelagic (oceanward) source directions, due to tectonic rotation and translation, but it does suggest a relatively consistent sense of, and gross magnitude of rotation and translation.

For comparison, we also plot the framework grain composition (quartz+chert v.s. feldspar v.s. total non-quartzose rock fragments) of one point-counted sample of lithic sandstone that overlies cherts at the Marin headlands (Fig. 4C). The framework composition of the sandstone of Marin Headlands falls along the edge of the field for sandstones of The Geysers terrane as a whole (compare with Jayko and Blake, this volume). We note that up to 3 percent detrital potassium feldspar occurs in the sandstones of the Marin Headlands terrane, whereas no potassium feldspar is present in the sandstones of The Geysers terrane. Relict microcline twinning, however, suggests that detrital potash feldspar has been albitized in The Geysers area.

Lenticular units of conglomerate associated with The Geysers terrane include zones of rounded pebbles with open frameworks within massive lithic sandstones; lenses of coarse, rounded to angular, poorly sorted polymict conglomerates or breccias locally with blocks more than 1 meter in length; and also some poorly sorted, monomict, pebble or cobble conglomerates composed of chert clasts, with a mudstone or siltstone matrix (i.e., pebbly or cobbly mudstones) (figs. 3 T, 3 U). The poorly sorted monomict and polymict conglomerates and breccias probably represent debris flows and slump deposits in the channelized inner region or the lower slope of submarine fan systems. Poorly-developed graded bedding and turbidite features in the sandstones associated with these conglomerates, and more abundant features suggestive of grain flow, slumping, and ripple lamination, further suggest the proximal channelized aspect of these coarse clastic rocks. Possibly these rocks represent the distributary part of a large submarine fan or canyon system.

Table 2, and Figure 6 illustrate the variable composition of 5 polymict conglomerate units in The Geysers terrane, determined from clast counts made in the field. We note that rare clasts of blueschist ("crocidolitic" metachert, glaucophane schist, and lawsonitic metasandstone) were found at localities 1, 2 and 4, suggesting recycled Franciscan rocks as the source of some conglomerate detritus. We also note the abundance of serpentinite detritus (including a lenticular boulder longer than 1 m) at locality 3 and the abundance of basalt or diabase clasts at all five localities.

The best exposed and most carefully studied conglomerate unit in the map area is a thick lens of pebble- to boulder-size conglomerate with a closed-framework pebble to cobble matrix, well exposed on the ridge above, and in Kelsey Creek (Localities 4 and 5, figs. 2, 6, and table 2). The conglomerate is about 230m thick and can be traced laterally for up to 3 km. At locality 4, mafic volcanic clasts are the most abundant clast population (23.4%), followed by sedimentary rock clasts, mostly of sandstone (17.1%), fine grained aplite and intermediate plutonic rocks (15.3%), black chert and argillite (14.4%), red and green chert (11.7%), silicic hypabyssal volcanic porphyry (6.3%), mafic plutonic rocks (2.7%), meta igneous rocks (2.7%), recrystallized limestone (1.8%), serpentinite (1.8%), metasedimentary rocks (1.8%), and vein quartz (0.9%).

The rounded clasts of intermediate plutonic rocks and hypabyssal porphyritic volcanic rocks are virtually ubiquitous both in polymict conglomerates of the Great Valley sequence and the Franciscan assemblage, as are dark, rounded chert pebbles. Seiders and others, (1979) have studied the chert clasts regionally in the Great Valley sequence and Franciscan, and determined that they contain Middle Triassic to Middle Jurassic radiolarian faunas. Seiders concluded that the conglomerates of the Great Valley sequence and Franciscan shared the same provenances. In this volume Seiders and Blome propose that there has been little if any latitudinal shift between the Franciscan and Great Valley sequence since the addition of these conglomerates to the Great Valley sequence and Franciscan. Even if so, however, large-scale translations between the Great Valley sequence and pelagic and oceanic components of the Central belt cannot be ruled out.

Although the ages of the chert clasts in conglomerates of the Great Valley sequence and Central belt of the Franciscan assemblage are well documented (Seiders and others, 1979), the radiometric age span of the porphyritic volcanic rock clasts is unknown. And although Paleozoic and Mesozoic rocks of the Sierran and Klamath basements have been proposed as sources for the chert and volcanic rock clasts, no attempts have been made to pin down specific source areas through trace element, isotopic, or geochemical fingerprinting. Thus, the sources of chert and volcanic rock clasts in these conglomerates may still be problematic.

Field relationships show that most blocks of chert-rich pebble conglomerate in the Franciscan Central belt of the map area are in melanges that also contain incorporated masses of the Coast Range ophiolite and overlying Great Valley sequence (previously discussed). Thus, at least some of the blocks of chert and volcanic-rich pebble conglomerate in the Franciscan Central belt melanges may have been faulted away from the west margin of the Great Valley sequence and translationally sheared into the Central belt with the masses of the Coast Range ophiolite during accretion of the Central belt to the California margin, as late as Paleogene time.

Seiders and others (1979) also studied clasts from the conglomerate localities (Locs. 4, 5) in the Kelsey Creek area, and found that some green tuffaceous meta-chert clasts contain conodonts that have color indices which indicate the source rocks have undergone thermal metamorphism to 400°-500°C (biotite grade), which further places an important constraint on conditions in the source area of these clasts.

Regionally metamorphosed sandstones of The Geysers terrane are typically reconstituted to textural zone 1 (Blake and others, 1967), but rocks of low textural zone 2 rarely are present. Sandstones of The Geysers terrane contain the metamorphic mineral assemblage quartz + albite + chlorite + white mica + pumpellyite ± lawsonite. Lawsonite is rare, and not well developed, and, more typically, these rocks are of pumpellyite-grade.

Relationship of The Geysers Terrane to Marin Headlands Terrane

In the previous discussion it was pointed out that the chert section of The Geysers terrane is nearly identical in age and relatively comparable in thickness to the chert section of the Marin Headlands terrane (Murchey, this volume). Furthermore, both chert sections are overlain by similar, thick terrigenous sections, locally with coaly debris, and underlain by thick sections of basaltic rocks. An ammonite of middle Cretaceous (Cenomanian age) has been reported from the sandstones overlying the chert of the Marin Headlands terrane (Blake and Jones, 1974), whereas sandstones of The Geysers terrane contain chert clasts with radiolarians of Cenomanian age (McLaughlin and Pessagno, 1978).

However, we also have pointed out, as have other authors in this volume, that there are distinct differences in the rocks of these two terranes. The basaltic rocks of The Geysers terrane (fig. 7) appear to be somewhat more alkalic, higher in SiO_2 and P_2O_5, and less iron-enriched than basaltic rocks from the Marin Headlands terrane. Karl (this volume) points out that cherts of the Marin Headlands section are depleted in both aluminum and potassium relative to iron and manganese compared to The Geysers section. We also note that prominent lenses of polymict and monomict conglomerates are present both with the sandstones and also interleaved with the basaltic rocks of The Geysers terrane, whereas no prominent conglomeratic units are reported for the Marin Headlands terrane.

From this data, we conclude that The Geysers terrane definitely has more continental margin- or arc-related affinities than the Marin Headlands terrane. However, it is questionable whether these differences really require completely different tectonic or depositional settings.

Distinct similarities and contrasts also exist between The Geysers terrane and the East Clear Lake terrane (=Yolla Bolly? terrane). The chert section of the East Clear Lake terrane to a large extent overlaps the age of the Marin Headlands and The Geysers terranes. Radiolarian assemblages from the cherts of the East Clear Lake terrane are as old as Tithonian and possibly as young as Aptian or Albian (C. D. Blome, written communication, 1983).

Table 2. Summary of pebble counts from conglomerates in The Geysers terrane. The Geyser-clear Lake Region

CLAST TYPE	LOCALITY NUMBER 1		2		3		4		5	
	f*/	%	f	%	f	%	f	%	f	%
-- Mafic Volcanic Rocks a/	21	17.8	27	25.2	21	19.1	26	23.4	13	12.9
-- Mafic plutonic rocks b/	7	59	1	0.9	1	0.9	3	2.7	2	1.9
-- Hypabyssal silicic, porphyritic and felsic rocks, and andesite	49	41.5	23	21.5	22	20.0	7	6.3	25	24.8
-- Fine-grained aplite and intermediate plutonic rocks	6	5.1	8	7.5	--	----	17	15.3	12	11.9
-- Serpentinite	--	---	1	0.9	11	10.0	2	1.8	--	---
-- Metaigneous rocks c/	--	---	-	----	11	10.0	2	1.8	--	---
-- Black chert, cherty argillite	16	13.6	17	15.9	27	24.5	16	14.4	11	10.9
-- Red and green chert	9	7.6	15	14.0	9	8.2	13	11.7	9	8.9
-- Recrystallized limestone	--	---	--	---	--	--	2	1.8	2	1.98
-- Other sedimentary rocks d/	8	6.8	8	7.5	14	12.7	19	17.1	18	17.8
-- Metasedimentary rocks e/	1	0.8	3	2.8	--	----	2	1.8	5	4.95
-- Calcareous concretion	--	---	1	0.9	--	----	--	----	--	---
-- Vein quartz	--	---	2	1.9	1	0.9	1	0.9	--	---
-- Quartzite	1	0.8	1	0.9	--	----	--	----	--	---
TOTALS	118	99.9	107	99.9	110	99.9	111	99.9	101	99.99

Notes:

*/ f = frequency

a/ Includes uralitized basalt and some diabase

b/ Includes gabbro and diabase

c/ Includes metatuff, metagabbro, and metadiabase

d/ Includes graywacke, shale, argillite, and chert-argillite breccia

e/ Includes quartz-mica schist, lawsonite- and glaucophane- bearing metagraywacke, slate, meta-argillite, and hornfels

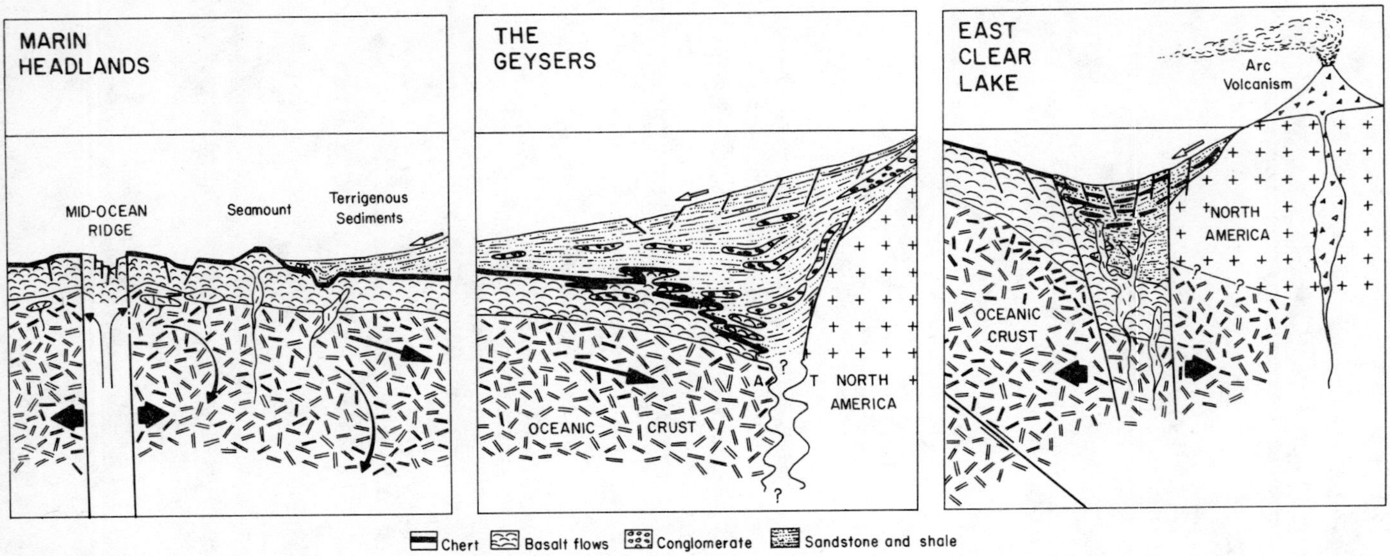

Figure 9. Possible depositional settings for the Marin Headlands, The Geysers, and East Clear Lake terranes. These depositional settings are not considered unique for each terrane.

To the north, at Hull Mountain, fossils from terrigenous sedimentary rocks of the Yolla Bolly terrane are of Cenomanian age (Blake and Jones, 1974), similar to the Marin Headlands and The Geysers sections. The basaltic-flow rocks of the East Clear Lake terrane are undated, and adequate dating of the full range of the metachert sections of the East Clear Lake and Yolla Bolly terranes is hampered by recrystallization brought about by textural metamorphism and poor preservation (Murchey, oral communication, 1983). Hence, it is reasonable to suspect that there is more complete age correspondence amongst these chert sections than the present data base demonstrates. However, the metacherts of the East Clear Lake (=Yolla Bolly?) terrane are distinctly tuffaceous compared to The Geysers and Marin Headlands terranes, are conspicuously interbedded with arc derived metasandstones, and have been shown elsewhere to be cut by mafic intrusive rocks, strongly suggesting an extensional depositional setting close to an arc.

These differences in field relations, petrography, and geochemistry, could be explained in terms of changes in the interaction of the North American plate margin with various spatially separated but interrelated depositional settings on the same adjacent oceanic plate (fig. 9). With respect to the North American margin, the Marin Headlands terrane represents the most outboard oceanic setting. The Geysers terrane seems to exhibit evidence of interaction of a more in-board part of the oceanic plate with a channeled slope setting; while the East Clear Lake terrane was in a near-arc extensional setting, possibly in a transform setting near a basin axis, where arc-derived sediment transgressed and interleaved with pelagic sediment as the basin deepened and/or the oceanic plate approached sediment sources along the North American margin (fig. 9). The pelagic sections of each of these terranes were overwhelmed by arc-derived terrigenous sediment in Cenomanian time. We suggest that the oceanic plate carrying the pelagic rocks of the Marin Headlands, The Geysers, and East Clear Lake (=Yolla Bolly?) terranes encountered (collided with) the continental margin at about Cenomanian time; at that time this plate began to break-up and the various pieces translated northward and rotated clockwise (McLaughlin and others, 1981; Gromme, this volume; Curry and others, 1983, this volume) forming the various terranes referred to here, that mixed into the Franciscan Central belt. The various original depositional oceanic settings of the siliceous pelagic sediments are poorly constrained, but may have been somewhere in the proximity of the Kula-Farallon Ridge, and/or in off-ridge seamount or plateau settings on the Farallon plate similar to the settings of pelagic limestones in the Franciscan Central belt (Sliter, this volume; Harbert and others, this volume).

The available evidence suggests that these more or less contemporaneous terranes were widely separated in location. However, it is not yet clear whether these rocks were derived from about the same belt of latitude or whether there was large longitudinal separation as well. Curry and others (this volume), present paleomagnetic data suggesting that the Marin Headlands terrane has been translated northward at least 7° relative to North America, while Gromme (this volume) suggests as much as 20° northward translation for the nearby cherts and basalts of the Nicasio area. Other paleomagnetic and paleontologic studies of foraminiferal limestones of the Franciscan assemblage suggest as much as 57° of northward translation for these elements of the Central belt (Alvarez and others, 1980) since Cenomanian or Turonian time; and perhaps 15° of northward drift of limestones of the Coastal terrane since Campanian or Maastrichtian time (Harbert and others, this volume). The faunas from the limestones contain foraminifers with equatorial affinities, further supporting large translations from equatorial paleolatitudes (Sliter, this volume). However, additional paleomagnetic data and paleoecologic data on radiolarian faunas of The Geysers, Marin Headlands, and East Clear Lake (=Yolla Bolly?) terranes will be needed before any final conclusions can be drawn regarding original continuity and connections between these terranes.

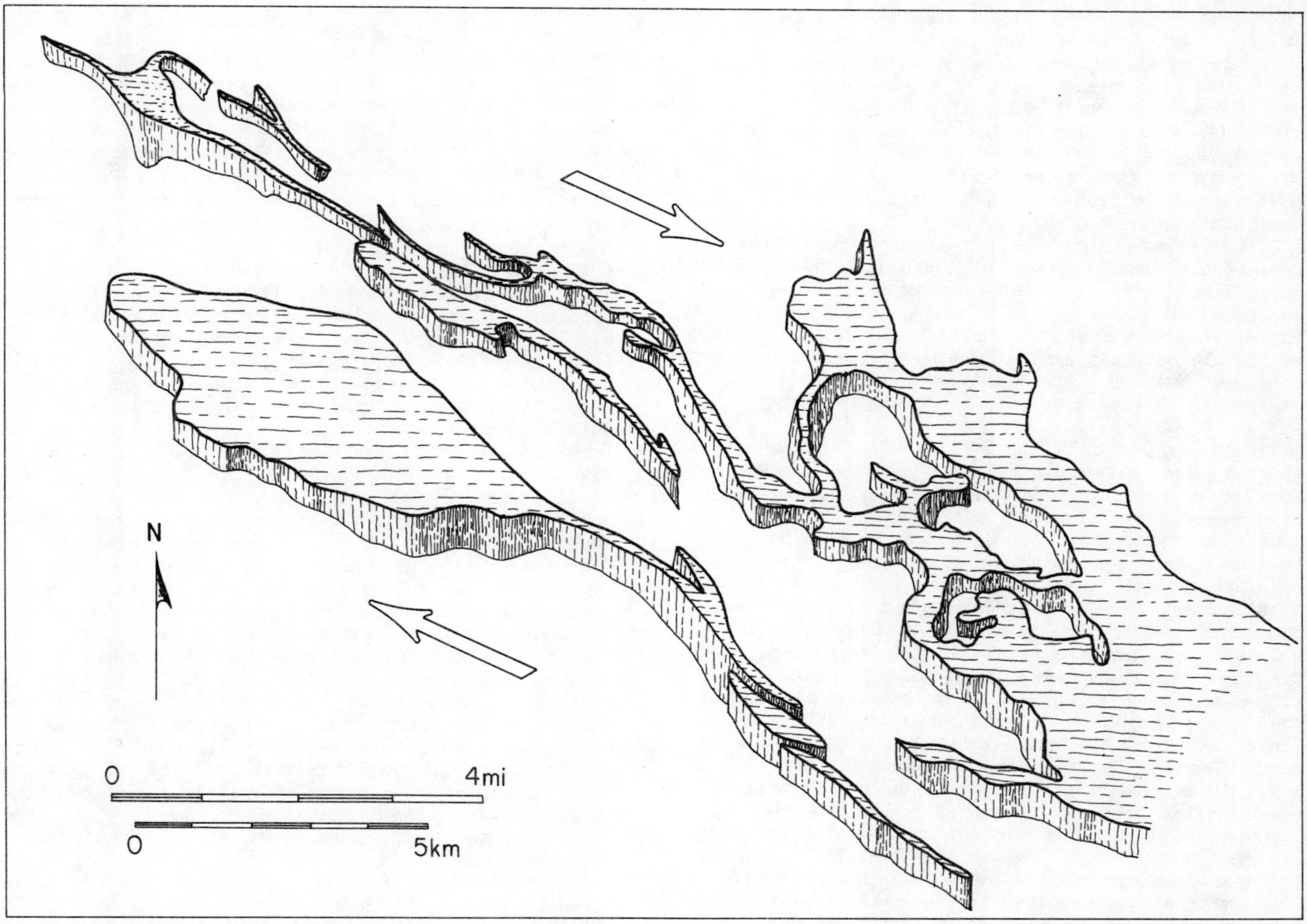

Figure 10. Schematic representation of the Coast Range ophiolite, showing its incorporation into melange on the southwest side of the Mayacmas Mountains, through imbrication, right shear and attenuation, and transposed folding.

Slabs of the East Clear Lake terrane within the Central belt

Two major slabs of lawsonitic metasandstone and several minor slabs reconstituted to textural zone 2 (Blake and others, 1967), locally with intercalated metachert and metabasalt, are interleaved with The Geysers terrane and zones of melange in the Central belt. The petrography and chemistry of the metasandstones of these slabs is similar to that of metasandstones in the East Clear Lake terrane (Figs. 4, 7), and the association of metasandstone and interbedded tuffaceous metachert is characteristic. Atypically, some of the more volcanic-rich lithic metasandstone contains jadeitic pyroxene, and some minor polymict stretched metaconglomerates are present. Highly folded metachert is most prevalent in a structurally-high large slab that composes Long Ridge southeast of Cobb Mountain and southwest of Collayomi Valley. Locally, elongate lenses of metadiabase are associated with the Long Ridge slab. Due to poor exposures, however, it is not clear whether or not the diabase was originally intrusive into the enclosing metasandstones of the Long Ridge slab.

Nearly all of the disrupted remnants of the Yolla Bolly (or East Clear Lake) terranes incorporated into the Central belt are attached to, or in contact with, disrupted rocks of the Coast Range ophiolite, which are also tectonically incorporated into the Central belt. This strongly suggests that disruption, imbrication and translational shearing of the Central belt largely post-dated any emplacement of the East Clear Lake terrane beneath the Coast Range ophiolite. It also should be emphasized that directly to the northeast, the East Clear Lake (=Yolla Bolly?) terrane is not directly juxtaposed with the Coast Range ophiolite beneath the Coast Range thrust; rather, the Little Indian Valley terrane intervenes. Thus, the incorporated remnants of the East Clear Lake (=Yolla Bolly?) terrane within the Central belt probably have been laterally translated within the Central belt. The orientation of sheared-out limbs of folds in some of the incorporated ophiolitic remnants in the Central belt (see fig. 10 and figs. 2 A, D, for style of shearing and folding of ophiolite into the Central belt) that are attached to the Yolla Bolly terrane suggests that these rocks have also been translated in a right-shear sense (see fig. 2).

Devils Den Canyon Terrane

The Devils Den Canyon terrane consists of a large intact slab of thin bedded to massive, arkosic to feldspathic sandstone and shale. This terrane forms an elongate, structurally low slab approximately 18 km long, exposed in the upper reaches of Little Sulphur Creek, Devils Den Canyon, and along the ridge line between Geyser Peak and Mercuryville. Data from geothermal wells demonstrate that the northeast side of the Devils Den Canyon terrane extends for some distance northeastward beneath The Geysers terrane, and probably comprises part of the reservoir rocks of The Geysers steam field (McLaughlin, 1981). The southwest side of the Devils Den Canyon terrane dips steeply to the southwest beneath melange and ophiolitic rocks of the Geyser Peak area. The terrane is enclosed by, or lenses-out, in melange to the northwest and southeast.

The sandstone of the Devils Den Canyon terrane is incipiently metamorphosed to textural zone 1 (Blake and others, 1967), and contains the mineral assemblage quartz + calcite + albite + white mica + chlorite + pumpellyite. There is no potash feldspar, although relict microcline twinning in some grains suggest that original detrital K-feldspar has been albitized, as also noted in sandstones of The Geysers terrane.

The sandstones (figs. 3 V, 4 and 6) are distinctly more quartzose relative to most sandstones from other parts of the Central belt, due mainly to high percentages of chert detritus in the framework grain compositions. The percentages of both monocrystalline quartz and chert are also generally increased relative to other terranes of the Central belt and the more eastern Franciscan terranes; and the abundance of chert detritus is up to almost 4 times that of the volcanic rock detritus (table 1 and also scatter plot, fig. 6). Although there is no distinct dominance of either mafic or intermediate to silicic detritus in the lithic component of sandstones in the Devils Den Canyon terrane, the ratio of mafic to siliceous igneous detritus tends to reach values of the order of 1.5. Sandstones of other terranes of the Central belt reach ratios above 2, up to 9, suggesting that volcanic detritus becomes more silicic to the southwest in the Central belt. Since the ratio of quartz to feldspar detritus in the Devils Den Canyon terrane is high (mean = .91) compared to other sandstones in the Geysers terrane of the Central belt (mean = .68), these data may reflect selective removal by mechanical erosion of less resistant and more labile rock fragments with distance from the source of terrigeneous detritus (i.e., in a down-fan or down-slope direction). By this interpretation, the Devils Den Canyon terrane might represent a portion of a dismembered submarine fan or slope complex that is more distal than the sandstones and conglomerates associated with The Geysers terrane and other structurally higher sandstones of the Central belt to the northeast. A down-fan increase in quartzose detritus at the expense of less resistant detritus has been noted in modern submarine fan systems such as the Astoria fan (Nelson and Nilsen, 1974).

The quartzose, chert-rich composition of the sandstone of the Devils Den Canyon terrane is also reflected in the chemical composition of these sandstones (fig. 7). A triangular plot (fig. 7 C) of SiO_2, Al_2O_3, $FeO + Fe_2O_3 + MgO + TiO_2 + P_2O_5 + MnO$, corresponds in a qualitative sense to a Q + C, F, L plot of the framework grains (fig. 4 C). On the plot of the chemical oxides, the Devils Den Canyon terrane falls within the field of composition of the Central belt but clusters toward the SiO_2 pole, similar to the Q + C, F, L plot.

On an AFM plot of sandstones of Franciscan terranes (fig. 7 B), those of the Devils Den Canyon terrane plot toward the alkali pole, reflecting their low content in mafic rock fragments (fig. 7). The alkalic composition of the sandstone of the Devils Den Canyon terrane probably reflects an arc-derivation for much of the detritus.

Mayacmas terrane

This disrupted terrane, herein named for the Mayacmas Mountains, is prominently exposed along the southwest side of the Mayacmas Range, northeast of Alexander Valley. The unit probably is continuous with the extensive zone of melange evident east and west of Highway 101 between Cloverdale and Ukiah. The Mayacmas terrane (fig. 2) is characterized by numerous large blocks and slabs of high-grade, coarse blueschist whose protoliths include ultramafic, basaltic, tuffaceous, and cherty rocks and sandstones (fig. 3 X). In addition, sporadic blocks of garnet-bearing amphibolite and eclogite that are partially to completely retrograded to glaucophane schist are also present. Other lower-grade blocks of basaltic volcanic rocks with intercalated radiolarian chert may be remnants of parts of the nearby The Geysers or Yolla Bolly terranes. In addition, numerous lenses of serpentinite, some associated with mafic rocks, have been sheared and incorporated into the melange from larger slabs of the Coast Range ophiolite (i.e., the Geyser Peak and Mount St. Helena ophiolite remnants of Hopson and others (1981)). Blocks (small car-size) of recrystallized gray (marble-like) limestone occur at two localities: a lime kiln was built at one of these localities. This relatively pure limestone was found to contain no recognizable microfossils or megafossils, in contrast to the gray Permanente quarry (=Calera Limestone) and pink Laytonville (informally designated Laytonville limestone of Alvarez and others, 1980) foraminiferal varieties present along the continuation of the Central belt to the northwest and southeast (Sliter, this volume). Also of note are a few large, massive blocks of gritty to conglomeratic sandstone with clasts that are similar to those from conglomerates of the lower part of the Great Valley sequence described by Seiders (1983). One of these blocks in melange south of the Geyser Peak ophiolite remnant (fig. 2 D) contains specimens of Buchia of Late Jurassic (Tithonian) or Early Cretaceous (Valanginian) age. We tentatively correlate these conglomerate blocks in the melange with the Great Valley sequence and suggest that they were ripped away from the Great Valley and tectonically incorporated into the melange, along with components of the Coast Range ophiolite, between middle Cretaceous and Paleogene time (see earlier discussions).

The scaly argillic "matrix" material (see earlier discussion) of this melange is reported to contain incipient metamorphic minerals such as phengitic white mica, pumpellyite, and locally, possible lawsonite (Cloos, 1982; 1983). Cloos has proposed that such melanges deform and become metamorphosed through deep ductile flow within a subduction zone under conditions of high pressure (between 5 and 10 kb) and low temperature (below about 250°C), P-T conditions similar to those associated with metamorphism of the so-called high-grade blueschist blocks included in the melange. Cloos also interpreted the distribution of high-grade blocks in the melange as resulting from large-scale density sorting during flowage within a presumed subduction zone. Although density sorting and ductile flowage may well have been important processes in formation of melange in the Central belt, it is difficult to reconcile the distribution, structure, and variable metamorphism of the numerous Central belt terranes without calling upon considerable right-shear as an intimate process in producing the melange fabric.

In particular, evidence for large-scale ductile right-shear can be seen in the structural style of deformed ophiolitic rocks incorporated into the melanges of the southwest and central Mayacmas Mountains (figs. 2, 10). We have attempted in Figure 10 to show schematically the style of transposed folding, ductile right-shear, and attenuation exhibited by exposures of ophiolitic rocks in this area. Zones of serpentinite are clearly entrained into the melange along the pervasive northwest-trending shear fabric. These zones of serpentinized ultramafic rocks may be traceable for tens of kilometers (for example, serpentinite in Mercuryville and Burned Mountain thrust zones) into major bodies of the dismembered Coast Range ophiolite (for example, the Geyser Peak and Mount St. Helena ophiolite remnants). Locally, the serpentinite in these northwest-trending shear zones exhibits tight folding, with one of the fold limbs transposed by northwest-trending shears (see, for example, northwest part of fig. 10). These data suggest that the serpentinite was incorporated into the melange through imbricate thrusting that was penecontemporaneous with, or followed by, tight isoclinal to recumbent folding. This folding turned into a pervasive, oblique right-shear fabric that locally transposed the limbs of the folds right-laterally and vertically along shear planes (see Fig. 10).

The timing of formation of the melange is difficult to constrain with a high degree of certainty, partly because it is likely that the melange formed over a long period of time and may have propagated northward with time (Fox, 1976). However, the ages of rocks in the various dismembered terranes of the Central belt do provide important constraints. The Geysers terrane contains radiolarian chert that records continuous pelagic sedimentation from middle Jurassic (Bajocian?) through early Late Cretaceous (Cenomanian) time. Thus, it is unlikely that the ocean floor on which these sediments were deposited was broken up prior to the late Cenomanian or Turonian, since such a major disruption probably would have severely disturbed the pelagic record. The Central belt contains rocks at least as young as Maastrichtian (Van Arsdale terrane) and probably as young as Paleocene, if rocks of the Middle Mountain terrane are considered. In northern California, northeast of Garberville, along the west side of the Central belt, there are numerous large blocks and slabs of arkosic sandstone (on the order of a kilometer or more long) that contain significant detrital potassium feldspar, and are metamorphosed to prehnite-laumontite grade. These K-feldspar-rich sandstone blocks are identical to some sandstones of the Franciscan Coastal terrane. The Coastal terrane in this area is outboard and to the west and is largely of Eocene age (McLaughlin and others, 1982, 1983). Thus it seems likely that tectonic mixing of rocks into the melange matrix of the Central belt continued from middle Cretaceous into Eocene time. Accretion of melanges of the Central belt to the California margin apparently was complete by the early and middle Miocene, as mildly deformed early and (or) middle Miocene sediments of the Eel River basin overlap the Central belt and Coastal terrane in northwestern California (McLaughlin and others, 1982; 1983a).

Pine Flat Road Terrane

South of Geyser Peak and in a small area to the northwest along Pine Flat Road (fig. 2 D) are intact slabs of well-bedded, folded, K-feldspar-bearing sandstone (fig. 3). This sandstone is intercalated with basaltic flows, tuff, and pillow breccia. A few lenses of radiolarian chert are also associated with basaltic rocks of the terrane. A sill-like body of diabase apparently intrudes the sandstones northwest of Geyser Peak. The terrane is underlain structurally by serpentinite in the area northwest and southeast of Pine Flat Road.

Sandstones of the Pine Flat Road terrane contain as much as 5 percent detrital potash feldspar, and are metamorphosed to prehnite-pumpellyite grade. Aside from their K-feldspar content and lower metamorphic grade, the framework grain composition of these sandstones falls within the compositional field of other sandstones within the Central belt (fig. 4); these sandstones are among the most siliceous of the Central belt (fig. 7 C). We have no data on the age of the rocks in this terrane. However, we speculate that these rocks might represent a less metamorphosed equivalent of The Geysers or the Marin Headlands terranes. Similar sandstone to the west along the Russian River has been called the Rio Nido terrane (see Jayko and Blake, this volume).

CONCLUSIONS REGARDING THE TIMING AND NATURE OF AMALGAMATION OF TECTONOSTRATIGRAPHIC TERRANES IN THE GEYSERS-CLEAR LAKE AREA

1. The framework grain compositions of sandstones (fig. 4) from various terranes of the Geysers-Clear Lake area span a wide range, and show distinct but overlapping compositional fields. With the exception of the Pine Flat Road terrane, a prominent increase in the ratio of chert to volcanic detritus from northeast-to southwest in the Franciscan assemblage (fig. 6) also corresponds with an increase in quartz (Qm, table 1). Tithonian and Valanginian sandstone in the Little Indian Valley terrane is the most lithic, with conspicuous intermediate volcanic rocks, plutonic rocks, and felsitic rock fragments. Sandstone of the Devils Den Canyon terrane, in a structurally low part of the Central belt of probable Late Cretaceous age, contains the highest percentage of chert and quartz grains. Possibly the westward-southwestward increase in quartz and chert reflects increasing influence of continent or arc-derived sediment in the more westward, younger sandstone terranes of the Franciscan.

2. Chemical data from the Franciscan assemblage sandstones of the East Clear Lake and Geysers terranes (fig. 7) strongly imply an arc derivation for the terrigenous sediment. Volcanic and intrusive rocks associated with The Geysers terrane are alkalic, possibly reflecting an off-ridge (ocean island or seamount), tectonic setting. Mafic rocks of middle Cretaceous age that intrude rocks related to the East Clear Lake terrane north and south of Clear Lake suggest a near-arc extensional setting.
3. Our observations of field relations in the Geysers-Clear Lake area together with aeromagnetic data indicate that rocks of the Little Indian Valley and East Clear Lake terranes probably were obductively accreted through east-directed thrusting onto an ophiolitic basement. This occurred concurrently with westward thrusting of the Coast Range ophiolite together with the overlying Great Valley sequence over these Franciscan rocks, about 95-92 m.y. ago. The Grizzly Creek melange terrane, involving the Coast Range ophiolite and Tithonian through Valanginian rocks in the upper plate of the Coast Range thrust, probably also formed at this time as the result of shearing, imbricate thrusting, and/or detachment between the Coast Range and structurally higher Stony Creek thrusts. All of these rocks (Little Indian Valley, East Clear Lake, and Grizzly Creek melange terranes) were appressed into southeast- plunging

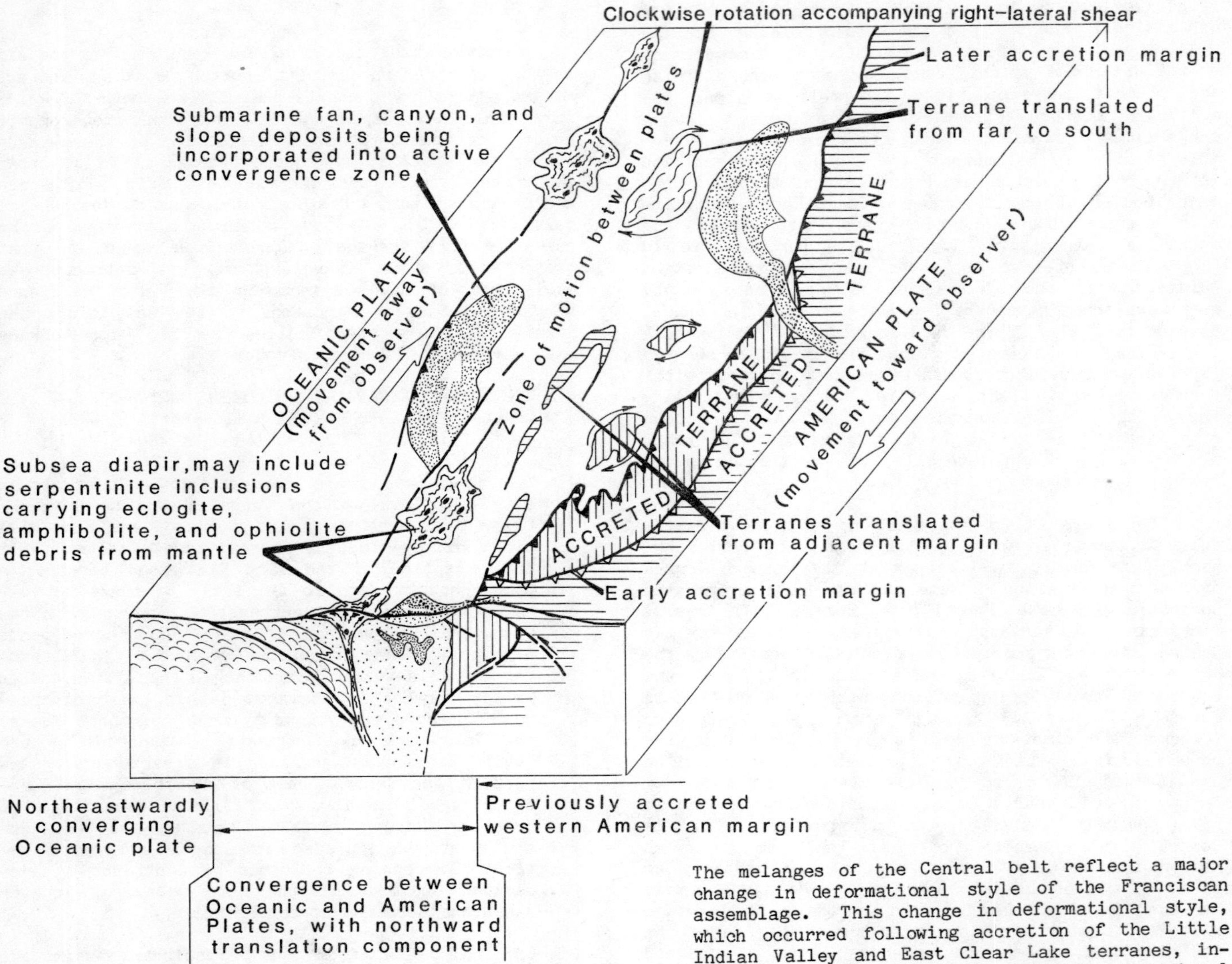

Figure 11. Schematic map and section (not to scale) depicting tectonic and sedimentologic processes contributing to formation of melange and terranes of the Central belt of the Franciscan assemblage. The Central belt is here envisioned as a wide zone of translational movement between the North American margin and an obliquely converging oceanic plate which formed about 90 m.y.a. during break-up of the Farallon plate along the Kula-Farallon Ridge.

folds prior to the emplacement of Late Cretaceous (Turonian or younger) strata of the Great Valley sequence above these rocks.

4. Melanges in the Franciscan Central belt probably were formed no earlier than Cenomanian-Turonian time (about 92 m.y.ago), since the pelagic record spanning at least the Middle Jurassic (Bajocian) through middle Cretaceous (Cenomanian), represented by cherts in the Geysers, the Marin Headlands, and East Clear Lake terranes within the Central belt, would probably have been severely disrupted by any nearby Late Jurassic or Early Cretaceous melange-forming tectonism of the oceanic plate on which the pelagic sediments were deposited. We know of no evidence for earlier melange-forming deformation in the Central belt.

The melanges of the Central belt reflect a major change in deformational style of the Franciscan assemblage. This change in deformational style, which occurred following accretion of the Little Indian Valley and East Clear Lake terranes, involved a change from complex folding and regional metamorphism exemplified in the largely intact Little Indian Valley and East Clear Lake terranes, to the melange deformation typical of the Central belt. We suggest that this change in deformational style corresponded with a major change in plate motion between North America and the Farallon or Kula plates (McLaughlin and others, 1981). The timing of this change in plate motion seems to correspond to inception of the Kula-Farallon spreading center. Structures in the Central belt, particularly those reflected in fragments of the Coast Range ophiolite, suggest that this change in plate motion resulted in significant subsequent northward translation of the various terranes of the Central belt. Available paleontologic data (Murchey; Murchey and Jones; Sliter; this volume) and paleomagnetic data (Alvarez and others, 1980; Curry and others, 1983; Gromme, this volume) suggest that the oceanic, seamount, and/or near-arc pelagic terranes represented in the Central belt by The Geysers, Marin Headlands, and East Clear Lake terrane equivalents, probably have been translated northward from equatorial (0°-20° N or S) paleolatitudes. Furthermore, translational tectonism of rocks in the Central belt appears to have occurred well into Paleogene time, as indicated by the youngest rocks present in melanges of the Central belt.

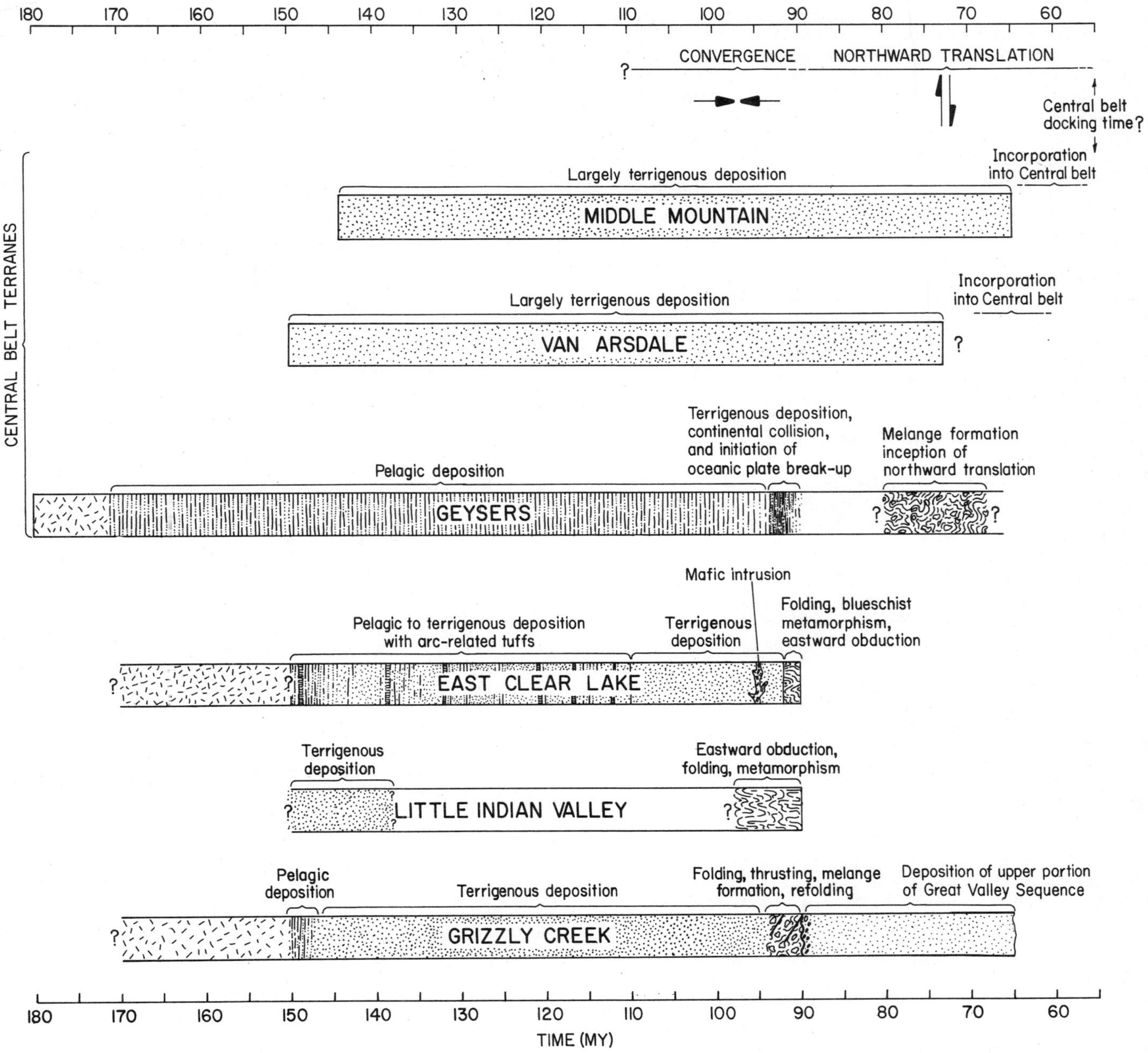

Figure 12. Flow chart depicting relative sequence and timing of deposition, metamorphism, and tectonic disruption for terranes (Middle Mountain, Van Arsdale, The Geysers, East Clear Lake, Little Indian Valley, Grizzly Creek terranes) of The Geysers-Clear Lake area. Note that beginning of melange formation in the Central belt of the Franciscan assemblage at about 90 m.y.a. marks a major change in deformational style in the Coast Ranges, corresponding to the inception of the Kula-Farallon Ridge.

Based on these arguments we suggest that the final "docking" or arrival time of the Middle Jurassic to middle Cretaceous pelagic rocks (and their associated basaltic basements) in northern California was about 55-60 m.y. ago.

The first contact, or collision of North America with the oceanic (Farallon?) plate on which cherts of The Geysers terrane were deposited is recorded by the Cenomanian or younger age of the first terrigenous sedimentary rocks deposited upon the chert (figs. 2, 11, 12). These sandstones represent overwhelming of the oceanic plate by continent or arc-derived sediment, as the chert-bearing part of the oceanic plate neared and first collided with the American plate margin about 90 m.y. ago. Subsequent to this first collision and break-up of the oceanic (Farallon?) plate, a northward component of plate motion (possibly on the newly formed Kula plate or an as yet unrecognized plate) relative to North America caused various pelagic elements of the former Farallon plate to be translated northward along the North American margin, within the pervasively sheared zone of contact with the oceanic plate. This translational tectonism probably propagated northward and continued until about 55 m.y.a., when large slabs of the Great Valley sequence and underlying Coast Range ophiolite were torn away from the western edge of the Great Valley province, drag-folded, imbricated, and right-laterally sheared-out into Central belt melanges, as the Central belt "docked" and accreted to northwestern California (figs. 11 and 12).

References Cited

Alvarez, Walter, Kent, D. V., Premoli-Silva, Isabella, Schweikert, R. A., and Larson, R. L., 1980, Franciscan complex limestone deposited at 17° south paleolatitude: Geological Society of America Bulletin, v. 91, p. 476-484.

Bailey, E. H., 1946, Quicksilver deposits of the western Mayacamas District, Sonoma County, California: California Journal of Mines and Geology, v. 42, no. 3, p. 199-230.

Bailey, E. H., Irwin, W. P., and Jones, D. L., 1964, Franciscan and related rocks and their significance in the geology of western California: California Division of Mines and Geology Bulletin 183, 177 p.

Bailey, E. H., and Blake, M. C., Jr., 1969, Tectonic development of western California during the Late Mesozoic (in Russian): Geotektonika, pt. 3, p. 17-30, pt. 4, p. 24-34.

Bailey, E. H., Blake, M. C., Jr., and Jones, D. L., 1970, On-land Mesozoic oceanic crust in California Coast Ranges: U.S. Geological Survey Professional Paper 700-C, p. C70-C81.

Berkland, J. O., 1969, Late Mesozoic and Tertiary sequence near the proposed Garett Tunnel, Mendocino and Lake counties, California: Geological Society of America, Abstracts with Programs for 1969, Pt. 3, p. 6-7.

-----1972, Paleogene "frozen" subduction zone in the Coast Ranges of northern California: International Geological Congress, 24th, Tectonics, Montreal, Sec. 3, p. 99-105.

-----1973, Rice Valley Outlier-New Sequence of Cretaceous-Paleocene Strata in northern Coast Ranges, California: Geological Society of America Bulletin, vol. 84, p. 2389-2406.

Berkland, J. O., Raymond, L. A. Kramer, J. C., Moores, E. M., and O'Day, M., 1972, What is Franciscan?: American Association of Petroleum Geologists Bulletin, v. 56, p. 2295-2302.

Blake, M. C., Jr., Jayko, A. S., and Howell, D. G., 1981, Geology of a subduction complex in the Franciscan assemblage of northern California: Proceedings, 26th International Geological Congress, Geology of Continental margins symposium, Paris, July 7-17, p. 267-272.

Blake, M. C., Jr., Howell, D. G., and Jones, D. L., 1982, Preliminary tectonostratigraphic terrane map of California: U.S. Geological Survey Open-File map 82-593.

Blake, M. C., Jr., and Jones, D. L., 1974, Origin of Franciscan melanges in northern California: Society of Economic Paleontologists and Mineralogists Special Publication 19, p. 345-357.

-----,1981, The Franciscan assemblage and related rocks in northern California: a reinterpretation: <u>in</u> The geotectonic development of California, W. G. Ernst, editor, Rubey Volume I, p. 306-328, Prentice Hall Inc., 706 p.

Blake, M. C., Jr., Irwin, W. P., and Coleman, R. G., 1967, Upside-down metamorphic zonation, blueschist facies, along a regional thrust in California and Oregon: U.S. Geological Survey Professional Paper 575-C, p. C1-C9.

Brice, J. C., 1953, Geology of the Lower Lake quadrangle, California: California Division of Mines Bulletin 166, 72 p.

Brown, R. D., Jrs., 1964, Thrust-fault relations in the northern Coast Ranges, California: <u>in</u>, the U.S. Geological Survey Professional Paper 475-D, p. D7-D13.

Carlson, Christine, 1981, Sedimentary serpentinites of the Wilbur Springs Area--A possible early Cretaceous Structural and Stratigraphic Link between the Franciscan Complex and the Great Valley sequence: unpublished M.S. thesis, Stanford University, 105 manuscript pages.

Cloos, Mark, 1982, Flow melanges: numerical modeling and geologic constraints on their origin in the subduction complex, California: Geological Society of America Bulletin, v.93, p. 330-345.

_______ 1983, Comparative study of melange matrix and metashales from the Franciscan subduction complex with the basal Great Valley sequence, California: Journal of Geology, v. 91, p. 291-306.

Curry, F., Cox, A., and Engebretson, D., 1983, Paleomagnetism of Franciscan rocks in the Marin headlands: EOS, Transactions, American geophysical Union, vol. 64, no. 3, p. 867.

Dickinson, W. R., Ingersoll, R. V., Cowan, D. S., Helmold, K. P., and Suczek, C. A., 1982, Provenance of Franciscan graywackes in coastal California: Geological Society of America Bulletin, v. 93, p. 95-107.

Dickinson, W. R., and Rich, E. I., 1972, Petrologic intervals and petrofacies in the Great Valley sequence, Sacramento Valley, California: Geological Society of America Bulletin, v. 83, p. 3007-3024.

Dickinson, W. R., and Seely, D. R., 1979, Structure and Stratigraphy of forearc regions: American Association of Petroleum Geologists Bulletin, v. 63, p.2-31.

Fox, K. F., Jr., 1976, Melanges in the Franciscan complex, a product of triple junction tectonics: Geology, v. 4, p. 737-740.

Gealey, W. K., 1951, Geology of the Healdsburg quadrangle, California: California Division of Mines and Geology Bulletin 161, 76 p.

Griscom, Andrew, 1983, The contact between the Great Valley sequence and the Franciscan assemblage, California, from magnetic data: EOS, transactions, American Geophysical Union, V. 64, no. 8, p. 867-868

Harland, W. B., Cox, A. V., Llewellyn, P. G., Pickton, C. A. G., Smith, A. G., and Walters, R., 1982, A geologic time scale, Cambridge University Press, 2nd edition.

Hearn, B. C., Jr., Donnelly-Nolan, J. M., and Goff, F. E., 1981, The Clear Lake Volcanics: Tectonic setting and magma sources: <u>in</u> U.S. Geological Survey Professional paper 1141, R. J. McLaughlin and J. M. Donnelly-Nolan, editors, p. 25-45.

Hopson, C. A., Mattinson, J. M., and Pessagno, E. A., Jr., 1981, Coast Range ophiolite, western California: <u>in</u> The geotectonic development of California, Rubey volume I, W. G. Ernst editor, Prentice Hall, New Jersey, p. 418-510.

Ingersoll, R. V., 1978, Paleogeogrpahy and paleotectonics of the Late Mesozoic forearc basin of northern and central California: <u>in</u> Mesozoic paleogeography of the western United States, Pacific coast paleogeography symposium 2, D. G. Howell and K. A. McDougall editors, Pacific Section, Society of Economic paleontologists and mineralogists, p. 471-482.

Isherwood, W. F., 1976, Gravity and magnetic studies of The Geysers-Clear Lake geothermal region, California, U.S.A., United Nations symposium on development and use of geothermal resources, 20-29, May, 1975, v. 2., p. 1065-1074.

Irvine, T. N., and Baragar, W. R. A., 1971, A guide to the chemical classification of the common volcanic rocks: Canadian Journal of Earth Science, v. 8, p. 523-548.

Irwin, W. P., 1960, Geologic reconnaissance of the northern Coast Ranges and Klamath Mountains, California: California Division of Mines and Geology, Bulletin 179, 79 p.

Karl, S. M., 1983, Geochemical and depositional environments of Upper Mesozoic radiolarian cherts from the Marin headlands, California, and from Pacific DSDP cores: EOS, Transactions, American Geophysical Union, vol. 64, no. 3, p. 867.

______ 1982, Geochemical and depositional environments of Upper Mesozoic radiolarian cherts from the northeastern Pacific rim and from Pacific DSDP cores: unpublished Ph.D. thesis, Stanford University, 245 p.

Lawton, J. E., 1956, Geology of the North half of the Morgan Valley quadrangle and the south half of the Wilbur Springs quadrangle: unpublished Ph.D. thesis, Stanford University, 223 p.

Layman, E. B., 1977, Intrusive Rocks in the near-trench environment: Two localities in the northern California Franciscan complex: unpublished M.Sc. thesis, Stanford University, 55 p.

Ludwig, W. J., Nafe, J. E., and Drake, C. L., 1970, Seismic Refraction, in, A. E. Maxwell, ed., The Sea, v. 4, p. 53-84.

MacDonald, G. A., and Katsura, T., 1964, Chemical composition of Hawaiian lavas: Journal of Petrology, v. 5, p. 82-133.

Mattinson, J. M., 1980, Jurassic ophiolitic belt of western North America: Age relationships and their tectonic implications: Geological Society of America, abstracts with programs, v. 12, p. 478.

Mattinson, J. M., and Echeverria, L. M., 1980, Ortigalita Peak gabbro, Franciscan Complex: U-Pb dates of intrusion and high-pressure-low-temperature metamorphism: Geology, v. 8, p. 589-593.

Maxwell, J. C., 1974, Anatomy of an orogen: Geological Society of America Bulletin, v. 85, p. 1195-1204.

McLaughlin, R. J., 1978, Preliminary geologic map and structural sections of the central Mayacmas Mountains and the Geysers steam field, Sonoma, Lake, and Mendocino Counties, California: U.S. Geological Survey Open-File Map 78-389, scale 1:24,000.

-----1981, Tectonic setting of pre-Tertiary rocks and its relation to geothermal resources in the Geysers-Clear Lake area: in U.S. Geological Survey Professional Paper 1141, R. J. McLaughlin and J. M. Donnelly-Nolan, editors, p. 3-23.

McLaughlin, R. J., and Pessagno, E. A., Jr., 1978, Significance of age relations above and below Upper Jurassic ophiolite in The Geysers-Clear Lake region, California: U.S. Geological Survey Journal of Research, v. 6, p. 715-726.

McLaughlin, R. J., and Stanley, W. D., 1976, Pre-Tertiary geology and structural control of geothermal resources, the Geysers steam field, California: United Nations symposium on development and use of geothermal resources, 2d, 20-29 May, 1975, Proceedings, v. 1, p. 475-486.

McLaughlin, R. J., Kling, S. A., Poore, R. Z., McDougall, K., and Beutner, E. C., 1982, Post-Middle Miocene accretion of Franciscan rocks, northwestern California: Geological Society of America Bulletin, v. 93, p. 595-605.

McLaughlin, R. J., Ellen, S. D., Miller, S. G., Lajoie, K. R., and Morrison, S. D, 1983a, Terrane boundary relations and tectonostratigraphic framework south of Eel River basin, northwestern California: Program and Abstracts, 58th Annual Meeting, Pacific Section, American Association of Petroleum, Geologists A.A.P.G., S.E.G., S.E.P.M., p. 112.

McLaughlin, R. J., Moore, D. E., Sorg, D. H., and McKee, E. H., 1983, Multiple episodes of hydrothermal circulation, thermal metamorphism, and magma injection beneath the Geysers steam field, California: Geological Society of America Abstracts with programs, v. 15, no. 5, p. 417.

McLaughlin, R. J., Ohlin, H. N., and Thormahlen, D. J., with contributions from Jones, D. L., Miller, J. W., and Blome, C. D., in press, Geologic map and structure sections of the Little Indian Valley-Wilbur Springs geothermal area, northern Coast Range, California: U.S. Geological Survey Miscellaneous Geological Investigations I- map, scale 1:24,000.

McLaughlin, R. J., Blake, M. C., Jr., and Jones, D. L., 1981, Franciscan melange in central northern California: Geological Society of America, Abstracts with programs, v. 13, no. 7, p. 508.

McNitt, J. R., 1968, Geology of the Kelseyville quadrangle Sonoma, Lake, and Mendocino counties, California: California Division of Mines and Geology, Map Sheet #9, 1:62,500.

Murchey, B. L., Jones D. L., and Couzin, R. I., 1983, Age and significance of chert in the Franciscan complex in the San Francisco Bay area: EOS, transactions, American Geophysical Union, vol. 64, no. 45, p. 867.

Nelson, C. H., and Nilsen, T. H., 1974, Depositional trends of modern and ancient deep-sea fans, in Dott, R. H., Jr., and Shaver, R. H., eds., Modern and ancient geocynclinal sedimentation: Society of Economic Paleontologists and Mineralogists Special Publication 19, p. 69-91.

Pessagno, E. A., Jr., 1977a, Upper Jurassic radiolaria and radiolarian biostratigraphy of the California Coast Ranges: Micropaleontology, v. 23, no. 1, p. 56-113.

-----1977b, Lower Cretaceous Radiolarian biostratigraphy of the Great Valley sequence and Franciscan Complex, California Coast Ranges: Cushman Foundation for foraminiferal research special publication 15, 87 p.,

Pessagno, E. A., Jr., Blome, C. D., and Longoria, J. F., in press, A revised radiolarian zonation for the Upper Jurassic of Western North America: Bulletin of American Paleontology, v. 85, no. 320.

Rich, E. I., 1971, Geologic map of the Wilbur Springs quadrangle, Colusa and Lake Counties, California: U.S. Geological Survey Miscellaneous Geological Investigations Map I-538, scale 1:48,000.

Seiders, V. M., 1983, Correlation and provenance of Upper Mesozoic chert-rich conglomerate of California: Geological Society af America Bulletin, v. 94, p. 875-888.

Seiders, V. M., Pessagno, E. A., Jr., and Harris, A. G., 1979, Radiolarians and conodonts from pebbles in the Franciscan assemblage and Great Valley sequence of the California Coast Ranges: Geology, v. 7, p. 37-40.

Suppe, J., and Foland, K. A., 1978, The Goat Mountain schists and Pacific Ridge complex: a re-deformed but still-intact Mesozoic Franciscan Schuppen complex: in Mesozoic paleogeography of the western United States, Pacific coast paleogeography symposium 2, D. G. Howell and K. A. McDougall, editors, Pacific Section, Society of Economic paleontologists and Mineralogists, p. 471-451

Swe, W., and Dickinson, W. R., 1970, Sedimentation and thrusting of Late Mesozoic rocks in the Coast Ranges near Clear Lake, California: Geological Society of America Bulletin, v. 81, p. 165-189.

Taliaferro, N. L., 1943, Franciscan-Knoxville problem: American Association of Petroleum Geologists Bulletin, v. 27, p. 102-219.

Worrall, D. M., 1981, Imbricate low-angle faulting in upper-most Franciscan rocks, south Yolla Bolly area, northern California: Geological Society of America Bulletin, v. 92, p. 703-729

Yates, R. G., and Hilpert, C. S., 1946, Quicksilver deposits of eastern Mayacmas district, Lake and Napa Counties, California: California Journal of Mines and Geology, v. 42, no. 3, p. 231-286.